W0259168

Informatik – Fachberichte

Band 122: Ch. Habel, Prinzipien der Referentialität. Untersuchungen zur propositionalen Repräsentation von Wissen. X, 308 Seiten. 1986.

Band 123: Arbeit und Informationstechnik. GI-Fachtagung. Proceedings, 1986. Herausgegeben von K. T. Schröder. IX, 435 Seiten. 1986.

Band 124: GWAI-86 und 2. Österreichische Artificial-Intelligence-Tagung. Ottenstein/Niederösterreich, September 1986. Herausgegeben von C.-R. Rollinger und W. Horn. X, 360 Seiten. 1986.

Band 125: Mustererkennung 1986. 8. DAGM-Symposium, Paderborn, September/Oktober 1986. Herausgegeben von G. Hartmann. XII, 294 Seiten, 1986.

Band 126: GI-16. Jahrestagung. Informatik-Anwendungen – Trends und Perspektiven. Berlin, Oktober 1986. Herausgegeben von G. Hommel und S. Schindler. XVII, 703 Seiten. 1986.

Band 127: GI-17. Jahrestagung. Informatik-Anwendungen – Trends und Perspektiven. Berlin, Oktober 1986. Herausgegeben von G. Hommel und S. Schindler. XVII, 685 Seiten. 1986.

Band 128: W. Benn, Dynamische nicht-normalisierte Relationen und symbolische Bildbeschreibung. XIV, 153 Seiten. 1986.

Band 129: Informatik-Grundbildung in Schule und Beruf. GI-Fachtagung, Kaiserslautern, September/Oktober 1986. Herausgegeben von E. v. Puttkamer. XII, 486 Seiten. 1986.

Band 130: Kommunikation in Verteilten Systemen. GI/NTG-Fachtagung, Aachen, Februar 1987. Herausgegeben von N. Gerner und O. Spaniol. XII, 812 Seiten. 1987.

Band 131: W. Scherl, Bildanalyse allgemeiner Dokumente. XI, 205 Seiten. 1987.

Band 132: R. Studer, Konzepte für eine verteilte wissensbasierte Softwareproduktionsumgebung. XI, 272 Seiten. 1987.

Band 133: B. Freisleben, Mechanismen zur Synchronisation paralleler Prozesse. VIII, 357 Seiten. 1987.

Band 134: Organisation und Betrieb der verteilten Datenverarbeitung. 7. GI-Fachgespräch, München, März 1987. Herausgegeben von F. Peischl. VIII, 219 Seiten. 1987.

Band 135: A. Meier, Erweiterung relationaler Datenbanksysteme für technische Anwendungen. IV, 141 Seiten. 1987.

Band 136: Datenbanksysteme in Büro, Technik und Wissenschaft. GI-Fachtagung, Darmstadt, April 1987. Proceedings. Herausgegeben von H.-J. Schek und G. Schlageter. XII, 491 Seiten. 1987.

Band 137: D. Lienert, Die Konfigurierung modular aufgebauter Datenbanksysteme. IX, 214 Seiten. 1987.

Band 138: R. Männer, Entwurf und Realisierung eines Multiprozessors. Das System „Heidelberger POLYP". XI, 217 Seiten. 1987.

Band 139: M. Marhöfer, Fehlerdiagnose für Schaltnetze aus Modulen mit partiell injektiven Pfadfunktionen. XIII, 172 Seiten. 1987.

Band 140: H.-J. Wunderlich, Probabilistische Verfahren für den Test hochintegrierter Schaltungen. XII, 133 Seiten. 1987.

Band 141: E. G. Schukat-Talamazzini, Generierung von Worthypothesen in kontinuierlicher Sprache. XI, 142 Seiten. 1987.

Band 142: H.-J. Novak, Textgenerierung aus visuellen Daten: Beschreibungen von Straßenszenen. XII, 143 Seiten. 1987.

Band 143: R. R. Wagner, R. Traunmüller, H. C. Mayr (Hrsg.), Informationsbedarfsermittlung und -analyse für den Entwurf von Informationssystemen. Fachtagung EMISA, Linz, Juli 1987. VIII, 257 Seiten. 1987.

Band 144: H. Oberquelle, Sprachkonzepte für benutzergerechte Systeme. XI, 315 Seiten. 1987.

Band 145: K. Rothermel, Kommunikationskonzepte für verteilte transaktionsorientierte Systeme. XI, 224 Seiten. 1987.

Band 146: W. Damm, Entwurf und Verifikation mikroprogrammierter Rechnerarchitekturen. VIII, 327 Seiten. 1987.

Band 147: F. Belli, W. Görke (Hrsg.), Fehlertolerierende Rechensysteme / Fault-Tolerant Computing Systems. 3. Internationale GI/ITG/GMA-Fachtagung, Bremerhaven, September 1987. Proceedings. XI, 389 Seiten. 1987.

Band 148: F. Puppe, Diagnostisches Problemlösen mit Expertensystemen. IX, 257 Seiten. 1987.

Band 149: E. Paulus (Hrsg.), Mustererkennung 1987. 9. DAGM-Symposium, Braunschweig, Sept./Okt. 1987. Proceedings. XVII, 324 Seiten. 1987.

Band 150: J. Halin (Hrsg.), Simulationstechnik. 4. Symposium, Zürich, September 1987. Proceedings. XIV, 690 Seiten. 1987.

Band 151: E. Buchberger, J. Retti (Hrsg.), 3. Österreichische Artificial-Intelligence-Tagung. Wien, September 1987. Proceedings. VIII, 181 Seiten. 1987.

Band 152: K. Morik (Ed.), GWAI-87. 11th German Workshop on Artificial Intelligence. Geseke, Sept./Okt. 1987. Proceedings. XI, 405 Seiten. 1987.

Band 153: D. Meyer-Ebrecht (Hrsg.), ASST'87. 6. Aachener Symposium für Signaltheorie. Aachen, September 1987. Proceedings. XII, 390 Seiten. 1987.

Band 154: U. Herzog, M. Paterok (Hrsg.), Messung, Modellierung und Bewertung von Rechensystemen. 4. GI/ITG-Fachtagung, Erlangen, Sept./Okt. 1987. Proceedings. XI, 388 Seiten. 1987.

Band 155: W. Brauer, W. Wahlster (Hrsg.), Wissensbasierte Systeme. 2. Internationaler GI-Kongreß, München, Oktober 1987. XIV, 432 Seiten. 1987.

Band 156: M. Paul (Hrsg.), GI – 17. Jahrestagung. Computerintegrierter Arbeitsplatz im Büro. München, Oktober 1987. Proceedings. XIII, 934 Seiten. 1987.

Band 157: U. Mahn, Attributierte Grammatiken und Attributierungsalgorithmen. IX, 272 Seiten. 1988.

Band 158: G. Cyranek, A. Kachru, H. Kaiser (Hrsg.), Informatik und „Dritte Welt". X, 302 Seiten. 1988.

Band 159: Th. Christaller, H.-W. Hein, M. M. Richter (Hrsg.), Künstliche Intelligenz. Frühjahrsschulen, Dassel, 1985 und 1986. VII, 342 Seiten. 1988.

Band 160: H. Mäncher, Fehlertolerante dezentrale Prozeßautomatisierung. XVI, 243 Seiten. 1987.

Band 161: P. Peinl, Synchronisation in zentralisierten Datenbanksystemen. XII, 227 Seiten. 1987.

Band 162: H. Stoyan (Hrsg.), Begründungsverwaltung. Proceedings, 1986. VII, 153 Seiten. 1988.

Band 163: H. Müller, Realistische Computergraphik. VII, 146 Seiten. 1988.

Band 164: M. Eulenstein, Generierung portabler Compiler. X, 235 Seiten. 1988.

Band 165: H.-U. Heiß, Überlast in Rechensystemen. IX, 176 Seiten. 1988.

Band 166: K. Hörmann, Kollisionsfreie Bahnen für Industrieroboter. XII, 157 Seiten. 1988.

Band 167: R. Lauber (Hrsg.), Prozeßrechensysteme '88. Stuttgart, März 1988. Proceedings. XIV, 799 Seiten. 1988.

Band 168: U. Kastens, F. J. Rammig (Hrsg.), Architektur und Betrieb von Rechensystemen. 10. GI/ITG-Fachtagung, Paderborn, März 1988. Proceedings. IX, 405 Seiten. 1988.

Informatik-Fachberichte 218

Herausgeber: W. Brauer
im Auftrag der Gesellschaft für Informatik (GI)

G. Stiege J. S. Lie (Hrsg.)

Messung, Modellierung und Bewertung von Rechensystemen und Netzen

5. GI/ITG-Fachtagung
Braunschweig, 26.-28. September 1989

Proceedings

Springer-Verlag
Berlin Heidelberg New York
London Paris Tokyo Hong Kong

Herausgeber

G. Stiege
J. S. Lie
Institut für Betriebssysteme und Rechnerverbund
TU Braunschweig
Postfach 3329, D-3300 Braunschweig

CR Subject Classification (1987): C.4

CIP-Titelaufnahme der Deutschen Bibliothek.
Messung, Modellierung und Bewertung von Rechensystemen und Netzen: ... GI/NTG-Fachtagung; proceedings. - Berlin; Heidelberg; New York; London; Paris; Tokyo: Springer
Bis 4 (1987) u.d.T.: Messung, Modellierung und Bewertung von Rechensystemen
NE: Gesellschaft für Informatik
5. Braunschweig, 26. - 28. September 1989. - 1989
(Informatik-Fachberichte; 218)

ISBN-13: 978-3-540-51713-9 e-ISBN-13: 978-3-642-75079-3
DOI: 10.1007/978-3-642-75079-3

NE: GT

2145/3140 - 543210 - Gedruckt auf säurefreiem Papier

Vorwort

Die Fachtagung "Messung, Modellierung und Bewertung von Rechensystemen" fand in der Vergangenheit alle zwei Jahre an wechselnden Orten statt. Sie wird von der gleichnamigen Interessengruppe des Fachbereichs 3 der Gesellschaft für Informatik (GI) und des Fachbereichs 4 der Informationstechnischen Gesellschaft (ITG) veranstaltet.

Die 5. Tagung mit dem erweiterten Titel "Messung, Modellierung und Bewertung von Rechensystemen und Netzen" findet vom 26. bis zum 28. September 1989 an der Technischen Universität Braunschweig statt. Ein eintägiges Tutorium mit vier Referenten geht ihr voraus, eine Ausstellung von Werkzeugen zur Messung und Modellierung wird sie begleiten.

Zahl und Qualität der eingereichten Vorträge gaben Anlaß zu Zufriedenheit. Über 50 Beiträge wurden eingereicht. Daraus wurden 18 nach anonymer Begutachtung für das wissenschaftliche Programm ausgewählt. Eine weitere Sitzung ist Kurzvorträgen aus der Praxis gewidmet. Vier eingeladene Vorträge, die ein weites Spektrum von Forschungsgebieten behandeln, vervollständigen das Programm.

Es ist sehr erfreulich, daß neben wichtigen Ergebnissen zur analytischen Modellierung, denen zwei Sitzungen gewidmet sind, gehaltvolle praxisnahe Beiträge eingereicht wurden, mit denen zwei Sitzungen über Betriebssysteme und je eine über Meßverfahren und Methodik eingerichtet werden konnten. Auch die Erweiterung des Gesamtthemas auf Netze erwies sich als richtig; aus diesem Bereich konnten zwei Sitzungen eingerichtet werden.

Eine wissenschaftliche Tagung läßt sich nicht ohne die Hilfe vieler durchführen. Hier ist vor allem die Arbeit der Mitglieder des Programmkomitees und weiterer Gutachter zu erwähnen. Auch das Entgegenkommen und die Unterstützung der TU Braunschweig und des Springer-Verlags waren hilfreich. Bei der lokalen Organisation gibt es viele helfende Hände, insbesondere sind die Damen S. Panzer, U. Dierks sowie die Herren H.-D. Brökelmann, E. Büscher, A. Gerns, K. Luck und D. Meier zu nennen. Ihnen allen sei herzlich gedankt. Für großzügige Unterstützung möchten wir uns bei den Firmen GEI Bonn, IVM Wolfsburg und Nixdorf Paderborn bedanken.

Braunschweig, im Juli 1989

G. Stiege
J. S. Lie

Programmkomitee

H. Beilner, Universität Dortmund
R. Bordewisch, Nixdorf Paderborn
G. Bolch, Universität Erlangen-Nürnberg
W. Gürich, KFA Jülich
H. L. Hartmann, TU Braunschweig
U. Herzog, Universität Erlangen-Nürnberg
L. Hieber, Datenzentrale Württemberg
W. Hoffmann, Siemens Erlangen
R. Klar, Universität Erlangen-Nürnberg
P. Kühn, Universität Stuttgart
H. Langendörfer, TU Braunschweig
F. Lehmann, Universität BuWe München
R. Lehnert, Philips Nürnberg
J. S. Lie, TU Braunschweig
M. Paterok, Universität Erlangen-Nürnberg
B. Schmidt, Universität Erlangen-Nürnberg
H. Schmutz, IBM Heidelberg
S. Schaßberger, TU Braunschweig
O. Spaniol, RWTH Aachen
P. P. Spies, Universität Oldenburg
G. Stiege, TU Braunschweig
B. Walke, Fernuniversität Hagen
S. Zorn, Siemens München
W. Zorn, Universität Karlsruhe

Inhaltsverzeichnis / Contents

Betriebssysteme I / Operating Systems I

Methodik / Methodology

Betriebssysteme II / Operating Systems II

Kurzvorträge aus der Praxis / Short Contributions

Netze I / Networks I

Netze II / Networks II

Mean Passage Times in Queueing Networks

Kwangho Kook

and

Richard F. Serfozo

Georgia Institute of Technology

Abstract

Major performance measures of a queueing network are the mean time a unit (i.e. customer) spends in a sector of the network and the mean time for a unit to move from one sector to another. We give expressions for these and other mean passage times on routes in Jackson queueing networks and in more general queueing networks with congestion-dependent processing and routing. In these networks, the units may overtake one another as they move.

1 Introduction

What is known about travel times of units in queueing networks? For a classical Jackson network, with the additional assumption that units cannot overtake each other, the total sojourn time of a unit in a set of nodes is the sum of independent exponential sojourn times at the nodes visited. This and related results appear in [1] - [5], [9] -[12], [15] - [16]. This is essentially the extent of our knowledge of travel times. The problem is that, without the overtake-free assumption, the single sojourn times of a unit at the nodes are typically not independent or exponentially distributed. Consequently, the total sojourn time in a set of nodes is generally intractable.

There are a variety of travel times in networks, in addition to sojourn times in sets of nodes, that are important for assessing the quality of a network. Two examples are:

- The time for a unit to go from one set of nodes to another set.
- The time that a unit spends as a certain type (or class) during its stay in a set of nodes.

Such passage times are the focus of this study.

We introduce a general notion of a "route" that units may take in a network. The "passage time" for a route is the time it takes for a unit to travel the route. We consider a variety of passage times for general Markovian queueing networks in which units may overtake one another. Our main results are expressions for the means, and in some cases, higher moments of these sojourn and passage times. We derive mean passage times as limits of laws of large numbers for regenerative processes. The difficulty is that it is not known whether a unit is undergoing a passage on a route until the route is completed. In other words, the number of units undergoing a passage at any time is a function of the future of the network process as well as its past. We overcome this difficulty by a subtle labeling device that helps us look into the future in a regenerative sense. The mean passage times are directly proportional to the average number of units in the network that are undergoing a passage. This relationship resembles the classical Little's law for single queues. Some of our results can therefore be viewed as Little-type laws for mean passage times in networks. Further details on the results and their proofs appear in [6], [7] and [14].

The rest of this paper is organized as follows. In section 2 we introduce the network and in section 3 we give two examples of our results. The expressions for mean passage times are in section 4. In section 5 we characterize higher moments of passage times in networks with single server nodes.

2 Preliminaries

We shall consider a queueing network in which discrete units move among the set of nodes $M = \{1, \ldots, m\}$. The evolution of the network is represented by the stochastic process $\{X(t) = (X_1(t), \ldots, X_m(t)) : t \geq 0\}$ that records the numbers of units at the respective nodes. This process X represents a closed network with N indistinguishable units in it. We point out later how the results extend to open networks and to multiple types of units. The state space of the process X is the set S_N of all non-negative integer-valued vectors $n = (n_1, \ldots, n_m)$ such that $n_1 + \ldots + n_m = N$, where n_j denotes the number of units at node j.

The units move one at a time between nodes. When the process X is in state n and one unit moves from node j to node k, the next state of X is $T_{jk}n$, where

$$T_{jk}n = \begin{cases} n - e_j + e_k & \text{if } n_j \geq 1 \\ n & \text{otherwise} \end{cases}$$

and e_j is the m-vector with 1 in position j and 0 elsewhere. We assume that the movements and processing is such that X is a continuous-time Markov chain with transition rates

$$q(n, n') = \begin{cases} p_{jk}\phi_j(n) & \text{if } n' = T_{jk}n \text{ for some } j, k \epsilon M \\ 0 & \text{otherwise,} \end{cases}$$

where $\phi_j : S_N \rightarrow R_+$ and $R_+ = [0, \infty)$. The $\phi_j(n)$ can be interpreted as the processing rate at node j which depends on the entire state vector n. This general rate function allows for a variety of congestion- dependent processing schemes [14]. The p_{jk} is the probability that a unit departing node j will move (instantly) to node k. We assume that $\phi_j(n) = 0$ if and only if $n_j = 0$, and that $\{p_{jk}\}$ is an irreducible Markov matrix. Then there is a unique distribution $w_1, \ldots, w_m$ (the stationary distribution of $\{p_{jk}\}$) that satisfies the routing-balance equations

$$w_j = \sum_{k=1}^{m} w_k p_k, \quad j = 1, \ldots, m.$$

Under these assumptions it follows that the network process X is irreducible.

The following result from [14] characterizes the equilibrium or stationary distribution of X.

Theorem 1.1 *Suppose that, for each n and j, k, ℓ,*

$$\phi_j(n)\phi_k(T_{j\ell}n)\phi_\ell(T_{k\ell}n) = \phi_k(n)\phi_j(T_{k\ell}n)\phi_\ell(T_{j\ell}n). \tag{1}$$

Then the equilibrium distribution of X is

$$\pi(n) = c_N \Pi_{j=1}^m w_j^{n_j} \Pi_{\nu=1}^{N-n_m} \frac{\phi_m(n - \sum_{s=1}^{\nu} e_{j_s} + \nu e_m)}{\phi_j(n - \sum_{s=1}^{\nu-1} e_{j_s} + (\nu-1)e_m)} \tag{2}$$

where $j_1, j_2, \ldots, j_{N-n_m}$ is any sequence in $\{1, \ldots, m-1\}$ such that $\sum_{s=1}^{\nu} e_{js} \leq n$ for each ν, and c_N is the normalizing constant.

The condition (1) ensures that the product in (2) is the same for any sequence $\{j_s\}$ as specified. Furthermore, condition (1) is sufficient and, with a slight weakening, is also necessary for X to be "partial balanced" in that its equilibrium distribution satisfies the partial balance equations

$$\pi(n)\sum_k q(n, T_{jk}n) = \sum_k \pi(T_{jk}n)q(T_{jk}n, n), \quad \text{each } j \text{ and } n.$$

These and other ramifications of (1) are discussed in [6], [7] and [14]. When the nodes of the network operate independently such that $\phi_j(n)$ is a function $\phi_j(n_j)$ of only n_j, then X is a Jackson network process for which (1) is automatically true and (2) reduces to the well-known expression

$$\pi(n) = c_N \Pi_{j=1}^m w_j^{n_j} \Pi_{\nu=1}^{n_j} \phi_j(\nu)^{-1}.$$

We are interested in various passage times for units moving in the network. A passage time, loosely speaking, is the time that a unit spends in a certain sector of the network, or the time it takes to travel a certain route. Before giving a formal definition, we cite some examples. Here J, K, S denote sets of nodes.

- The time it takes for a unit to pass through J (i.e, the time between entering and exiting J, or the "sojourn" time in J).

- The time it takes for a unit to travel from J to K.

- The time it takes for a unit to pass through the fixed path $(j_1, \ldots, j_\nu)$ (in that order), or to pass through the node sets $(J_1, \ldots, J_\nu)$.

- The sojourn time in S of a unit that enters S from J and exits into K.
- The time between which a unit makes a one-step transition from J to K and its next one-step transition from J' to K'.

For this study, we define routes and passage times for them as follows.

Definition 1.2. *A* **simple route** *of the network is a finite sequence* $r = (r_1, \ldots, r_\ell)$ *in* M. *A unit* **traverses the route** r *if it enters the node* r_1 *and proceeds through the nodes* $r_2, \ldots, r_\ell$ *in that order in the next* $\ell - 1$ *moves. For a unit that traverses the route* r*, its* **passage time** *through a designated subsequence of the route is the total time spent at those nodes in the subsequence.*

Typically, one would be interested in a passage time for a "designated" part of a route and not all of it. An example for $r = (r_1, \ldots, r_\ell)$ is the passage time over the " interior" nodes $r_2, \ldots r_{\ell-1}$: the times at r_1 and r_ℓ are disregarded.

Definition 1.3. *A* **route** $\mathfrak{R}$ *is a finite or infinite collection of simple routes such that one is not a contiguous segment of another. A unit* **traverses the route** $\mathfrak{R}$ *if it traverses one of the simple routes in* $\mathfrak{R}$*, and its* **passage time** *on* $\mathfrak{R}$ *is its passage time on the simple route.*

Be mindful that a route $\mathfrak{R}$ is a non-random set, but a unit traversing it takes a random simple route selected by the evolution of the network process X. The random passage time through $\mathfrak{R}$ is a function of the random simple route selected and the random queueing and service times at its nodes. The condition that any route in $\mathfrak{R}$ cannot be a segment of another route ensures that a single unit cannot traverse two simple routes in $\mathfrak{R}$ simultaneously. However, the routes in $\mathfrak{R}$ may overlap in other ways and several units may traverse $\mathfrak{R}$, or a simple route in $\mathfrak{R}$ simultaneously. Some examples of passage times for routes are as follows.

- The time it takes for a unit to travel from J to K is the passage time for the route
$$\mathfrak{R} = \{(j, i_1, \ldots, i_n, k) : j \in J, i_1, \ldots, i_n \notin J \cup K, k \in K\}$$
on the interior nodes (i.e. the passage time on $r = (j, i_1, \ldots, i_n, k)$ is the total time at nodes $i_1, \ldots, i_n$).
- The time it takes for a unit to pass through the node sets $J_1, \ldots, J_\nu$ is the passage time for the route
$$\mathfrak{R} = \{(i, j_{11}, \ldots, j_{1n}, j_{21}, \ldots, j_{2n}, \ldots, j_{\nu 1}, \ldots, j_{\nu n}, k) : i \notin J_1, k \notin J_\nu$$
$$\text{and } j_{s1}, \ldots, j_{sn} \in J, \text{ for } s = 1, \ldots \nu\}$$

along the interior nodes.

There are other kinds of routes and passage times of interest that do not fall in our setting such as (a) the sojourn time of a unit in the set J at those nodes in which the unit encountered more than 10 units when it arrived at the node, and (b) the time between two occurrences of the event that a unit enters a node with more than 10 units. These routes are functions of the network state and cannot be defined by a fixed set like our $\Re$.

Throughout this study, we consider a fixed route $\Re$. Let T_k denote the passage time for the kth unit that "begins" a traverse through $\Re$, $k = 1, 2, \ldots$. These times are dependent and generally overlap since several units may be traversing $\Re$ at once. Keep in mind that the index k is the kth unit that begins the route $\Re$, but this unit may not be the kth one to complete $\Re$. Our aim is to establish the existence of the average or mean passage time

$$T = \lim_{n \to \infty} n^{-1} \sum_{k=1}^{n} T_k \quad \text{a.s.}$$

and to find a way to compute it. This limit is with probability one and the T will be a constant. In our results, the T is given by the Little-type equation $U = \lambda T$ where U will be the average number of units undergoing a passage on the route $\Re$ and λ is the rate at which units enter the route.

Before presenting our main results, we will give two examples of mean passage times.

3 Examples of Mean Passage Times

The simplest passage times are sojourn times. Here and in the following sections, we let $\pi_j(\cdot)$ denote the jth marginal distribution of π. The π_j is the equilibrium distribution of the number of units at node j. We also let U_j denote the mean of π_j.

Example 3.1 Sojourn Time in the Node Set J The mean sojourn time T in the node set J is given by $U = \lambda T$, where

$$U = \sum_{j \epsilon J} U_j,$$

which is the expected number of units in J in equilibrium, and

$$\lambda = \sum_{i \notin J} \sum_{j \epsilon J} c_N c_{N-1}^{-1} w_i p_{ij},$$

which represents the rate of flow of units into J. It is of interest to note that this mean sojourn time T in J is not the sum $\sum_{j\in J} T^{(j)}$ of the mean sojourn times at the single nodes in J.

This result, which follows from Theorem 4.2, can be proved directly - we did it several years ago using standard applications of laws of large numbers. However, we have not seen this in the literature, aside from the special case with no overtaking. Other passage times such as the next one are more involved.

Here and later, we will use certain exit probabilities defined as follows. Suppose that a single unit moves in discrete time on the set M such that its location over time is a discrete-time Markov chain with transition probabilities $\{p_{jk}\}$, which are the routing probabilities for the network. For a fixed subset J of M, let η_{jk} denote the probability that the unit starting at j in J exits J into the node k in J^c (the k is the first node visited outside J). By conditioning on the first step of the unit, it follows that

$$\eta_{jk} = p_{jk} + \sum_{\ell\epsilon J} p_{j\ell}\eta_{\ell k} \quad j \in J, k \in J^c$$

These equations can be used to compute $\{\eta_{jk}\}$. We call $\{\eta_{jk}\}$ the **exit probabilities** for J.

Example 3.2. The Time It Takes for a Unit to Travel from J to K. Consider the time it takes for a unit to travel from J to K, where J and K are disjoint. For the sake of simplicity, assume that a unit visits at least one node in $(J\cup K)^c$ between the movement from J to K.

Theorem 3.3 *The average time T it takes for a unit to travel from J to K is given by $U = \lambda T$, where*

$$\lambda = \sum_{i\in J} \sum_{j\notin J\cup K} \sum_{k\in K} \frac{c_N}{c_{N-1}} \tilde{\omega}_i p_{ij}\eta_{jk}.$$

and

$$U = \sum_{j\in K^c} \omega^{-1} U_j \sum_{i\in J} \sum_{\delta\in J\cup K} \sum_{k\in K} \omega_j(i,\delta)\eta_{jk}. \tag{3}$$

Here $\{\eta_{jk}\}$ are the exit probabilities for K^c, and $w_j(i,\delta)$ is the solution to

$$\omega_\delta(i,\delta) = \tilde{\omega}_i p_{i\delta} + \sum_{s\notin K} \omega_s(i,\delta)p_{s\delta}$$

$$\omega_\ell(i,\delta) = \sum_{s\notin K} \omega_s(i,\delta)p_{s\ell} \quad \ell \notin K, \ell \neq \delta.$$

and

$$\tilde{\omega}_j = w_j - \sum_{i \in J} \sum_{\delta \in J \cup K} w_j(i, \delta) \quad j \in J.$$

4 Expressions for Mean Passage Times

We are now ready to present our main results. To proceed, we need to introduce a labeling scheme for the units. A unit is a "special" unit if it is traversing a route $r = (r_1, \ldots, r_\ell)$ in $\Re$ and it is in the designated segment of the route. Otherwise a unit is a "non-special" unit. Let $X^*(t) = (X_1^*(t), \ldots, X_m^*(t))$ denote the numbers of special units at the respective nodes at time t. Let $N(t)$ denote the number of units that begin a traverse of the route in $\Re$ during the time period $(0, t]$. the $N(t)$ and $X^*(t)$ may not be determined entirely by the history $\{X(s) : 0 \leq s \leq t\}$ of X: they may depend on the future $\{X(u) : u \geq t\}$ as well. This is true even if $\Re$ were the route for the passage time from J to K. Consequently, $\{(X(t), N(t), X^*(t)) : t \geq 0\}$ is not Markovian. The dependency of $X^*(t)$ and $N(t)$ on the future of X is a major difficulty. To handle these dependencies, we shall consider a broader Markov process that tracks all the units through the network.

Associated with $\Re$, we define $\tilde{\Re}$ as the set of all simple routes $r = (r_1, \ldots, r_\ell)$ such that $r \in \Re$, or $(r_1, \ldots, r_{\ell-1})$ is the initial segment of some simple route in $\Re$ but r is not a route or a segment of a route in $\Re$. In other words, $\tilde{\Re}$ contains $\Re$ and all simple routes whose first $\ell - 1$ elements are initial segments of routes in $\Re$. We will classify the units according to the route they are traversing in $\tilde{\Re}$ and the stage on that route. Specifically, if a unit is traversing on the route $r = (r_1, \ldots, r_\ell)$ and it is currently at node r_s for some $s = 1, \ldots, \ell$, then the unit is of class $\alpha = sr$. That is, the unit is at stage s on the route r. Otherwise, we say that the unit is of class $\alpha = 0$. The set of all classes is therefore

$$C = \{0\} \cup \{sr : r = (r_1, \ldots, r_\ell) \in \tilde{\Re}, s = 1, \ldots, \ell\}.$$

We let $\tilde{X}_{\alpha j}(t)$ denote the number of α-units at node j at time t. The stochastic process $\{\tilde{X}_{\alpha j}(t) : \alpha \in C, j \in M\}$ has states $\tilde{n} = (n_{\alpha j} : \alpha \in C, j \in M)$. Although the process $\tilde{X}(t)$ is not a function of $\{X(s) : s \leq t\}$, its evolution can be described by the basic parameters of X. This is true because the route $\Re$ is defined solely by sequences of nodes, independent of the dynamics of X.

Because of the Markovian nature of X and the definition of $\Re$, it follows that $\tilde{X}$

is a Markov process. The transition rates of $\tilde{X}$ are

$$\tilde{q}(\tilde{n}, T_{\alpha j\beta k}\tilde{n}) = p_{\alpha j\beta k}\frac{n_{\alpha j}}{n_j}\phi_j(n)$$

where $n_j = \sum_\alpha n_{\alpha j}$ is the total number of units at node j and $\frac{n_{\alpha j}}{n_j}$ represents the probability that a unit being served at node j is a type α unit. The $p_{\alpha j\beta k}$ is the probability that an α-unit exiting node j enters node k as a β-unit. To specify these probabilities, we will use the probability

$$p(r) = \Pi_{i=1}^{\ell-1} p_{r_i r_{i+1}}$$

that a unit arriving at node r_1 will traverse the route $r = (r_1, \ldots, r_\ell)$. Then from the preceding description

$$p_{\alpha j\beta k} = \begin{cases} p_{jk}p(r) & \text{if } (\alpha,\beta) = (0, 1r) \qquad k = r_1 \\ p_{jk} & (\alpha,\beta) = (0,0) \\ 1 & (\alpha,\beta) = (sr, (s+1)r) \quad (j,k) = (r_s, r_{s+1}) \\ p_{jk}p(r') & (\alpha,\beta) = (\ell r, 1r') \quad (j,k) = (r_\ell, r'_1) \\ p_{jk} & (\alpha,\beta) = (\ell r, 0) \qquad j = r_\ell \\ 0 & \text{otherwise.} \end{cases}$$

Here r and r' are in $\Re$, the α, β are in C and the unspecified j, k are any nodes in M for which the transitions are feasible. This completes the definition of the process $\tilde{X}$. We will assume, without loss of generality, that the route $\tilde{\Re}$ is such that $\tilde{X}$ is irreducible and positive recurrent.

Note that the process X^* of special units can be expressed as

$$X_j^*(t) = \sum_{\alpha \in C_j} \tilde{X}_{\alpha j}(t)$$

where $C_j = \{sr : r \in R, r_s = j$, and r_s is in the designated segement of $r\}$. The next result gives the equilibrium distributions of $\tilde{X}$ and X^*.

Theorem 4.1. *The equilibrium distribution of* $\tilde{X}$ *is*

$$\ddot{\pi}(\ddot{n}) = c\Phi(\ddot{n})\Pi_\alpha\Pi_j\frac{n_j!}{n_{\alpha j}!}\omega_{\alpha j}^{n_{\alpha j}}$$

where c is a normalization constant and $\omega_{\alpha j}$ is the solution of the equations

$$\omega_{\alpha j} = \sum_{\beta k} \omega_{\beta k} p_{\beta k\alpha j} \quad \alpha \in C, j \in M.$$

The equilibrium distribution of X_j^* *is*

$$\pi_j^*(\nu)\sum_{k=\nu}^{\infty}\pi_j(k)\frac{k!}{\nu!(k-\nu)!}(\frac{\omega_j^*}{\omega_j})^{\nu}(1-\frac{\omega_j^*}{\omega_j})^{k-\nu}$$

where $\omega_j = \sum_{\alpha\in C}\omega_{\alpha j}$, *and* $\omega_j^* = \sum_{\alpha\in C_j}\omega_{\alpha j}$.

The following theorem is our main result. It says that the mean passage time

$$T = \lim_{n\to\infty} n^{-1}\sum_{k=1}^{n} T_k \quad \text{a.s.}$$

exists and is a constant given by $U = \lambda T$, where

$$\lambda = \lim_{t\to\infty}\frac{N(t)}{t} \quad \text{a.s.}$$

is the rate of passages through $\Re$, and

$$U = \lim_{t\to\infty} t^{-1}\int_0^t \sum_j X_j^*(s)ds \quad \text{a.s.}$$

is the average number of units undergoing a passage in $\Re$. The λ and U are non-random constants. The relation $U = XT$ is similar to Little's law $L = \lambda W$ for a queueing system relating the average queue length L to the average waiting time W in the system.

For the following, we let $\tilde{n}$ be a fixed state for $\tilde{X}$ and let $0 < \tau_1 < \tau_2 \ldots$ denote the successive times at which $\tilde{X}$ enters $\tilde{n}$.

Theorem 4.2. *Under the preceding assumptions, the limits* T, λ *and* U *exist and are constants of the form*

$$\begin{aligned}
\lambda &= \frac{E[N(\tau_2)-N(\tau_1)]}{E[\tau_2-\tau_1]} \\
&= \sum_j \sum_{r\in\Re}\frac{c_N}{c_{N-1}}\omega_{0j}p_{jr_1}p(r) \\
&\quad + \sum_{r\in\tilde{\Re}}\sum_{r'\in\Re}\frac{c_N}{c_{N-1}}\omega_{\ell r r_\ell}p_{r_\ell}r_1'p(r') \\
U &= \frac{E[\int_{\tau_1}^{\tau_2}\sum_j X_j^*(s)ds]}{E[\tau_2-\tau_1]}
\end{aligned}$$

$$= \sum_{j\in M} \frac{\omega_j*}{\omega_j} U_j$$

$$T = \frac{E[\sum_{k=N(\tau_1)}^{N(\tau_2)} T_k]}{E[N(\tau_2) - N(\tau_1)]}.$$

Furthermore, $U = \lambda T$.

The proof [7] uses Theorem 4.1 and three applications of strong laws of large numbers for regenerative processes. The following are some extensions and related results.

Interpretation of T as an Expected Passage Time Suppose $\{\tilde{T}_k\}$ is a stationary version of the passage times $\{T_k\}$. Then it follows that

$$E\tilde{T}_k = T.$$

In other words, T is the expected passage time for the stationary times.

Convergence of Expected Passage Times The limits λ, U and T in Theorem 4.2 are also the limits of expectations of the variables. That is,

$$\lim_{t\to\infty} E[t^{-1}N(t)] = \lambda, \lim_{t\to\infty} E[t^{-1}\int_0^t \sum_j X_j^*(s)ds] = U$$

and

$$\lim_{t\to\infty} E[n^{-1}\sum_{k=1}^{n} T_k] = T = \lambda^{-1}U.$$

These follow by using the key renewal theorem for regenerative processes.

Central Limit Theorem We have shown that the average $n^{-1}\sum_{k=1}^n T_k$ converges to $T = \lambda^{-1}U$ with probability one. The deviation between this average and its limit can be described by the following central limit theorem. Define

$$Z_n = \frac{1}{A\sqrt{n}}\sum_{k=1}^{n}(T_k - T)$$

where

$$A^2 = \frac{E[\sum_{k=N(\tau_1)+1}^{N(\tau_2)} T_k - T_{\tau_1}]^2}{E[\tau_2 - \tau_1]}.$$

Theorem 4.3. *Suppose* $E(\tau_1 - \tau_2)^2$ *and* $E[(\sum_{k=N(\tau_1)+1}^{N(\tau_2)} T_k^2]$ *are finite. Then* A *is finite and* Z_n *converges in distribution to a standard normal random variable as* $n \to \infty$.

This and more general functional central limit theorems follow by such theorems for regnerative processes (see for instances [13]).

Regenerative and Semi-Stationary Network Processes We have assumed that X is a Markovian network process. The results above will carry over in an obvious way to a non- Markovian process $\tilde{X}$ such that $\tilde{X}$ is regenerative over $\tau_1, \tau_2, \ldots$. Another generalization is that $\tilde{X}$ is semi-stationary and ergodic over these times as in [13].

Open Networks and Networks with Multiple Types of Units The results above also apply, with appropriate changes in notation, to open networks and to closed or open networks with multiple types of units, where the units may change types as they move. For instance, one may be interested in the mean time that a unit spends in J as a certain type of unit (it may change its type several times while in J).

5 Passage Times for an Open Network with Single Server Nodes

In certain networks in which the input flow to a route is a Poisson process, the passage time distribution or its moments may be more tractable than when the input flow to the route is not Poisson. In this section, we discuss such a network and derive expressions for the moments of a passage time. Throughout this section, we assume that $\{X(t) : t \geq 0\}$ is an open version of the network process described above. The actual form of the transition rates and equilibrium distribution π is not required and so we will not display them (see [14]). A special case is that X is an open Jackson network. We assume that each node has a single server that serves units on a first-come-first-serve basis. Furthermore, we assume, for simplicity, that the network is acyclic: a unit never visits a node more than once. Units may overtake one another.

We shall consider a passage time for the route $\Re$. A unit traversing $\Re$ takes a random simple route selected by the evolution of the network process. Let $p(r)$ be the probability that a unit traversing the route $\Re$ takes the simple route r. If one can get the passage time distribution $F_r(t)$ for the simple route r, then the passage time distribution for $\Re$ has the mixed distribution

$$F_{\Re}(t) = \sum_{r \in \Re} p(r) F_r(t).$$

The characterization of $F_{\Re}$ therefore reduces to the characterization of F_r for a simple route r. Generally r may not be an overtake free route, and its passage time distribution is intractable. We will, therefore, restrict our attention to obtaining expressions for the moments of the passage time distribution F_r.

We label the units as in section 4. That is, a unit is a special unit if it is traversing r and it is in the designated segment of the route. Otherwise a unit is a non-special unit. Our discussion is for passage times when the system is in equilibrium, and so we assume that X is stationary. Let U be the random variable denoting the number of special units undergoing a passage on r. This U corresponds to the number of special units $\sum_j X_j^*(t)$ in section 4 whose limiting mean we denoted as U. Let T be the random variable denoting the passage time for a unit on the designated segment of the route r. The T corresponds to T_k in section 4 when $\{T_k : k \geq 1\}$ is stationary. As before, let λ denote the rate at which passages on r begin. The following result is a relation between the passage time moment ET^ℓ and the ℓth factorial moment of U, which is defined by

$$E[U^\ell] = E[\frac{U(U-1)\dots(U-\ell+1)}{\ell!}].$$

Theorem 5.1. *For the open acyclic queueing network described above with single-server nodes,*

$$E[T^\ell] = \lambda^{-\ell} \ell! E[U^{(\ell)}]. \tag{4}$$

Proof As in section 4, we can construct a Markovian network process $\tilde{X}$ for keeping track of traversing units. First note that from [14], we know that $\tilde{X}$ has the property that "a moving unit sees a time average". Therefore, the distribution of the state of the network process $\tilde{X}$ left behind by a unit departing from the designated segment of the route r is the same as the equilibrium distribution of $\tilde{X}$. Now the special units don't pass each other on the simple route r since the nodes serve units one at a time. Then the number of special units left behind by the unit departing from the designated segment of the route r is the number of special units that arrived during the passage time of that unit on r. Also, since the open network is acyclic, the stream of units entering r is a Poisson process [14] with rate λ. From these observations, it follows that

$$P\{U = i\} = P\{N(T) = i\}$$

where $\{N(t) : t \leq 0\}$ is a Poisson process with rate λ independent of T. This says that the equilibrium distribution of U equals the distribution of the number of special units left behind on the designated segment of r by the unit departing there. Then the generating function of U is

$$\begin{aligned} E[z^U] &= E[z^{N(T)}] \\ &= E[E(z^{N(t)} \mid T)] \\ &= E[exp^{-\lambda T(1-z)}]. \end{aligned}$$

Repeated differentiation of this with respect to z yields (4).□

To compute the factorial moments of U, one can use the following expression. Here X_j is a random variable denoting the number of units at node j in equilibrium.
Theorem 5.2 *Under the assumptions of Theorem 5.1,*

$$E[U^{(\ell)}] = \sum_{\alpha \in A_\ell} \Pi_{j=1}^{m} (\frac{w*_j}{w_j})^{\alpha_j} E[X_j^{(\alpha_j)}],$$

where A_ℓ is the set of all vectors $(\alpha_1, \ldots, \alpha_m)$ with non-negative integer entries such that $\alpha_j = 0$ when j is not on the route r and $\sum \alpha_j = \ell$.

References

[1] Daduna, H. (1982). Passage Times for Overtake-Free Paths in Gordon-Newell Networks. *Adv. Appl. Prob. 14*, 672- 686.

[2] Daduna, H. (1986). Cycle Times in Two-Stage Closed Queueing Networks: Applications to Multiprogrammed Computer Systems with Virtual Memory. *Operations Res. 34*, 281-288.

[3] Fayolle, G., Iasnogorodski, R. and Mitrani, I. (1983). The Distribution of the Sojourn Time in a Queueing Network with Overtaking: Reduction to a Boundary Value Problem. In *Performance '83* (eds. Agrawala, A.K. and S.K. Tripathi), North Holland, Amsterdam.

[4] Kelly, F.P. (1979). Reversibility and Stochastic Networks. John Wiley and Sons.

[5] Kelly, F.P. and Pollett, P.K. (1983). Sojourn Times in Closed Queueing Networks. *Adv. Appl. Prob. 15*, 638-656.

[6] Kook, K. (1989). Equilibrium Behavior of Markovian Network Processes. Ph.D. Thesis, Georgia Institute of Technology.

[7] Kook, K. and R.F. Serfozo (1989). Mean Passage Times in Markovian Network Processes. Technical report Georgia Institute of Technology (in preparation).

[8] Kuehn, P.J. (1979). Approximate Analysis of General Queueing Networks by Decomposition. *IEEE Trans. Comm. COM-27*, 113-126.

[9] Lemoine, A.J. (1979). Total Sojourn Time in Networks of Queues. TR No. 79-020-1, Systems Control, Inc., Palo Alto, California.

[10] Mckenna, J. (1989). A Generalization of Little's Law to Moments of Queue Lengths and Waiting Times in Closed, Product-Form Queueing Networks, *J. Appl. Prob. 26* 121-133.

[11] Melamed, B. (1982), Sojourn Times in Queueing Networks, *Math. Oper. Res. 7*, 223-244.

[12] Reich, E. (1957). Waiting Times When Queues are in Tandem. *Ann. Math. Statist. 28* 768-773.

[13] Serfozo, R.F. (1975). Functional Limit Theorems for Stochastic Processes Based on Embedded Processes. *Adv. Appl. Prob. 1*, 125-139.

[14] Serfozo, R.F. (1989). Markovian Network Processes: Congestion-Dependent Routing and Processing. To appear in *Queueing Systems.*

[15] Shassberger, R. and H. Daduna (1983). The Time for a Round Trip in a Cycle of Exponential Queues. *J. ACM 30*, 146-150.

[16] Walrand, J. and Varaiya, P. (1980). Sojourn Times and Overtaking Condition in Jacksonian Networks. *Adv. App. Prob. 12*, 1000-1018.

[17] Whittle, P. (1986). Systems in Stochastic Equilibrium, John Wiley and Sons.

MEAN VALUE ANALYSIS FOR THE DURATION OF HEAVY TRAFFIC PERIODS IN SUBNETWORKS OF A QUEUEING NETWORK

Hans Daduna

Institut für Mathematische Stochastik, Universität Hamburg,
Bundesstraße 55, D-2000 Hamburg

Summary: We investigate under equilibrium conditions the duration of times where in a prescribed subnetwork of a product-form network at each node of the subnetwork at least a fixed number of customers is present. For the open network case we obtain closed formulas, for the closed network case recursion formulas and computational algorithms to compute mean values for such times intervals.

I. Introduction

Congestion control is one of the most important problems in networks of queues.

In planning such networks, however, the standard procedure is modelling the system as a product-form network (REISER [82]) and then (a) identifying bottlenecks and overloaded subnetworks, (b) changing parameters (e.g., channel capacities, work rates), if possible, (c) deciding where additional congestion control has to be implemented.

This paper contributes to the problem of characterizing the behaviour of overloaded subnetworks in product-form networks; to be more precise:

Fix a prescribed subnetwork $\tilde{I}$ and a congestion level $c \geq 0$ (where c is thought to be large). Call a time period during which each node of $\tilde{I}$ has at least c customers present a "heavy traffic period of order c for $\tilde{I}$".

We compute the mean duration of such periods.

The discussion of these and similar problems was opened by KÜHN [83 a,b] who investigated busy periods for nodes in stochastic networks and introduced the notion of "simultaneous busy periods" (which in our terminology would be a "heavy traffic period of order 1") for a set of prescribed nodes. For some simple networks he derived first-entrance-equations for the Laplace-Stieltjes transform of the residual simultaneous busy period distribution, which can be solved numerically. The computational difficulties which arise in solving this and similar problems became appearent from KÜHN's own developements, and from the solution of a simple example in DADUNA [85]. Only recently some related problems were investigated by MASSEY [87] (simultaneous busy period for the nodes of an open tandem of $M/M/1/\infty$ queues), BACCELLI/MASSEY/WRIGHT [88] (the same for a closed cyclic queue) and BACCELLI/MASSEY [88] (busy periods and simultaneous busy periods for an open two-stage tandem queue). In these papers residual (simultaneous) busy period distributions are expressed in terms of lattice Besselfunctions which can be used to compute asymptotic expansions of the probabilities.

While in these tandem networks the service rates at the nodes are assumed to be state-independent, in DADUNA [88], [89] the problem of computing mean (simultaneous) busy periods for a prescribed network in (general) product-form networks with state-dependent service rates is addressed and computational algorithms are derived which resemble the usual algorithms of performance analysis.

In the present paper for ease of notation we deal with single chain GORDON-NEWELL and JACKSON networks with state-dependent service times.

In section II for the closed network case we derive a recursion and a computational algorithm, in section III a series representation for the open network case, which all give mean heavy traffic period durations of general order.

In section IV we discuss obvious generalizations.

II. Heavy traffic periods in closed networks

We consider a GORDON-NEWELL network of single server nodes $\{1,2,\ldots,J\} = \tilde{J}$ under first-come-first-served (FCFS) discipline with state-dependent service rates. If at node j there are $n_j > 0$ customers present, the service rate at node j is $\mu_j(n_j) > 0$, $\mu_j(0) = 0$, $j=1,\ldots,J$.

The network is closed with $N \geq 1$ identical customers cycling in it. The migration of the customers is governed by an irreducible Markovian routing matrix $R = (r_{ij}: i,j=1,\ldots,J)$.

Let $\alpha = (a_1,\ldots,\alpha_J)$ be the stochastic solution of the traffic equation $x = x \cdot R$. Then the steady-state distribution of the Markovian joint queue length process $X = (X(t):t\geq 0)$ with state space

$$S = \{n_1,\dots,n_J) : n_j \in \mathbb{N}\ ,\ j=1,\dots,J,\ n_1 +\dots+n_J = N\}$$

(where n_j is the queue length at node j including the customer just in service if any,) is $\pi = (\pi(\bar{n}) : \bar{n} \in S)$, given by

$$(2.1)\qquad \pi(n_1,\dots,n_J) = \left[\prod_{j=1}^{J}\ \prod_{k_j=1}^{n_j} \frac{\alpha_j}{\mu_j(k_j)}\right]\cdot G(N,J)^{-1},\ (n_1,\dots,n_J) \in S,$$

where $G(N,J)^{-1}$ is the norming constant. We assume henceforth that the network is in equilibrium.

In this network we define for the subnetwork $\tilde{I} = \{1,2,\dots,I\}$, $1 \leq I \leq J-1$, and for $c \geq 1$ with $c\cdot I \leq N$, the heavy traffic period of order c for $\tilde{I}$ to be a maximal connected time interval during which at each of the nodes $1,\dots,I \in \tilde{I}$ at least c customers are present. (For definiteness we assume such period to be continued if at the same moment a departure, say, from node $i \in \tilde{I}$ would let it end, but due to a new arrival at node i a new such period would commence.)

Let $T(\tilde{I},c)$ be a generic $\mathbb{R}_+$-valued random variable distributed like the heavy traffic period of order c for $\tilde{I}$ in equilibrium, and denote by $t(\tilde{I},c)$ its mean.

We have the following

<u>2.1 Theorem</u>

Set $\tilde{R} = \tilde{J} - \tilde{I}$ and for $\tilde{L} \subseteq \tilde{J}$, $M \geq 0$,

$$S(\tilde{L},M) = \{(n_j : j\in\tilde{L}) : n_j\in\mathbb{N},\ j\in\tilde{L},\ \sum_{j\in\tilde{L}} n_j = M\},$$

and for $\tilde{K} \subseteq \tilde{I}$ and $M \geq |\tilde{K}|\cdot c$

$$h(\tilde{K},M,c) = \sum_{\substack{(n_j:\, j\in\tilde{K}\cup\tilde{R})\,\in S(\tilde{K}\cup\tilde{R},M)\\ n_j\,\geq\, c,\ j\in\tilde{K}}} \ \prod_{j\in\tilde{K}\cup\tilde{R}} \left[\prod_{k_j=1}^{n_j} \frac{\alpha_j}{\mu_j(k_j)}\right].$$

Then the equilibrium mean heavy traffic period of order c for $\tilde{I}$ is

$$t(\tilde{I},c) =$$

$$(2.2) \quad = h(\tilde{I},N,c) \cdot \left[\sum_{i=1}^{I} \alpha_i (1-r_{ii}) \left[\prod_{k_i=1}^{c-1} \frac{\alpha_i}{\mu_i(k_i)} \cdot h(\tilde{I}-\{i\},N-c,c) \right]\right]^{-1}.$$

Proof: We apply a relation known from renewal theory which connects the lifetime distribution and the steady state residual lifetime distribution. This relation was proved to hold for suitably defined "ergodic sojourn time distributions" and "ergodic exit time distributions" in a subset of the state space for any Markov process by KEILSON [79], p. 93 .

Let

$$G = \{(n_1,\ldots,n_J) : n_j \in \mathbb{N},\ j \in \tilde{J},\ n_i \geq c,\ i \in \tilde{I},\ n_1+\ldots+n_J = N\} \subseteq S$$

and B = S - G. Then a heavy traffic period of order c for $\tilde{I}$ is a sojourn time of X in G.

By standard Markov process arguments (KELLY/POLLETT [83] it follows that under equilibrium conditions the probability that a heavy traffic period of order c for $\tilde{I}$ is started in state

$$(n_1,\ldots,n_I,\ldots n_J) \in G ,$$

(where for at least one $i_o \in \tilde{I}$: $n_{i_o} = c$)

is

$$(2.3) \quad p(n_1,\ldots,n_I,\ldots,n_J) =$$

$$= \sum_{i=1}^{I} 1_{(n_i=c)} \sum_{\substack{j=1 \\ j\neq i}}^{J} \pi(n_1,\ldots,n_i-1,\ldots,\ n_j+1,\ldots,n_J)\cdot\mu_j(n_j+1)\cdot r_{ji}\cdot$$

$$\cdot \left[\sum_{\substack{(m_1,\ldots,m_J)\in G \\ \exists k\in\tilde{I}: m_k = c}} \sum_{i=1}^{I} 1_{(m_i=c)} \cdot \right.$$

$$\left. \cdot \sum_{\substack{j=1 \\ j\neq i}}^{J} \pi(m_1,\ldots,m_i-1,\ldots,\ m_j+1,\ldots,m_J) \cdot \mu_j(m_j+1) \cdot r_{ji} \right]^{-1}$$

The numerator in (2.3) is just the ergodic flow rate (KEILSON [79], p.86, 87) or probability flux from B to $(n_1,\ldots,n_I,\ldots,n_J)$ (KELLY/POLLETT [83]).

Therefore, conditioning the distribution of $T(\tilde{I},h)$(on the state of the network just after the heavy traffic period of order c for $\tilde{I}$ commences) shows that we have to compute the "ergodic" sojourn time in G" as KEILSON

[79], p. 89, defines it. KEILSON proved as a corollary to his result on the exit and sojourn time distribution that the "mean ergodic sojourn time in G" can be computed from

(2.4) $\pi(G) = h_{GB} \cdot t(I,c)$,

where h_{GB} is the ergodic flow rate from G to B, and $\pi(G)$ the stationary probability for X to be in G.

To solve this equation for $t(\tilde{I},c)$ we write explicitely:

$$\sum_{\substack{(n_1,\dots,n_J)\in S \\ n_i \geq c,\, i\in\tilde{I}}} \prod_{j=1}^{J} \left[\prod_{k_j=1}^{n_j} \frac{\alpha_j}{\mu_j(k_j)}\right] \cdot G(N,J)^{-1} =$$

$$= \sum_{\substack{(n_1,\dots,n_J)\in S \\ n_k \geq c,\, k\in\tilde{I} \\ \exists i\in I: n_i = c;}} \prod_{j=1}^{J} \left[\prod_{k_j=1}^{n_j} \frac{\alpha_j}{\mu_j(k_j)}\right] \cdot G(N,J)^{-1} \cdot$$

$$\cdot \left[\sum_{i=1}^{I} 1_{(n_i = c)}\, \mu_i(c) \cdot (1-r_{ii})\right] \cdot t(\tilde{I},c) ,$$

where we used $h_{GB} = h_{BG}$.

This yields $t(\tilde{I},h) =$

$$= \left[\sum_{\substack{n_1+\dots+n_J = N \\ n_i \geq c,\, i\in\tilde{I}}} \sum_{j=1}^{J} \left[\prod_{k_j=1}^{n_j} \frac{\alpha_j}{\mu_j(k_j)}\right]\right] \cdot$$

$$\cdot \left[\sum_{i=1}^{I} \left[\sum_{\substack{n_1+\dots+n_J = N \\ n_i = c, \\ n_k \geq c,\, k\in\tilde{I}-\{i\}}} \prod_{j=1}^{J} \left[\prod_{k_j=1}^{n_j} \frac{\alpha_j}{\mu_j(k_j)}\right]\right] \cdot \mu_i(c) \cdot (1-r_{ii})\right]^{-1} =$$

$$= \left[\sum_{\substack{n_1+\dots+n_J = N \\ n_i \geq c,\, i\in\tilde{I}}} \prod_{j=1}^{J} \left[\prod_{k_j=1}^{n_j} \frac{\alpha_j}{\mu_j(k_j)}\right]\right] \cdot$$

$$\cdot \left[\sum_{i=1}^{I} \left(\prod_{k_i=1}^{c-1} \frac{\alpha_i}{\mu_i(k_i)} \right) \left\{ \sum_{\substack{n_1+\ldots+n_J = N-c \\ n_k \geq c, k \in \tilde{I}-\{i\} \\ n_i=0}} \prod_{\substack{j=1 \\ j \neq i}}^{J} \left(\prod_{k_j=1}^{n_j} \frac{\alpha_j}{\mu_j(k_j)} \right) \right\} \alpha_i (1-r_{ii}) \right]^{-1} .$$

A special case of importance follows from theorem 2.1 as

2.2 Corollary

Assume the service rates at nodes $1,\ldots,I$ of the subnetwork $\tilde{I}$ are state--independent, i.e.: $\mu_j(n) = \mu_j$, $n > 0$, and let for $\tilde{L} \subseteq \tilde{J}$, $M \geq 0$

$$g(\tilde{L},M) = \sum_{(n_j : j\in\tilde{L}) \in S(\tilde{L},M)} \prod_{j\in\tilde{L}} \left\{ \prod_{k_j=1}^{n_j} \frac{\alpha_j}{\mu_j(k_j)} \right\} .$$

(2.5) Then $t(\tilde{I},c) =$

$$= \left(\prod_{i=1}^{I} \frac{\alpha_i}{\mu_i} \right) g(\tilde{J}, N-c\cdot I) \cdot$$

$$\cdot \left[\sum_{i=1}^{I} \alpha_i (1-r_{ii}) \left(\prod_{\substack{j=1 \\ j\neq i}}^{I} \frac{\alpha_j}{\mu_j} \right) \cdot g(\tilde{J}-\{i\}, N-c\cdot I) \right]^{-1} .$$

2.3 Remarks:

(1) Note that for the result of Corollary 2.2 only the nodes of subnetwork $\tilde{I}$ must have state-independent service rates.

(2) (2.5) reveals a noteworthy invariance property, which agrees with intuition: $t(\tilde{I},c)$ depends on N, the number of customer in the network, and c, the order of the high traffic period, only through the difference $d(I) = N-c\cdot I$.

It follows that under the assumptions of 2.2 the mean heavy traffic period of order c with N customer equals the mean "heavy traffic period order 1" with $N-(c-1)\cdot I$ customers in the network.

The latter case was investigated by DADUNA [89], Corollary 2.2, denoted as mean "simultaneous busy period" for the subnetwork $\tilde{I}$ according to the definitions used by KÜHN [83a,b].

2.4 Corollary

Assume that the nodes $1,\dots,I$ of the subnetwork $\tilde{I}$ are exponential multi-server nodes with m_i service channels at node i, each having service intensity μ_i, $i = 1,\dots,I$, and let for $\tilde{L} \subseteq \tilde{J}$ and $M \geq 0$

$\bar{g}(\tilde{L},M) =$

$$= \sum_{(n_j: j\in\tilde{L})\in S(\tilde{L},M)} \left[\prod_{j\in\tilde{L}\cap\tilde{I}} \left(\frac{\alpha_j}{\mu_j\cdot m_j}\right)^{n_j} \right] \left[\prod_{j\in\tilde{L}\cap\tilde{R}} \prod_{k_j=1}^{n_j} \frac{\alpha_j}{\mu_j(k_j)} \right] .$$

(2.6) Then for $c \geq \max(m_i : i\in I)$ $t(\tilde{I},c) =$

$$= \left[\prod_{i=1}^{I} \frac{\alpha_i}{\mu_i\cdot m_i} \right] \bar{g}(\tilde{J},N-c\cdot I) \cdot$$

$$\cdot \left[\sum_{i=1}^{I} \alpha_i(1-r_{ii}) \left[\prod_{\substack{j=1\\ j\neq i}}^{I} \frac{\alpha_j}{\mu_j m_j} \right] \bar{g}(\tilde{J}-\{i\},N-c\cdot I) \right]^{-1} .$$

2.5 Remarks:

(1) Remarks 2.3 (1),(2) apply as well.

(2) Under the assumptions of Corollary 2.4 the mean heavy traffic period of order c for $\tilde{I}$ equals the mean "simultaneous busy period" for $\tilde{I}$ in a network where the multiservers at nodes $i\in\tilde{I}$ are substituted by exponential single server nodes with state-independent service rates $(\mu_i\cdot m_i)$. This is clear from the observation that nodes $i\in\tilde{I}$ in the original network have all channels continuously working during a heavy traffic period of order $c \geq m_i$.

For the special cases of Corollaries 2.2 and 2.4 we have derived the mean heavy traffic period of order c for $\tilde{I}$ in terms which are well known in performance analysis of queueing networks. In fact, $g(\tilde{L},M)$ and $\bar{g}(\tilde{L},M)$ are norming constants (partition functions) for the steady state probabilities of closed queueing networks, and several algorithms exist to compute them in an efficient way, e.g.: convolution algorithm, LBANC algorithm (see BRUELL/BALBO [80], LAVENBERG [83]), extended mean value analysis (see AKYILDIZ/BOLCH [83]), RECAL-recursion (see CONWAY/GEORGANAS [86]).

In these cases we therefore have transformed the problem of computing $t(\tilde{I},c)$ to performing well known procedures. On the other hand the terms $h(\tilde{K},M,c)$ which are needed to be evaluated in (2.2) in general are not partition functions. But, obviously algorithms similar to those mentioned above can be derived to deal with that case. We present a recursion similar to BUZEN's [73] algorithm.

2.6 Lemma We have

$$(2.7)\quad h(\emptyset,M,c) = g(\tilde{R},M)\ ,\quad M \geq c,\ c \geq 1\ ,$$

$$(2.8)\quad h(\tilde{K},|\tilde{K}|\cdot c,c) = \prod_{j\in\tilde{K}} \left[\prod_{k_j=1}^{c} \frac{\alpha_j}{\mu_{j(k_j)}}\right]\ ,\quad c \geq 1,\ \tilde{K} \subseteq \tilde{I}\ ,$$

and for $\tilde{K} \subseteq I,\ |\tilde{K}|\cdot c < M,\ i\in\tilde{K}$

$$(2.9)\quad h(\tilde{K},M,c)\ \sum_{m=(|\tilde{K}|-1)\cdot c}^{M-c} h(\tilde{K}-\{i\},m,c)\cdot\left[\prod_{n=1}^{M-m}\frac{\alpha_i}{\mu_i(n)}\right].$$

Proof: Follows immediately from the definition.

Using the formulas of Lemma 2.6 we obtain from (2.2) by successive applications of (2.9) eventually expressions like (2.7) and (2.8), where (2.8) is given above, while (2.7) is a norming constant in the traditional form.

This proves the following:

2.7 Algorithm

"Computing the mean heavy traffic period of order c for subnetwork $\tilde{I}$ in a GORDON-NEWELL network."

Step 1: Compute $g(\tilde{R},M)$ $M = 0,1,\ldots,N-c\cdot I$ by BUZEN's algorithm.

Step 2: For $k = 1,\ldots,I-1$ do:
< for all $\tilde{K} \subseteq \tilde{I}$ with $|\tilde{K}| = k$
compute $h(\tilde{K},M,c)$, $M\in\{k\cdot c,\ldots,N-(I-k)\cdot c\}$,
by (2.8),(2.9) and using the results of step 1 >

Step 3: Compute $h(\tilde{I},N,c)$
by (2.9) using the result of step 2.

Step 4: Compute $t(\tilde{I},c)$
by (2.2) using the results of step 3 and step 2.

2.8 Remark:

The algorithm has a linear structure in $k = |\tilde{K}|$, $k = 1,\dots,I-1$. But on each of these stages in step 2 an inherent parallelismn allows to compute the $h(\tilde{K},M,c)$, $\tilde{K} \subseteq I$, $|\tilde{K}| = k$, $M\in\{k\cdot c,\dots,N-(I-k)\cdot c\}$ (for fixed k) concurrently. The data needed to perform the computations are provided from the computations at stage k-1.

III. Heavy traffic periods in open networks

In this section we state the counterpart to the results of section II for open networks of queues.

We shall use (without explicit definition) the notation of section II.2 with the following additional conventions: The network is now totally open. Identical customers arrive at node j in a Poisson stream of intensity $\gamma_j \geq 0$, $j = 1,\dots,J$. The arrival processes are independent and independent from the service processes. Any customer eventually leaves the network in finite time with probability 1. Routing is governed by $R = (r_{ij} : i,j = 0,1,\dots,J)$, where 0 is absorbing and r_{io} is the probability that a customer leaving node i leaves the network.

$\alpha = (\alpha_1,\dots,\alpha_J)$ is the solution of the equations

$$(3.1)\quad x_j = \gamma_j + \sum_{i=1}^{J} x_i r_{ij}\,, \quad j = 1,\dots,J.$$

The system is assumed to be ergodic and to be in its equilibrium

$\pi = (\pi(\bar{n}) : \bar{n} \in \mathbb{N}^J)$, given by

$$\pi(n_1,\dots,n_J) = \prod_{j=1}^{J} \left[\prod_{k_j=1}^{n_j} \frac{\alpha_j}{\mu_j(k_j)}\right] \cdot G(J)^{-1}, \quad (n_1,\dots,n_J) \in \mathbb{N}^J ,$$

where G(J) is the norming constant.

The definition of a heavy traffic period of order c for $\tilde{I} = \{1,\dots,I\} \subseteq \tilde{J}$ as given in section II is carried over directly and we have the following

3.1 Theorem

For $c \geq 1$ the mean heavy traffic period of order c for the subnetwork $\tilde{I}$ in the open JACKSON network described above is

$$(3.2)\quad t(\tilde{I},c) = \left[\sum_{i=1}^{I} \frac{\alpha_i(1-r_{ii})}{\sum_{n=c}^{\infty} \left(\prod_{k=c}^{n} \frac{\alpha_i}{\mu_i(k)} \right)} \right]^{-1} .$$

Proof: KEILSON's equation (2.4) applies again and can be evaluated directly.

As in the closed network case we have some simple corollaries.

<u>3.2 Corollary</u> If in the JACKSON network described above we have at nodes in $\tilde{I}$ only multiserver nodes with m_i service channels and service rates μ_i, $i\in\tilde{I}$, then for $c \geq \max(m_i : i\in\tilde{I})$

$$(3.3)\quad t(\tilde{I},c) = \left[\sum_{i=1}^{I} \alpha_i \,(1-r_{ii})\, (\mu_i m_i - \alpha_i) \right]^{-1} .$$

For the case of state-independent service rates at the nodes of the sub-network $\tilde{I}$, i.e., $m_i = 1$ for all $i\in\tilde{I}$, we obtain

$$(3.4)\quad t(\tilde{I},c) = \left[\sum_{i=1}^{I} \alpha_i \,(1-r_{ii})\, (\mu_i - \alpha_i) \right]^{-1} , \quad c \geq 1 .$$

<u>3.3. Remarks:</u>

(1) Formulas (3.2), (3.3), (3.4) show a remarkable form of insensitivity: If the rest of the network, i.e., nodes $\tilde{R} = \{I + 1,\ldots,J\}$ is changed in any way, only leaving the partial solution $(\alpha_1,\ldots,\alpha_I)$ of the traffic equation (3.1) invariant, then the mean heavy traffic period of order c for $\tilde{I}$ remains unchanged.

(2) (3.3) and (3.4) reveal the invariance property discussed in Remark 2.3 (2). In (3.4) this implies again that $t(\tilde{I},c)$, $c \geq 1$, equals the mean "simultaneous busy period for nodes 1,...,I" as it was computed in DADUNA [89], Corollary 3.2, while (3.3) can be interpreted similar to Remark 2.5 (2) as mean "simultaneous busy period for nodes in $\tilde{I}$ " in a network with suitably defined state-independent service rates for nodes in $\tilde{I}$.

IV. Discussion

We have computed the mean heavy traffic period of order n for the nodes of a subnetwork of product-form queueing networks.
The results are simple consequences of a general mean value formula and show that for open networks we obtain an extremly simple expression in terms of series which have to be evaluated. For closed networks even in the simplest cases it is not possible to find general closed form solutions due to the occurence of expressions similar to partition functions. But this - as we have shown - allows to apply efficient algorithms similar to those which are well known in performance analysis.

As the proof of theorem 2.1 shows the results hold for much more general networks. With suitable modifications it can be shown that the analogon of theorem 2.1 and 3.1 hold in any (open, closed or mixed) product-form network where all the usual modifications are allowed:
multiple chains of customers, type dependend non-exponential service time distributions at symmetric servers (see KELLY [79], BASKETT/CHANDY/MUNTZ/PALACIOS [76]), etc.. Again suitable algorithms are already available (see e.g. LAVENBERG [83]).

Because the proof depends only on evaluation of steady-state probabilities and ergodic flow rates the mean heavy traffic periods remain unchanged as long as these quantities do not change. This leads to the phenomenon of insensitivity for mean heavy traffic periods under variation of service time distribution at symmetric servers as long as the mean service time is fixed.

The methods used here can be easily applied to investigate heavy traffic periods of "mixed order", i.e. for each of the nodes is prescribed the minimal number of customers staying there during such a period.

In case of a closed network it can be seen that for a parametric analysis of the mean heavy traffic period for the nodes $1,\ldots,I$ the "rest $I+1,\ldots,J$ of the network" can be lumped together into one node with state-dependent service rates by NORTON's theorem (for a discussion see e.g. BALSAMO/IAZEOLLA [84].

The result of the theorem is stated under the assumption of stationary of the investigated busy periods. It is easy to see that it can be stated as a result on limiting mean busy periods as well (KELLY/POLLETT [83].

V. References

AKYILDIZ, I.F.; BOLCH, G. [83]: Erweiterung der Mittelwertanalyse zur Berechnung der Zustandswahrscheinlichkeiten für geschlossene und gemischte Netze, in: KÜHN, P.J.; SCHULZ, K.M.(eds.): Messung, Modellierung und Bewertung von Rechensystemen, Proc. 2. GI/NTG-Fachtagung, Stuttgart 1983, Informatik-Fachberichte 61, 267-276, 1983.

BACELLI, F.; MASSEY, W.A. [88]: A transient analysis of the two-node series Jackson network, Report INRIA, 1988.

BACELLI, F.; MASSEY, W.A.; WRIGHT, P.E. [88]: Determining the exit time distribution for a closed cyclic network, Report INRIA, 1988.

BALSAMO, S.; IAZEOLLA, G. [84]: Aggregation and Disaggregation in Queueing Networks: The Principle of Product-Form-Synthesis, in: IAZEOLLA, G.; COURTOIS, P.J.; HORDIJK, A.(eds.): Mathematical Computer Performance and Reliability 95-109, North-Holland, Amsterdam, 1984.

BASKETT, F.; CHANDY, K.M.; MUNTZ, R.R.; PALACIOS, F.G. [1975]: Open, closed, and mixed networks of queues with different classes of customers, J. Assoc. Comput. Mach. 22, 248-260, 1975.

BRUELL, S.C.; BALBO, G. [80]: Computational algorithms for closed queueing networks, North-Holland, New York 1980.

BUZEN, J.P. [73]: Computational algorithms for closed queueing networks with exponential servers. Communications ACM 16, 527-531, 1973.

CONWAY, A.E.; GEORGANAS, N.D. [86]: A new method for computing the normalization constant of multiple chain queueing networks, INFOR 24(3), 1986.

DADUNA, H. [85]: The busy period distribution in a closed tandem of queues, Optimization 16, 755-766, 1985.

DADUNA, H. [88]: Busy periods for subnetworks in stochastic networks: Mean value analysis, J. Ass. Comput. Mach. 35, 668-674, 1988.

DADUNA, H. [89]: Simultaneous busy periods for nodes in a stochastic network, Performance Evaluation 9, 103-109, 1989.

KEILSON, J. [79]: Markov chain models - Rarity and exponentiality, Springer-Verlag, Heidelberg, 1979.

KELLY, F.P. [79]: Reversibility and stochastic networks, Wiley, Chichester, 1979.

KELLY, F.P.; POLLETT, P.K. [83]: Sojourn times in closed queueing networks, Adv. Appl. Prob. 15, 638-656, 1983.

KÜHN, P.J. [83a]: Analysis of busy periods and response times in queueing networks by the method of first passage times, in: AGRAWALA, A.K.; TRIPATHI, S.K. (eds.): Performance '83, 437-455, North-Holland, Amsterdam, 1983.

KÜHN, P.J. [83b]: Analysis of busy period and response time distributions in queueing networks, in KÜHN, P.J.; SCHULZ, K.M.(eds.): Messung, Modellierung und Bewertung von Rechensystemen, Inform. Fachber. 61, Springer-Verlag, Heidelberg, 135-149, 1983.

LAVENBERG, S. (ed.) [83]: Computer performance modelling handbook, Academic Press, New York, 1983.

MASSEY, W.A. [87]: Calculating exit times for series Jackson networks, J. Appl. Prob. 24, 226-234, 1987.

REISER, M. [82]: Performance evaluation of data communication systems, Proc. of the IEEE 70, 171-196, 1982.

ASSEMBLY-LIKE QUEUES AND CLOSED-LOOP SYSTEMS WITH ADMISSION CONTROL

Manfred Kramer
Konstanz

Abstract

This paper studies a queueing system with a double queue of customers and permits fed by independent Markovian input processes. Customers can only be served when a permit for service is available. Their service times are governed by a PH-distribution. This system is isomorphic to a closed two-loop queueing network. We propose a computational approach to the stationary distribution of the queue length. The model was motivated by the sliding window flow-control mechanism widely used in communcation networks.

1. Introduction

Processes requiring simultaneous possession of several resources are ubiquitous in communication networks, operating systems and dataflow computers. A virtual channel connection in a packet switching network controlled by the sliding window protocol may also be seen as such a process, because a packet is transmitted only if a permit to do so can be withdrawn from a limited pool. The number of permits increases by one whenever the correct receipt of a packet is acknowledged [9].

Queueing networks with passive resources and synchronization delays are not covered by the BCMP-theorem. A simple instance is the assembly-like queue, where a single server processes pairs of matching items. Related systems known as semaphore and fork/join queues are descriptive tools for synchronization mechanisms in parallel processing computers [1].

Earlier investigations into assembly-like queues assume that items of different types arrive according to Poisson processes and that their processing times are exponentially distributed. Since supply of items is unlimited , the queue is only stable if a bound on their excess number exists. A matrix-geometric solution for the queue length process on that condition was given in [6].

A recent study focuses on the output process of a similar model with equal bounds on the waiting rooms for both types of items [2]. For vanishing service times the system degenerates to the familiar double-ended queue, which is essentially a single server system with state-dependent arrival rates of two types of customers [8, p. 301]. This fact is exploited in the papers [5] and [9], which provide rather detailed case-studies of communication networks .

In this paper we study a closed queueing network with a main server protected by an admission control policy similar to the sliding window mechanism. This system is shown to be isomorphic to an assembly-like queue with two queues of finite capacity, one for customers and one for permits , fed by state-dependent Markovian input processes. Services are subject to a general PH-distribution. Our main objective is an efficient algorithmic solution to the stationary queue length distribution.

2. The queueing network and the isomorphic single server queue

Let us consider the queueing system defined as follows:
(i) There is a fixed population of N customers and M permits circulating endlessly in the network. Customers and permits join their own admission queues in front of a synchronization barrier. As soon as a customer and a permit are simultaneously present, the customer is transferred immediately to the server queue taking the permit with him.
(ii) The service times of customers (in possession of a permit) are independent and identically distributed according to a common PH-distribution $A(\cdot)$ with irreducible representation (α, T) of order K. Service is rendered by a main server according to the FIFO rule.
(iii) Customer release their permit after service. Both are returned to their admission queues by exponential servers operating at rates $\lambda_n > 0$, $0 \le n < N$, and $\mu_m > 0$, $0 \le m < M$, respectively, when a total of n customers and m permits are waiting for the main server.

This system of queues and servers is depicted in Fig. 1 with symbols for join and fork transitions similarly defined as in stochastic Petri nets.

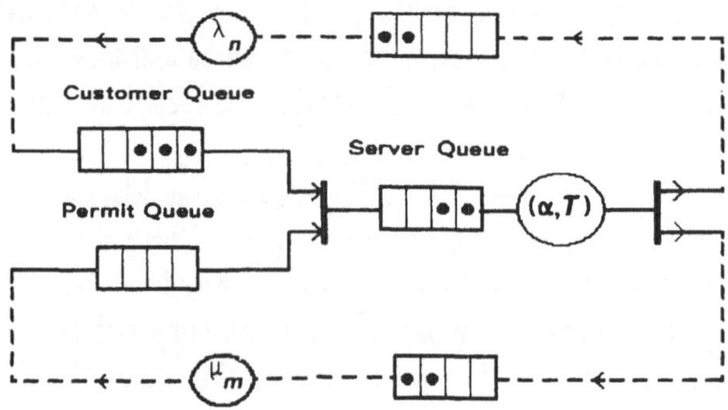

Fig. 1 The Closed-Loop Queueing System

A continuous-time PH-distribution of order K generates a random variable with the same distribution as the distribution of the time until absorption in a Markov process on the set $\{0,...,K\}$, where states or phases $0,...,K$-1 are transient and state K is absorbing. Every PH-distribution can be represented by a pair (α, T), where the vector α and the matrix T are both of order K-1 and specify the initial probabilities of starting in particular transient states and the transition intensities between these states respectively. For more details on PH-distributions see [8, p. 46].

This closed two-loop network represents a generic model of a virtual channel connection controlled by a sliding window, the size of which is given by the maximum number of permits. The PH-distribution allows flexible modelling of network delays, possibly following the guidelines established in [4]. State dependent interarrival intensities mirror the effect of finite packet sources or of throttling the network input on overload. An infinite or Markovian server for permits results from an aggregation step according to Norton's Theorem and represents the reverse channel through which successful transmissions are acknowledged [9].

It is well-known that a closed cyclic loop with a constant customer population served in turn by a general and an exponential server is isomorphic to the familiar M/G/1 model with finite waiting room [3]. This concept of isomorphism is based on the stochastic equivalence of corresponding queueing processes in the steady state.

By virtue of this principle our model can be reduced to an isomorphic single server queueing system by substituting equivalent Markovian input processes with state-dependent intensities for the exponential servers. It suffices also to model the synchronization mechanism only by a queue of customers and one of permits, because the length of the server queue is always equal to the length of the shorter queue.

This assembly-like queueing system is pictorially represented by the interior part of Fig. 1 drawn in solid lines. It is conceptually simpler than the network model and therefore referred to in all subsequent derivations, if not stated otherwise.

3. State-dependent Markovian input

We now introduce the generic term of a service cycle to denote the period betweeen two successive service completions. A service cycle starting with $i \geq 0$ customers and $j \geq 0$ permits may then be called an (i, j)-service cycle.

Let $c_{n,m}^{(i,j)}$, $0 \leq n \leq N\text{-}i$, $0 \leq m \leq M\text{-}j$, be the probability that n customers and m permits arrive during an (i,j)-service cycle. Let $\mathbf{c}_{n,m}^{(i,j)}$ for $i > 0$ and $j > 0$ be the vector whose k-th element is the probability $c_{n,m}^{(i,j,k)}$ of the same event, given the service starts in phase $k, 0 \leq k \leq K\text{-}1$. As a general convention we define quantities with indices outside their specified ranges to be zero.

The duration of a service cycle depends primarily on the initial number of customers i and permits j:

(i) If $i = 0$ or $j = 0$, the server is initially idle. The ensuing service cycle can be decomposed into the time interval until the first arrival and either into a subsequent $(i+1, j)$- or $(i, j+1)$-cycle, depending on whether a customer or a permit arrives at that instant. Conditioning on that event we obtain the following relation for the arrival probabilities.

$$(\lambda_i + \mu_j) c_{n,m}^{(i,j)} = \lambda_i \, c_{n-1,m}^{(i+1,j)} + \mu_j \, c_{n,m-1}^{(i,j+1)} , \tag{1}$$

for $i = 0$ or $j = 0$.

(ii) If $i > 0$ and $j > 0$, service starts immediately. The service process during an (i,j)-service cycle augmented by the number of customers and permits is an absorbing Markov process as the service process itself. Its increments during a first passage from the starting state to the set of absorbing states determine the conditional arrival probabilities.

With the vector of service completion rates $\boldsymbol{t} = -\boldsymbol{T}\mathbf{1}$, where $\mathbf{1}$ denotes the column vector with all its components equal to 1, the standard equation for the first passage probabilities in finite Markov processes given in [11, Theorem 9.3.2] is readily adapted to our problem and rewritten as

$$((\lambda_i + \mu_j)\boldsymbol{I} - \boldsymbol{T})\,\boldsymbol{c}_{n,m}^{(i,j)} = \lambda_i \, \boldsymbol{c}_{n-1,m}^{(i+1,j)} + \mu_j \, \boldsymbol{c}_{n,m-1}^{(i,j+1)} + \boldsymbol{t}\,\delta_{n,0}\,\delta_{m,0} , \tag{2}$$

for $i > 0$ and $j > 0$. The matrix on the left-hand side is a nonsingular M-matrix, since $\boldsymbol{T}$ has negative entries on the diagonal, nonnegatives otherwise, and is invertible under nondegeneracy assumptions [8, p.45] .

The unconditional arrival probabilities are then given by

$$c_{n,m}^{(i,j)} = (1 - \boldsymbol{\alpha}\cdot\mathbf{1})\,\delta_{n,0}\,\delta_{m,0} + \boldsymbol{\alpha}\,\boldsymbol{c}_{n,m}^{(i,j)} , \tag{3}$$

for $i > 0$ and $j > 0$.

The foregoing equations relate the arrival probabilities of the (i,j)-service cycle to those of the $(i+1, j)$- and $(i,j+1)$-service cycles. Thus, starting with the (N,M)-service cycle, the conditional arrival probabilities of each (i,j)-service cycle can be successively determined by solving $N-i$ times $M-j$ linear equations (2) of order K, for descending indices $i > 0$ and $j > 0$. The unconditional probabilities are then obtained from (3) and finally from (1). This amounts to $O(N^2M^2K^3)$ floating point operations in total.

4. Arrivals from Poisson or finite sources

In the following two cases this computational work can be substantially reduced:

(i) Customers from a Poisson source enter a queue of finite capacity N. A customer not accepted is turned away and lost.

(ii) Customers originate from N sources. Each source emits a customer after an exponentially distributed think time, and is blocked until that customer is served.

First we note that in both cases the length of an $(i+1, j)$- and of an (i, j)-service cycle are identically distributed for $i > 0$ and $j \geq 0$.

In case (i) of a Poisson source the number of customers emitted during each of these cycles is also distributed according to the same law. But $N-i-1$ customers at most enter the queue during an $(i+1,j)$-service cycle in contrast to $N-i$ customers during an (i,j)-service cycle, because one waiting position more is available in the latter case.

Hence we obtain the following relationship between the arrival probabilities of both service cycles

$$c_{n,m}^{(i+1,j)} = c_{n,m}^{(i,j)} + \delta_{n,N-i-1}\, c_{n+1,m}^{(i,j)} \quad , \tag{4}$$

In case (ii) of finite sources one of the sources, which may be tagged, is blocked during an $(i+1,j)$-, but not during an (i,j)-service cycle. Thus, the probability that n customers arrive during an $(i+1,j)$-service cycle is the same as the probability that n and $n+1$ customers respectively arrive during an (i,j)-service cycle, depending on whether this tagged source is silent or not.

Given n arrivals from $N-i$ sources during an (i,j)-cycle, the probability that the tagged source has been silent is $(N-i-n)/(N-i)$, since all sources behave according to an identical law. Likewise, given the total number of $n+1$ arrivals, the probability that the tagged source has emitted a customer is $(n+1)/(N-i)$.

Removing the conditioning on the tagged source we obtain the following relationship between the arrival probabilities of both cycles

$$(N-i)\,c_{n,m}^{(i+1,j)} = (N-i-n)\,c_{n,m}^{(i,j)} + (n+1)\,c_{n+1,m}^{(i,j)} \quad . \tag{5}$$

The arrival intensities are given by $\lambda_n = \lambda(1-\delta_{n,N})$ in case (i) of a Poisson source, and by $\lambda_n = (N-n)\lambda$ in case (ii) of finite sources, when n customers are in the queue, $0 \le n \le N$. A few calculations reveal that for both kinds of arrival processes equation (1) can be rewritten in the common form

$$(\lambda_{n+i} + \mu_0)\,\mathbf{c}_{n,m}^{(i,0)} = \lambda_{n+i-1}\,\mathbf{c}_{n-1,m}^{(i,0)} + \mu_0\,\mathbf{c}_{n,m-1}^{(i,1)} \quad , \tag{6}$$

for $i > 0$.

By symmetry, analogous simplifications are possible, if permits arrive with intensities $\mu_m = \mu(1-\delta_{m,M})$ or $\mu_m = (M-m)\mu$, $0 \le m \le M$. Thus we have

$$(\lambda_0 + \mu_{m+j})\,\mathbf{c}_{n,m}^{(0,j)} = \lambda_0\,\mathbf{c}_{n-1,m}^{(1,j)} + \mu_{m+j-1}\,\mathbf{c}_{n,m-1}^{(0,j)} \quad , \tag{7}$$

for $j > 0$.

Equations (4) and (5) remain valid for conditional probabilities too, if we replace $c_{n,m}^{(i,j)}$ by the vector $\mathbf{c}_{n,m}^{(i,j)}$. With the resultant relations and some algebra, equation (2) can again be rewritten in both cases as

$$((\lambda_{n+i}+\mu_{m+j})I-T)\,c^{(i,j)}_{n,m}=\lambda_{n+i-1}\,c^{(i,j)}_{n-1,m}+\mu_{m+j-1}\,c^{(i,j)}_{n,m-1}+t\,\delta_{n,0}\,\delta_{m,0} \tag{8}$$

for $i>0$ and $j>0$.

Equations (4) or (5) and their analogues for permits reduce the arrival probabililites of the $(i+1\,,\,j)$- and $(i\,,\,j+1)$- service cycle to the corresponding probabilities of the $(i\,,\,j)$-service cycle . The computational work needed for this reduction is negligible. After the ultimate step only N-1 times M-1 linear systems (8) for $i=1$ and $j=1$ remain. In the best case the matrix T is bidiagonal as for the generalized Erlang distribution [8, p.46]. Then the computational work amounts to $O(NMK)$ floating point operations .

5. The embedded Markov chain

Consider the bivariate process of the number of customers and permits at departure epochs . The set of states with a common number of customers may be called a level. Let $\pi_{n,m}$ be the stationary probability that a departing customer leaves behind a queue of n customers and m permits, and let $\boldsymbol{\pi}_n$ be the vector with m-th component $\pi_{n,m}$, $0\le m\le M-1$.

The queue length at departure instants is readily shown to be given by the following system of equilibrium equations

$$\boldsymbol{\pi}_n=\sum_{i=0}^{n+1}\boldsymbol{\pi}_i\,C^{(i)}_{n+1-i}\quad , \tag{9}$$

for $0\le n\le N-1$ and $\boldsymbol{\pi}_N=\mathbf{0}$, where $C^{(i,j)}_{n+1-i}$ is the matrix with (j,m)-entry $c^{(i,j)}_{n+1-i,m+1-j}$, $0\le j,\, m\le M-1$. The particular structure of these equations with a coefficient matrix of the M/G/1 type reflects the fact that both a customer and a permit leave simultaneously the system at service completion epochs. Once a solution to these equations is known, the probabilities $\pi_{n,m}$ are obtained uniquely by normalization.

Based on entirely probabilistic arguments we outline now a general procedure to solve systems with a coefficient matrix of the M/G/1 type. A similar approach to G/M/1 type systems was reported in [7].

Let $P_{i,n}$, $0\le i\le n$, be the matrix of probabilities for a first passage from level i to level n without hitting any taboo level $<n$. The matrices $P_{i,n}$ are substochastic for $n\ge 1$, because there is always a positive probability for touching one of the taboo levels $<n$ before reaching the final level n . In the same way the matrix $P_{0,0}$ is seen to be stochastic.

Since the queue length process is skip-free in negative direction, any path from level i to level n-1 which avoids lower levels must either visit the final level n-1 directly after the first service or pass through level n. This decomposition leads to

$$P_{i,n-1} = C^{(i)}_{n-i} + P_{i,n} P_{n,n-1} \tag{10}$$

for $0 \le i \le n$ and $1 \le n \le N$. Of course, we have $P_{i,N} = 0$ for $0 \le i \le N$.

We now show by an inductive proof that the system of equations (9) is equivalent to

$$\pi_n = \sum_{i=0}^{n} \pi_i P_{i,n} \quad , \tag{11}$$

for $0 \le n \le N$.

Equations (9) and (11) are obviously identical for $n = N-1$. Suppose now that the induction hypothesis (11) is true for a fixed index $n \le N-1$. Summing equation (10) premultiplied by the vector π_i for $0 \le i \le n$, replacing the first and second sum arising on the right by π_{n-1} and π_n as in the original equations (9) and the hypothesis respectively we obtain another instance of (11) with n replaced by $n-1$. This completes the proof.

Since $P_{n,n}$ is substochastic for $n \ge 1$, the matrix $I - P_{n,n}$ is a nonsingular M-matrices with a LU decomposition obtained by natural pivoting. Once the matrices $P_{i,n}$ for level n are known, equation (10) can be solved for $P_{n,n-1}$, which in turn determines the remaining matrices $P_{i,n-1}$, $i \le n-1$, for level $n-1$. Starting with $P_{i,N} = 0$, the complete array of matrices $P_{i,n}$ can be generated in descending order of n.

$P_{0.0}$ is the transition matrix of the Markov chain obtained if the embedded process is observed only at queue length level 0. Its invariant vector π_0 is uniquely determined up to a multiplicative constant by equation (11) taken for $n = 0$. Each of the vectors π_n for $n \ge 1$ can then be recursively computed by backsubstituting $\pi_0,\dots,\pi_{n-1}$ into (11) and solving the resultant linear systems with coefficient matrix $I - P_{n,n}$, whose previously stored LU factorization may be reused.

The task dominating all is to set up an almost triangular array of matrices of order M by evaluating equations (10). This requires $O(N^2M^3)$ floating point operations. Because only (non-negative) probabilities are involved, numerically stable computations can be expected.

6. The queue length

Let c be the mean overall length of a service cycle in the steady state. We recall that a service cycle includes not only the time spent in serving a customer , but also a possible idle time of the server.

This initial idle time, if any, is equal to the maximum of the interarrival time of a customer and a permit, if neither a customer nor a permit is left by the customer served last, otherwise is it equal to the interarrival time of the complementing permit or customer. Hence it follows

$$c = \pi_{0,0}\left(\frac{1}{\lambda_0} + \frac{1}{\mu_0} - \frac{1}{\lambda_0+\mu_0}\right) + \sum_{n=1}^{N-1} \pi_{n,0}\frac{1}{\mu_0} + \sum_{m=1}^{M-1} \pi_{0,m}\frac{1}{\lambda_0} + a \quad , \tag{12}$$

where $a = -\alpha T^{-1}\mathbf{1}$ is the mean service time.

Let Π_n be the stationary probability for a queue of n customers at arbitrary instants. We must relate it to the corresponding probability at embedded departure instants given by $\pi_n = \boldsymbol{\pi}_n \mathbf{1}$.

This is possible by virtue of a basic equation given in [10, eq. (4.59)]. Rewritten in our notation and with $1/c$ in place of of the mean throughput it states

$$\Pi_n = c^{-1}\frac{\pi_n}{\lambda_n} , \tag{13}$$

for $0 \le n \le N-1$. Furthermore, we have

$$\Pi_N = 1 - c^{-1}\sum_{n=0}^{N-1}\frac{\pi_n}{\lambda_n} , \tag{14}$$

since probabilities sum up to 1 .

Equation (13) is a fundamental relationship between time- and customer- averages and holds for single server queueing systems with state-dependent Markovian input processes. Its proof is given in [10, p.284] and translates literally to our model.

Little's law enables us to compute the mean actual waiting time of a customer from his admission to the queue until service completion

$$W = Nc - \sum_{n=0}^{N-1}(N-n)\frac{\pi_n}{\lambda_n} . \tag{15}$$

This reduces to the simple formula $W = Nc - 1/\lambda$ in case of arrivals from finite sources. Going back to the isomorphic queueing network, equation (15) expresses the mean sojourn time spent in the interior system as the difference between the mean cycle time of a customer around the loop and a term accounting for the mean delay at the exponential server.

7. Numerical example

A simple numerical example may illustrate the feasibility of the proposed approach. Let us consider the following system:

Each of $N = 6$ sources emits a customer after a think time which is exponentially distributed with parameter $\lambda = 0.2$. A pool accommodating $M = 6$ permits is refilled in a Poisson stream at rate $\mu = 0.5$. Service times are distributed according to a common PH-distribution with representation (α, T) given by $\alpha = \{0.3, 0.1, 0.2, 0.3, 0.1\}$ and

$$T = \begin{vmatrix} -2.0 & 0.0 & 0.2 & 0.0 & 0.0 \\ 1.5 & -3.0 & 0.9 & 0.6 & 0.0 \\ 0.0 & 0.5 & -1.0 & 0.0 & 0.2 \\ 0.0 & 0.0 & 1.0 & -1.2 & 0.0 \\ 0.2 & 0.0 & 0.0 & 1.0 & -2.0 \end{vmatrix}$$

Table 1 shows the joint distribution of the number of customers and permits at departure points in the steady-state.

$\pi_{n,m}$	0	1	2	3	4	5
0	.015	.013	.011	.007	.004	.002
1	.027	.025	.023	.018	.011	.005
2	.041	.038	.039	.037	.029	.015
3	.048	.044	.045	.049	.050	.036
4	.037	.034	.035	.037	.050	.056
5	.015	.013	.014	.014	.017	.043

Tab.1 The stationary joint distribution of the queue length and of the number of available permits at departure instants

In Table 2 the distributions of the queue length at embedded and arbitrary instants are presented.

n	0	1	2	3	4	5	6
π_n	.051	.111	.200	.272	.249	.117	.000
Π_n	.018	.047	.107	.193	.266	.249	.120

Tab.2 The stationary queue length distribution at departure and arbitrary instants

The mean length of a service cycle is $c = 2.34$ and the mean waiting time is $W = 9.07$.

References

[1] F. Baccelli, A.M. Makowski, A. Shwartz, Simple Computable Bounds and Approximations for the Fork-Join Queue
in : *T. Hasegawa , H. Takagi , Y. Takahashi (eds.) , Computer Networking and Performance Evaluation*
North-Holland, Amsterdam- New York - Oxford 1986

[2] U. N. Bhat, Finite Capacity Assembly-like Queues
Queueing Systems **1** (1986), 85-101

[3] J.L Carroll, A. van de Liefvoort, L. Lipsky, Solutions of M/G/1//N-type Loops with Extensions to M/G/1 and GI/M/1 Queues
Oper. Res. **30** (1982), 490-514

[4] S.D. Hohl, P.J. Kühn , Approximate Analysis of Flow and Cycle Times in Queueing Networks
in : *L.F.M. de Moraes , E. de Souza e Silva, L.F.G. Soares (eds.) , Data Communications and their Performance* , 471-485
North-Holland, Amsterdam- New York - Oxford 1988

[5] U. Körner, H.G. Perros, S. Fdida, G. Shapiro , End to End Delays in a Catenet Environment
in : *L.F.M. de Moraes , E. de Souza e Silva, L.F.G. Soares (eds.) , Data Communications and their Performance* , 441-452
North-Holland, Amsterdam- New York - Oxford 1988

[6] G.Latouche, Queues with Paired Customers
J. Appl. Prob. **18** (1981), 684-696

[7] G.Latouche, P.A. Jacobs, D.P. Gaver, Finite Markov Chain Models Skip-Free in One Direction
Nav. Res. Log. Quart. **31** (1984), 571-588

[8] M.F. Neuts, *Matrix-Geometric Solutions in Stochastic Models*
Johns Hopkins University Press, Baltimore-London 1981

[9] M. Reiser, Admission Delays on Virtual Routes with Window Flow Control
in : *G. Pujolle (ed.) Performance of Data Communication Systems and their Applications*
North-Holland, Amsterdam- New York - Oxford 1981

[10] H.C. Tijms, *Stochastic Modeling and Analysis*
J. Wiley, New York 1986

[11] H. Zhenting, G. Quingfeng, *Homogeneous Denumerable Markov Processes*
Springer, Berlin-Heidelberg and Science Press , Beijing 1988

Aktuelle Probleme und Lösungen zur Leistungsanalyse von modernen Rechensystemen mit Hardware-Meßwerkzeugen

Wilhelm Föckeler
Nixdorf Computer AG
Abt. Leistungsanalyse
Pontanusstr. 55
4790 Paderborn

Norbert Rüsing
Universität-GH-Paderborn
FB 14 / FG Datentechnik
Pohlweg 47 - 49
4790 Paderborn

1. Einleitung

Messungen zur Analyse des Leistungsverhaltens eines Rechensystems dienen dem Vergleich der seine Leistung beschreibenden Parameter mit jenen Leistungsparametern, die entweder im Rahmen einer Systemoptimierung angestrebt werden oder die in der Entwurfsphase definiert worden sind. Die Meßinstrumente, die man zur Erkennung und Untersuchung der Ursachen einsetzt, werden in Hardware- und Software-Monitore unterschieden. Das Zusammenspiel von Hard- und Software-Monitor-Funktionalitäten ist unter dem Begriff *Hybrid-Monitoring* zusammengefaßt /1-5/.

Neue Architekturmerkmale und ein hoher Integrationsgrad der eingesetzten Halbleitertechnologie erschweren die Anwendung der bekannten Hardware-Meßverfahren /3, 5/. Handelt es sich um Multi-Tasking-Systeme, ist es nicht ohne weiteres möglich, Meßdaten einem bestimmten Prozeß zuzuordnen. Beim direkten Hardware-Zugriff stoßen auch hybride Meßverfahren an den Schutzkonzepten moderner Betriebssysteme (UNIX, OS/2[1]) und Mikroprozessoren (M680xx, 80x86) auf nicht ohne weiteres zu überwindende Hürden. Bedingt durch die zunehmende funktionelle Integrationsdichte der am Markt angebotenen Mikroprozessoren wird es immer schwieriger, beliebige Meßpunkte galvanisch zu erreichen. Monitore lassen sich daher nicht mehr in herkömmlicher Weise einsetzen.

Um diesen einer allgemeinen Meßfähigkeit entgegenwirkenden Entwicklungen der Rechnertechnik gegenüberzutreten, wurde ein neues hybrides Monitor-Konzept entwickelt /6/. Auf dessen Basis werden seit nunmehr zwei Jahren bei der Nixdorf Computer AG neue prozeßselektive Meßverfahren eingeführt, über die in der vorliegenden Arbeit berichtet wird.

2. Monitoring-Verfahren

Hardware-Monitore sind externe Meßgeräte und in der Regel systemunabhängig. Von ihnen gehen keine Rückwirkungen auf das jeweils zu beobachtende System aus. Diese Monitore "veralten" mit der fortschreitenden Rechner-Technologie. Um eine maximale Flexibilität bei der Anwendung der verschiedenen Meßverfahren zu erhalten, sollte ein Hardware-Monitor immer "schneller" als das Prüflings-System sein.

Bei Software-Monitoren handelt es sich um Programme, die dem Betriebssystem eines Rechensystems angegliedert oder in dieses eingebettet sind und abwechselnd mit den zu messenden Objektprogrammen ablaufen. Daher stellt im Gegensatz zu einem Hardware-Monitor ein sich im aktiven Zustand befindender Software-Monitor eine zusätzliche Systemlast dar. Software-Monitore können nur Informationen verarbeiten und sammeln, die der Software zugänglich sind. Sie eignen sich daher gut zur Beobachtung von Systemzuständen innerhalb der Software-Schichten eines Rechensystems, sie sind aber an einzelne Rechensysteme bzw. Rechnerfamilien und deren Betriebssysteme gebunden.

Die zunehmende Integration in der Halbleiter-Technologie schränkt die Möglichkeiten, auf Hardware-Ebene Informationen über das jeweils beobachtete Rechensystem zu gewinnen, ein. Daher wird die Bedeutung des reinen Hardware-Monitorings für Messungen an Rechensystemen abnehmen. Somit ist es notwendig, konventionelle Verfahren zur Meßdatengewinnung weiterzuentwickeln.

Beim Hybrid-Monitoring werden die Vorteile von Hard- und Software-Monitoren vereint: Mit einfachen, das Zielsystem möglichst gering belastenden Meß- oder Sammelroutinen (*Software-Sensoren*) werden der Software zugängliche Meßdaten in vereinbarte und dem Hardware-Monitor zugängliche Speicherzellen geschrieben (vgl. Abb. 1). Diese können Bestandteil des Arbeitsspeichers, aber auch speziell innerhalb oder außerhalb des Zielsystems implementierte Register sein. Die Vorteile liegen auf der Hand: Ein Hardware-Monitor erhält nun Zugang zu applikations- und betriebsmittelorientierten Informationen der höheren Software-Schichten. Das Zielsystem wird nicht mit der regelmäßigen Abspeicherung der anfallenden Meßdaten belastet, und die Zeitetikettierung der Meßdaten erfolgt nun außerhalb des Zielsystems und läßt sich daher auch mit der gewünschten Genauigkeit durchführen. Informationen, die an chipinternen Meßpunkten eines Mikroprozessors anstehen und einem Hardware-Monitor nicht, wohl aber einem Software-Monitor zugänglich sind, kann man ebenfalls in extern zugängliche Speicherzellen laden (vgl. Kap. 5).

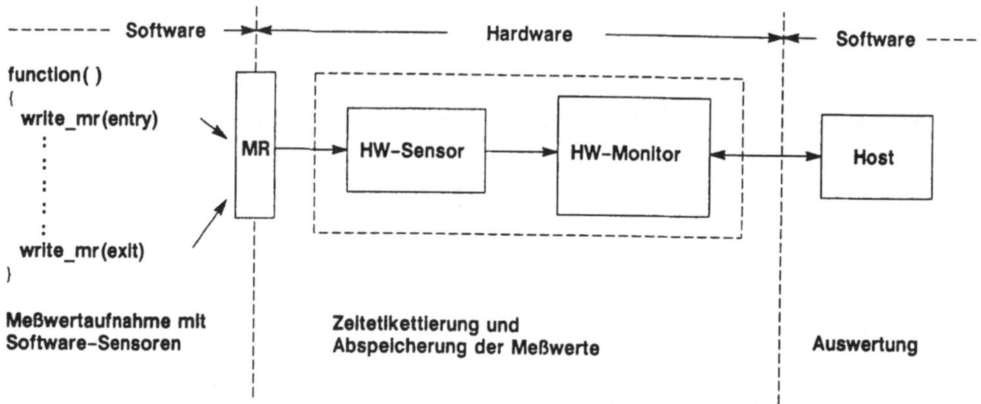

Abb. 1 Ausgabe von SW-Informationen an einen HW-Monitor über ein Monitor-Register (MR)

3. Beschränkungen von Hardware-Meßverfahren bei ihrer Anwendung in Rechensystemen mit modernen Mikroprozessoren

Die von Großrechnern bekannten Architekturen sind mittlerweile auf der Ebene der 32 Bit-Mikroprozessoren verfügbar. Sie schränken die Wirksamkeit der konventionellen Monitoring-Methoden ein. Diese Erschwernisse rühren nicht nur von den eigentlichen Architekturmerkmalen her. Sie werden auch durch die hohe Integration von Funktionalitäten auf den CPU-Chips verursacht.

So sind logische Adreßräume nutzbar, welche den Umfang eines wirtschaftlich sinnvoll realisierbaren physikalischen Arbeitsspeichers übersteigen. Auf solchen Prozessoren basierende Multi-Tasking-Systeme verwenden virtuelle Speicherkonzepte. Diese bewirken, daß die mit einem Hardware-Monitor von Speicherzugriffen abgeleiteten Meßdaten nicht mehr einem bestimmten Prozeß zugeordnet werden können.

Zur Durchsatzsteigerung sind Mikroprozessoren in den meisten Fällen mit internen Pipeline-Strukturen versehen. Diese Eigenschaft hat u.a. zur Folge, daß vorgeholte Instruktionen nach einer ausgeführten Verzweigung nicht mehr abgearbeitet werden. Insbesondere bei RISC-Architekturen manipulieren die Compiler die Reihenfolge der abzuarbeitenden Befehle, um die durch Datenspeicherzugriffe bedingten Verzögerungen zu kompensieren. Diese Beispiele zeigen, daß es in vielen Fällen nicht mehr möglich ist, nur durch Beobachten der Bus-Aktivitäten direkt auf den tatsächlichen Programmablauf schließen zu können.

Auf einem Mikroprozessor-Chip implementierte Daten- und Befehls-Caches bewirken, daß im Falle erfolgreicher Cache-Zugriffe keine Informationen über die ablaufende Befehlsfolge an die Außenwelt gelangen.

In Rechensystemen mit auf dem Mikroprozessorchip integrierter Speicherverwaltungseinheit (Memory Management Unit, MMU) ist der Bezug zwischen physikalischer und zugehöriger virtueller Adresse nicht nachvollziehbar. Ein sicherer Schluß auf den tatsächlichen Programmablauf ist aus dem durch eine Aufzeichnung der Busaktivitäten gewonnenen Befehlsstrom nicht mehr möglich.

Die zunehmenden Taktraten (> 30 MHz) verlangen die Nutzung entsprechend schneller Technologien in den Meßgeräten. Bei auf RISC-Architekturen basierenden Systemen verschärft sich die Situation, weil die meisten Befehle innerhalb eines Taktzyklus ausführbar sind.

Untersucht man die physikalische Struktur von Rechensystemen bezüglich der Zugänglichkeit ihrer relevanten Meßpunkte für die Sensoren von Hardware-Monitoren, so kann man sie in verschiedene Klassen einteilen und damit eine in /7/ vorgestellte Systematik verfeinern:

I. Der ausgenutzte logische Adreßraum ist galvanisch zugänglich und identisch mit dem physikalischen Adreßraum. Die Adreßräume sind direkt (1:1) aufeinander abgebildet.

II. Der ausgenutzte logische Adreßraum ist nicht immer galvanisch zugänglich[2] und kleiner oder gleich dem physikalischen Adreßraum. Der physikalische Adreßraum wird durch entsprechende Umsetzverfahren in feste Segmente eingeteilt.

III. Der ausnutzbare logische Adreßraum ist galvanisch zugänglich und größer als der verfügbare physikalische Adreßraum, welcher vollständig oder zum Teil dem logischen Adreßraum zugeordnet wird.

IV. Rechensysteme (III), bei denen der logische Adreßraum *nicht* galvanisch zugänglich ist. Aus der physikalischen Adresse kann nicht auf die logische geschlossen werden, weil der *offset* zwischen logischer und physikalischer Adresse nicht konstant bleibt.

Rechensysteme (I) findet man u.a. in *embedded applications* wie Controller, digitale Regler oder Filter. Als Beispiel für Rechensysteme (II) sei exemplarisch die System-Familie 8860/62/64 genannt. Die Systeme (I) und (II) basieren zumeist auf 8-Bit und 16-Bit-Mikroprozessoren. Bei Rechensystemen (III) und (IV) findet man vornehmlich 32-Bit-, aber auch 16-Bit-Mikroprozessoren. Zu ihnen zählen Multi-User- und Multi-Tasking-Systeme mit virtueller Speicherverwaltung (z.B. UNIX). Systeme vom Typ (IV) unterscheiden sich von Systemen vom Typ (III) durch die auf dem CPU-Chip integrierte Speicherverwaltungseinheit (z.B. MC68030), wodurch die Zugriffe auf die virtuellen, logischen Adressen der "Außenwelt verborgen" bleiben.

4. Meßstrategien

Es ist kennzeichnend für Rechensysteme (I) mit einem Single-Tasking-Betriebssystem, daß durch die Beobachtung der CPU-Signale auf den Programmablauf zurückgeschlossen werden kann. Das Hardware-Monitoring von Rechensystemen mit einem Multi-Tasking-Betriebssystem wirft dagegen besondere Probleme auf, weil der zu beobachtende Befehlsablauf ohne zusätzliche Informationen nicht einem bestimmten Prozeß (Task) zugeordnet werden kann. Darum werden besondere Eigenschaften der Prozeß- und Speicherverwaltung durch das Betriebssystem für eine prozeßselektive Meßdatenübernahme zur Vereinfachung des Meßablaufs herangezogen.

Handelt es sich bei Rechensystemen (I) um Multi-Tasking-Systeme, so sind den verschiedenen Prozessen Speicherbereiche exklusiv zugeordnet. Es ist daher auch möglich, durch Beobachtung der CPU-Signale Leistungsmaße wie z.B. die Auftrittshäufigkeit bestimmter Ereignisse oder die Laufzeiten von Programmen zu bestimmen. Voraussetzung hierzu ist die Verfügbarkeit entsprechender Informationen über das zu beobachtende Programm wie z.B. die *entry/exit*-Adressen einzelner Prozeduren.

Greifen in Rechensystemen (II) verschiedene Prozesse auf gemeinsame Code- und Datensegmente zu, ist ein prozeßselektives Monitoring wie bei Systemen (I) nicht möglich. Die CPU-Signale sind nur dann richtig zu interpretieren, wenn sie vom Systemstart an durchgängig beobachtet und ausgewertet werden[3]. Ein solches Verfahren ist natürlich sehr aufwendig und zeitraubend. Es ist daher sinnvoller, das Meßverfahren auf die Methode für Rechensysteme (I) zurückzuführen, indem man per Software in eine vereinbarte Speicherzelle (Monitor-Register, s.w.u.) ein Merkmal (z.B. Prozeßidentifikation (PID) oder Segmentregister-Inhalt) ablegt, welches der Steuerung eines Hardware-Meßwerkzeugs dient. Sind gemeinsame Code- und Datensegmente dagegen nicht zugelassen, ist durch die Interpretation der führenden Adreß-Bits des physikalischen Adreßraums eine Prozeßidentifikation prinzipiell möglich.

In Rechensystemen (III) ist es für prozeßselektive Messungen ebenfalls notwendig, die aktuelle PID dem Hardware-Monitor bekanntzumachen. Die Ausgabe der PID kann dazu ebenfalls über ein Monitor-Register erfolgen.

Rechensysteme (IV) lassen die prozeßbezogene Messung und Interpretation der CPU-Signale nicht zu, weil trotz des Zugangs zur PID der Zugriff auf physikalische Adressen nicht den Rückschluß auf die adressierten logischen Speicherzellen erlaubt. Hier existiert zur Zeit eine Lücke im Bereich der verfügbaren Meßtechnologie.

5. Ein hybrides Monitor-Konzept

Da es insbesondere bei modernen Rechensystemen mit hochintegrierten Mikroprozessoren und virtueller Speicherverwaltung (Systeme (III) und (IV)) nicht mehr möglich sein wird, relevante prozeßbezogene Meßdaten mit Hardware-Meßwerkzeugen abzuleiten, wurde ein neues Meßkonzept entwickelt. Das Blockschaltbild in Abb. 2 erläutert die Architektur dieses *hybriden Monitor-Konzepts* /6/.

Der Prüfling ist über eine systemspezifische Adaptions-Einheit mit dem Datenvorverarbeitungs-Modul (DVM) des Meßsystems verbunden. Ein zweiter Informationskanal verbindet eine im Prüfling vereinbarte und als *Monitor-Register* bezeichnete Speicherzelle mit dem DVM.

Das Monitor-Register kann entweder in der Form einer dafür reservierten Zelle des Arbeitsspeichers (virtuelles Monitor-Register) oder durch ein separates Hardware-Register (physikalisches Monitor-Register) realisiert sein. Im ersten Fall erfolgt die physikalische Anbindung des Monitor-Registers an das DVM über die systemspezifische Adaption, im zweiten Fall über eine zweite Hardware-Schnittstelle.

Die systemspezifische Adaption wertet die Buszyklen der CPU bezüglich der Speicherzugriffe (Code, Daten, etc.) aus und übergibt die so entstandenen Informationen dem DVM. Der Inhalt des Monitor-Registers wird durch das Betriebssystem kontrolliert. Über eine Software-Schnittstelle ist auch Anwenderprogrammen der Zugriff auf das Monitor-Register möglich. Über dieses erhält das DVM programmspezifische Informationen, die z.B. der Beeinflussung seines Speichermechanismus (SW-Trigger) dienen können.

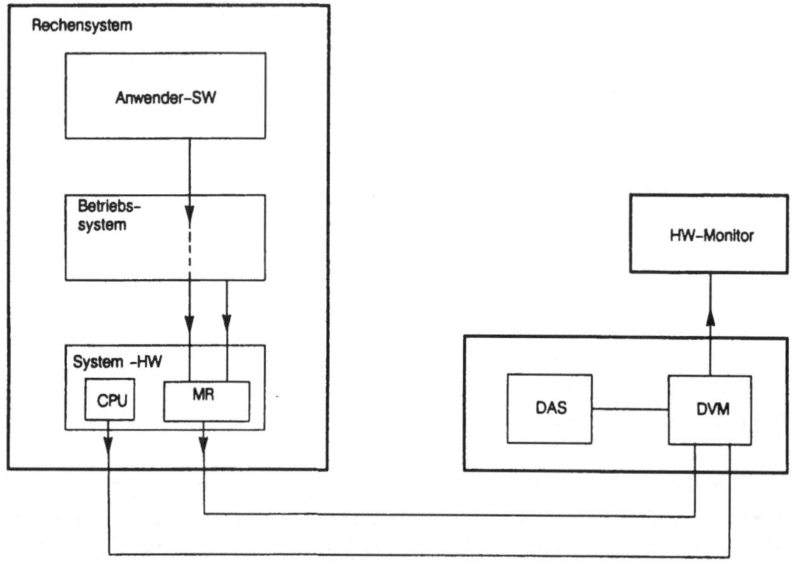

Abb. 2 Architektur des hybriden Monitorkonzepts (MR: Monitor-Register, DAS: Datenanalyse-System, DVM: Datenvorverarbeitungs-Modul)

Das DVM ist wiederum mit einem Daten-Analyse-System (DAS) und bei Bedarf auch mit einem Hardware-Monitor verbunden. Letzterer stellt flexibel konfigurierbare Meß- und Verarbeitungsverfahren zur Verfügung. Das DVM dient in diesem Fall als Datenkonzentrator. Mit einer Konfiguration aus Datenkonzentrator und Hardware-Monitor werden bei Nixdorf schon seit über sieben Jahren hardwarenahe Messungen erfolgreich durchgeführt.

Bei Rechensystemen (II) können beispielsweise über ein Monitor-Register gesendete Informationen (z.B. aktuelle Speicherbank, PID) das DVM steuern. Im Fall von Rechensystemen (III) schreibt das Betriebssystem die jeweils aktuelle PID in das Monitor-Register. Das DVM triggert auf die ihm bekannt gemachten PID's und kann so prozeßselektiv Meßdaten aufzeichnen.

Mit der zur Zeit verfügbaren Technologie sind Hardware-Messungen an Rechensystemen (IV) nur auf symbolischer Ebene (vgl. Abschnitt 6.2) möglich, indem die Anwenderprogramme an interessierenden Positionen zum Beschreiben eines Monitor-Registers instrumentiert werden (vgl. Abschnitt 6.3).

6. Realisierung

Der im Kapitel 5 vorgestellte Ansatz ist primär als konzeptionelle Grundlage sowohl zum Ausbau bestehender als auch zur Entwicklung neuer Meßverfahren zu interpretieren. Als Zielsysteme stehen Rechner mit den Betriebssystemen UNIX, OS/2 und MS-DOS[4] im Vordergrund. Um die anfallenden Entwicklungskosten gering zu halten, werden am Markt angebotene Werkzeuge beschafft und den meßtechnischen Bedürfnissen angepaßt.

6.1 Monitoring der CPU-Signale

So wurde auf ein kommerzielles Produkt des Herstellers CADRE/MicroCASE, die Software Analyse Workstation (SAW, Abb. 3, /8, 9/), zurückgegriffen, welche Funktionalitäten von DVM und DAS besitzt. Im Gegensatz zu diesen ist die SAW ein autark arbeitendes System mit graphischer Benutzeroberfläche, welches die funktionellen Eigenschaften eines Logikanalysators (Interactive State Analyzer, ISA) mit denen eines Hardware-Monitors (Softanalyst, SFA) verbindet.

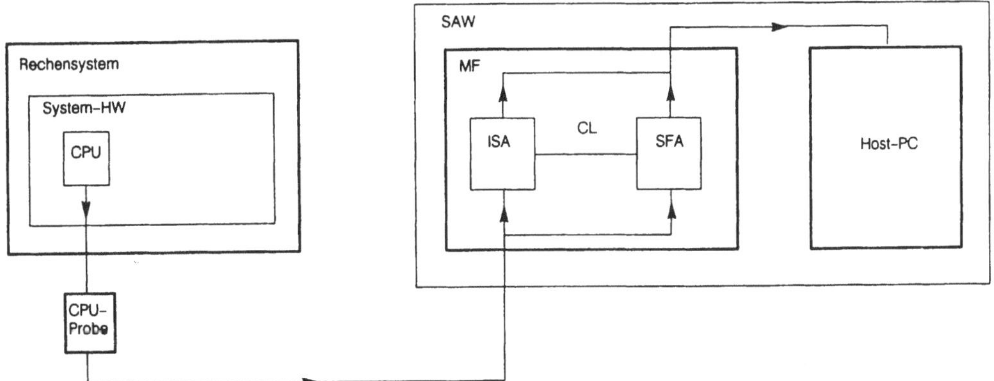

Abb. 3 Software Analyse Workstation (SAW; MF: Mainframe, ISA: Interactive State Analyzer, SFA: Softanalyst, CL: Cross Link)

Die SAW besteht in ihrer Minimalkonfiguration aus drei Hardware-Einheiten: einem MS-DOS-PC als Host, dem sogenannten Mainframe als Baugruppenträger und einer Probe zur Anpassung an die CPU des jeweiligen Zielsystems. Für die meisten Standard-Mikroprozessoren sind spezifische Probes verfügbar. Alternativ ist eine Adaptierung an beliebige Meßpunkte eines Prüflings mit Sensorleitungen möglich.

Es sind die folgenden Betriebsarten implementiert: Software-Trace (*MicroTrace*) auf Assembler-Ebene (ISA), Symbolischer Trace (*SymTrace*) auf Hochsprachen-Ebene (SFA), Performance-Analyse (SFA), Programm-Verifikation(*code coverage*, SFA). Die Betriebsart Performance Analyse erlaubt die Überprüfung von Programm-Aktivitäten im Adreßraum (*event mapping*, *event timing*) und der Abhängigkeiten zwischen aufrufenden und aufgerufenen Prozeduren. Weiterhin gestattet eine anzuwählende Option (*cross link*) die Steuerung (*enable / disable*) der verschiedenen Betriebsarten des SFA durch den ISA.

Allen Betriebsarten ist gemeinsam, daß sie die Informationen, welche ein Linker in die MAP-Datei schreibt, verwerten können und damit zur Transparenz der Meßergebnisse beitragen.

Die SAW ist in der Lage, die von einem Mikroprozessor mit Befehls-Pipeline vorgeholten Befehle, welche nach einer Programmverzweigung nicht mehr abgearbeitet werden, als solche zu erkennen und unberücksichtigt zu lassen. Bei einigen 32-Bit-Mikroprozessoren werden auch gezielt nicht dokumentierte Signalkombinationen mit ausgewertet. So können z.B. bei den CPU-Bausteinen MC68020/30 die bei *cache hits* adressierten Speicherzellen erfaßt werden.

Eine prozeßselektive Messung erfolgt zum Beispiel durch die Beobachtung und Interpretation der Zugriffe auf ein Monitor-Register, in welches das Betriebssystem die aktuelle PID ablegt. Durch entsprechende Einstellung der Trigger-Bedingungen in der Betriebsart MicroTrace werden die CPU-Zyklen nur dann aufgezeichnet, wenn die aktuelle PID mit der des relevanten Prozesses übereinstimmt. Ergänzend dazu steuert der ISA die gewählte SFA-Betriebsart. So ist z.B. mit dieser Konfiguration und der Betriebsart Performance Analyse ein prozeßselektives Address-Mapping /3/ durchführbar. Durch Beifügen weiterer Triggerbedingungen kann man die gewählte Analyse-Methode auf bestimmte Prozeduren innerhalb eines Prozesses beschränken.

6.2 Symbolisches Monitoring auf Hochsprachenebene

Der vorhergehende Abschnitt behandelte das Monitoring der CPU-Signale mit der SAW. Dazu reicht ein virtuelles Monitor-Register zur Übergabe von ergänzenden Informationen wie z.B. der vom Betriebssystem bereitgestellten PID aus. Sind die mit einer CPU-spezifischen Probe erzielbaren Prozessorzyklen nur von untergeordneter Relevanz und ist ein physikalisches Monitor-Register vorhanden, so kann die SAW über passive Sensoren direkt mit diesem verbunden werden.

Die zu messenden Programme müssen dazu mit Schreibaufträgen in das Monitor-Register instrumentiert werden. Der Zugriff auf dieses erfolgt in der Programmiersprache C über eine Funktion, welche als Parameter einen Integer-Wert übergibt. Der Overhead beschränkt sich bei einer aktuellen Implementierung auf wenige µs.

Um Anwenderprogrammen den Schreibzugriff auf das Monitor-Register zu gestatten, ist eine Unterstützung durch das Betriebssystem des jeweiligen Prüflings notwendig, weil die Schutzkonzepte moderner Rechnerarchitekturen einen direkten, schnellen Zugriff auf Hardware-Ressourcen nicht zulassen. Dieser ist bei einem Betriebssystem relativ einfach zu implementieren, dessen *Kernel*-Quellen vorliegen und entsprechende Erweiterungen zulassen (z.B. UNIX).

Die Programm-Instrumentierung sei an einem Beispiel erläutert. Angenommen, man interessiere sich für die Laufzeit einer Prozedur, so werden der Anfang (*entry*) und das Ende (*exit*) der Prozedur mit Anweisungen zum Beschreiben des Monitor-Registers instrumentiert. Zur Unterscheidung der Meßpunkte werden verschiedene Informationen übergeben, die zur Identifikation der gespeicherten Ereignisse bei der Auswertung dienen. Bei umfangreichen Programmen ist eine Organisation der Übergabeparameter in Nummernkreise sinnvoll, um mögliche Doppelbelegungen der Ereignisse zu vermeiden. Die Instrumentierung des Codes zur Markierung von *entry*/*exit*-Ereignissen kann manuell oder mittels Präprozessor erfolgen. Darüber hinaus ist es möglich, beliebige Ereignisse manuell zu instrumentieren. Die in das Monitor-Register geschriebenen Informationen werden von der SAW ausgelesen und im Falle der Betriebsarten MicroTrace oder SymTrace mit einer Zeitmarke versehen und anschließend abgespeichert.

So wurden mit Hilfe dieses Meßprinzips in den letzten Monaten bei Anwendungen aus dem Banken- und dem Bürobereich Schwachstellen aufgedeckt, die das Antwortzeitverhalten negativ beeinflußten. Es zeigte sich oft, daß die erhöhten Programmlaufzeiten durch Zugriffe auf Prozeduren und Bibliotheksfunktionen verursacht wurden, deren Zeitbedarf der jeweilige Programmierer gelegentlich um mehrere Größenordnungen unterschätzte. Die Antwortzeiten ließen sich reduzieren, indem durch Residenthalten von Daten Mehrfachabfragen vermieden wurden.

Der notwendige Investitionsaufwand von ungefähr DM 60000,00 verhindert eine breite Streuung der SAW unter den Software-Entwicklern eines Unternehmens. Praktische Erfahrungen mit der SAW und anderen Werkzeugen haben gezeigt, daß in vielen Fällen ein einfach zu bedienendes Trace-Werkzeug mit genauer Zeitetikettierung der gespeicherten Ereignisse für die symbolische Ebene (Hochsprachen) gefragt ist. Ein Trace-Werkzeug als reine Software-Lösung zu realisieren, scheidet jedoch aus, weil bei aufwendig instrumentierten Programmen der entstehende *overhead* zur Verfälschung der Meßergebnisse führt. Dieser ist bei der Fehlersuche mit Sicherheit von untergeordneter Bedeutung. Gilt es jedoch, den Nachweis zu erbringen, daß die geforderten Spezifikationen (z.B.: Antwortzeiten) eingehalten werden, sind die entstehenden Meßwertverfälschungen nicht zu tolerieren.

Bei Nixdorf wird die SAW vornehmlich für aufwendige Meßaufgaben auf der Assembler- und auf der symbolischen Ebene eingesetzt. Für allgemeine Trace-Messungen wird zur Zeit auf ein HW-Werkzeug älterer Generation zurückgegriffen, welches den Anforderungen noch genügt. Um die Verbreitung hybrider Werkzeuge zur Bearbeitung einfacher Meßaufgaben deutlich zu erhöhen, erfolgt zur Zeit die Realisierung einer universellen Lösung in Form eines Zusatzboards für PC-Produkte. Dieses ist für UNIX-, OS/2- und MS-DOS-Systeme anwendbar und zeichnet sich durch einen relativ geringen Preis, einen hohen Grad an Mobilität und eine einfache Bedieneroberfläche aus.

6.3 Monitoring hochintegrierter Mikroprozessoren am Beispiel des Projekts LOPY

Der Umstand, daß es in Rechensystemen, welche auf Mikroprozessoren mit integrierter MMU basieren, nicht möglich ist, mit Hardware-Meßwerkzeugen einen Programmablauf auf Assemblerebene direkt zu verfolgen (vgl. Kap. 4), führte zu Arbeiten mit dem Ziel, diese Einschränkung auszugleichen.

Ziel des Projekts LOPY[5] (*Logical from Physical Address Translator*) ist die Ergänzung der SAW mit einem Hardware-Modul für Messungen an UNIX-Systemen mit der CPU MC68030 /10/. Die Festlegung auf das Betriebssystem UNIX erlaubt im Hinblick auf ein vereinfachtes Verfahren zur Code-Instrumentierung die Auswertung von systemspezifischen Eigenschaften.

LOPY ist eine Hardware-Baugruppe zur Umsetzung der von einer aktiven Probe für die CPU MC68030 gelieferten physikalischen Adressen in logische Adressen. LOPY wird zwischen die Mikroprozessor-Probe und die SAW geschaltet.

Die CPU MC68030 ist entsprechend der Systematik in Kapitel 3 Rechensystemen der Klasse (IV) zuzuordnen. Neben der integrierten MMU besitzt sie zwei *on chip caches* für Befehle und Daten.

Die Cache-Speicher bewirken, daß zwischen der externen Busaktivität und den ausgeführten Instruktionen kein direkter kausaler Zusammenhang besteht. Die Methode, jede ausgeführte Instruktion anhand ihrer Busaktivität zu identifizieren, funktioniert daher ohne besondere Hilfsmittel nur sehr lückenhaft.

Im Falle eines *cache hit* liegt die Speicheradresse für wenige ns auch am externen CPU-Bus an. Die MC68030-Probe der SAW erfaßt *cache hits* und leitet die anliegenden Adreßinformationen an die SAW weiter. Das zur Adresse gehörende Datum kann mittels eines Postprozessors im nachhinein ergänzt werden, sofern der Inhalt des Cache-Speichers zu einem bestimmten Startzeitpunkt extern bekannt ist.

Ständen keine weiteren Informationen über ein zu messendes Programm zur Verfügung, müßte, da der einzige nicht von Vorzuständen abhängige Prozessorstatus der Reset-Status ist, eine Messung immer nach einem CPU-Reset, also mit dem Systemstart beginnen, um alle für eine Auswertung notwendigen Informationen zu erhalten. Zur Vereinfachung dieses wenig praktikablen Meßablaufs bietet sich die Ausnutzung von Eigenschaften des Betriebssystems UNIX an. Bei prozeßselektiven Messungen muß jeder Task-Wechsel und damit jede Änderung der aktuellen PID von LOPY erkannt werden. Durch eine Änderung im UNIX-Kernel wird bei jedem *context switch* die nun aktuelle PID in ein von der Probe und damit von LOPY lesbares Monitor-Register geschrieben. Gleichzeitig bedingt die Prozeßumschaltung auch, daß mit dem Übergang einer Task in den Zustand "rechnend" die bisherigen Inhalte der CPU-internen Cache-Speicher für die nun aktive Task nicht mehr verwendbar sind. Noch bestehende Einträge aus früheren aktiven Phasen der Task sind bereits protokolliert worden und somit bekannt. Cacheinhalte aus *shared memory regions* sind bekannt, falls auf die PID´s aller am "Sharing" beteiligten Prozesse getriggert wird. Die für eine sich anschließende Rekonstruktion der im Trace enthaltenen *cache hits* erforderlichen Informationen sind somit in der Trace-Datei vorhanden.

Die durch die MMU vorgenommene Abbildung des logischen Adreßraums auf den physikalischen Adreßraum ist von den Umsetztabellen und der aktuellen CPU-internen MMU-Konfiguration als auch dem Inhalt des *address translation cache* abhängig. Der Rückschluß von physikalischen auf logische Adressen kann nur dann sinnvoll durchgeführt werden, wenn der Umsetzlogik die aktuellen Inhalte CPU-interner Register bekannt sind. Eine programmierbare Umsetzlogik ermöglicht die Erzeugung einer inversen Abbildung der Adreßräume. Auf CPU-interne Umkonfigurierungen, z.B. des *translation-control-* oder des *root-pointer*-Registers der MMU, wird sofort reagiert. Sie führen daher nicht zu Fehlinterpretationen des an die SAW weitergeleiteten Datenstroms.

Das UNIX-Prozeßkonzept ermöglicht auf einfache Weise eine Initialisierung von LOPY. Dazu werden die aktuellen Konfigurationsdaten der MMU vor dem Start eines Anwenderprozesses LOPY übermittelt. In UNIX-Systemen kann dieser Ablauf durch einen Initialisierungsprozß bewerkstelligt werden. Der Prozeß übergibt LOPY die relevanten Registerinhalte der CPU sowie die PID's der Kindprozesse und wartet auf deren Beendigung.

Die hier vorgestellte Meßumgebung erlaubt, Zugriffe auf ein Monitor-Register zu vermeiden. Folglich sollte auf eine Codeinstrumentierung des Anwenderprogramms verzichtet werden können. Da die Dauer eines Traces nicht von vornherein begrenzt ist, liegen dem Entwurf von LOPY Echtzeitbedingungen zugrunde.

7. Zusammenfassung und Ausblick

Der zunehmende Integrationsgrad von Mikroprozessoren und die damit verbundene Zunahme von *on chip*-Funktionalitäten machen ein prozeßorientiertes Hardware-Monitoring von Rechensystemen zunehmend aufwendiger. Hybride Verfahren, bei denen die Software Informationen an einen Hardware-Monitor übergibt, finden ihre Grenzen an den Schutzkonzepten der Betriebssysteme, welche auch von der Hardware-Seite vermehrt unterstützt werden.

Innerhalb dieser Grenzen bieten sich zwei Auswege an. Für Untersuchungen auf Hochsprachenebene können Anwenderprogramme mit Schreibanweisungen in ein Monitor-Register instrumentiert werden. Dagegen kann die gezielte Ausnutzung der internen Strukturen der Zielsysteme zu einer Vereinfachung der Meßproblematik auf Assembler-Ebene führen. Dieses wurde am Beispiel der CPU MC68030 gezeigt.

Für die Zukunft ist zu erwarten, daß das reine Hardware-Monitoring nicht mehr uneingeschränkt möglich sein wird. Der für das Hybrid-Monitoring notwendige Software-Aufwand wird durch die Schutzkonzepte der Betriebssysteme und der Mikroprozessoren bestimmt. Daraus erhebt sich die Forderung, künftige Mikroprozessoren mit integrierten Monitor-Funktionen zu versehen (*on chip monitoring*).

[1] UNIX ist ein eingetragenes Warenzeichen der AT&T. OS/2 ist ein eingetragenes Warenzeichen der Microsoft Corporation.

[2] Z.B. wird bei den Mikroprozessoren 8088/86 die physikalische Adresse durch die Addition der logischen Adresse zum Inhalt des Segmentregisters erzeugt, welcher CPU-extern nicht bekannt ist. Weil der Segmentregister-Inhalt die führenden externen Adreßstellen des Prozessors beeinflußt, ist im vorliegenden Fall diese Einschränkung unerheblich.

[3] Diese Aussage verliert ihre Gültigkeit, wenn der CPU-Chip mit einem aktivierten *on chip cache* für Instruktionen oder Daten betrieben wird.

[4] MS-DOS ist ein eingetragenes Warenzeichen der Microsoft Corporation.

[5] Eine Kooperation zwischen der Universität-GH-Paderborn und der Nixdorf Computer AG.

Literatur

/1/ Beilner, H. Leistungsanalyse von Rechensystemen
Vorlesung, Fernuniversität-GH in Hagen, 1988

/2/ Klar, R. Hardware/Software-Monitoring
Informatik-Spektrum, **8**(1), 37 - 38, 1985

/3/ Bordewisch, R. Messung und Bewertung von Betriebssystem-Komponenten
In: B. Mertens (Hrsg.): Messung, Modellierung und Bewertung von Rechensystemen, Reihe Informatik-Fachberichte, Bd. 41, 14 - 28, Springer-Verlag, Berlin, Heidelberg, New York 1981

/4/ Ferrari, D. A Hybrid Measurement Tool for Minicomputers
In: Ferrari, D.; Spadoni, M. (Hrsg.): Experimental Computer Performance and Evaluation, North-Holland, Amsterdam 1981

/5/ Terplan, K. Leistungsoptimierung von Computersystemen und Rechnernetzen
Reihe Kontakt & Studium (Hrsg.: Bartz, W.J.), Expert Verlag, Grafenau 1982

/6/ Hagenhoff, H.J., Kooke, W., Totzauer, G., Vering, M. Vorschlag für ein integriertes hybrides Monitorkonzept
Internes Dokument, Nixdorf Computer AG, Paderborn, 1987

/7/ Föckeler, W. Hardware-Monitoring zur Leistungsmessung von Rechensystemen mit modernen Mikroprozessoren
Nachrichtentechnisches Kolloquium, Universität-GH-Paderborn, 7.10.1988

/8/ Ableidinger, B., Agrawal, N., Nobles, C. Real-time analyzer furnishes high-level look at software operation
Electronic Design, Sept. 19, 117 - 131, 1985

/9/ Sundermeier, B. Software Analysis Aids Prototype Evaluation
Electronics Test, 49 - 51, Sept. 1987

/10/ MC68030 Enhanced 32-Bit Microprocessor User´s Manual
Datenbuch, Fa. Motorola, 1987

Ein Monitorsystem zur verzögerungsfreien Überwachung von Multiprozessoren

Thomas Bemmerl, Robert Lindhof, Thomas Treml
Institut für Informatik der TU München
Lehrstuhl für Rechnertechnik und Rechnerorganisation
Arcisstr. 21, D-8000 München 2
Tel.: 089 / 21 05 - 82 47 oder - 23 82
DFN: bemmerl@infovax.informatik.tu-muenchen.dbp.de

Detaillierte Informationen über das Laufzeitverhalten eines parallelen Systems werden in fast allen Phasen der Benutzung, Konfigurierung, Programmierung und Bewertung von Multiprozessoren benötigt. Insbesondere werden Laufzeitinformationen über das dynamische Ablaufverhalten von Werkzeugen wie Leistungsmeßsystemen, Debugging-Systemen, Testsystemen, Animationssystemen und Lastbalanzierern ausgewertet. Der vorliegende Artikel beschreibt ein Monitorsystem für Multiprozessoren, mit welchem Laufzeitinformationen für die erwähnten Werkzeuge gewonnen werden können. Besondere Eigenschaften dieses Monitors sind die verzögerungsfreie Überwachung, die zeitrichtige Korrelation von Ereignissen unterschiedlicher Verarbeitungselemente und die Verfügbarkeit verteilter und zentraler Zeitmodelle.

1. Einführung und Motivation

Laufzeitinformationen sind zum Verständnis des dynamischen Ablaufs von Rechensystemen unerläßlich. Insbesondere gilt dies für Multiprozessoren mit mehreren Verarbeitungselementen und parallelen Abläufen. Aus diesem Grund ist es in vielen Phasen der Benutzung, Konfigurierung, Programmierung, Analyse und Bewertung von parallelen Systemen notwendig, geeignete Laufzeitinformationen über den dynamischen Ablauf eines Multiprozessors zu erhalten. Geeignete Hilfsmittel (Werkzeuge) für die erwähnten Phasen der Multiprozessor-Benutzung werten die gewonnene Laufzeitinformation entsprechend der Werkzeugeigenschaften aus. Werkzeugumgebungen bestehend aus Systemen zur Gewinnung von Laufzeitinformationen (Monitore) und Systemen zur Auswertung dieser Informationen (Werkzeuge) sind die Voraussetzung für den verstärkten Einsatz von Multiprozessoren und verteilten Systemen. Durch den Einsatz dieser Werkzeugumgebungen ist die Produktivität bei der Nutzung von Multiprozessoren wesentlich zu steigern. Im Rahmen des Projekts TOPSYS (TOols for Parallel SYStems) am Institut für Informatik wird eine solche integrierte Werkzeugumgebung für den Multiprozessor iPSC/2 mit Hypbercubetopologie entworfen und prototypisch implementiert [Bem88a]. In dieser Werkzeugumgebung werden Werkzeuge für die Fehlersuche (Debugging), die Leistungsmessung, den Test, die Visualisierung und den dynamischen Lastausgleich auf die von einem Monitorsystem gewonnenen Laufzeitinformationen abgestützt. Die Werkzeuge werten die Laufzeitinformationen entsprechend des beabsichtigten Untersuchungszwecks (Fehlersuche, Lestungsmessung, Lastausgleich, etc.) aus und bereiten sie zur graphischen Darstellung auf. Abbildung 1 zeigt dieses Integrationsmodell. Werkzeuge und Monitorsystem von TOPSYS bilden eine integrierte und interaktive Werkzeugumgebung zur Analyse und Bewertung des dynamischen Ablaufverhaltens von Multiprozessoren.

Das Monitorsystem von TOPSYS verwendet optional unterschiedliche Instrumentierungstechniken zur Überwachung des Multiprozessors. In der jetzigen Phase des Projekts werden Hardware-, Software- (Objektcode-), Hybrid-Instrumentierung und Simulation betrachtet. Die Entwicklung unterschiedlicher Monitore dient der Untersuchung der Wechselwirkungen zwischen Instrumentierungstechnik und Meßqualität im Bezug auf Verzögerungsfreiheit, Implementierungsaufwand und Meßgenauigkeit. Der Rest des Artikels beschreibt die Konzepte des Monitorsystems von TOPSYS unter besonderer Berücksichtigung des verzögerungsfreien Hardware-Monitors.

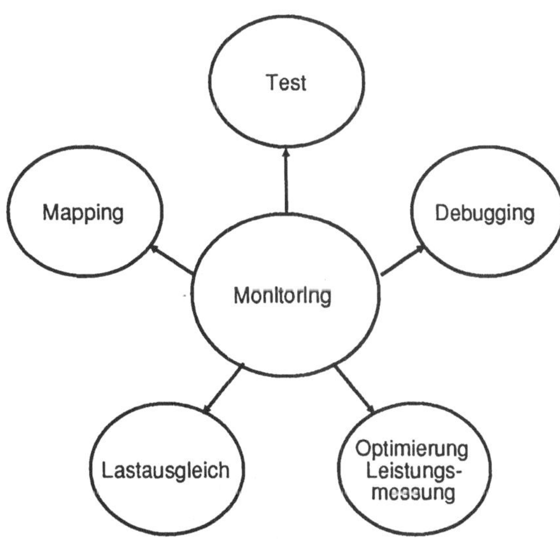

Abb. 1: Integration von Werkzeugen und Monitorsystem

Forschungsarbeiten über die Beobachtung und Überwachung von Multiprozessoren und verteilten Systemen wurden in den letzten Jahren aus den genannten Gründen verstärkt vorangetrieben. Die wichtigsten Arbeiten auf diesem Gebiet werden zum Zwecke der Einordnung des TOPSYS-Projekts bzgl. des Monitorsystems nachfolgend kurz vorgestellt. Die erste Klasse von Arbeiten beschäftigt sich mit spezifischen theoretischen Grundlagen bei der Überwachung von parallelen und verteilten Systemen. Dabei wird hauptsächlich die Problematik einer zeitlichen Ordnung und eines konsistenten globalen Zustands behandelt. Zu dieser Klasse zählen die Arbeiten von Chandy/Lamport [Cha85], [Lam78] und Spezialetti/Kearus [Spe88]. Eine zweite Klasse von Arbeiten versucht Multiprozessoren und verteilte Systeme aus Gründen der höheren Flexibilität mit Hilfe von Software-Monitoren zu überwachen. In vielen Fällen werden bei diesen Arbeiten keine realen parallelen Systeme betrachtet, sondern lediglich Simulationen von solchen. Mit dieser Technik arbeiten Mühlenbein [Müh86], Segall [Seg85], Bailey [Bai88], Miller [Mil88], Joyce [Joy87], Eichholz [Eic88], Bates [Bat88], Curtis/Wittie [Cur82] und Harter [Har85]. Bezüglich der Verzögerung des zu überwachenden Systems werden häufig nur Schätzungen angegeben, so daß es keine definierten Aussagen über die Rückwirkung dieser Monitortechnik auf das zu untersuchende Objekt gibt. Eine weitere Klasse von Arbeiten beschäftigt sich weniger mit den Überwachungstechniken von parallelen Abläufen, sondern mehr mit der Reproduzierbarkeit dieser Abläufe. Dazu gehören die Arbeiten von LeBlanc [LeB85], [LeB87]. Mit dem Problem einer möglichst verzögerungsfreien Überwachung von Multiprozessoren und verteilten Systemen beschäftigen sich nur recht wenige Forschungsgruppen. In den Arbeiten von Hofmann/Klar [Hof87] und Haban/Wybranietz [Wyb88] werden Hardware- bzw. Hybrid-Monitore zur Leistungsmessung verwendet. Diese Monitore sind jedoch nur für diese Art der Messung ausgelegt. In den Projekten von Burkhart [Bur88] und Gregoretti [Gre86] beschränkt man sich auf speichergekoppelte Multiprozessoren, was vieles bei der Überwachung durch die Verfügbarkeit zentraler Betriebsmittel vereinfacht.

Das Motiv für die Entwicklung eines Hardware-Monitors innerhalb des TOPSYS-Projekts ist unserer Meinung nach die absolute Notwendigkeit einer möglichst verzögerungsfreien Überwachung bei Multiprozessoren. Gerade bei der Beobachtung asynchroner paralleler Abläufe können die Meßverfälschungen einer verzögerungsbehafteten Meßmethodik nicht hingenommen werden. Frühere Untersuchungen am Institut für Informatik haben gezeigt, daß selbst Software-Monitore bei der Überwachung von nur einzelnen Knotenprozessoren den Ablauf bis zu 500% verlangsamen können [Bei88].

Der Artikel versucht in Abschnitt 2 auf der Basis einer Modellvorstellung von parallelen Abläufen grundlegende Anforderungen an Monitorsysteme zu erarbeiten. Abschnitt 3 beschreibt das Grundkonzept des verteilten Monitorsystems von TOPSYS. Abschnitt 4 und 5 erläutern die Entwurfskonzepte und die Implementierung des lokalen Hardware- bzw. Hybrid-Monitors.

2. Parallele Abläufe und deren Überwachung

Bei den vorausgesetzten parallelen Systemen betrachten wir den allgemeinsten Fall der MIMD-Systeme ohne gemeinsame Betriebsmittel. Dies bedeutet, es werden insbesondere Multiprozessoren mit verteiltem Speicher betrachtet. Auf dem Multiprozessor werden Programme zur Ausführung gebracht, welche auf dem Prozeßmodell kommunizierender Prozesse basieren. Dabei ist auch zugelassen, daß mehrere Prozesse quasiparallel auf einem Knotenprozessor ausgeführt werden. Die Zustandsänderungen der Prozesse erzeugen entsprechend der operationellen Semantik, Zustandsfolgen über der Zeit. Die Zustandsübergänge stellen die durch den Monitor zu überwachenden Vorgänge dar. Aus Konsistenzgründen mit der Literatur bezeichnen wir die Zustandsübergänge als Ereignisse und die Folgen von Zustandsübergängen als Ereignisfolgen [Lam78], [Mil88], [Spez88]. Zur Illustration der Ereignisfolgen über der Zeit verwenden wir sog. Raum/Zeit-Diagramme in denen der Austausch von Nachrichten zwischen Prozessen als Pfeile ausgedrückt werden (Abbildung 2) [Lam78]. Zur Vereinfachung werden in diesen Darstellungen nicht interessierende Ereignisse nicht aufgeführt. Über das Vorhandensein einer zentralen Zeitbasis oder von nur lokalen Uhren wird in dieser Darstellung nichts ausgesagt.

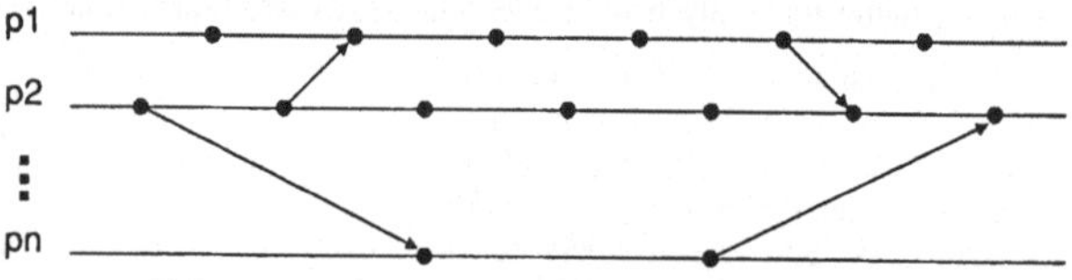

Abb. 2: Prozessmodell im Raum/Zeit-Diagramm

Die Aufgabe des Monitorsystems ist es nun, das Auftreten von Ereignissen zu erkennen bzw. zu überwachen. Die Erkennung eines Ereignisses wird als Trigger bezeichnet. Ein Trigger ist dann in der Lage, das Monitorsystem zu veranlassen, die unterschiedlichsten Aktionen durchzuführen. Ein Monitorsystem ist hauptsächlich dadurch charakterisiert, welche Ereignistypen überwacht und welche Aktionen ausgelöst werden können. Die nachfolgenden Absätze beschreiben die von Monitorsystemen für parallele Systeme zu überwachenden Ereignistypen und auszuführenden Aktionen. Zu den jeweiligen Ereignis- und Aktionsklassen werden erklärende Beispiele angegeben. Weitere Ereignis- und Aktionsbeschreibungen von Monitorsystemen finden sich in [Bem88b], [Hab88], [Mil88].

Primitive Ereignisse

Die primitiven Ereignisse lassen sich in drei Klassen einteilen. In der ersten Klasse von Ereignissen, den Kontrollfluß-Ereignissen KE wird geprüft, ob ein Prozeß eine bestimmte Ausführungsstelle erreicht.

Beispiele:

- Prozeß p1 führt Anweisung 113 aus.
- Prozeß p2 betritt Prozedur alfa.

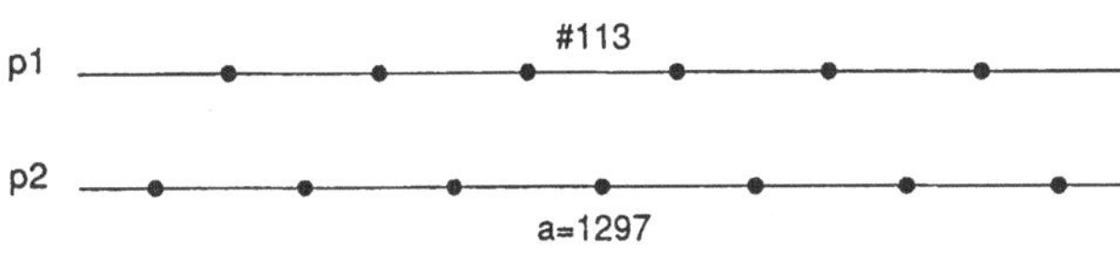

Abb. 3: Daten- und Kontrollereignisse

Die zweite Klasse von Ereignissen beschreibt Aussagen über den Datenfluß. Die Datenfluß-Ereignisse (DE) überwachen die Zustandsänderung von Datenobjekten.

Beispiele:

- Variable a von Prozeß p2 erhält den Wert 1297.
- Prozedurparameter x der Prozedur p von Prozeß p1 wird gelesen.

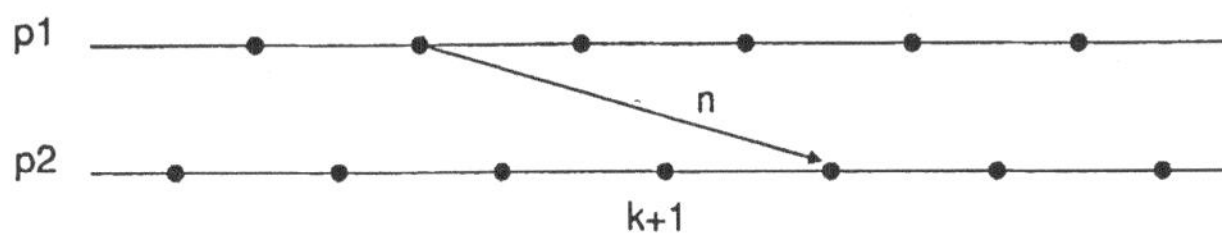

Abb. 4: Parallelitätsereignisse

Charakteristisch für DE's und KE's ist, daß sie sich auf einem ganz bestimmten (oder keinen) Prozeß beziehen und deshalb immer auf einen Knotenprozessor beschränkt bleiben. Im Raum/Zeit-Diagramm werden DE's und KE's durch Beschriftung einzelner oder mehrerer Ereignisse dargestellt (Abbildung 3).

Die Parallelitäts-Ereignisse (PE) werden zwar implementierungstechnisch meist aus DE's und KE's zusammengesetzt; aus Übersichtlichkeitsgründen werden sie hier jedoch als eigenständige Ereignisklasse aufgeführt. Die PE's formulieren Aussagen über die Kommunikation von Prozessen und über verschiedene Implementierungsaspekte des Prozeßmodells, wie Nachrichtenpufferlängen und Prozeßwarteschlangen des darunterliegenden Betriebssytemkerns.

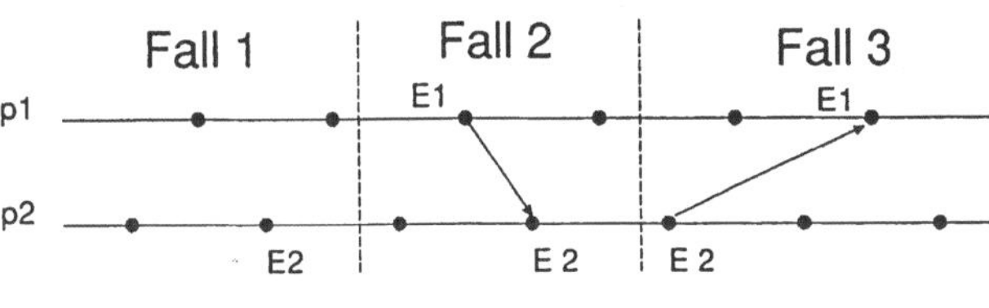

Abb. 5: Ereignisverküpfungen

Beispiele:

- Prozeß p1 sendet Nachricht n an Prozeß p2.
- Die Anzahl der wartenden Prozesse k wird um eins erhöht.

Charakteristisch für PE's ist, daß auch sie noch mit den Informationen eines Prozessorknotens erkannt werden können.

Ereignisverknüpfungen

Bei der Untersuchung paralleler Abläufe ist das Zusammenspiel der einzelnen Prozesse äußerst kritisch. Ein adäquates Monitorsystem muß deshalb die Verknüpfung von primitiven Ereignissen erlauben. Von besonderer Bedeutung sind hierbei die Formulierung von Verknüpfungen mit Aussagen über die zeitliche Ordnung von Ereignissen. Drei Möglichkeiten der Verknüpfung sind notwendig:

Fall 1: Ereignis E1 oder Ereignis E2 tritt auf (E1 v E2).

Fall 2: Ereignis E1 tritt zeitlich vor Ereignis E2 auf (E1$\rightarrow$E2).

Fall 3: Sowohl Ereignis E1 als auch Ereignis E2 tritt auf, wobei die zeitliche Reihenfolge beliebig ist (E1 $\wedge$ E2). Dies wird manchmal auch als Gleichzeitigkeit in verteilten Systemen definiert [Lam78], [Hab88].

Wichtig dabei ist, daß mehr als zwei Ereignisse verknüpfbar sind und daß die einzelnen Ereignisse unterschiedliche Prozesse und damit Knotenprozessoren betreffen können. Beispiele für die Fälle 1 bis 3 sind in Abbildung 5 dargestellt.

Aktionen

Beim Auftreten von Ereignissen und Ereignisverknüpfungen ist es notwendig, eine Anzahl von Aktionen anstoßen zu können. Die Aktionen können in drei Klassen eingeteilt werden.

- Die Aktionen zum Anhalten einer oder mehrerer paralleler Abläufe vor Erreichen des nächsten kritischen Ereignisses. Nach dem Anhalten kann der Zustand der einzelnen Prozesse von den Werkzeugen genauer analysiert und verändert werden. Wichtig dabei ist, daß auch Prozesse auf anderen Prozessorknoten angehalten werden können. Diese Aktion wird üblicherweise als Haltepunkt oder Breakpoint bezeichnet und unterstützt hauptsächlich die Phasen Fehlersuche, Test und Visualisierung.
- Aufzeichnen von Zustandsübergängen zusammen mit der Zeit des Auftretens. Diese Klasse von Aktionen bezeichnen wir als Trace. Aufgezeichnet werden können dabhei hauptsächlich Ausführungspunkte von Prozessen, Werte von Datenobjekten und Kommunikations- bzw. Synchronisationsparameter. Die Traceaufzeichnung wird von Werkzeugen für die Fehlersuche, den Test, die Visualisierung, die Leistungsmessung und den Lastausgleich verwendet.

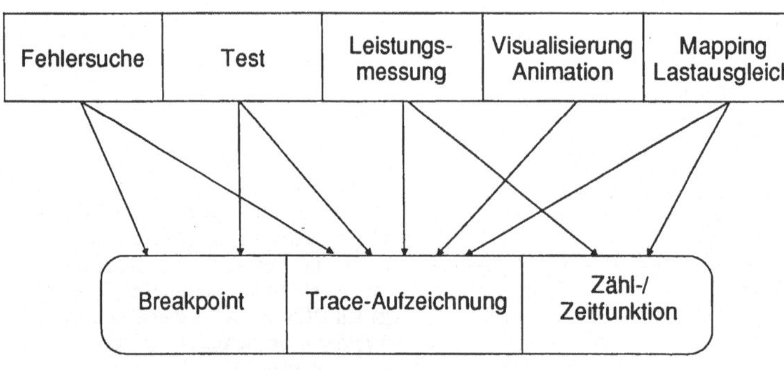

Abb. 6: Zuordnung von Werkzeugen zu Monitoraktionen

- Die dritte Klasse von Aktionen stellen die Zähl- und Zeitfunktionen dar. Hier können beim Auftreten spezieller Ereignisse einzelne Zähler erhöht bzw. erniedrigt werden und Uhren gestartet bzw. angehalten werden. Die Zähl-/Zeitfunktionen werden insbesondere in den Phasen Leistungsmessung und Lastausgleich verwendet.

Zur Spezifikation der Ereignisse, Aktionen und deren Verknüpfung werden Monitor-/Debugging- und Ereignisdefinitonssprachen verwendet, wie sie in [Bem88b], [Hof87], [Hab88] und [Bat83] beschrieben sind. Damit können Ereignisse und Aktionen auf unterschiedlichen Abstraktionsebenen spezifiziert werden. In Abb. 6 ist die Zuordnung von Monitor-Aktionsklassen und Werkzeugen zusammengefasst. Die einzelnen Ereignisklassen werden selbstverständlich von allen Werkzeugen benutzt. Ausserdem ist eine beliebige Zuordnung von Ereignisklassen zu Aktionsklassen programmierbar.

Neben diesen, die Funktionalität eines Monitorsystems betreffenden Eigenschaften, müssen einige allgemeingültige Anforderungen aufgestellt werden. Nachfolgend seien diese allgemeinen Anforderungen kurz angesprochen.

- Wie bereits erwähnt, müssen Ereignisse über Prozessorgrenzen hinweg verknüpfbar sein und Aktionen auch auf anderen Prozessorknoten als dem eigenen auslösbar sein.
- Die Ereigniserkennung, Aktionsauswertung und Weiterleitung der Daten an die Werkzeuge hat zur Laufzeit zu geschehen (on-line), um den interaktiven Charakter der Werkzeuge zu unterstützen.
- Eine ebenfalls aus der interaktiven Art der Werkzeuge abgeleitete Anforderung ist die nach Monitortechniken, welche keine Rekompilierung des Objektsystems erforderlich machen. Insbesondere ist damit jegliche Art der Quellcodeinstrumentierung ausgeschlossen.
- Nicht zuletzt steht die bereits erwähnte Forderung nach einer möglichst verzögerungsfreien Monitortechnik.

3. Das Monitor-Konzept von TOPSYS

Nach Hofmann/Klar [Hof87] ist das Überwachen von Multiprozessoren und verteilten Systemen nur durch ein verteiltes Monitorsystem möglich. Aus diesem Grund wird durch jeweils einen Monitor der Ablauf eines Knotenprozessors überwacht. Die einzelnen Monitore stehen mit einem zentralen Entwicklungsrechner in Verbindung, über welchen sie programmiert werden und auf welchen auch die Werkzeuge zur Auswertung der durch die Monitore gewonnenen Laufzeitinformation zur Verfügung stehen. Diese verteilte Monitorstruktur mit zentralem Entwicklungsrechner entspricht auf natürliche Weise den Host/Target-Entwicklungsumgebungen der meisten heute verfügbaren Multiprozessoren.

Für modelltheoretische Untersuchungen betrachtet man die Monitore bzw. Werkzeuge auf dem Entwicklungsrechner ebenfalls als Prozesse mit zugehörigen Zustands- bzw. Ereignisfolgen. Monitorprozesse und zu überwachende Prozesse kommunizieren durch den Austausch von Nachrichten. Bei Objektcode-Instrumentierung wird diese Kommunikation durch Software-Interrupts (Traps) realisiert, bei Hardware-/Hybrid-Monitoren durch die Überwachung des Knotenprozessorbusses. Außerdem kommunizieren die Monitore selbstverständlich mit den Werkzeugen auf dem Entwicklungsrechner und umgekehrt. Für die Realisierung von globalen Haltepunkten, konsistenten globalen Zuständen und globalen Zeiten ist jedoch auch ein Nachrichtenaustausch der Monitore untereinander notwendig. Über die Formulierung von Zeitbedingungen und Zeitschranken der zu überwachenden Prozesse und Monitorprozesse können Aussagen über die Qualität der Monitortechnik im Bezug auf die Verzögerungsfreiheit getroffen werden. Abbildung 7 zeigt diese zeitbehaftete Modellvorstellung des Monitorsystems im Raum/Zeit-Diagramm und gibt die Zeitbedingung für die Realisierung eines globalen Haltepunkts an.

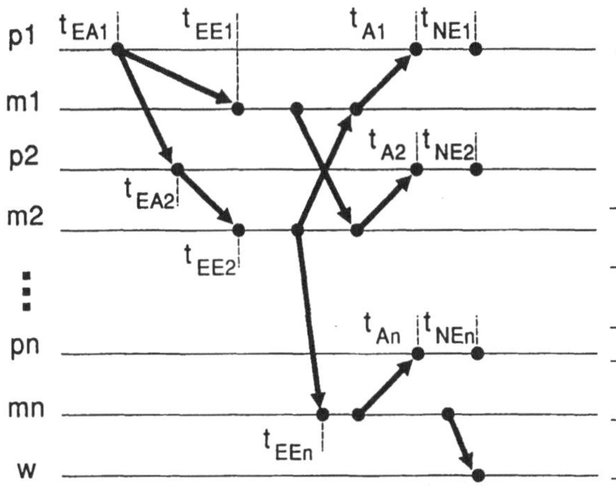

Abb. 7: Monitormodell im Raum/Zeit-Diagramm

Globaler Breakpoint genau dann, wenn:

$\forall p_i, m_i: |t_{Ai} - t_{EEi}| < |t_{NEi} - t_{EEi}|$

Bedeutungen der Zeiten:

- t_{EAi}: Relevantes zu überwachendes Ereignis tritt auf.
- t_{EEi}: Relevantes zu überwachendes Ereignis wird erkannt.
- t_{Ai}: Zeitpunkt zu dem Prozeß i angehalten wird.
- t_{NEi}: Zeitpunkt zu dem nächstes kritisches Ereignis in Prozeß i auftritt.
- pi: Ereignisfolge des zu überwachenden Prozesses i.
- mi: Ereignisfolge des Monitorprozeß i.
- w: Ereignisfolge des Werkzeugprozeß.

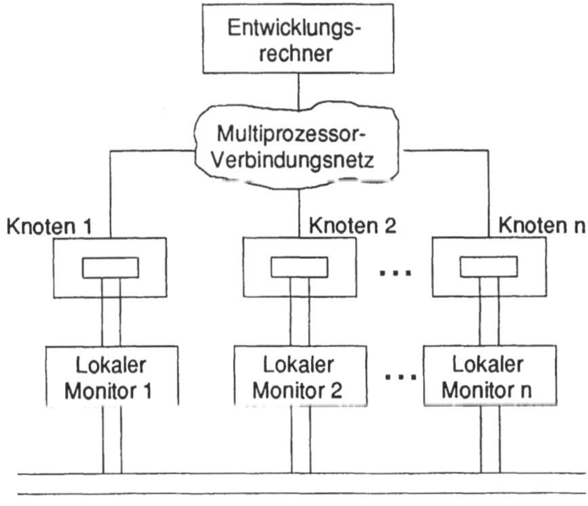

Abb. 8: Verteiltes Monitorkonzept von TOPSYS

Die beeinflussenden Parameter für die Zeitschranken hängen sehr stark von der Implementierungstechnik des Monitorsystems ab. Es ist ein Hauptinteresse des Projekts TOPSYS, zumindest die Größenordnungen dieser Zeitschranken für zwei konträre Instrumentierungstechniken (Hardware-/Hybrid- und Software-Monitoring) festzulegen.

Beim jetzigen Implementierungskonzept des Monitorsystems werden lokale Monitore an die Knotenprozessoren adaptiert. Um dieses Konzept auch für Multiprozessoren mit eine großen Anzahl von Knotenprozessoren verwenden zu können, kommunizieren die lokalen Monitore über das normale Verbindungsnetz des Multiprozessors mit dem Entwicklungsrechner. Optional könnte jedoch auch das Diagnose-Netzwerk des Multiprozessors sowie ein extra Hypercube-Netzwerk niederiger Dimension verwendet werden [Int88].

Zur Realisierung der Ereignisverknüpfung über Prozessorgrenzen hinweg und der Implementierung des globalen Haltepunktes haben wir in Abschnitt 2 festgestellt, dass eine Kommunikation der lokalen monitore untereinander notwendig ist. Bei der Implementierung von Softwaremonitoren wird diese Kommunikation mit Hilfe der Übertragung von Nachrichten über das normale Kommunikationssystem erledigt. Die relativ lange Laufzeit dieser Nachrichten führt selbstverständlich dazu, daß globale Haltepunkte und konsistente Zustände nur auf sehr grobgranularer Ebene herzustellen sind. Um globale Haltepunkte und Konsistente Zustände auf der Ebene von Hochsprachen-Anweisungen oder Maschinenbefehlen realisieren zu können, ist es notwendig, die Inter-Monitor-Kommunikation durch ein extra Verbindungsnetzwerk zu implementieren. Im Rahmen von TOPSYS wird dafür ein Crosstrigger-Bus verwendet, welcher alle lokalen Hardware-Monitore über neun bidirektionale Leitungen verbindet. Die 8 Crosstrigger-Leitungen werden von den einzelnen Monitoren im Zeitmultiplex zur Übermittlung von Triggern an andere Monitore verwendet. Neben den Triggern kann über eine Synchronisationsleitung auch ein zentraler Takt für alle Knoten geführt werden. Darauf wird in Abschnitt 4 noch genauer eingegangen. Abbildung 8 zeigt ein Blockschaltbild des beschriebenen verteilten Hardware-Monitor-Konzepts.

4. Der Lokale Hardware/Hybrid-Monitor

Nachdem das Zusammenwirken aller Monitore des Multiprozessors beschrieben wurde, widmen wir uns nun der Struktur eines lokalen Hardware/Hybrid-Monitors. Dabei beschreiben wir zunächst die Grobstruktur des Monitors und erläutern dann die, nach unserer Meinung nach neuartigen, Konzepte dieses Entwurfs.

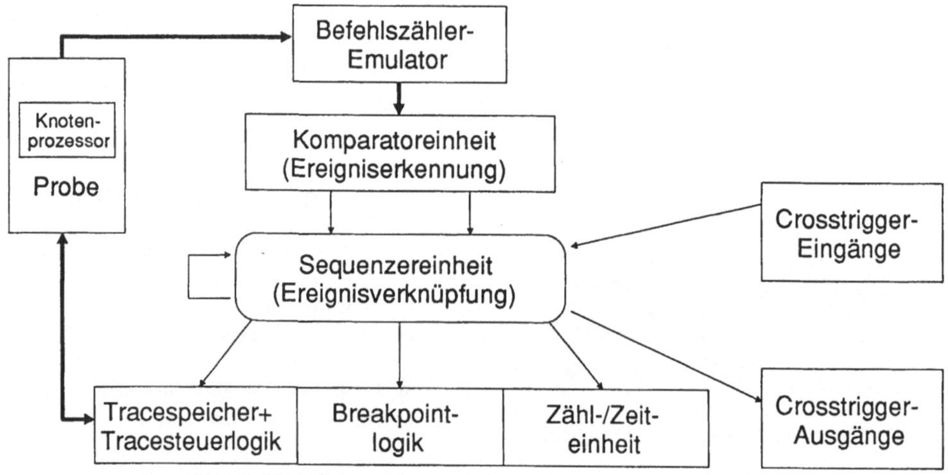

Abb. 9: Blockdiagramm des lokalen Hardware-/Hybrid-Monitors

Grobstruktur und Leistungsdaten

Der Hardware/Hybrid-Monitor wird über die sog. Probe an den Knotenprozessor adaptiert. Im Rahmen der jetzigen Implementierung ist dies der 32-Bit-Prozessor iAPX386. Auf der Probe befinden sich Treiber, welche die Signale des Prozessors verstärken. Die Adressen-, Daten- und Steuersignale des Knotenprozessors werden über die Probe in die Komparatoreinheit geführt. Dabei werden die Befehlsadressen noch im Befehlzähler-Emulator (BZE) umgewandelt. Auf die Funktion und Bedeutung des BZE wird weiter unten noch detaillierter eingegangen. Ab der Komparatoreinheit wurde versucht, den Hardware-/Hybrid-Monitor möglichst prozessorunabhängig zu gestalten. Durch dieses Vorgehen ist bei einer Adaption an einen anderen Prozessortyp lediglich die Anpassung der Probe und des BZE erforderlich.

Die Komparatoreinheit realisiert die Überwachung von primitiven Daten- und Kontrollereignissen (siehe Abschnitt 2). Dabei kann die implementierte Komparatoreinheit parallel 14 primitive Ereignisse überwachen. Durch die parallele Überwachung wird bereits gleichzeitig die in Abschnitt 2 geforderte "Oder-Verknüpfung" realisiert (Fall 1). Die sog. "UND-Verknüpfung" (Fall 3) wird beim Hardware-Monitor in der Komparatoreinheit exakt implementiert. Die von der Komparatoreinheit erzeugten Trigger können zusammen mit den Crosstrigger-Leitungen in der Sequenzereinheit weiter miteinander verknüpft werden. Dadurch wird die in Abschnitt 2 geforderte sequentielle Verknüpfung realisiert (Fall 2). Die implementierte Sequenzereinheit kann gleichzeitig 14 Trigger mit einer Sequenztiefe von 4 verknüpfen.

Die Ausgänge der Sequenzereinheit werden verwendet, um die geforderten Aktionen, Breakpoint, Trace-Aufzeichnung und Zähl- bzw. Zeitmessungen anzustoßen. Außerdem ist es natürlich möglich, die auf dem lokalen Monitor verknüpften Ereignisse wieder über den Crosstrigger-Bus an andere Monitore zu melden. Die Sequenzereinheit kann parallel 18 Aktionen gleichzeitig anstoßen. Die Breakpoint-Logik löst eine nicht maskierbare Unterbrechung beim Knotenprozessor aus, worauf die auf diesem Knotenrechner anzuhaltenden Prozesse gestoppt werden können. Die Zähl- und Zeiteinheit kann einfache und integrierende Zähloperationen über programmierbare Zeitintervalle ausführen. Außerdem ist es möglich, absolute Zeitmessungen mit unterschiedlicher Genauigkeit (programmierbare Zeitbasis von 1µs-1ms, 32 bit) durchzuführen. Die Zeitbasis kann optional von lokalen Uhren oder von einer über die Synchronisationsleitung eingebrachten zentralen Uhr abgegriffen werden. Dieses Vorgehen wurde gewählt, um mit lokalen und zentralen Zeiten bei Fertigstellung des Monitors experimentieren zu können. Insbesondere sollen die Auswirkungen beider Alternativen auf die Meßqualität untersucht werden.

Der Tracespeicher erlaubt die Aufzeichnung von allen, über den Adreß-, Daten- und Kontrollbus des Prozessors transportierten Informationen mit zugehörigen Zeitstempeln der Zähl-/Zeiteinheit. Der Tracespeicher kann also wiederum lokale oder zentrale Zeiten enthalten. Außerdem kann auch der Crosstrigger-Bus im Tracespeicher mitprotokolliert werden, um Ereignisfolgen von anderen Prozessoren zu rekonstruieren. Der Tracespeicher ist 8K Einträge tief und 64 Bit breit. Die 8K Einträge werden durch die Programmierung der Aufzeichnungsart zu unterschiedlich großen Aufzeichnungsrahmen (Frames) konfiguriert.

Komparator- und Triggertechnik

Die Erkennung von primitiven Ereignissen kann prinzipiell mit zwei Techniken erledigt werden. Auf dem Markt verfügbare Komparatorbausteine vergleichen das Sollbit-Muster mit dem Istbit-Muster. Diese Technik wurde bei einem früheren Hardware-Monitor-Entwurf verwendet und hat sich als sehr unflexibel herausgestellt [Bem86]. Für jedes zu überwachende primitive Ereignis ist ein Komparatorbaustein notwendig. Auf Grund dieser Erfahrung fand beim Entwurf des beschriebenen Monitors die sogenannte RAM-Komparatortechnik ihre Anwendung [Wol84]. Dabei werden schnelle Speicherbausteine als sogenannte TAG-RAMs verwendet. Wird eine Speicherzelle des Tag-RAM's aktiviert und ist das Tag gesetzt, so ist der Trigger erfüllt. Diese Technik hat den Vorteil, daß bei gleicher Bausteinanzahl parallel viele Ereignisse überwacht werden können und der RAM-Komparator wesentlich flexibler programmiert werden kann. Ausserdem ist ein Vorteil dieser Technik, dass sehr einfach mit Hilfe von sog. "Don't Care-Bits" Bereichserkennungen durch Ausblendung realisiert werden können.

Monitorunterstützung durch den Knotenprozessor

Hochleistungs-Knotenprozessoren von heutigen Multiprozessoren verwenden sehr intensiv durchsatzsteigernde Konzepte wie überlappte Verarbeitung (Pipelining) und Cachespeicher. Außerdem wird versucht durch zunehmende Transistoranzahlen immer weitere Funktionseinheiten (z.B. Speicherverwaltungseinheit, Numerik-Koprozessor) in den Prozessorbaustein zu integrieren. Dies führt dazu, daß eine Monitorschaltung an den externen Prozessoranschlüssen den internen Ablauf des Prozessorbausteins nicht mehr exakt beobachten kann. Auf lange Sicht wird dies sicher dazu führen, daß Monitorschaltungen in die Prozessoren integriert werden. Bis sich jedoch die Prozessorhersteller zu dieser Entscheidung durchgerungen haben, gibt es nur die Möglichkeit spezielle Testversionen eines Prozessors für die Überwachung zu verwenden. Bei diesen speziellen Testversionen werden Informationen über den internen Ablauf des Prozessors auf zusätzlichen Statusleitungen nach außen geführt. Die Testversionen der Prozessoren bezeichnet man als Bondout-Bausteine.

Bei dem beschriebenen lokalen Monitor wird diese Technik angewendet und ein Bondout-Baustein des iAPX386-Prozessors eingesetzt. Auf Grund der zusätzlichen Statusleitungen ist der erwähnte Befehlszähler-Emulator (BZE) in der Lage die Arbeitsweise der internen Prozessorpipeline nachzubilden und den genauen Ausführungszeitpunkt von Operationen zu bestimmen. Abbildung 10 zeigt die Pinbelegung des Bondout-Bausteins, wobei die zusätzlichen Statusleitungen besonders gekennzeichnet sind.

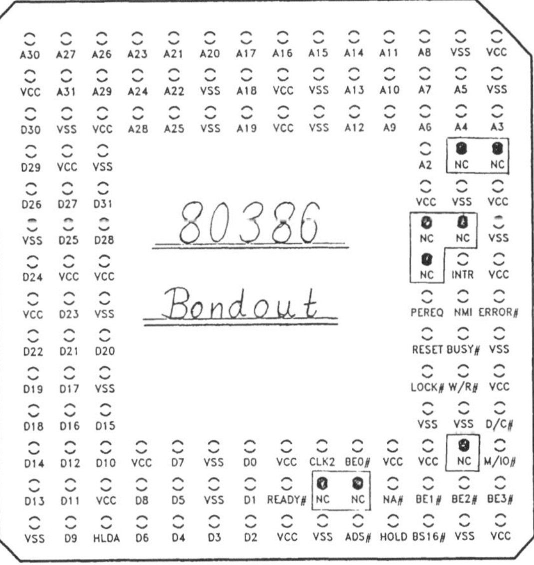

Abb. 10: Bondout des iAPX386

Monitorunterstützung durch das Betriebssystem

Parallelitäts-Ereignisse (PE) können, wie in Abschnitt 2 erwähnt, durch eine kombinierte Überwachung von Kontroll- und Daten-Ereignissen über den Ablauf von Implementierungsroutinen des Betriebssystems ausgewertet werden. Insbesondere wird mit den PE's die Interaktion zwischen Prozessen überwacht. Dieses Vorgehen hat jedoch den Nachteil, daß viele Betriebsmittel des Hardware-Monitors durch die PE's belegt werden. Eine alternative Technik ist, den Hardware-Monitor durch den Einbau monitorunterstützender Konzepte in das Betriebssystem bei der Erkennung von Parallelitäts-Ereignissen zu unterstützen. Dabei werden zur Spezifikationszeit der PE's Sensoren im Betriebssystem aktiviert, welche notwendige Informationen (z.B. Prozeßidentifikatoren, Nachrichtenpointer, Warteschlangenlängen,

Prozeßzustände) zur Laufzeit der Applikationsprozesse in die Register des Hardware-Monitors schreiben. Wichtig ist, daß die Sensoraktivierung dynamisch zur Laufzeit geschehen kann, was sich positiv auf die Flexibilität dieser Monitortechnik auswirkt. Eine detailliertere Beschreibung dieser, in der Literatur häufig auch als Hybrid-Monitortechnik bezeichnete, Instrumentierungstechnik findet sich in [Bei88]. Dort wird auch durch Messungen belegt, daß die Sensoren den Programmablauf im schlechtesten Fall um 1% verzögern.

5. Implementierungsdetails

Der Prototyp des beschriebenen lokalen Hardware-Monitors wird auf einer 2-fach Doppeleuropakarte realisiert. Der Monitor hat die gleiche Größe wie das normale Knotenboard des iPSC/2 und kann neben einem Knotenboard über die normale Rückwandverdrahtung (Backplane) in das Gehäuse des iPSC/2 Multiprozessors integriert werden. Die jetzigen Schaltzeiten des Monitors sind für einen 16 MHz iAPX386 Knotenprozessor ausgelegt. Zu diesem Zweck werden für die RAM-Komparatoren statische Speicherbausteine mit einer Zugriffszeit von 15 ns eingesetzt. Neben Programmierbaren Logikbausteinen (PAL's) für verschiedene Verknüpfungslogiken, werden Logic Cell Arrays (LCA) mit 1200 Gattern und einer Taktfrequenz von 70 MHz für die Realisierung der Tracespeicher- und Zähl-/Zeitsteuereinheit benutzt.

6. Stand der Arbeiten und Zukunft

Die beschriebene TOPSYS Werkzeugumgebung befindet sich zur Zeit in der Test- und Integrationsphase. Parallel zu dem Hardware-Monitor-Prototyp werden der Softwaremonitor und die in Abschnitt 1 geforderten Werkzeuge entwickelt. Ein erster Prototyp einer Werkzeugumgebung inklusive Monitorsystem für den verfügbaren Multiprozessor iPSC/2 mit 32 Prozessorknoten wird Mitte 1989 zur Verfügung stehen.

Da der beschriebene Hardware-Monitor noch relativ viel Platz benötigt, wird zur Zeit in Zusammenarbeit mit dem Lehrstuhl für Rechnergestütztes Entwerfen eine integrierte Version des Hardware-Monitors in Form eines VLSI-Bausteins entwickelt. Die Integration des Hardware-Monitors wird dadurch erleichtert, daß bereits der beschriebene Monitor mit Hilfe einer CAE-Station entwickelt wurde und so die Schaltpläne und Simulationsergebnisse in maschinenlesbarer Form vorliegen.

Bibliographie

[Bai88] M.L. Bailey, D. Socha, D. Notkin: Debugging Parallel Programs using Graphical Views; Int. Conf. on Parallel Processing, p. 46 - 49, Aug. 1988

[Bat83] P.C. Bates, J.C.Wileden: High Level Debugging of Distributed Systems: the Behavioral Abstraction Approach; Journal of Systems and Software 4 (3), p. 255 - 264, Dec. 1983

[Bat88] P. Bates: Distributed Debugging Tools for Heterogeneous Distributed Systems; 8th Int. Conf. on Distr. Comp. Systems, p. 308 - 316, June 1988

[Bei88] H.J. Beier, T. Bemmerl: Software Monitoring of Parallel Programs; CONPAR 88, p. 71 -78, Sept. 1988

[Bem86] T. Bemmerl: Realtime High Level Debugging in Host/Target Environments; EUROMICRO'86 Symposium, p. 387 - 400, Sept. 1986

[Bem88a] T.Bemmerl: An Integrated and Portable Tool Environment for Parallel Computers; Int. Conf. on Parallel Processing, p. 50 -53, Aug. 1988

[Bem88b] T. Bemmerl: Quellbezogenes Debugging von Multimikroprozessoren ; GI-Jahrestagung 1988, p. 615 - 629, Okt. 1988

[Bur88] H. Burkhart, R. Millen: Techniken und Werkzeuge der Programmbeobachtung am Beispiel eines Modula-2 Monitorsystems; Informatik Forschung und Entwicklung, 3, p. 6 - 21, 1988

[Cha85] K.M. Chandy, L. Lamport: Distributed Snapshots: Determining Global States of Distributed Systems; ACM Trans. on Comp. Syst. 3 (1), p. 63 - 75, Feb. 1985

[Cur82] R. Curtis, L. Wittie: BUGNET: a debugging system for parallel programming environments; 3th Int. Conf. on Distr. Comp. Syst. p. 18 - 22, 1982

[Eic88] S. Eichholz, F. Abstreiter: Runtime Observations of Parallel Programs; CONPAR 88, p. 63 - 70, Sept. 1988

[Gar84] Garcia-Molina: Debugging a Distributed Computing System; IEEE Trans. on Soft. Eng., Vol.-SE-10, No. 2, p. 210 - 219, 1984

[Gre86] F. Gregoretti, F. Maddaleno, M. Zamboni: Monitoring Tools for Multimicroprocessors; EUROMICRO'86 Symp., p. 409 - 416, Sept. 1986

[Hab88] D. Haban, W. Weigel: Global Events and Global Breakpoints in Distributed Systems; 21st Hawaii Conf. on System Science, Jan. 1988

[Har85] P.K. Harter, D.M. Heimbigner, R. King: IDD: An Interactive Distributed Debugger; 5th Conf. on Distr. Comp. Syst., p. 498 - 506, May 1985

[Hof87] R. Hofmann, R. Klar, N. Luttenberger, B. Mohr: ZÄHLMONITOR 4: Ein Monitorsystem für das Hardware- und Hybrid-Monitoring von Multiprozessor- und Multicomputer-Systemen; Messung und Modellierung von Rechensystemen, p. 79 - 99, 1987

[Int88] Intel: iPSC/2 Users Manual; Intel Corp., 1988

[Joy87] J. Joyce, G. Lomow, K. Slind, B. Unger: Monitoring Distributed Systems; ACM Trans. on Comp. Syst. 5 (2), p. 121 - 150, 1987

[Lam78] L. Lamport: Time, Clocks and Ordering of Events in a Distributed System: Com. of the ACM 21 (7), p. 558 - 565, July 1978

[LeB85] R.J. LeBlanc, A.D. Robbins: Event-Driven Monitoring of Distributed Programs; 5th Int. Conf. on Distr. Comp. Syst., p. 515 - 522, May 1985

[LeB87] T.J. LeBlanc, J.M. Mellor-Crummey: Debugging Parallel Programs with Instant Replay; IEEE Trans. on Comp. 36 (4), April 1987

[Mil88] B.P. Miller, J.D. Choi: Breakpoints and Halting in Distributed Programs; Int. Conf. on Distr. Comp. Syst., p. 316 - 323, 1988

[Müh86] H. Mühlenbein, F. Limburger, S. Streitz, S. Warnhaut: MUPPET, A Programming Environment for Message-Based Multiprocessors; Fall Joint Comp. Conf., Nov. 1986

[Seg85] Z. Segall, L. Rudolph: PIE: A Programming and Instrumentation Environment for Parallel Processing; IEEE Software, p. 22 - 37, Nov. 1985

[Spe88] M. Spezialetti, J.P. Kearus: A General Approach to Recognizing Event Occurrences in Distributed Computations; Int. Conf. on Distr. Comp. Systems, p. 300 - 307, 1988

[Wol84] A. Wolper: An Emulator Architecture; Intel Technology Journal, p. 9 - 13, Fall 1984

[Wyb88] D. Wybranietz, D. Haban: Monitoring and Performance Measuring Distributed Systems during Operation; GI/ITG Tagung Org. und Betrieb von Rechensystemen, p. 308 - 323, März 1988

MESSUNGEN ZUR BEURTEILUNG DER REALZEITEIGENSCHAFTEN EINES KOMMUNIKATIONSKONTROLLERS FÜR EIN MAP-NETZ

A.E. Elnakhal, H. Rzehak
Universität der Bundeswehr München
Werner-Heisenberg-Weg 39, D-8014 Neubiberg

Zusammenfassung
In dem Beitrag wird über Zeitmessungen des Transportdienstes eines MAP-Netzes berichtet. Hierbei wird insbesondere der Einfluß der Quittierungsverfahren und der Fenstergröße des ISO-Transportprotokolls herausgearbeitet. Ferner wird die Bedeutung der Interprozeßkommunikation und der Prüfsummenberechnung für eine Implementierung analysiert.

Abstract
The paper reports on measurements of time properties of the transport service of a MAP-network. Especially the influence of acknowledgement policies and the window size of the ISO Transport Protocol is studied. Some of the implementation aspects such as interprocess communication and checksum computation are discussed.

1. Motivation und Meßumgebung

1.1 Allgemeines

Verschiedene Realisierungen von MAP-LANs (Manufacturing Automation Protocol - Local Area Networks) im Rahmen des ISO-Reference Modell unterscheiden sich zur Zeit zwar durch die vom Hersteller aus Interessengründen eingesetzten Hard- und Software Technologien, haben aber eine gemeinsame Struktur, die sich in drei Komponenten untergliedern läßt: Host oder Gerät, Kommunikationskontroller und das physikalische Netz.

Jeder Knoten in einem LAN verfügt über den sogenannten Kommunikationskontroller (KK), der als eine Schnittstelle zwischen den auf dem Host laufenden Anwenderprozessen und dem physikalischen Netz angesehen werden kann. Er ist in der Regel eine mit einem Prozessor ausgerüstete intelligente Komponente und hat die Entlastung des Host-Prozessors von der Abwicklung aller (oder eines Teils der) Kommunikationsprotokolle (KP) zur Aufgabe vgl. Bild 1.1 .

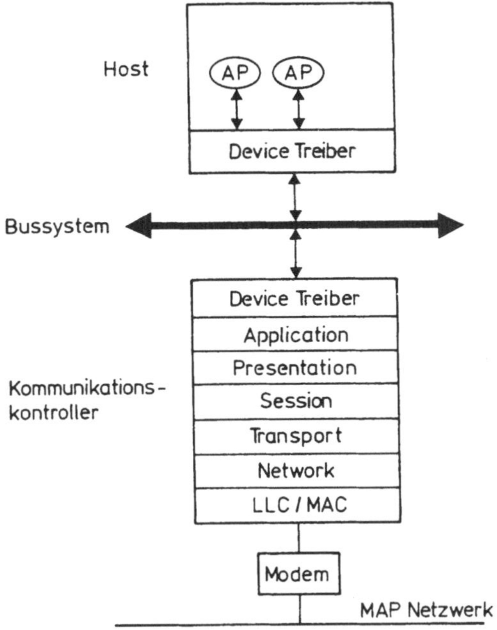

Bild 1.1 Aufbau eines MAP-Knotens

In diesem KK erfolgt die Realisierung der Zugriffsschicht des ISO-Modelles teilweise durch einen integrierten Baustein, der die Dienste von IEEE 802.4 abwickelt. Die übrigen Protokolle des ISO-Modells sind in Form von Tasks bzw. Prozessen implementiert, deren Verwaltung und Synchronisation von einem Realzeit-Betriebssystem erfolgt (z.B. VRTX von Ready Systems, IRMX von Intel). Jede Task hat eine Warteschlange für Nachrichten, mit deren Hilfe die Tasks durch Senden und Empfangen von Nachrichten kommunizieren.

1.2 Leistungsmerkmale eines Kommunikationskontrollers

Die Leistungsmerkmale des KK werden durch die Güte der Software zur Protokollabwicklung, softwaremäßige Umgebung (Betriebssystem, Interprozeßkommunikation) und durch die Fähigkeit der Hardware (Prozessor, Speicher) ausgeprägt. Um die Rolle, die der KK bei der Nachrichtenverzögerung spielt, genauer zu betrachten, wurden Leistungsmessungen des Transportdienstes an einem bestehenden MAP-Netz durchgeführt.

In der Arbeit, über die hier berichtet wird, wurden im zu untersuchenden KK nur die unteren vier Schichten des ISO-Modelles (Phy, MAC+LLC, Netzwerk, Transport) benutzt, da z.Zt. nur eine Implementierung von MAP 2.1 zur Verfügung steht und damit in den höheren Schichten keine Übereinstimmung mit dem zukünftigen Standard MAP 3.0 besteht. Die Transportdienste werden von einer darüberliegenden Schicht zur Lasterzeugung und Messung angefordert. Diese Schicht absorbiert auch auf der Empfangsschicht die angelieferten Protokolldateneinheiten (PDU).

Die ersten Zeitmessungen an MAP Implementierungen wurden in [STR 88] und [MAR 88] veröffentlicht. In [MAR 88] wurden die Ausführungszeiten der Anwenderdienste, bei im übrigen unbelastetem Netz, für den Kommunikationskontroller der Firma Industrial Networking Incorporated gemessen. In [STR 88] wurden Zeitmessungen der Transportdienste, bei im übrigen unbelastetem Netz, für den Kommunikationskontroller der Firma Intel in Abhängigkeit von der Nachrichtenlänge durchgeführt, so daß der Segmentierungsaufwand deutlich wurde.

Weiterhin wurden Messungen des Durchsatzes in Abhängigkeit von der Fenstergröße vorgenommen, die frühere Modellierungsergebnisse [TAK 87] hinsichtlich der Steigerung des Durchsatzes bei Vergrößerung der Fenstergröße bestätigen.

In dieser Studie werden Zeitmessungen des Transportdienstes auf dem Kommunikationskontroller der Firma Motorola sowohl beim unbelasteten Fall, vgl. 2.3, als auch unter poissonischer Last für unterschiedliche Verbindungstopologien durchgeführt. Dabei werden zwei zulässige Quittierungsverfahren der Transportschicht betrachtet (Nth Acknowledgement und Explicit Acknowledgement, siehe 2.2). Die Ursache des Zeitaufwandes zur Abwicklung der Transportdienste wird näher verfolgt und analysiert.

1.3 Das Versuchsnetz

Das Netz besteht aus fünf Workstations der Firma Motorola, die hostseitig unter dem Betriebssystem UNIX System V.3 auf VME-Bus Basis laufen. Jeder Knoten ist mit einem MAP-Kommunikationskontroller MVME 372 der Firma Motorola ausgerüstet, der über einen eigenen Prozessor (MC 68020) und Speicher verfügt und sich des Betriebssystems VRTX der

Firma Ready Systems bedient. Die Dienste der Zugriffsschicht (IEEE 802.4) auf dem KK werden durch den VLSI-Baustein MC68824 abgewickelt, zusätzlich dient eine MAC-Task als Treiber für diesen Baustein.

Der Anschluß des Knoten an das physikalische Netz erfolgt über ein Carrierband-Modem MCV-20 der Firma Computrol, 5 Mbps. Die Kommunikation zwischen Host und KK geschieht mit Hilfe des Host-Device Treiber Protokoll BPP (Buffered Pipe Protocol).

2. Meßvorrichtung

2.1 Allgemeiner Ablauf und Meßgrößen

Aufgrund der fehlenden Realzeitfähigkeit des Betriebssystems UNIX wurden die Messungen direkt auf dem KK vorgenommen. Auf dem KK eines jeden Knoten wurde eine zusätzliche Task kreiert (Initiator oder Responder), die an Stelle der Sitzungsschicht die Dienste der Transportschicht anfordern und deren Ausführungszeiten erfassen.

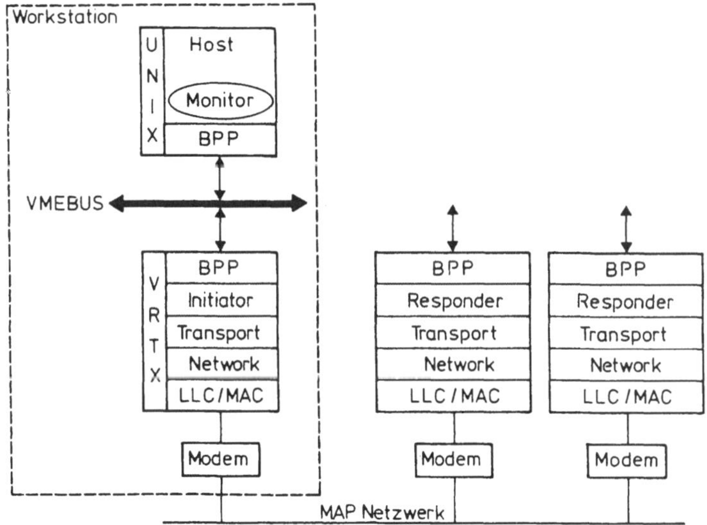

Bild 2.1 Aufbau des Versuchsnetzes

Der Initiator baut wahlweise Verbindungen zu dem gewünschten Responder auf und belastet diese Verbindungen mit Anforderungen gemäß der vorgegebenen Last. Der Responder bekommt die vom Initiator abgeschickte Anforderung, absorbiert sie und beantwortet sie ebenfalls mit einer

Anforderung. Ein Monitorprogramm auf dem Host dient der Koordinierung des Testvorgangs (Verbindungswünsche, Lastmuster, Anzahl der Anforderungen, Parameter einstellen (Fenstergröße), Start bzw. Stop des Testes), vgl. Bild 2.1. Zur Übermittlung der Parameterdaten vom Initiator an Responder KK werden die im Connect-request Primitive zulässigen 32 Octett User_Data ausgenutzt.

Initiator und Responder bilden jeweils einen Transportbenutzer, indem sie sich der Service-Schnittstelle der Transportschicht bedienen. Gemessen werden soll die Antwortzeit und die Durchlaufzeit. Unter Antwortzeit des Transportbenutzers T_a versteht man die Zeit, die zwischen Absenden einer Nachricht und Empfang der entsprechenden Antwort vergeht, vgl. Bild 2.2. Unter Durchlaufzeit T_d des Transportsystems versteht man die Laufzeit von dem Transportbenutzer der Sendestation bis zum Transportbenutzer der Empfangsstation.

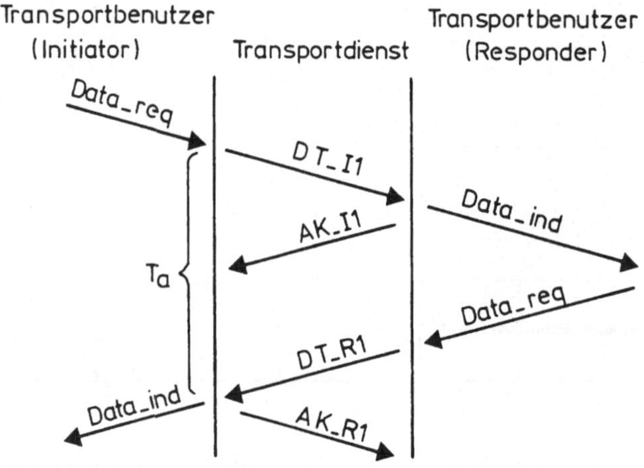

Bild 2.2 Antwortzeit T_a des Transportbenutzers

2.2 Protokolloptionen

Für alle Messungen wurden folgende Optionen für die Transportschicht gleich eingestellt:

- **Datenlänge:** um eine Segmentierung zu vermeiden, deren Effekt hier nicht weiter verfolgt wird, und bedingt durch die max. Pufferlänge von 1100 Byte und max. TPDU-Länge von 1024 Octett, haben wir eine TSDU-Länge von 1000 Octett festgelegt.

- **keine Prüfsumme**: der durch die Prüfsumme verursachte Zeitaufwand hängt von der Fähigkeit der CPU und der Nachrichtenlänge ab. Die Messungen an unserem System ergaben einen CPU-Zeitbedarf von etwa 4 μs pro Octett. Somit ist eine Erhöhung der Durchlaufzeit T_d einer Nachricht der Länge 1000 Octett von etwa 8 ms zu erwarten.

- **Priorität**: Alle Nachrichten haben die gleiche Priorität, es werden also keine Vorrangdaten generiert.

Der Verbindungsaufbau und -abbau erfolgte vor bzw. nach den Messungen. Die Auswirkung dieser Optionen, so wie andere interne Parameter der Transportschicht, wurden in [NBS 85] für ein über Satellite-Network operierendes Transportsystem (als WAN-Umgebung) simulativ untersucht.

Zwei Quittierungsverfahren der Transportschicht wurden berücksichtigt:

A) Nth acknowledgement (NACK)

Der Sender schickt N Dateneinheiten ohne auf Quittung zu warten, der Empfänger quittiert nur die N-te Nachricht, dabei charakterisiert N in unserem System die Fenstergröße n. Mit der Quittierung der N-ten Nachricht werden die vorigen Nachrichten implizit mitquittiert (acknowledges accumulation).

B) Explicit acknowledgement (SACK)

Der Sender verhält sich wie bei NACK, der Empfänger quittiert jede empfangene Dateneinheit.

Erwartungsgemäß ist die Quittierungsstrategie NACK günstiger als SACK, da bei SACK zusätzlicher Verkehr durch die Quittungen verursacht wird, dessen Umfang durch die Messungen deutlich wird.

2.3 Lasterzeugung und Erfassung der Meßgrößen

Unter dem unbelasteten Fall werden die Anforderungen so generiert, daß zu jeder Zeit maximal eine einzige Anforderung sich im Netz befindet. Es wird also die minimale Verzögerung gemessen. Im belasteten Fall werden die Anforderungen gemäß dem vorgegebenen Ankunftsprozess erzeugt.

Die gewünschten Zwischenankunftszeiten der Anforderungen werden vom Monitorprogramm (Hostseitig) off-line generiert und in Tabellen abgelegt, die mittels BPP an die Initiatortask abgeschickt werden. Der erste Wert der Zwischenankunftszeiten wird in der Variable X zugewiesen. Diese Variable wird jede Millisekunde dekrementiert. Erreicht sie den Wert Null, so wird eine Anforderung bzw. TSDU (Transport Service Data Unit) generiert und die Uhrzeit im Datenteil vermerkt. Die Variable X wird nun mit dem nächsten Wert der Tabelle aktualisiert, und die TSDU wird auf der entsprechenden Verbindung zum Responder abgeschickt. Erhält der Responder diese TSDU, so generiert er eine TSDU als Antwort, in welcher der Zeitvermerk aus der erhaltenen TSDU eingetragen ist und schickt sie auf der entsprechenden Verbindung zum Initiator zurück, der sie absorbiert und die Antwortzeit durch die Differenz der aktuellen Zeit und dem Zeitvermerk bildet.

Eine meßtechnische Erfassung der Durchlaufzeit T_d ist ohne auf den Netzknoten synchron laufende Uhren nicht durchführbar.

T_d läßt sich folgendermaßen zerlegen:

$$T_d = T_s + T_e + T_m$$
$$T_a = 2\, T_d + T_q$$

T_s - Verarbeitungszeit der Daten beim sendenden Knoten

T_e - Verarbeitungszeit der Daten beim empfangenden Knoten

T_m - Frameverzögerung von MAC-User bis MAC-User entlang des phys. Medium

T_q - Zeitaufwand durch Quittung (falls eine Quittung generiert wird)

T_s bzw. T_e besteht aus der Verzögerung des PDU beim Durchlaufen der Softwarepfade, des Senders bzw. des Empfängers der Schichtenprotokolle (ab MAC-Schicht) plus des Zeitaufwands zur Verwaltung der einzelnen Prozesse.

T_m besteht aus der Verabeitungs- und Sendezeit der sendenden MAC-Schicht, inklusive Netzwerkverzögerung T_{ms} und der Empfangs- und Verarbeitungszeit der empfangenden MAC-Schicht T_{me}.

3. Einflußgrößen auf den Zeitbedarf

3.1 Grundsätzliche Betrachtungen

Die Zeit für die Ausführung von Diensten der KPs hängt hauptsächlich von folgenden Parametern ab:

- Umfang und Qualität der zu erbringenden Dienste, sowie Anzahl der kommunizierenden Prozesse;
- Hardwarekonfiguration (Fähigkeit der CPU, vorhandene Ressourcen);
- Interprozeßkommunikation und Pufferverwaltungsstrategie (Betriebssystem);
- Leistung, Kapazität und Zuverlässigkeit des Transportsystems;
- interne Parameter der einzelnen ISO-Schichten und die Güte der Implementierung.

3.2 Auswirkung des Betriebssystems auf die Zeiten zur Protokollabwicklung

Wie schon erwähnt, wird jede ISO-Schicht normalerweise softwaremäßig durch einen Prozeß (Task) realisiert, die durch Nachrichtenaustausch untereinander kommunizieren, um Dienste anzufordern bzw. zu erbringen. Es ist die Aufgabe des Betriebssystems (BS), die Prozesse zu verwalten. Dies geschieht durch BS-Aufrufe, deren Aufwand nicht nur von der Art des eingesetzten Betriebssystems, sondern auch von der Fähigkeit des Kommunikationsprozessors abhängig ist. Im Folgenden betrachten wir <u>nur</u> die BS-Aufrufe, die einen Taskwechsel bewirken, um den Aufwand durch die Task-Switching-Time abzuschätzen.

Der Transportprozeß P_4 wird gerade durch eine vom Transportbenutzer ausgelöste Anforderung aktiv und will einen Dienst vom Netzwerkprozeß P_3 anfordern. Dazu schickt P_4 eine Nachricht an P_3 und wird anschließend passiviert. Daraufhin bekommt P_3 die Nachricht und wird aktiviert.

Kalkuliert man den durch die Betriebssystemaufrufe zur Abwicklung dieser Aktivitäten verursachten Zeitaufwand, so ergibt sich z.B. für das BS VRTX der Firma "Ready Systems" gemäß Firmenangaben für Speicher ohne Wait State einmal 357 μs bei 10 MHz-68010 CPU und 153 μs bei 16,67 MHz-68020 CPU, vgl. Anhang I. Für unser System (12,5 MHz-68020

CPU) wurde ein Zeitaufwand von etwa 270 μs gemessen. T_s bzw. T_e enthalten das dreifache dieser Zeit, da jedesmal 4 Schichten durchlaufen werden. T_m enthält das zweifache dieser Zeit (sendende und empfangende MAC-Driver).

Der Umfang und die Typen, der zur Abwicklung interner Protokoll-Aktivitäten benötigten BS-Aufrufe, sind unterschiedlich für die verschiedenen Protokollaufgaben und den Zustand der Finite State Maschine des Protokolls. (Belegen und Freigeben von Ressourcen, Timerverwaltung.)

4. Ergebnisse und Bewertung

4.1 Messungen im unbelasteten Fall

Unter Einhaltung der in 2.3 definierten Regel für den unbelasteten Fall wurden für die beiden Strategien unterschiedliche minimale Antwortzeiten gemessen. Für Strategie SACK wurde immer eine Antwortzeit T_{SACK} = 22 ms festgestellt.

Für Strategie NACK ergab sich die Antwortzeit bei einer Fenstergröße n für die ersten n-1 Anforderungen zu $T_{NACK(n-1)}$ = 18 ms, für die nte Anforderung wurde eine Antwortzeit von $T_{NACK(n)}$ = 22 ms = T_{SACK}.

Der Zeitunterschied zwischen T_{SACK}, $T_{NACK(n)}$ und $T_{NACK(n-1)}$ ergibt sich daher, daß die Quittung Ak-I1 im Bild 2.2 bei den ersten (n-1) Anforderungen der Strategie NACK entfällt, d.h. es gilt T_q. = 4 ms.

Nun läßt sich die Durchlaufzeit des Transportsystems im unbelasteten Fall T_d abschätzen:

$$T_d = T_{NACK(n-1)}/2 = 9 \text{ ms}.$$

In [RZE 89] wurde die Sendezeit in einem unbelasteten Netz zu 3.3 ms meßtechnisch erfaßt. Diese erfaßte Zeit enthält allerdings den Aufwand durch die Ergänzung eines confirm von TBC an MAC und vom MAC-Driver an LLC-Schicht, wo die gesendete Nachricht entlang des phy. Medium schon gesendet wurde, und deshalb nur einen Teil dieser Zeit zu T_{ms} beiträgt. Der Empfang und die Verarbeitung der Nachricht bis hin zur LLC-Schicht T_{me} kann mit der Zeit, die zur Erzeugung eines Confirms von TBC an MAC und von MAC an LLC benötigt wird verglichen werden.

Daher stellt die in [RZE 89] gemessene Zeit eine sehr gute Näherung für die Zeitanteile der Zugriffsschicht T_m an die Durchlaufzeit T_d. Die verbliebene Zeit von T_d nach T_m und Taskswitching (vgl. Anhang I) ist also 9-(3.3+1.6) = 4.1 ms. Diese 4.1 ms sind den Betriebssystemaufrufen und Softwarepfaden zur Durchführung der Protokollaktivitäten zuzurechnen.

4.2 Messungen unter poissonscher Last

Es wird weiterhin die Antwortzeit gemessen, wobei die Anforderungen gemäß einem Poisson Ankunftsprozeß mit der Rate LAM generiert und auf die aufgebauten M Verbindungen gleich verteilt werden, so daß jede bestehende Verbindung mit einer poissonschen Last der Rate LAM/M belastet wird. Die Messungen wurden für unterschiedliche Verbindungstopologien und Fenstergrößen vorgenommen und über die Anforderungsrate LAM in $(100 \text{ ms})^{-1}$ gezeichnet.

Zur meßtechnischen Erfassung der einzelnen Kurven wurde die Zwischenankunftsrate LAM schrittweise vergrößert, bis ein Systemzusammenbruch zustande kommt. Ein Systemzusammenbruch (breack down) kommt dann vor, wenn die Ankunftsrate LAM derart groß ist, daß mehr Anforderungen eintreffen, als das System bedienen kann, was einen Pufferüberlauf zur Folge hat (Stabilitätsbedingung Angebot<1 wird nicht mehr erfüllt). Aus diesem Grund hören die aufgenommenen Kurven zu unterschiedlichen Werten des Angebotes auf. Damit verdeutlichen die Kurven nicht nur den Unterschied bei den Antwortzeiten, sondern zeigen auch quantitativ die Grenze der Belastbarkeit in den einzelnen Fällen.

4.3 Mehrfachquittungen (Fenstergröße)

Bild 4.1 zeigt die Antwortzeit über die Ankunftsrate LAM einer einzigen Transportverbindung (M=1) für die beiden Quittierungsmechanismen SACK und NACK, die Kurven sind mit der Fenstergröße n=1,8,15 parametrisiert, dabei bedeutet der Schlüssel SACK8 eine Verbindung der Strategie SACK mit der Fenstergröße n=8.

Für die Fenstergröße n=1 geht das Quittierungsverfahren NACK in SACK über, deshalb gibt es nur eine Kurve SACK1.

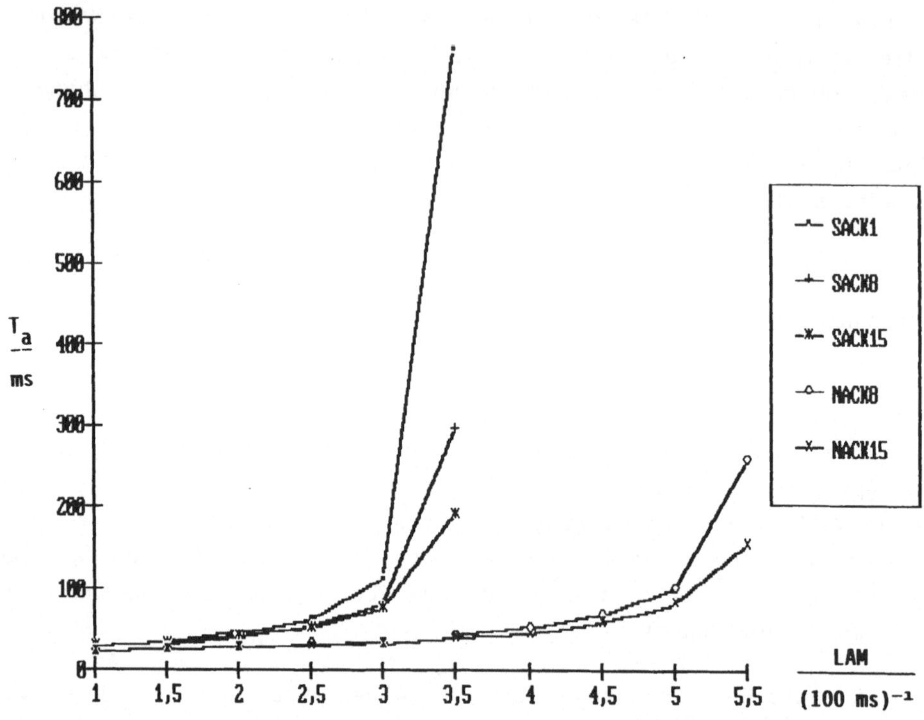

Bild 4.1 Antwortzeit T_a über dem Angebot einer einzigen Transportverbindung zwischen zwei Knoten bei unterschiedlichen Fenstergrößen n=1,8,15 für die Quittierungsstrategien NACK und SACK

Die Kurven bestätigen nicht nur den erwarteten Vorteil des Verfahrens NACK mit größerem n gegenüber den anderen (vgl. 2.2), sondern zeigen auch quantitativ den erheblichen Unterschied bei steigender Last. Eine spürbare Verbesserung der Antwortzeiten der einzelnen Quittierungsstrategien mit steigendem n tritt erst dann ein, wenn man genügend Last zum Abbauen hat.

4.4 Mehrere Transportverbindungen zwischen zwei Knoten

Dieser Versuch dient für Aussagen über eine mögliche Entlastung einer Verbindung zwischen zwei Knoten durch zusätzliche Verbindungen zwischen beiden Knoten. Die Antwortzeitmessungen wurden jeweils für M = 1,2 Verbindungen aufgenommen. Dabei wurden die Anforderungen mit der Rate LAM generiert und auf die M bestehenden Verbindungen gleichverteilt. Bild 4.2 zeigt die Antwortzeiten für die unterschiedlichen Fälle über der Anforderungsrate LAM. Dabei bedeutet SACKMV M-fache Verbindungen zwischen zwei Knoten der Quittierungsstrategie SACK. Die Fenstergröße für alle Messungen wurde auf n=8 festgehalten.

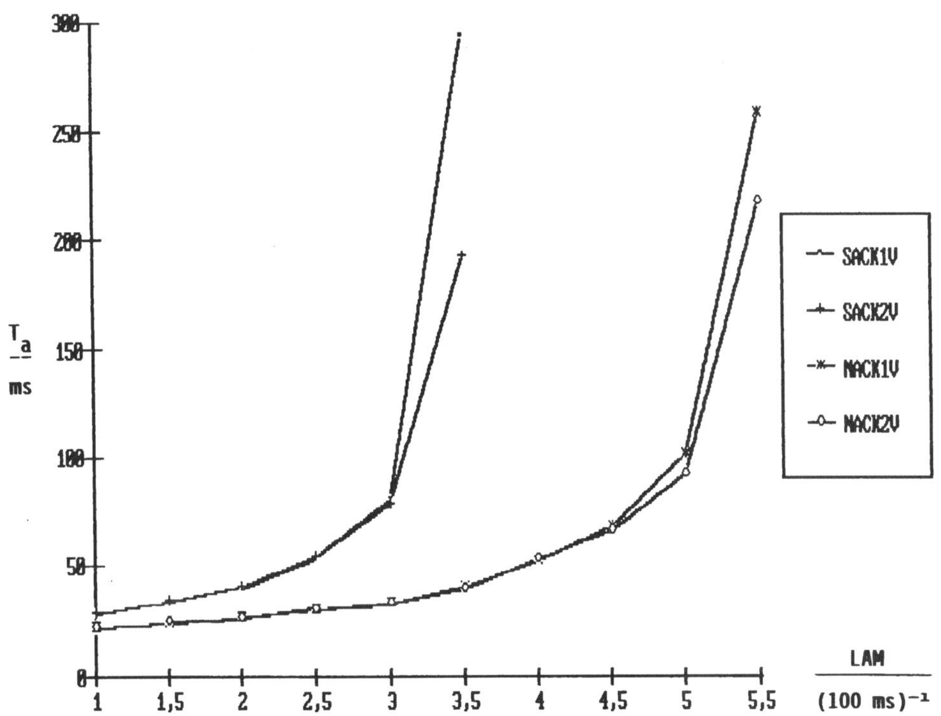

Bild 4.2 Antwortzeit T_a über dem Angebot bei M-fachen Transportverbindungen (M=1,2) zwischen zwei Knoten für die Quittierungsstrategien NACK und SACK

Eine spürbare Verbesserung der Antwortzeiten tritt beim großen Angebot auf. Dies erklärt sich dadurch, daß man bei zweifachen Verbindungen die praktische Fenstergröße 2*8 hat, deren Vorteil erst dann zustande kommt, wenn man genügend Last zum Abbauen hat.

Diese Entlastung kann man also bei einer einfachen Transportverbindung durch eine größere Fenstergröße erzielen, gleichzeitig spart man sich hierbei den Verwaltungsaufwand für zusätzliche Transportverbindungen, der beim Vergleich der Kurven NACK2V Bild 4.2 und NACK15 Bild 4.1 deutlich wird. Bei den Kurven NACK2V Bild 4.2 hat man zwar die praktische Fenstergröße 2*8, allerdings werden zwei Quittungen für 16 Anforderungen (zwei Verbindungen) in jede Richtung ausgelöst. Außerdem hat die Transportmaschine zwei Verbindungen zu verwalten. Bei der Kurve NACK15 Bild 4.1 wird nur eine Quittung für 15 Anforderungen in jede Richtung ausgelöst und die Transportmaschine hat nur eine einzige Transportverbindung zu verwalten.

Bei Strategie SACK sieht man zwar den deutlichen Vorteil von zwei Verbindungen mit n=8 gegenüber einer Verbindung mit n=8, jedoch ist keine Verbesserung gegenüber einer Verbindung mit n=15 zu verzeichnen, da der Quittierungsaufwand in beiden Fällen fast der gleiche ist.

4.5 Verbindungen eines Knoten zu mehreren Empfänger-Knoten

Bei diesen Messungen handelt es sich um eine einseitige Entlastung des Transportsystems auf der Responderseite, indem der Initiator zu zwei Knoten jeweils eine Verbindung aufbaut. Die mit der Rate LAM poissonisch generierte Last wird auf die zwei Verbindungen gleich verteilt. Die Fenstergröße wurde für alle Messungen auf n=8 festgehalten.

Wir erwarten in diesem Fall einen deutlichen Abfall der Antwortzeiten gegenüber dem Fall in 4.4. Das liegt darin begründet, daß der auf Initiatorseite generierte Verkehr Responderseitig von zwei KKs getragen wird, die jeweils nur mit der Hälfte von diesem Angebot belastet werden. Dies führt zu kleineren Wartezeiten auf der Responderseite, bedingt durch kleine Verkehrsaufkommen, und damit zu günstigeren Antwortzeiten.

In Bild 4.3 sieht man die Antwortzeiten für die beiden Quittierungsstrategien SACK und NACK über die Anforderungsrate LAM für zwei Verbindungen, bei entlastetem Responder und zwecks Vergleich für zwei Verbindungen bei einem nicht entlasteten Responder, wobei NACK1R für Strategie NACK, ein Responder-Knoten und zwei Verbindungen steht, und NACK2R, die Strategie NACK, zwei Responder-Knoten und zwei Verbindungen bedeutet.

Die Kurven bestätigen quantitativ die erwartete Verbesserung der Antwortzeiten bei Entlastung des Responders, so wie die dadurch erzielte höhere Grenze der Belastbarkeit des Transportsystems.

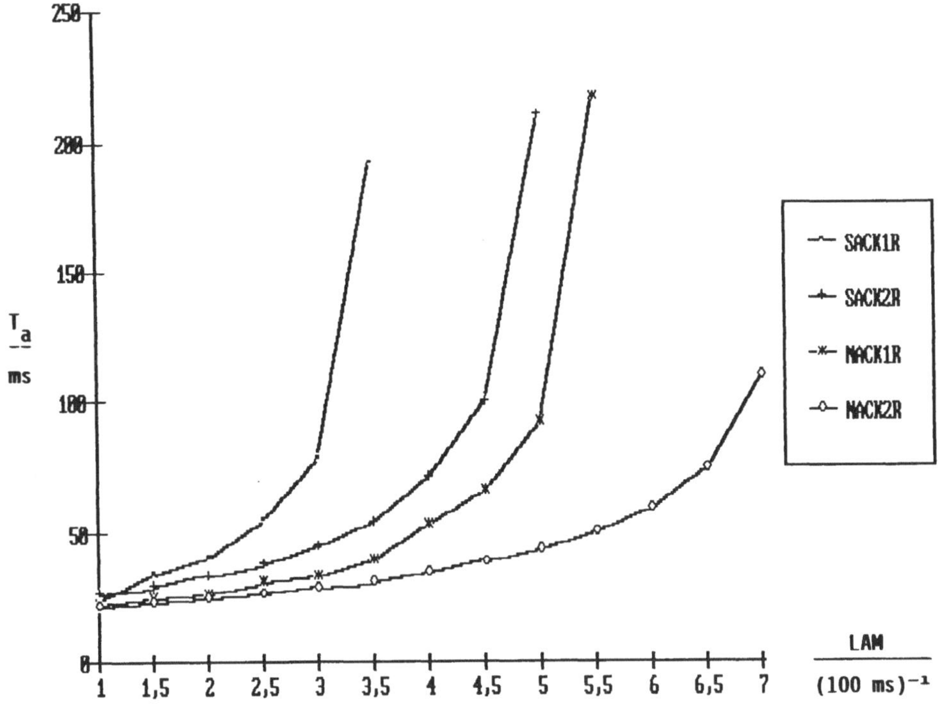

Bild 4.3 Antwortzeit T_a für zwei Transportverbindungen mit einem Initiator-Knoten und einem Responder-Knoten (1R) bzw. zwei Responder-Knoten (2R) bei den Strategien SACK und NACK

5. Ausblick und weitere Arbeiten

Aufgrund fehlender synchroner Uhren auf dem Knoten, konnte man nur die Antwortzeiten messen, die sich im belasteten Fall nicht mehr zerlegen lassen. Trotzdem hat man durch die vorgenommene Untersuchung am MAP-Transportsystem einen Eindruck über die Leistungserwartung der KKs und deren Abhängigkeit, sowohl von protokollinternen Parametern bzw. Mechanismen (Datenflußkontrolle) als auch von der Implementierung der Protokolle (SW, HW, Umgebung) und dem Verkehrsaufkommen auf dem Sender- bzw. Empfängerknoten gewonnen. Für die meisten Anwendungen ist die Ausführungszeit der von der Schicht 7 angebotenen Dienste (z.B. MMS) bei unterschiedlichem Lastmuster auf dem Netzknoten maßgebend. Um genaue Aussagen bzgl. den Ausführungszeiten der Anwenderdienste zu erzielen, wurden auf den Netzknoten synchron laufende Uhren mit einer Auflösung von 10 μs implementiert.

Auf dieser Basis laufen momentan Meß- und Modellierungsarbeiten mit dem Ziel, Aussagen über die unterschiedlichen Implementierungen von MAP zu treffen, sowie Möglichkeiten zur Optimierung hinsichtlich der Abwicklungszeiten der Anwenderdienste zu entwickeln.

Anhang I

Aktivität	Zeitaufwand [μs]	
	68010 10MHz	68020 16,657MHz
Tansport Layer is activ, after pending for user_Messages		
Post Message To Network (no rescheduling occurs)	133	58
Pend on Queue (no message is available, the calling task (Transport) suspends, rescheduling occurs, network task is activ.)	224	95
	Σ 357	Σ 153

Literaturverzeichnis

[MAR 89] Marathe, M.; Smith, R.A.: Performance of a MAP Network Adapter; IEEE Network, May 1988

[NBS 85] COMSAT/NBS Experiment Plan for Transport Protocol; National Bureau of Standards, Report No. 3141, March 1985

[RZE 89] Rzehak, H.; Jäger, R.: Ein Meß- und Monitorkonzept zur Beurteilung der Realzeitfähigkeit eines MAP-Netzes; Kommunikation in verteilten Systemen; ITG/GI-Fachtagung, Stuttgart, FEb. 89

[TAK 87] Takagi, H.; Murata, M.: Two-Layer Modelling for Local Area Networks,; IEEE 1987, INFOCOM

[STR 88] Strayer, W.T.; Weaver, A.C.: Performance Measurement of Data Transfer Services in MAP; IEEE Network, Vol. 2, No. 3, May 1988

[VRTX1] Versatile Real-Time Executive - Users Guide; Ready Systems, Software Release 1, Document-No. 542101001, April 1987

[VRTX2] VRTX32/680X0 Timing Reference; Ready Systems, Document-No. 540011001, May 1987

An Approach to the Numerical Analysis of Multiple-Queue, Cyclic Service Systems

Christoph Strelen
Berthold Bärk
Institut für Informatik
University of Bonn, Germany
Römerstr. 164, D-5300 Bonn 1

Abstract

A cyclic service system with multiple, nonidentical finite queues and gated limited service is considered. The model is an imbedded Markov chain whose state space is reduced by applying the usual independence assumption. The analysis is done by the power method. Expectations or even probability distributions of cycle lengths, transmission lengths, buffer utilization, waiting times, throughput, and blocking probabilities are computed. The results are compared with those obtained from a simulation model.

1 Introduction

This is a study in numerical modelling of polling systems. By the term "numerical model" we mean a set of mathematical formulae, often based on probability theory and stochastic processes, whose evaluation requires quite a lot of computation.

The modelled object of our investigation is a token passing satellite access system. The earth stations receive messages for transmission according to Poisson processes, which may have different parameters. The stations are inhomogeneous and have only finite buffers as intermediate storage. A reservation scheme prevents collisions and monopolization by means of limited service. Gating is observed.

After we had got the task to model the system, we tried analytical and event driven simulation models. But the first were not satisfying because of the complexity of the system - limited buffers, limited service, inhomogenous transmission stations - and we were once again shocked by the enormous need of computing time to do the simulations, for we tried to get insight into the system's behaviour by trying different values for a fair number of independent system parameters. Thus we decided to build a numerical model which seems to have essential advantages compared to the classical analytical and simulation models: 1) Many details of the real system can be modelled, which are in this case limited buffers, a complicated service discipline with reservations and limited service, and inhomogeneity of the stations. 2) The transient behaviour is also analyzed. 3) There are detailed results, not only mean values, but probability distributions for some performance parameters.

A drawback of this technique is not quite obvious: the difficulty to build and verify numerical models like the proposed one. In contrast to the model's concise representation in this description, which is mainly based on random variables, the program and its more abstract mathematical formulation is rather lengthy and less clear, for it is concerned with probabilities and probability distributions instead. It was a cumbersome and time consuming job to realize and test the model. But this problem seems not to be unseparably connected with the technique; highly abstract algorithms can be automatically translated into executable programs, provided they are precise. A similar approach is used in modelling tools.

Our model of the system is an imbedded Markov chain with a finite discrete state space. The buffer contents at the transmission starts define the states. The stations' states are assumed to be independent in order to avoid a state space explosion.

The Markov chain is numerically analyzed: Beginning with an initial state distribution - here according to an empty system - the Chapman-Kolmogorov equations are evaluated successively. This classic method is due to R. v. Mises [16] and is sometimes called *power method*, see Stewart [26]. The evaluation can be continued until equilibrium is nearly reached. Thus not only the equilibrium state distributions are obtained, but also the transient behaviour is analyzed.

Compared with simulation models, the considered "numerical models"

- give results without statistical deviations,
- provide not only mean values, but also probability distributions,
- provide performance measures of the transient behavior,
- save much computing time.

Compared with analytical models, they

- allow to model a system in more detail,
- provide in general more information of the system, e.g. of its transient behavior, or distributions of quantities of interest,
- need some more computing time.

The power method is equivalent to doing an infinite number of simulation runs in parallel, each until equilibrium is reached.

In general, it may take quite a long time until equilibrium accomplishing the power method, but we did never observe this with our model. Such a slow convergence can coincide with a long transient phase of the real system. In this case it seems to be more important to obtain the transient distributions than the equilibrium ones, and it is not necessary to compute until they are reached.

Stewart compares in [26,27] some numerical techniques which apply to our model, and he exhibits other methods which converge much faster (the lopsided simultaneous iteration), but these do not apply to the transient phase. In another model with continuous state space ([29]), we also tried the power method and accelerated convergence by Aitken's δ^2-method, see [28] for example. Thus we obtained both, faster convergence and the transient phase analysis.

The computed numerical results are compared with those obtained from the simulation model mentioned above. Measured figures were not at our disposal, and analytical models, which consider finite buffers and limited service, are not available.

Models of multiqueue systems are of great importance for the performance analysis of computer and communication systems. There are numerous articles on the analysis of them, [1]–[25], considering different polling mechanisms, such as

- exhaustive, non-exhaustive, limited, gated service disciplines,
- cyclic or priority order services,
- finite or infinite buffers in the stations,
- switchover times or none,
- symmetric or asymmetric stations and loads.

Surveys on this topic are given by Takagi [18,20]. Some of these models are numerical. In [24], the power method is applied to solve an imbedded Markov chain, where an exponential message length distribution is assumed, or a continuous two moment approximation is taken for it, and the service discipline is non-exhaustive, i.e. one message per station is transmitted within a cycle, if the corresponding queue is not empty. The states of the queues are assumed to be independent in some sense. In [23], the service discipline is gated, and at most one packet of a message is transmitted by every station within a cycle. The dependency between the stations is considered at the price of high computational costs. Here, a direct solution method is applied: the Markov chain's equilibrium probabilities are calculated by solving systems of $O(2^K)$ linear equations, where K denotes the number of queues. In both models, asymmetric stations and loads, and finite buffers are considered; this is a merit of the numerical models.

The paper is organized as follows. In Sec. 2, the system is defined and the Markov chain is modelled. In Sec. 3, performance measures are derived from the state probabilities. In Sec. 4, sample results are discussed and compared with simulation results. More details of the treatment are contained in [1].

2 The System and its Model

The satellite access system under consideration uses time multiplex; collisions do not occur. The access right is passed by an implicit token between the stations in a predefined fixed order, thus forming cycles. All transmission of data must be announced by the stations one cycle in advance. Limited service is observed in order to prevent monopolization: the number of packets per transmission are limited, and if data are left behind they are transmitted in subsequent cycles. The buffer space in the stations is limited. Arriving data which cannot be stored are lost. Messages to be transmitted arrive at the stations according to Poisson processes; the lengths of the messages are random. The stations are inhomogeneous: the parameters of the arrival processes, the service limits, and the buffer space sizes may be different. Our analysis is confined to just different arrival process rates but a generalization is straightforward.

All stations are organized in a logical ring. Each of them is identified by an integer $k \in \mathcal{K} = [1 : K]$, where K denotes the number of stations in the system. The stations transmit only in their own time slices. TS_t, $t \in I\!N$, is the tth time slice, and it is attached to station $k = k(t) \triangleq (t-1) \bmod K + 1$. This station k has the choice to transmit

- a request message RM_t,
- nothing,
- a data burst DB_t, or
- a request message RM_t and a data burst DB_t

in TS_t.

A request message RM_t serves the purpose to accomplish a reservation for transmitting data in TS_{t+K}. All data bursts must be reserved by request messages, which announce their lengths. A data burst consists of some packets, which are the basic unit of data. The number of packets in a data burst is limited by $\xi \in I\!N$, i.e. the protocol has limited service. We call ξ the *transmission* or *service limit*.

The unit of time is the time which is necessary to transmit one packet.

The r.v. Y_t denotes the length of the time slice TS_t. It is y if y packets are transmitted, and t_{RM} otherwise. We assume that t_{RM} is a constant multiple of the time unit.

T_t denotes the time at which the time slice TS_t begins. The period of the time slices $TS_t, \ldots, TS_{t+K-1}$ is called *cycle t*; it consists of one time slice for every station. The length of cycle t is denoted by the r.v. $Z_t = Y_t + \ldots + Y_{t+K-1} = T_{t+K} - T_t$.

The buffer space in the stations for storing data between arrival and transmission is limited: ν packets can be stored, at maximum. ξ of these storage units are always designated for outgoing packets.

The buffer limit ν and the transmission limits ξ could be different for different stations in our analysis, but for the sake of simplicity, we choose them all equal.

Messages with data for transmission arrive at the stations according to Poisson processes with rate λ_k, $k \in \mathcal{K}$, at station k. A message consists of L packets. We assume the distribution of L to be the same for all messages, but again we could consider different distributions for different stations.

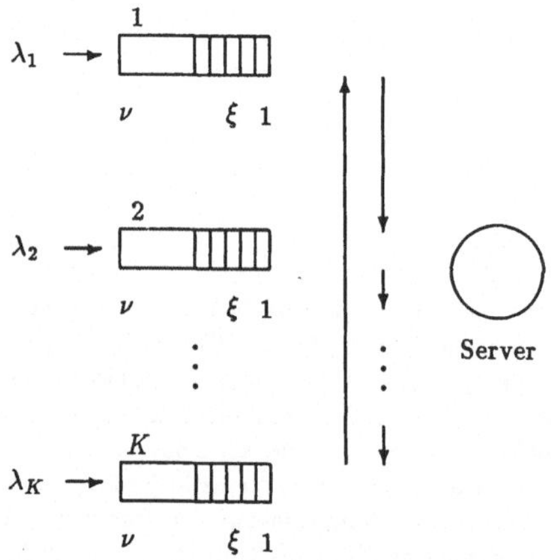

Fig. 1 Cyclic Server Multiple Queue Model

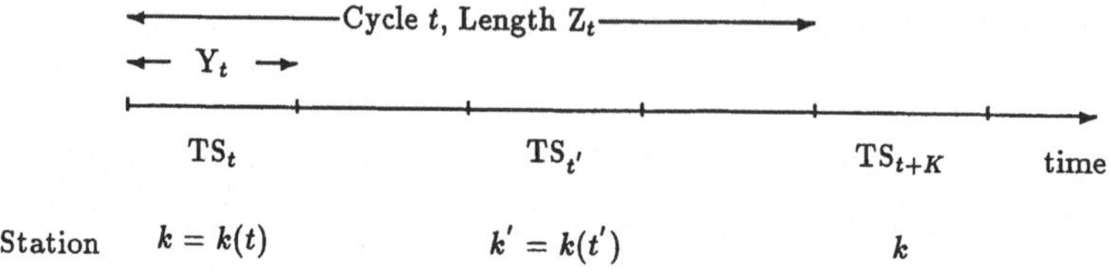

Fig. 2 Time Slices and Cycles

Now we are going to propose the model of the satellite access system, a Markov chain. We begin with some general notation and the definition of two functions. Let X be any integer random variable. $\Omega_X \subset \mathbb{N}$ denotes its sample space and x_i, $i \in \Omega_X$, its distribution, i.e. $x_i = P\{X = i\}$. $\min(X, m)$ defines the minimum of any r.v. X and an integer m:

$$P\{\min(X, m) = j\} = \begin{cases} x_j, & j \in \Omega, \\ 1 - \sum_{i \in \Omega} x_i, & j = m, \end{cases}$$

where $\Omega = \{i \mid i \in \Omega_X, i < m\}$. The sample space of the resulting r.v. is $\Omega \cup \{m\}$.

$A = \alpha(\lambda, Z)$ with $\Omega_A = [0 : \nu + 1]$ is a r.v. which describes the number of packets arriving in a period of length Z according to a Poisson process with rate λ messages per unit of time: a_i is the probability of arriving i packets, $i \leq \nu$, and $a_{\nu+1}$ the probability of arriving more than ν packets. $\alpha(\lambda, Z)$ is defined as follows. We assume L, the number of packets in a message, to be uniformly distributed over $\Omega_L = [\alpha : \beta]$. Other distributions could be chosen as well. Let M' denote the number of arriving messages in the period, and L_j denote their respective lengths, $j = 1, \ldots, M'$. The L_j are assumed to be i.i.d. like L. Thus, the distribution of M' is

$$m'_j = \sum_{i \in \Omega_Z} e^{-\lambda i} \frac{(\lambda i)^j}{j!} z_i, \quad j \in \mathbb{N}_0,$$

and with

$$M \triangleq \min(M', \lfloor \nu/\alpha \rfloor + 1)$$

and

$$A' \triangleq L_1 \oplus \ldots \oplus L_M,$$

we get

$$A = \min(A', \nu + 1).$$

Here, $\oplus$ denotes the sum of two independent r.v. whose distribution is calculated by convolution.

Now we are prepared to define the Markov chain. Let N_t denote the number of packets in the buffer of station $k(t)$ at time T_t except those packets which are announced for transmission in time slice TS_t. The state of the system at T_t is given by $(N_{t-K+1}, \ldots, N_t)$. We regard all arrivals of messages at station $k(t)$ to take place a short moment before T_t. Thus, the transitions $(N_{t-K}, \ldots, N_{t-1}) \rightarrow (N_{t-K+1}, \ldots, N_t)$ happen at time $T_t, t \in \mathbb{N}$. Thereby the component N_{t-K} disappears, N_t is added, and the other components remain unchanged.

The new component is given by

$$N_t = \min(U_{t-K} + A_{t-K}, \nu - \xi),$$

where U_{t-K} is that part of N_{t-K} which could not be scheduled by the request message RM_{t-K} for transmission in the data burst DB_t because of the limited service, and $A_{t-K} = \alpha(\lambda_k, Z_{t-K})$ counts the packets which arrived within cycle $t - K$ at station k.

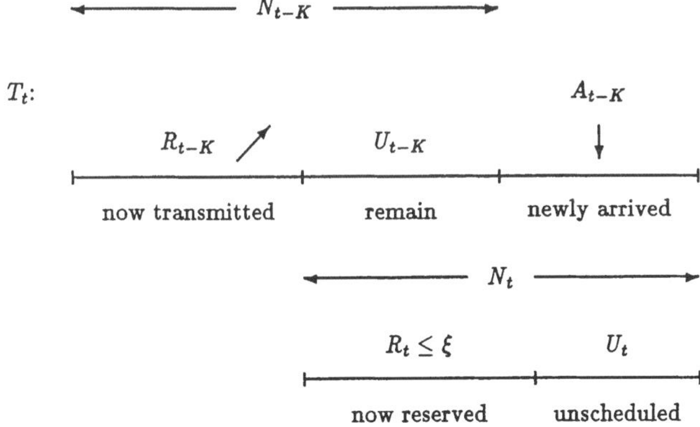

Fig. 3 State Transition for Station $k(t)$ at Time T_t, Buffer Contents

We somewhat simplify the expression for N_t by assuming the r.v. U_{t-K} and A_{t-K} being independent, $N_t = \min(U_{t-K} \oplus A_{t-K}, \nu - \xi)$; this does not seem to bring in a great error. More serious is the following independence assumption: When calculating the length of cycle $t-K$, we assume the time slice lengths being independent, i.e. we take $Z_{t-K} = Y_{t-K} \oplus \ldots \oplus Y_{t-1}$ instead of the correct $Z_{t-K} = Y_{t-K} + \ldots + Y_{t-1}$. This is to avoid state space explosion.

From the new N_t we calculate

$$R_t = \min(N_t, \xi),$$

the number of packets scheduled for transmission in TS_{t+K},

$$U_t = N_t - R_t = \begin{cases} 0 & \text{if } N_t \leq \xi, \\ N_t - \xi & \text{otherwise,} \end{cases}$$

the number of packets remaining unscheduled, and

$$Y_{t+K} = \begin{cases} t_{RM} & \text{if } R_t = 0, \\ R_t & \text{otherwise,} \end{cases}$$

the length of time slice TS_{t+K}. The r.v.

$$B_{t-K} \triangleq \min(U_{t-K} \oplus A_{t-K}, \nu - \xi + 1)$$

is calculated to determine the blocking probability: $b_{t-K,\nu-\xi+1}$ is the probability that packets arrive at station k within cycle $t-K$ and are lost on account of buffer storage shortage.

Now we are prepared to write down the algorithm for the calculation of the introduced random variables; this algorithm evaluates the Chapman-Kolmogorov equations.

Algorithm

Initialization
for $t := 1$ **to** K **do**
  $N_t := 0$.
  $Y_t := t_{RM}$.
od.
repeat

t	$:= t + 1$.	{Counts the time slices}
k	$:= (t-1) \bmod K + 1$.	{Station number}
Z	$:= Y_{t-K} \oplus \ldots \oplus Y_{t-1}$.	{Cycle $t-K$ length, Z_{t-K}}
A	$:= \alpha(\lambda_k, Z)$.	{Arrivals at station k within cycle $t-K$}
U	$:= \begin{cases} 0 & \text{if } N_{t-K} \leq \xi, \\ N_{t-K} - \xi & \text{otherwise.} \end{cases}$	$\{U_{t-K}\}$
B	$:= \min(U \oplus A, \nu - \xi + 1)$.	
$p_{b,t}$	$:= b_{\nu-\xi+1}$.	{Blocking probability of station k in cycle $t-K$}
R	$:= \min(N_{t-K}, \xi)$.	$\{R_{t-K}\}$
Y_t	$:= \begin{cases} t_{RM} & \text{if } R = 0, \\ R & \text{otherwise.} \end{cases}$	{Time slice TS_t length}
N_t	$:= \min(U \oplus A, \nu - \xi)$.	{New state}

until equilibrium is nearly reached.

Remark 1 The distributions of the r.v. N_t, N_{t-K}, and Y_t are stored in vectors n_k and y_k, $k \in \mathcal{K}$.

Remark 2 The results of the convolutions for the calculation of Z can be applied more than once, thus saving operations.

The number of operations which must be executed for one time slice in the loop is

$$O(\nu^2 + \nu K \xi + K^2 \xi^2),$$

if the convolutions $L_1 \oplus \ldots \oplus L_j$, $j \in [1 : \lfloor \frac{\nu}{\alpha} \rfloor + 1]$ are done once outside the loop, if no fast Fourier transform is applied, and if the idea of remark 2 is not considered.

3 Performance Measures

From the results of the algorithm which evaluates the Markov chain, performance measures of interest can be obtained: Moments of all computed distributions, the probability that station k transmits a data burst in cycle $t + K$,

$$1 - n_{t,0},$$

the mean number of data bursts in cycle $t + K$,

$$K - \sum_{t'=t}^{t+K-1} n_{t',0},$$

the mean number of packets in DB_{t+K},

$$E[R_t]/(1 - n_{t,0}),$$

throughput of station k in cycle $t + K$,

$$E[R_t]/E[Z_{t+K}] \quad \text{[packets / unit of time]},$$

which also equals ρ_k, the server utilization by station $k(t)$, i.e. the fraction of time the server is busy transmitting data of station $k(t)$ in cycle $t + K$, and the total server utilization ρ, i.e. the fraction of time used for transmission of data in cycle $t + K$,

$$\Big(\sum_{k \in \mathcal{K}} E[R_{t+k-1}]\Big)/E[Z_{t+K}].$$

We define the time slice utilization of station $k(t)$ in cycle t,

$$\sigma_k = E[R_t]/\xi,$$

and of the whole system,

$$\sigma = \sum_{\tau=t}^{t+K-1} E[R_\tau]/(K\xi).$$

It is not so easy to find an expression for the waiting time of a packet. In the following analysis, some simplifying assumptions are made, mainly independece assumptions.

The waiting time W of a packet $\mathcal{D}$ between its arrival and its transmission consists of three components:

$$W = W' + W'' + W'''.$$

W' is the time between the arrival of $\mathcal{D}$ at station $k(t)$ and the beginning of the next time slice of this station, say T_t. W'' is the time between T_t and T_{t+K} i.e. the length Z_t of cycle t. $\mathcal{D}$ cannot be transmitted within TS_t, because it must be scheduled before. W''' is the time between T_{t+K} and the start of its transmission.

For W',

$$E[W'] = E[Z_{t-K}^2]/(2E[Z_{t-K}]), \tag{1}$$

holds because more packets arrive in longer cycles. For W'' we take

$$E[W''] = E[Z_t^2]/E[Z_t]; \tag{2}$$

this accounts approximately for the dependency of the lengths of subsequent cycles, which are positively correlated.

$\mathcal{D}_1$, the first packet which arrives at station k in cycle $t - K$, has $N_{t-K} = R_{t-K} + U_{t-K}$ packets in front of itself, which are transmitted before $\mathcal{D}_1$: the R_{t-K} packets in cycle t, and then U_{t-K} packets in $\lfloor U_{t-K}/\xi \rfloor$

"long" cycles $\tau = t + K, t + 2K, \ldots$. Then, in the next cycle, $\mathcal{D}_1$ is transmitted behind the remaining data, i.e. $U_{t-K} \bmod \xi$ packets.

In a "long" cycle, station k transmits ξ packets, $Y_\tau = \xi$. We assume

$$Z_\tau^{(\text{long})} = \xi \oplus Y_{\tau+1} \oplus \ldots \oplus Y_{\tau+K-1}. \tag{3}$$

Now we consider any packet, chosen with equal probability from those which arrived within cycle $t - K$, say $\mathcal{D}_J$. The index denotes the order of arrival, $\mathcal{D}_2$ arrives after $\mathcal{D}_1$ etc. Because of the random selection,

$$P\{J = j | \tilde{A} = i\} = 1/i, \quad j \in [1 : i],$$

where $\tilde{A} \stackrel{\Delta}{=} A|A > 0$ is the number of packets which arrived within cycle $t - K$, conditioned that there is at least one arrival. It follows

$$P(J = j) = \sum_{i \in \Omega_{\tilde{A}}, i \geq j} \frac{1}{i} \tilde{a}_i, \; j \in \Omega_{\tilde{A}}. \tag{4}$$

Clearly $\mathcal{D}_J$ has $U_{t-K} + R_{t-K} + J - 1$ packets in front of itself, if it is not lost because of buffer storage shortage. As mentioned for $\mathcal{D}_1$, the R_{t-K} packets are transmitted in cycle t. Thus

$$F' = U_{t-K} \oplus J - 1 \tag{5}$$

packets remain to be transmitted in the "long" cycles $\tau = t + K, t + 2K, \ldots$, and in order to take into account the limited buffer, we consider

$$F = F' | F' \leq \nu - \xi - 1 \tag{6}$$

instead.

Thus we get

$$E[W'''] = \sum_{j=0}^{\nu-\xi-1} f_j \Big(\sum_{i=1}^{\lfloor j/\xi \rfloor} E[Z_{t+i\,K}^{(\text{long})}] + j \bmod \xi \Big). \tag{7}$$

Our basic independence assumption leads to an underestimation of the variances of the buffer contents N_t, and of the cycle lengths Z_t. By result of this, $E[W']$ and $E[W'']$ are underestimated by (1) and (2), respectively.

There is a positive correlation between subsequent transmission lengths, $Y_\tau, Y_{\tau+1}, \ldots$. Therefore $Z_t^{(\text{long})}$ is underestimated by (3).

A similar analysis as for the waiting times can be accomplished for the sojourn times of messages in the stations.

The time complexity of the evaluation of (3) is $O(K^2\xi^2)$, and of (4), (5), and (7) $O(\nu^2)$. (3) must be evaluated $O(\nu)$ times. Thus the overall complexity is

$$O(\nu^2 + \nu K^2 \xi^2).$$

The space requirements are of order $O(\nu K)$, and they are determined by the representations of the distributions of the r. v. N, Y, and A for all K stations and for the r.v. Z.

4. Examples

We analyzed 53 systems and compared the calculated values with simulation results. The systems have $K = 3$, 5, or 10 stations. The message lengths are uniformly distributed between α and β, where $\alpha = 1$. We considered three triples of β, ξ (transmission limit), and ν (buffer limit) values.

$$\beta = 1, \quad \xi = 4, \quad \nu = 20,$$
$$\beta = 4, \quad \xi = 4, \quad \nu = 20,$$
$$\beta = 10, \quad \xi = 10, \quad \nu = 40.$$

Each system was tested with different arrival rates, λ_k messages per unit of time.

Figures 5 to 9 show graphically the differences between numerical and simulation results in equilibrium:

Fig. 5: mean cycle lengths (Z_t)
Fig. 6: mean waiting times (W)
Fig. 7: mean number of data bursts per cycle
Fig. 8: mean number of packets per data burst
Fig. 9: blocking probabilities

All the differences are relative compared to the simulation results, except for the blocking probabilities in Fig. 9, which shows absolute deviations.

As can be seen in the figures, the accuracy of the results of our numerical model is good or reasonable. The errors are greater, in general, if

- the time slice utilization is medium,
- the messages are long,
- the number of stations are small.

In contrast to this, the results are more accurate, if the utilization is low or high, the messages are short and the number of stations are large.

The results are more accurate for the mean cycle length $E[Z_t]$, the mean number of packets in the buffer $E[N_t]$, and the mean number of packets per data burst. The expected waiting time $E[W]$, the blocking probability p_b, Var Z and Var N are less accurate.

The main reason for the errors is the independence assumption: the stations do not behave independently! Consequently the modelled behaviour is more uniform than that of the real system; the computed variances of the cycle lengths and of the buffer contents are too small, just as the computed probabilities of little or high buffer charge, see Figs. 4, 4', 4". As we observed in simulations, the following is more realistic: greater variances of Z_t and N_t; phases of many long cycles alternate with phases of empty cycles (without transmissions); most of the messages arrive at the stations during long cycles. These tendencies are even stronger if messages are long.

Under light load conditions, the probability that messages arrive during a transmission is small; most of the messages arrive within empty cycles. Thus, the correlation between the stations is small, e.g. their respective buffer contents. If the system is heavily loaded, the buffers are almost always filled up, almost all time slices are of maximum length ξ, all stations send data bursts in nearly every time slice. In this case, too, the stations behave independently. Thus the independence assumption is less disadvantageous if the time slice utilization is low or nearly one. Then the errors are smaller as can be verified in Figs. 6, 6', 6", for example.

But under very light load conditions, subsequent cycle lengths Z_t and Z_{t+K} are nearly uncorrelated. Thus the approximation (2) overestimates $E[W'']$; $E[W''] = E[Z_t]$ would be more appropriate. Indeed, the two leftmost entries in Fig. 6 reduced to less than 0.05 if calculated accordingly.

The underestimation of mean waiting times in many examples is due to the dependency of the stations' states as was remarked at the end of Sec. 3.

The errors of the blocking probabilities result from the underestimation of the buffer contents' variances: the probability of a buffer contents near to the mean number of packets in the buffer, $E[N_t]$, is overestimated, but in contrast to this, the computed probabilities of little or high buffer utilization are too small. This can be seen in Figs. 4, 4', and 4". Thus, if the mean buffer contents $E[N_t]$ is only slightly smaller than the buffer limit ν, i.e. if σ is close to one, the blocking probability $P\{N_t + A > \nu\}$ is estimated too large; see Fig. 4'. But if σ is smaller, and $E[N_t]$ is much smaller than ν, we get too small a value for the blocking probability, see Fig. 4".

The computations were performed on an PC-AT-type personal computer with 80286/80287 processors running at 8 MHz. The time required for one iteration step for all stations varied from less than 1 s to about 25 s, depending on the values of K, ν and ξ. The number of iteration steps until near equilibrium rises with increasing utilization from 10 to about 80. In none of our examples did the space requirements for data exceed 20 KByte.

As the résumé of this section, we conjecture that the independence assumption is the essential reason for the observed errors.

Conclusion

A numerical model of a multiqueue system with one server, reservations, limited service, gating, finite buffers, and asymmetric stations is proposed. The model is a discrete state Markov chain, which is solved by the power method. By this technique, many details of the modelled system can be considered, and the transient behaviour is also investigated, not only the equilibrium case. The computational costs are lower than with simulation models, and the results are more detailed, e.g. probability distributions can be evaluated. It is true that building such a model can be cumbersome and error prone, but this could be alleviated by a modelling tool. A modification of the proposed model for other polling systems, e.g. other service disciplines or other asymmetries can be accomplished easily. The main drawback of our model is that the dependencies of the stations are not considered appropriately; this will be done in a further development. Compared with the power method, lopsided simultaneous iteration provides faster convergence towards equilibrium distributions; this technique can be modified such that the transient distributions and a sequence of modified distributions, which approach the equilibrium distribution rapidly, are obtained in parallel. Probably, fast Fourier transforms could be applied, thus speeding up the convolutions. Numerical models are not confined to discrete probability distributions; continuous distributions can be handled as well if suitable representations for them are choosen.

References

[1] B. Bärk, Numerische Analyse eines zyklischen Wartesystems, Diploma Thesis, University of Bonn, 1988.

[2] O. J. Boxma, Two symmetric queues with alternating service and switching times, Proc. Performance '84, Paris, 409-431.

[3] O. J. Boxma and B. Meister, Waiting-time approximations for cyclic-service systems with switchover times, ACM Perf. Eval. Rev. 14,1 (1986) 254-262.

[4] W. Bux, Local-area subnetworks: A performance comparison, IEEE Trans. Comm. COM-29 (1981) 1465-1473.

[5] R. B. Cooper and G. Murray, Queues served in cyclic order, Bell Syst. Tech. J. 48 (1969) 675-689.

[6] R. B. Cooper, Queues served in cyclic order: Waiting times, Bell Syst. Techn. J. 49 (1970) 399-413.

[7] M. Eisenberg, Two queues with changeover times, Oper. Res. 19 (1971) 386-401.

[8] M. Eisenberg, Queues with periodic service and changeover times, Oper. Res. 20 (1972) 440-451.

[9] S. W. Fuhrmann and Y. T. Wang, Mean waiting time approximations of cyclic service systems with limited service, Performance '87, P.-J. Courtois and G. Latouche (eds.) (North-Holland, Amsterdam, 1988) 253-265.

[10] O. Hashida, Analysis of multiqueue, Rev. El. Commun. Lab. 20 (1972) 189-199.

[11] O. Hashida and K. Ohara, Line accommodation capacity of a communication control unit, Rev. El. Commun. Lab 20 (1972) 231-239.

[12] O. C. Ibe and X. Cheng, Analysis of polling systems with single-message buffers, Proc. of IEEE GLOBECOM '86, Houston, TX (1986) 939-943.

[13] A. R. Kaye, Analysis of a distributed control loop for data transmission, Proc. of 22nd Int. Symp. on Comp. Comm. Networks and Teletraffic, Polytechnic Institute of Brooklyn, NY.(1972) 47-58.

[14] P. J. Kuehn, Multiqueue systems with nonexhaustive cyclic service, Bell. Syst. Tech. J. 58 (1979) 671-699.

[15] M. A. Leibowitz, An approximate method for treating a class of multiqueue problems, IBM J. Res. Develop. 5 (1961) 204-209.

[16] R. v. Mises and H. Pollaczek-Geiringer, Verfahren zur Gleichungsauflösung, ZAMM 9 (1929) 152-164.

[17] H. Takagi, On the analysis of a symmetric polling system with single-message buffers, Perf. Evaluation, Vol. 5, No. 3 (1985) 149-157.

[18] H. Takagi, Analysis of polling systems, (The MIT Press, Cambridge, MA, 1986).

[19] H. Takagi, Analysis and applications of a multiqueue cyclic service system with feedback, IEEE Trans. Comm., Vol. COM-35, No. 2 (1987) 248-250.

[20] H. Takagi, A survey of queueing analysis of polling models, Proc. of 3rd Intern. Conf. on Data Comm. Systems and their Performance, Rio de Janeiro, Brasil (1987) 277-296.

[21] H. Takagi, Exact analysis of round-robin scheduling of services, IBM Journal of Research and Development, Vol. 31, No. 4 (1987) 484-488.

[22] T. Takine, Y. Takahashi and T. Hasegawa, Exact analysis of asymmetric polling systems with single buffers, IEEE Trans. on Comm., Vol. COM-36, (1988).

[23] T. Takine, Y. Takahashi and T. Hasegawa, Average message delay of an asymmetric single-buffer polling system with round-robin scheduling of services, Proc. of the 4th International Conference on Modelling Techniques and Tools for Computer Performance Evaluation, Palma de Mallorca (1988) 233-243.

[24] P. Tran-Gia and T. Raith, Approximation for finite capacity multiqueue systems, Proc. of the 3rd GI/NTG-Fachtagung, Dortmund (1985) 332-345.

[25] R. M. Wu and Y.-B. Chen, Analysis of a loop transmission system with round-robin scheduling of services, IBM Journal of Research and Development, Vol. 19, No. 5 (1975) 486-493.

[26] W. J. Stewart, A comparison of numerical techniques in Markov modelling, Comm. ACM 21,2 (1978) 144-152.

[27] W. J. Stewart, A direct numerical method for queueing networks, Proc. of the 4th International Symposium on Modelling and Performance Evaluation of Computer Systems, Vienna (1979).

[28] J. Stoer, Einführung in die Numerische Mathematik (Springer Verlag, Berlin, 1976).

[29] J. Ch. Strelen, Piecewise approximation of densities applying the principle of maximum entropy: Waiting times in G/G/1-systems, Proc. of the 4th International Conference on Modelling Techniques and Tools for Computer Performance Evaluation, Palma de Mallorca (1988) 493-512.

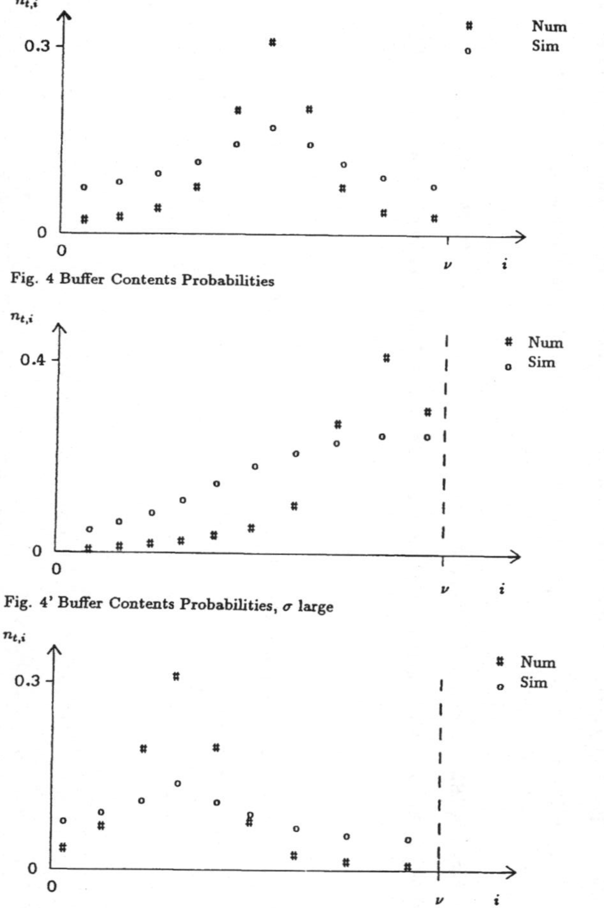

$n_{t,i}$
0.3
0
Num
o Sim
ν
i

Fig. 4 Buffer Contents Probabilities

$n_{t,i}$
0.4
0
Num
o Sim
ν
i

Fig. 4' Buffer Contents Probabilities, σ large

$n_{t,i}$
0.3
0
Num
o Sim
ν
i

Fig. 4" Buffer Contents Probabilities, σ small

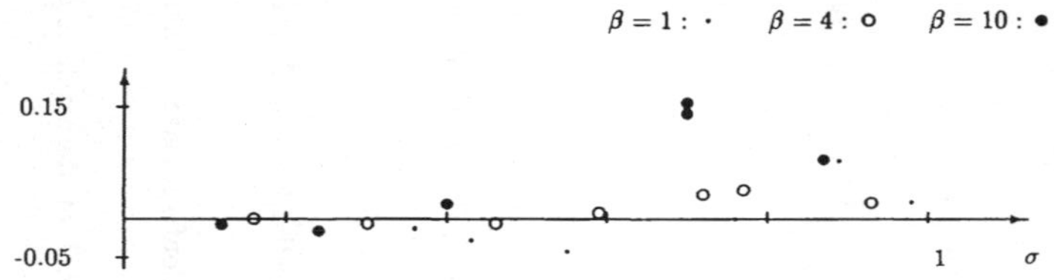

$\beta = 1 : \cdot$
$\beta = 4 : \circ$
$\beta = 10 : \bullet$
0.15
-0.05
1
σ

Fig. 5 Mean Cycle Lengths, Relative Differences, $K = 3$

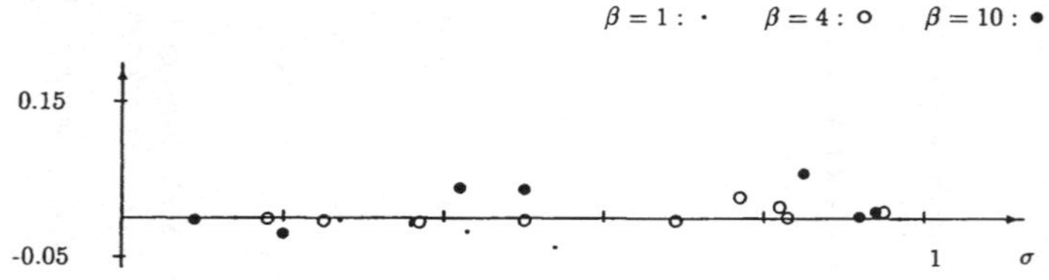

$\beta = 1 : \cdot$
$\beta = 4 : \circ$
$\beta = 10 : \bullet$
0.15
-0.05
1
σ

Fig. 5' Mean Cycle Lengths, Relative Differences, $K = 5$

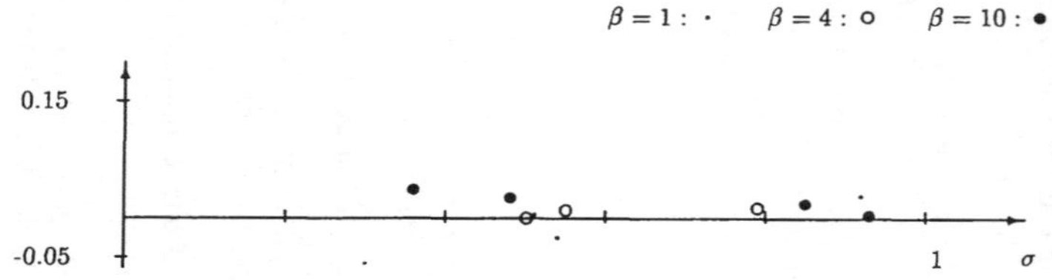

$\beta = 1 : \cdot$
$\beta = 4 : \circ$
$\beta = 10 : \bullet$
0.15
-0.05
1
σ

Fig. 5" Mean Cycle Lengths, Relative Differences, $K = 10$

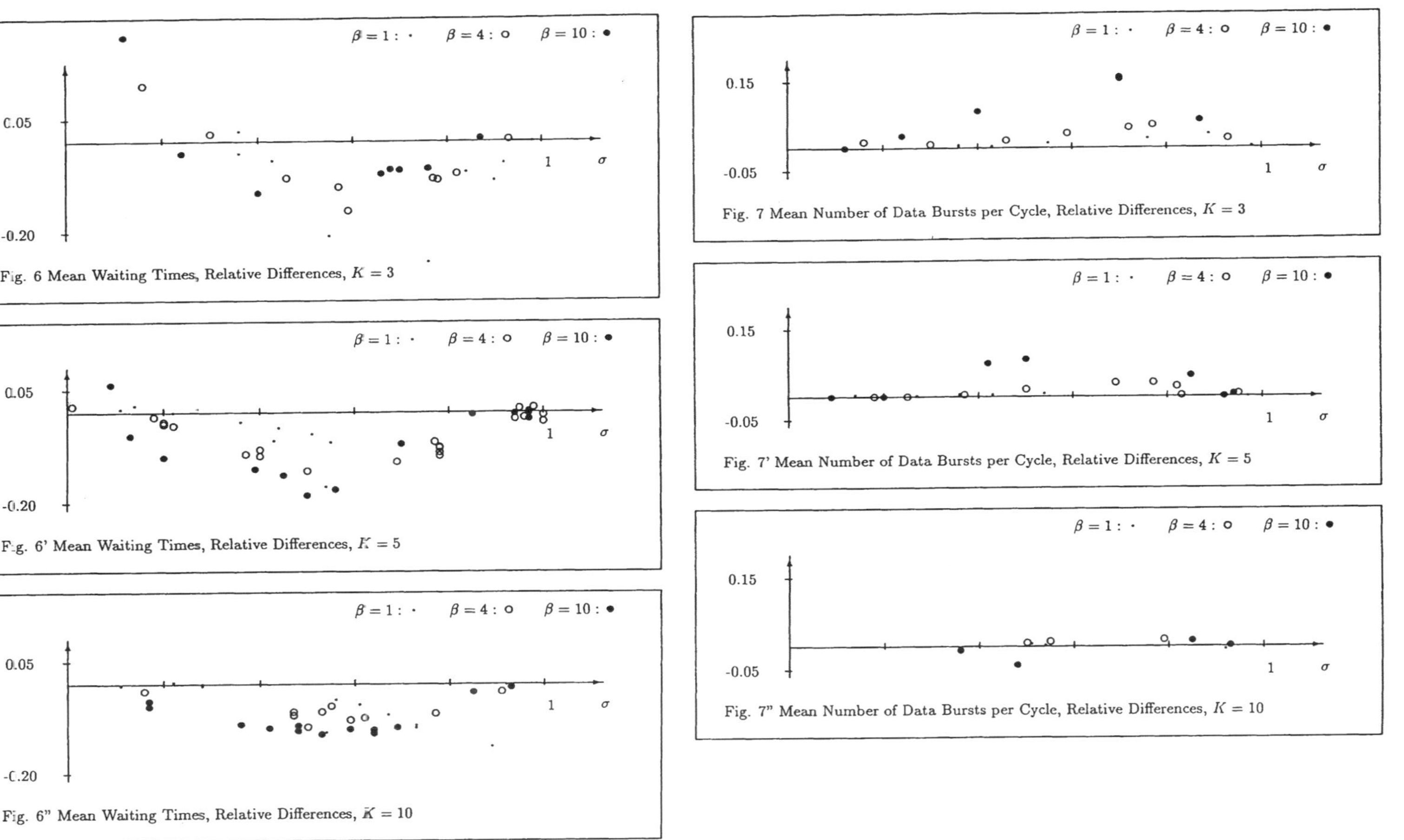

Fig. 6 Mean Waiting Times, Relative Differences, $K = 3$

Fig. 6' Mean Waiting Times, Relative Differences, $K = 5$

Fig. 6" Mean Waiting Times, Relative Differences, $K = 10$

Fig. 7 Mean Number of Data Bursts per Cycle, Relative Differences, $K = 3$

Fig. 7' Mean Number of Data Bursts per Cycle, Relative Differences, $K = 5$

Fig. 7" Mean Number of Data Bursts per Cycle, Relative Differences, $K = 10$

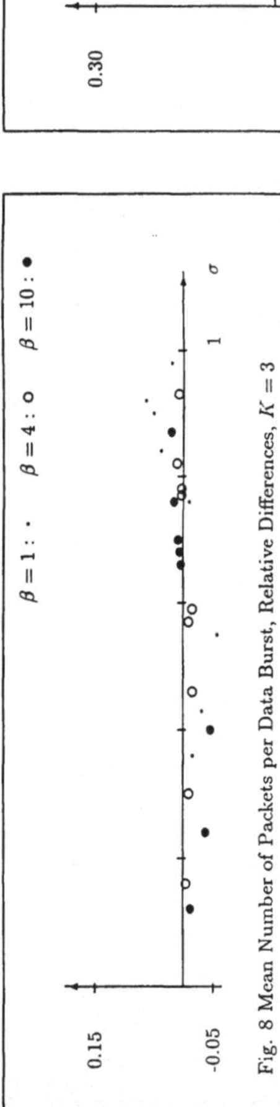

Fig. 8 Mean Number of Packets per Data Burst, Relative Differences, $K = 3$

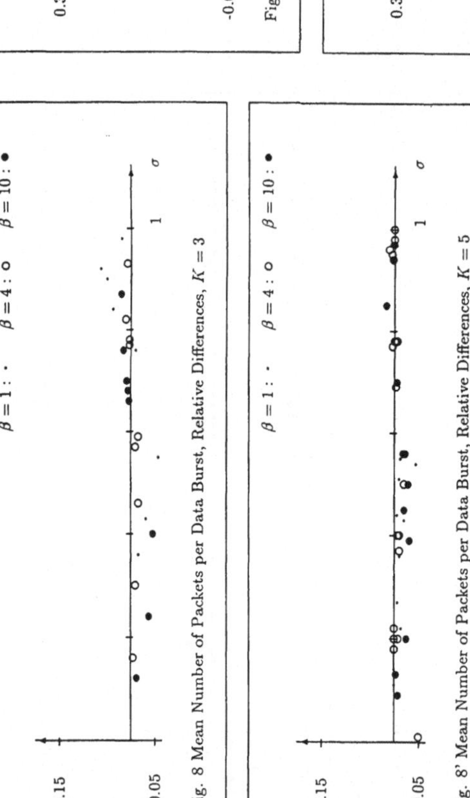

Fig. 8' Mean Number of Packets per Data Burst, Relative Differences, $K = 5$

Fig. 8" Mean Number of Packets per Data Burst, Relative Differences, $K = 10$

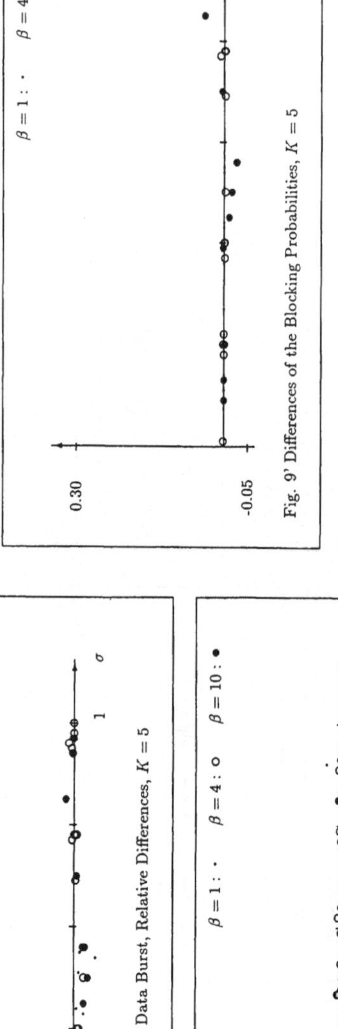

Fig. 9 Differences of the Blocking Probabilities, $K = 3$

Fig. 9' Differences of the Blocking Probabilities, $K = 5$

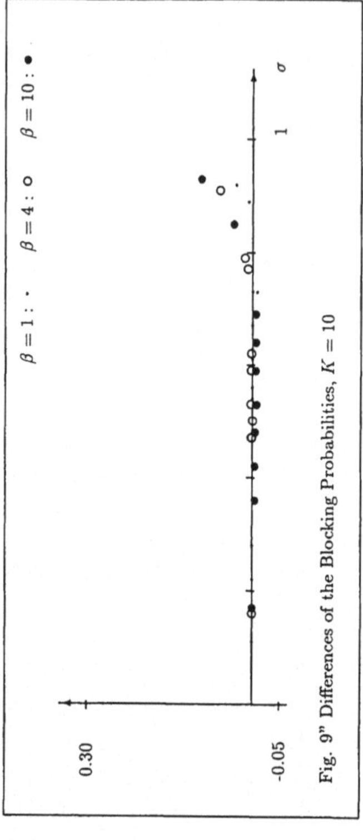

Fig. 9" Differences of the Blocking Probabilities, $K = 10$

WAITING TIMES IN POLLING SYSTEMS WITH MARKOVIAN SERVER ROUTING

O.J. Boxma
Centre for Mathematics and Computer Science
P.O. Box 4079, 1009 AB Amsterdam, The Netherlands
Faculty of Economics, Tilburg University
P.O. Box 90153, 5000 LE Tilburg, The Netherlands

J.A. Weststrate
Faculty of Economics, Tilburg University
P.O. Box 90153, 5000 LE Tilburg, The Netherlands

ABSTRACT
This study is devoted to a queueing analysis of polling systems with a probabilistic server routing mechanism. A single server serves a number of queues, switching between the queues according to a discrete time parameter Markov chain. The switchover times between queues are nonnegligible. It is observed that the total amount of work in this Markovian polling system can be decomposed into two independent parts, viz., (i) the total amount of work in the corresponding system without switchover times and (ii) the amount of work in the system at some epoch covered by a switching interval. This work decomposition leads to a pseudoconservation law for mean waiting times, i.e., an exact expression for a weighted sum of the mean waiting times at all queues. The results generalize known results for polling systems with strictly cyclic service.

1. INTRODUCTION

A system in which one server visits a set of queues, in some order, is commonly referred to as a polling system. A large number of queueing theoretic studies about polling systems has been published. The vast majority of these studies considers polling systems in which the server serves the queues in a strictly cyclic order. Several service strategies at the queues have been investigated and implemented in actual computer-communication networks; these strategies range from exhaustive (a queue is served until it is empty) to 1-limited (when the queue is non-empty, the server serves exactly one customer).

The main performance measures of cyclic polling systems are the mean waiting times at the various queues. When all queues have an exhaustive service strategy, the exact mean waiting times at the queues can be determined by solving a system of linear equations. For most other service strategies exact mean waiting times are only known in special cases; see the surveys of Takagi [1986,1988] for detailed results and further references.

Recently some generalizations have been considered, which encompass a much larger class of cyclic polling systems. One generalization of purely cyclic polling is a polling system with a service order table, i.e., a list of stations which the server must successively visit. Stations can be given higher priority by listing them more often in the table. See Boxma et al. [1988]. The polling table [1,2,...,N] gives the purely cyclic case, and [1,2,1,3,...,1,N] represents the important star polling scheme.

Another generalization concerns polling systems with a random polling scheme. In a random polling scheme, the polling order is not fixed but determined by some random mechanism. In a recent study, Kleinrock and Levy [1988] analyzed the behaviour of a random polling system in which the next station polled will be the jth station with probability p_j, independent of the present station. A large p_j corresponds to a high priority for the jth station. Kleinrock and Levy [1988] consider three different systems. In each one, all stations have the same service strategy: exhaustive, gated or 1-limited. For the exhaustive and gated strategies, they give the individual mean response times (waiting times plus service times) as the solution of a system of linear equations. For the 1-limited strategy they determine the mean response time for the special case of a completely symmetric system. They state that their results can be used to predict the expected delay in an exhaustive slotted ALOHA system; the random polling mechanism represents the random scheme according to which it is decided which station will transmit during the next slot.

In this study we also consider a polling system with probabilistic server routing. In our case the next station polled will be determined by a discrete time parameter Markov chain. We shall sometimes speak of *Markovian polling.* This includes cyclic polling and the purely random polling scheme of Kleinrock and Levy as special cases. The service strategies at the various queues may be different (exhaustive at one queue, 1-limited at the next one, etc.).

The switchover times of the server between queues are assumed to be nonnegligible. Hence there is no *work conservation.* However, the work conservation principle can be extended to a *work decomposition* principle (Boxma and Groenendijk [1987], Boxma [1989]). This work decomposition principle states that the amount of work in a polling system with switchover times can be decomposed into two independent parts, viz., (i) the amount of work in the same polling system but without switchover times and (ii) the amount of work in the system at some epoch covered by a switching interval.

Boxma and Groenendijk [1987] have used the work decomposition principle for cyclic polling systems to derive a *pseudoconservation law* for such systems, viz., an exact expression for a weighted sum of the mean waiting times at the queues. These results yield new insight into the behaviour of polling systems and can be used to obtain approximations for the individual mean waiting times. In the present study a pseudoconservation law will be derived for Markovian polling.

This study has been motivated by various considerations. Firstly, Markovian polling provides a theoretically interesting generalization of cyclic polling, and therefore we have considered it worthwhile to try and generalize the work decomposition principle and pseudoconservation law of Boxma and Groenendijk [1987] to the case of Markovian polling. Secondly, Markovian polling appears to have some interesting practical applications. While cyclic polling has been successfully used to model systems where a central controller polls many stations, Markovian polling may be used to model *distributed systems* (Kleinrock and Levy [1988]). As an example, Levy [1984] uses a random polling model to predict the mean delay in a slotted Aloha system. We believe that a second example may be found in the Orwell ring protocol (Mitrani, Adams and Falconer [1986]). In this protocol, c slots of equal length rotate around a ring. Each slot can accommodate one packet. A packet in a slot filled by a station Q_i is addressed to station Q_j with a certain probability. Q_j empties the slot and passes it on empty to the next downstream station. This is a major departure from other slotted ring protocols, where a slot can be released only by the station which filled it. For the case of $c = 1$ slot, it seems an interesting possibility to use a Markovian polling model to approximate the performance of the Orwell ring protocol. Such a model captures the stochastic character of the order of service of the stations, although it ignores the fact that the server transition probabilities in reality depend on whether or not a packet is waiting for transmission (whether or not a queue is non-empty).

A final reason for studying Markovian polling (and polling tables, for that matter) is that they open possibilities for optimization by considering various choices of the transition probabilities (and of the polling table). As a first step towards obtaining insight into this matter, we have compared the performance of polling systems with either cyclic or star polling and the performance of Markovian polling systems with the same server visit frequencies.

The paper is organized as follows. In Section 2 a model description is presented. Section 3 contains some preliminaries concerning the mean visit times of the server at the various queues, and a brief discussion of ergodicity conditions. Section 4 gives the work decomposition for Markovian polling systems. Section 5 is devoted to the main result of this study, the derivation of the pseudoconservation law. The determination of the mean interdeparture times of the server between queues will be central in that analysis. It will be shown that a system of N^2 equations in the N^2 unknown mean interdeparture times of the server, can be related to a system of N^2 equations in the N^2 unknown mean entrance times of the underlying reversed Markov chain (in fact, the system can be decomposed into N sets of N linear equations). For some special cases, explicit expressions for the mean interdeparture times are derived.

2. Model description

In this paper we consider a (continuous time) queueing system with N stations (queues), $Q_1,...,Q_N$, where each station has an infinite buffer capacity to store waiting messages (customers).

Message arrival process.

Customers arrive at all queues according to independent Poisson processes. The arrival intensity at Q_i is λ_i, $i = 1,...,N$. The total arrival rate is given by:

$$\Lambda := \sum_{i=1}^{N} \lambda_i.$$

Customers who arrive at Q_i are called type-i customers.

Service process.

The service times of type-i customers are independent, identically distributed stochastic variables. Their distribution $B_i(.)$ has first moment β_i and second moment $\beta_i^{(2)}$. The offered traffic, ρ_i, at Q_i is defined as:

$$\rho_i := \lambda_i \beta_i, \quad i = 1,...,N,$$

and the total offered traffic, ρ, as:

$$\rho := \sum_{i=1}^{N} \rho_i.$$

Polling strategy.

The N stations are served by a single server S who visits the stations according to a Markovian polling scheme (cf. Fig. 1).

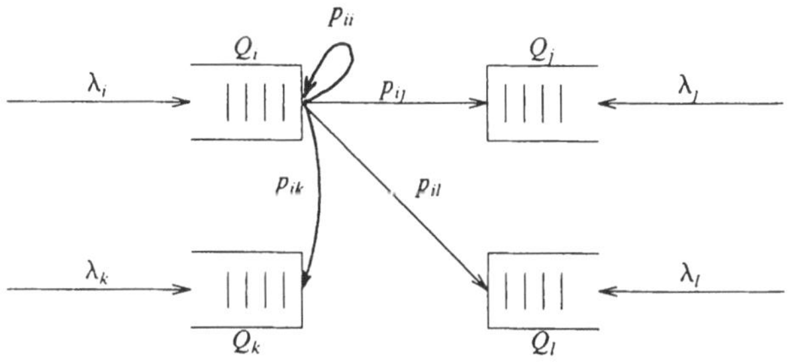

Figure 1

The next station to be polled is determined according to an irreducible positive recurrent discrete time parameter Markov chain $M = \{\mathbf{d}_n, n = 0,1,...\}$ with state space $I = \{1,...,N\}$. With $\{\mathbf{d}_n = i\}$ we denote the event that the nth station polled after $t = 0$ is station Q_i, $i \in I$.

We assume that the Markov chain M has stationary one-step transition probabilities, i.e., the conditional probabilities $Pr\{\mathbf{d}_{n+1} = j | \mathbf{d}_n = i\}$, $i,j \in I$, are independent of n.

Define:

$$p_{ij} := Pr\{\mathbf{d}_{n+1}=j|\mathbf{d}_n=i\}, \quad i,j\in I, \quad n=0,1,\ldots, \tag{2.1}$$

$$q_i := \lim_{n\to\infty} Pr\{\mathbf{d}_n=i\}, \quad i\in I, \quad n=0,1,\ldots. \tag{2.2}$$

For the waiting time analysis in Section 5 it will turn out to be essential to consider the *time reversed process* of the Markov chain M, $\tilde{M} = \{\tilde{\mathbf{d}}_n, n=0,1,...\}$, obtained from M by reversing the time parameter. The following theorem is proved in Kelly [1979], pp. 28,29:

THEOREM 2.1
If M is a stationary discrete time parameter Markov chain with state space I, one-step transition probabilities p_{ij}, $i,j\in I$, and with equilibrium distribution $\{q_j, j\in I\}$, then the reversed process $\tilde{M}$ is a stationary discrete time parameter Markov chain with state space I, one-step transition probabilities

$$\tilde{p}_{ij} := Pr\{\tilde{\mathbf{d}}_{n+1}=j|\tilde{\mathbf{d}}_n=i\} = \frac{q_j}{q_i}p_{ji}, \quad i,j\in I, \tag{2.3}$$

and with the same equilibrium distribution $\{q_j, j\in I\}$.

In the sequel $\tilde{M}$ will be called the reversed Markov chain.

Service strategy.
For the service strategies at the stations there are various possibilities, which differ in the number of customers who may be served in a queue during a visit of S to that queue. Assume that S visits Q_i. If Q_i is not empty S acts as follows, depending on the service strategy at Q_i:

I Exhaustive service (E): S serves type-i customers until Q_i is empty;
II Gated service (G): S serves exactly those type-i customers present upon his arrival at Q_i (a gate closes upon his arrival);
III 1-Limited service (1-L): S serves exactly one type-i customer.

In the sequel we will allow mixed service strategies (e.g., exhaustive at Q_1, 1-limited at Q_2 and Q_4, gated at Q_3, etc.).
After the visit period at Q_i (which has length zero when Q_i is empty) S switches with probability p_{ij} to Q_j, $i,j\in I$.

REMARK 2.1
We have restricted ourselves to the above-mentioned three main disciplines in polling systems. We could have included other strategies; in fact we show in Section 5 how 1-limited service in Markovian polling leads to the Bernoulli service discipline (cf. Keilson and Servi [1986]) in cyclic polling.

Switching process.
A switchover time is needed to switch from Q_i to Q_j, $i,j\in I$. The switchover times of the server between Q_i and Q_j are independent, identically distributed stochastic variables with mean s_{ij} and second moment $s_{ij}^{(2)}$.

The message arrival processes, the service demand processes and the switching processes are assumed to be mutually independent.

3. PRELIMINARY RESULTS

First some definitions. We define the visit time of the server at Q_i, $\mathbf{V}_i$, as:

$$\mathbf{V}_i := \text{ the time between the arrival of the server at } Q_i \text{ and its subsequent departure from } Q_i, \quad i \in I. \tag{3.1}$$

NOTE

If $p_{ii}>0$ and if, moreover, the switchover time from Q_i to Q_i is zero with positive probability, then a visit to Q_i may immediately be followed by another such visit.

We also define

$$\mathbf{u}_n := \text{ the time between the departure of the server from the (n-1)th station polled and its departure from the nth station polled after t=0 ;}$$

$\mathbf{u}_n$ is the sum of the switchover time from the $(n-1)$th station polled after $t=0$ to the nth station, and the visit time to the latter station. The process $SM = \{(\mathbf{d}_n,\mathbf{u}_n),\ n=0,1,...\}$, with $M = \{\mathbf{d}_n,\ n=0,1,...\}$ the discrete time parameter Markov chain described in Section 2, is a semi-Markov process (cf. Cinlar [1975], Chapter 10, Section 5). Note that when the $(n-1)$th station polled after $t=0$ is Q_i and the nth station polled after $t=0$ is Q_j:

$$E\mathbf{u}_n = s_{ij} + E\mathbf{V}_j.$$

We also have to introduce the server's interdeparture time between Q_i and Q_j, $\mathbf{T}_{ij}$, $i,j \in I$:

$$\mathbf{T}_{ij} := \text{ the time between a departure of S from } Q_j \text{ and its last previous departure from } Q_i. \tag{3.2}$$

So during the time span $\mathbf{T}_{ij}$, for $j \neq i$, S has not returned to Q_i but there may have been several visits to Q_j.
From a work balancing argument it follows that

$$E\mathbf{V}_i - \rho_i E\mathbf{T}_{ii}, \quad i \in I. \tag{3.3}$$

Another balancing argument yields:

$$\frac{q_i E\mathbf{V}_i}{q_j E\mathbf{V}_j} = \frac{\rho_i}{\rho_j}, \quad i,j \in I, \tag{3.4}$$

(cf. Cinlar [1975], p. 341), i.e. the ratio of the average amount of time S spends at Q_i and the average amount of time S spends at Q_j equals the ratio of the average traffic loads at Q_i and Q_j.
Combining (3.3) and (3.4) gives:

$$q_i E\mathbf{T}_{ii} = q_j E\mathbf{T}_{jj}, \ i,j \in I, \tag{3.5}$$

hence $q_i E\mathbf{T}_{ii}$, $i \in I$, is a constant.

Ergodicity conditions

A necessary condition for ergodicity of the system is $\rho<1$. When the service strategy at each queue is either exhaustive or gated this condition can be shown to be also sufficient. Without proof we observe that a necessary condition for ergodicity for a 1-limited station Q_i is:

$$\lambda_i E\mathbf{T}_{ii} < 1. \tag{3.6}$$

Indeed, $\lambda_i E\mathbf{T}_{ii}$ equals the mean number of arrivals to Q_i between two successive visits (and potential services) of the server at Q_i. For the mixed service strategies that we allow, condition (3.6) should be added to the stability condition $\rho<1$ for those queues at which we have a 1-limited service strategy. In the sequel it will be assumed that the ergodicity conditions are fulfilled, and that the system is in equilibrium.

4. Work decomposition

In this section we state that the amount of work in the Markovian polling system with switchover times can be decomposed into two independent terms, one of which is the amount of work in the same system but *without* switchover times. Before we give the work decomposition theorem we introduce the notion of 'corresponding M/G/1-system'. The corresponding M/G/1-system indicates a single-server system with exactly the same arrival processes, service demand processes and scheduling disciplines (i.e. a procedure for deciding which customer, if any, should be in service at any time) as the Markovian polling system under consideration, but without switchover times. The principle of work conservation implies that the amount of work in the latter system is independent of the service discipline: Markovian polling and FCFS service for all customers, irrespective of the queue they join, give rise to identical amounts of work at all times.
Define:

$\mathbf{V}_{MP}$:= steady-state amount of work in the Markovian polling system,
$\mathbf{V}$:= steady-state amount of work in the corresponding M/G/1-system,
$\mathbf{Y}$:= steady-state amount of work in the Markovian polling system at some epoch covered by a switching interval.

We relate these quantities in the next theorem.

Theorem 4.1
Consider a single-server multi-queue Markovian polling system as described in Section 2. Suppose the system is ergodic and stationary. Then the steady-state amount of work in this system, $\mathbf{V}_{MP}$, *is distributed as the sum of the steady-state amount of work in the corresponding M/G/1-system,* $\mathbf{V}$, *and the steady-state amount of work at some epoch covered by a switching interval,* $\mathbf{Y}$:

$$\mathbf{V}_{MP} \overset{D}{=} \mathbf{V} + \mathbf{Y}, \tag{4.1}$$

where $\overset{D}{=}$ *stands for equality in distribution. Furthermore,* $\mathbf{V}$ *and* $\mathbf{Y}$ *are independent.*

Proof
See Boxma [1989].

5. The pseudoconservation law

In this section we use Theorem 4.1 to derive an expression for a weighted sum of the mean waiting times. As a consequence of Theorem 4.1:

$$E\mathbf{V}_{MP} = E\mathbf{V} + E\mathbf{Y}, \tag{5.1}$$

and hence from M/G/1 theory, cf. Cohen [1982]:

$$E\mathbf{V}_{MP} = \frac{\sum_{i=1}^{N} \lambda_i \beta_i^{(2)}}{2(1-\rho)} + E\mathbf{Y}. \tag{5.2}$$

On the other hand, when $\mathbf{X}_i$ denotes the number of waiting type-i customers and $\mathbf{W}_i$ the waiting time of a type-i customer in the Markovian polling system with switchover times:

$$E\mathbf{V}_{MP} = \sum_{i=1}^{N} \beta_i E\mathbf{X}_i + \sum_{i=1}^{N} \rho_i \frac{\beta_i^{(2)}}{2\beta_i} = \sum_{i=1}^{N} \rho_i E\mathbf{W}_i + \frac{1}{2}\sum_{i=1}^{N} \lambda_i \beta_i^{(2)}, \tag{5.3}$$

the first equality following from the fact that service is non-preemptive, and the second equality following from Little's formula. Combination of (5.2) and (5.3) yields:

$$\sum_{i=1}^{N} \rho_i E\mathbf{W}_i = \rho \frac{\sum_{i=1}^{N} \lambda_i \beta_i^{(2)}}{2(1-\rho)} + E\mathbf{Y}. \tag{5.4}$$

To obtain an expression for this weighted sum of mean waiting times it remains to determine $E\mathbf{Y}$, the mean amount of work at some epoch covered by a switching interval. This determination, to be presented below, will in Theorem 5.1 finally lead to an exact expression for a weighted sum of the mean waiting times (a pseudoconservation law).

Denote by $\mathbf{Y}_{ij}$ the amount of work in the Markovian polling system at some epoch covered by a switchover from Q_i to Q_j, $i,j \in I$. $E\mathbf{Y}$ can be expressed as a weighted sum of all $E\mathbf{Y}_{ij}$, averaging over the switchover durations and the frequencies with which transitions between queues occur:

$$E\mathbf{Y} = (1/\sigma)\sum_{i=1}^{N} q_i \sum_{j=1}^{N} p_{ij} s_{ij} E\mathbf{Y}_{ij}, \tag{5.5}$$

with

$$\sigma := \sum_{i=1}^{N} q_i \sum_{j=1}^{N} p_{ij} s_{ij}, \tag{5.6}$$

the average mean switchover time. Note that in the purely cyclic case ($p_{i,i+1}=1$, $i \in I$): $\sigma = (1/N)\sum_{i=1}^{N} s_{i,i+1}$, with $\sum_{i-1}^{N} s_{i,i+1}$ the mean total switchover time in one cycle.

It remains to determine $E\mathbf{Y}_{ij}$, $i,j \in I$. $E\mathbf{Y}_{ij}$ is composed of three terms, only one of which depends on j:

$E\mathbf{M}_i^{(1)}$:= the mean amount of work in Q_i at a departure epoch of S from Q_i,

$E\mathbf{M}_i^{(2)}$:= the mean amount of work in $Q_1,\ldots,Q_{i-1},Q_{i+1},\ldots,Q_N$, at a departure epoch of S from Q_i,

$\rho\frac{s_{ij}^{(2)}}{2s_{ij}}$:= the mean amount of work that arrived in the system during the past part of the switching interval (from Q_i to Q_j) under consideration.

So we can write:

$$E\mathbf{Y}_{ij} = E\mathbf{M}_i^{(1)} + E\mathbf{M}_i^{(2)} + \rho\frac{s_{ij}^{(2)}}{2s_{ij}}. \tag{5.7}$$

We shall first consider $E\mathbf{M}_i^{(2)}$, the mean amount of work in $Q_1,\ldots,Q_{i-1},Q_{i+1},\ldots,Q_N$ at a departure epoch of the server from Q_i. Q_k ($k \neq i$) can make two contributions to $E\mathbf{M}_i^{(2)}$:

- the mean amount of work left behind in Q_k by S at his last departure from Q_k,

- the mean amount of work that has arrived in Q_k during $\mathbf{T}_{ki}$, the server's interdeparture time between Q_k and Q_i (cf. (3.2)).

We obtain the following relation:

$$E\mathbf{M}_i^{(2)} = \sum_{k \neq i} E\mathbf{M}_k^{(1)} + \sum_{k \neq i} \rho_k E\mathbf{T}_{ki}, \quad i \in I. \tag{5.8}$$

Substitution of (5.8) in (5.7) gives:

$$E\mathbf{Y}_{ij} = \sum_{k=1}^{N} E\mathbf{M}_k^{(1)} + \sum_{k \neq i} \rho_k E\mathbf{T}_{ki} + \rho \frac{s_{ij}^{(2)}}{2s_{ij}}, \quad i,j \in I. \tag{5.9}$$

$E\mathbf{M}_k^{(1)}$ and $E\mathbf{T}_{ki}$ still have to be determined.

Determination of $E\mathbf{M}_k^{(1)}$, $k \in I$

The $E\mathbf{M}_k^{(1)}$ are derived below for an exhaustive, gated or 1-limited service strategy at Q_k:

(i) Q_k has an exhaustive service strategy: Q_k is left behind empty by S, so

$$E\mathbf{M}_k^{(1)} = 0. \tag{5.10}$$

(ii) Q_k has a gated service strategy: $E\mathbf{M}_k^{(1)}$ equals ρ_k times the mean visit time of S at Q_k, hence (cf. (3.3)),

$$E\mathbf{M}_k^{(1)} = \rho_k E\mathbf{V}_k = \rho_k^2 E\mathbf{T}_{kk}. \tag{5.11}$$

(iii) Q_k has a 1-limited service strategy: a similar derivation as in Boxma and Groenendijk [1987] leads to,

$$E\mathbf{M}_k^{(1)} = \lambda_k E\mathbf{T}_{kk}[\rho_k(E\mathbf{W}_k + \beta_k)] + (1-\lambda_k E\mathbf{T}_{kk})0 = \rho_k \lambda_k E\mathbf{T}_{kk} E\mathbf{W}_k + \rho_k^2 E\mathbf{T}_{kk}, \tag{5.12}$$

($\lambda_k E\mathbf{T}_{kk}$ is the fraction of visits of S to Q_k that result in a service, and $\rho_k(E\mathbf{W}_k + \beta_k)$ equals the mean amount of work that has arrived during the sojourn time of a departing customer).
Substituting (5.10),...,(5.12) in (5.9) gives:

$$E\mathbf{Y}_{ij} = \sum_{k \in g, 1-l} \rho_k^2 E\mathbf{T}_{kk} + \sum_{k \in 1-l} \rho_k \lambda_k E\mathbf{T}_{kk} E\mathbf{W}_k + \sum_{k \neq i} \rho_k E\mathbf{T}_{ki} + \rho \frac{s_{ij}^{(2)}}{2s_{ij}}, \quad i,j \in I, \tag{5.13}$$

with g and $1-l$ denoting the group of queues with gated and 1-limited service strategies respectively.

Determination of $E\mathbf{T}_{ki}$, $k,i \in I$

We first introduce the event

B_{ji} := 'the last visit before a visit of S to Q_i was to Q_j'.

For all $k,i \in I$:

$$E\mathbf{T}_{ki} = \sum_{j=1}^{N} E\{\mathbf{T}_{ki} | B_{ji}\} Pr\{B_{ji}\}. \tag{5.14}$$

Determination of $E\mathbf{T}_{ki}$ requires looking backwards in time (cf. Theorem 2.1). We can write for all $i,j \in I$:

$$Pr\{B_{ji}\} = Pr\{\mathbf{d}_{n-1}=j|\mathbf{d}_n=i\} = Pr\{\tilde{\mathbf{d}}_{n+1}=j|\tilde{\mathbf{d}}_n=i\} = \tilde{p}_{ij}, \tag{5.15}$$

the one-step transition probabilities of the reversed Markov chain $\tilde{M}$. It easily follows that:

$$E\{\mathbf{T}_{ki}|B_{ji}\} = E\{\mathbf{T}_{kj}\} + s_{ji} + E\mathbf{V}_i \quad \text{if } j \neq k, \tag{5.16}$$
$$E\{\mathbf{T}_{ki}|B_{ji}\} = s_{ki} + E\mathbf{V}_i \quad \text{if } j=k.$$

Substitution of (5.15) and (5.16) into (5.14) gives, for $k,i \in I$:

$$E\mathbf{T}_{ki} = \sum_{j \neq k} [E\mathbf{T}_{kj} + s_{ji} + E\mathbf{V}_i]\,\tilde{p}_{ij} + [s_{ki} + E\mathbf{V}_i]\,\tilde{p}_{ik}. \tag{5.17}$$

If we define

$$f(i) := E\mathbf{V}_i + \sum_{j=1}^{N} s_{ji}\,\tilde{p}_{ij}, \quad i \in I, \tag{5.18}$$

then we can rewrite (5.17) as:

$$E\mathbf{T}_{ki} = f(i) + \sum_{j \neq k} E\mathbf{T}_{kj}\,\tilde{p}_{ij}, \quad i,k \in I. \tag{5.19}$$

Clearly, the set of N^2 linear equations (5.19) can be decomposed into N sets of N linear equations. In the next lemma it will be shown that the N^2 unknown mean interdeparture times $E\mathbf{T}_{ki}$, $k,i \in I$ can be expressed in the N^2 mean entrance times between queues in the underlying reversed Markov chain $\tilde{M}$. The mean entrance time between Q_i and Q_j, $\tilde{v}_{ij}$, in the reversed Markov chain $\tilde{M}$ is defined as:

$$\tilde{v}_{ij} := E\{\#\text{ steps required for the first entrance into } Q_i \text{ starting from } Q_j\}, \quad i,j \in I, \tag{5.20}$$

(cf. Cohen [1982], p. 33).
Note that from the theory of Markov chains and from Theorem 2.1 we have,

$$\tilde{v}_{ii} = \frac{1}{q_i}, \quad i \in I. \tag{5.21}$$

We now formulate:

Lemma 5.1
For all $i,k \in I$:

$$E\mathbf{T}_{ki} = f(i) + \sum_{l \neq k} f(l)\frac{\tilde{v}_{ik} + \tilde{v}_{kl} - \tilde{v}_{il}}{\tilde{v}_{ll}}. \tag{5.22}$$

Proof
Denote by $\tilde{S}$ the server for the reversed Markov chain $\tilde{M}$. We start with two observations:
(i) $E\mathbf{T}_{ki}$ is in the reversed Markov chain $\tilde{M}$ the average time between an arrival of $\tilde{S}$ at Q_i and his first subsequent arrival at Q_k.
(ii) $f(i)$ is in the reversed Markov chain $\tilde{M}$ the average time between an arrival of $\tilde{S}$ at Q_i and his arrival at the next station to be visited after Q_i (possibly again Q_i).
Using these observations we can write for $i,k \in I$:

$$E\mathbf{T}_{ki} = f(i) + \sum_{l \neq k} f(l)\, E\{\#\text{ times } \tilde{S} \text{ visits } Q_l \text{ before it visits } Q_k \text{ starting from } Q_i\}. \tag{5.23}$$

Further on we can write for $i,k,l \in I$ and $k \neq l$ (see Chung [1967], p.46):

$$E\{\#\text{ times } \tilde{S} \text{ visits } Q_l \text{ before it visits } Q_k \text{ starting from } Q_i\} = \sum_{n=1}^{\infty} {}_k\tilde{p}_{il}^{(n)},$$

with

$${}_k\tilde{p}_{il}^{(n)} = Pr\{\tilde{\mathbf{d}}_n = l,\ \tilde{\mathbf{d}}_m \neq k,\ m = 1,...,n-1 | \tilde{\mathbf{d}}_0 = i\},$$

i.e., the probability of going from Q_i to Q_l in n steps without visiting Q_k. Using Corollary 2 on page 65 of Chung [1967] we find:

$$\sum_{n=1}^{\infty} {}_k\tilde{p}_{il}^{(n)} = [\tilde{v}_{ik} + \tilde{v}_{kl} - \tilde{v}_{il}] / \tilde{v}_{ll}.$$

So we obtain:

$$E\{\#\text{ times } \tilde{S} \text{ visits } Q_l \text{ before it visits } Q_k \text{ starting from } Q_i\} = \tag{5.24}$$
$$[\tilde{v}_{ik} + \tilde{v}_{kl} - \tilde{v}_{il}] / \tilde{v}_{ll}, \quad i,k,l \in I,\ k \neq l.$$

Combining (5.23) and (5.24) gives relation (5.22).

REMARK 5.1
An alternative way to prove Equation (5.22) is by means of matrix manipulations. The following steps are required:
(1) Denote by $\overline{T} = (E\mathbf{T}_{11}, \ldots, E\mathbf{T}_{1N}, E\mathbf{T}_{21}, \ldots, E\mathbf{T}_{NN})'$ the N^2-dimensional column vector of the unknown mean interdeparture times. Then we can write (5.19) in the following form:

$$\overline{T} = A\overline{T} + \overline{b},$$

with obvious definitions of the N^2 by N^2 matrix A and the N^2-dimensional vector $\overline{b}$.
(2) Because the eigenvalues of A are all less than one (Seneta [1981]), $\overline{T}$ can be written as:

$$\overline{T} = [I - A]^{-1}\overline{b} = [\sum_{n=0}^{\infty} A^{(n)}]\overline{b} = [I + \sum_{n=1}^{\infty} A^{(n)}]\overline{b}.$$

(3) It appears that $\sum_{n=1}^{\infty} A^{(n)}$ is a blockdiagonal matrix with N blocks of size $N \times N$. Denoting the (i,l)th element of the kth block by $C_k(i,l)$, it can be shown that for $i,l \in \{1,...,N\}$, $k \in \{1,...,N\}$,

$$C_k(i,l) = 0, \ l = k, \qquad C_k(i,l) = \sum_{n=1}^{\infty} {}_k\tilde{p}_{il}^{(n)}, \ l \neq k.$$

Combination of steps 2 and 3 now yields (5.22).

So, to determine $E\mathbf{T}_{ki}$, $k,i \in I$ we can determine the mean entrance times, $\tilde{v}_{ik}$, $i,k \in I$, of the underlying reversed Markov chain $\tilde{M}$. It is known that these are the solution of

$$\tilde{v}_{ik} = 1 + \sum_{j \neq k} \tilde{p}_{ij}\tilde{v}_{jk}, \quad i,k \in I; \tag{5.25}$$

cf. Cohen [1982].

From a theoretic point of view it is interesting to make the link between interdeparture times of S, and entrance times of the underlying reversed Markov chain (the more so because the semi-Markov process and its underlying (reversed) Markov chain arise so naturally in the present queueing model). From a numerical point of view it constitutes no real advantage to solve the set of equations (5.25) instead of the set of equations (5.19).

For $k=i$ Lemma 5.1 yields, with $\tilde{v}_{ll} = 1/q_l$, $l \in I$,

$$q_i E\mathbf{T}_{ii} = \sum_{l=1}^{N} q_l f(l), \ i \in I. \tag{5.26}$$

This demonstrates the fact that $q_i E\mathbf{T}_{ii}$ does not depend on i (cf. (3.5)). We now determine $C := q_i E\mathbf{T}_{ii}$, successively using (5.26), (5.18), (2.3), (5.6) and (3.3):

$$C = \sum_{l=1}^{N} q_l f(l) = \sum_{l=1}^{N} q_l \sum_{m=1}^{N} \tilde{p}_{lm} s_{ml} + \sum_{l=1}^{N} q_l E\mathbf{V}_l =$$
$$\sum_{m=1}^{N} q_m \sum_{l=1}^{N} p_{ml} s_{ml} + \sum_{l=1}^{N} q_l E\mathbf{V}_l = \sigma + \sum_{l=1}^{N} \rho_l q_l E\mathbf{T}_{ll} = \sigma + \rho C.$$

Hence

$$C = \frac{\sigma}{1-\rho}, \tag{5.27}$$

so

$$E\mathbf{T}_{ii} = \frac{1}{q_i} \frac{\sigma}{1-\rho}, \tag{5.28}$$

$$E\mathbf{V}_i = \frac{1}{q_i} \frac{\rho_i \sigma}{1-\rho}. \tag{5.29}$$

Substituting (5.28) in (5.13) gives:

$$E\mathbf{Y}_{ij} = \frac{\sigma}{1-\rho} \sum_{k \in g, 1-l} \frac{\rho_k^2}{q_k} + \frac{\sigma}{1-\rho} \sum_{k \in 1-l} \frac{\rho_k}{q_k} \lambda_k E\mathbf{W}_k + \tag{5.30}$$
$$\sum_{k \neq i} \rho_k E\mathbf{T}_{ki} + \rho \frac{s_{ij}^{(2)}}{2 s_{ij}}, \ i,j \in I,$$

with $E\mathbf{T}_{ki}$ as in Lemma 5.1.

Combining (5.4), (5.5) and (5.30) gives our main result which is formulated in Theorem 5.1 below. As before, denote by g and $1-l$ the group of queues with gated and 1-limited service strategies, and further denote by e the group of queues with exhaustive service strategies.

THEOREM 5.1

Consider an ergodic and stationary single-server multi-queue Markovian polling system with mixed service strategies as described in Section 2. Then:

$$\sum_{k \in e,g} \rho_k E\mathbf{W}_k + \sum_{k \in 1-l} \rho_k \left[1 - \frac{\lambda_k}{q_k} \frac{\sigma}{1-\rho}\right] E\mathbf{W}_k = \tag{5.31}$$

$$\rho \frac{\sum_{i=1}^{N} \lambda_i \beta_i^{(2)}}{2(1-\rho)} + \frac{\sigma}{1-\rho} \sum_{k \in g, 1-l} \frac{\rho_k^2}{q_k} + \frac{\rho}{2\sigma} \sum_{i=1}^{N} q_i \sum_{j=1}^{N} p_{ij} s_{ij}^{(2)} + \frac{1}{\sigma} \sum_{i=1}^{N} q_i \sum_{j=1}^{N} p_{ij} s_{ij} \sum_{k \neq i} \rho_k \mathbf{ET}_{ki},$$

with $\mathbf{ET}_{ki}$ *as in Lemma 5.1.*

REMARK 5.2
In the purely cyclic case, $p_{i,i+1}=1$, $i \in I$, (5.31) reduces to (3.22) in Boxma and Groenendijk [1987]. Note that $(\sigma / q_k) = n\sigma$ in (5.31) corresponds to the total mean switchover time, s, in (3.22) of that publication, cf. below (5.6).

REMARK 5.3
Theorem 5.1 can be generalized to the case of a batch arrival process with correlated sizes of the batches simultaneously arriving at the various queues (cf. the cyclic polling model of Levy and Sidi [1988]).

REMARK 5.4
Kleinrock and Levy [1988] restrict themselves to the special case that $p_{ij} = p_j$ (random polling) and $s_{ij} = s_i$, $s_{ij}^{(2)} = s_i^{(2)}$ for all $i,j \in I$. In this case $q_k = p_k$, $k \in I$, and (5.17) reduces to:

$$E\mathbf{T}_{ki} = \frac{\sigma}{1-\rho} [\frac{\rho_i}{q_i} - \frac{\rho_k}{q_k} + \frac{1}{q_k}], \quad k,i \in I. \tag{5.32}$$

To interpret this formula, note that

$$E\mathbf{T}_{ki} - E\mathbf{V}_i + E\mathbf{V}_k = \frac{1}{q_k} \frac{\sigma}{1-\rho} = E\mathbf{T}_{kk}, \quad k,i \in I; \tag{5.33}$$

and observe that in this case M is reversible, so that $E\mathbf{T}_{ki} - E\mathbf{V}_i + E\mathbf{V}_k$ also equals the mean time between a departure from Q_i (or Q_k, as $p_{ij}=p_j$ for all i) and the first subsequent departure from Q_k.
Formula (5.31) reduces to:

$$\sum_{k \in e,g} \rho_k E\mathbf{W}_k + \sum_{k \in 1-l} \rho_k [1 - \frac{\lambda_k}{p_k} \frac{\sigma}{1-\rho}] E\mathbf{W}_k = \tag{5.34}$$
$$\rho \frac{\sum_{i=1}^{N} \lambda_i \beta_i^{(2)}}{2(1-\rho)} + \frac{\sigma}{1-\rho} \sum_{k \in g, 1-l} \frac{\rho_k^2}{p_k} - \frac{\sigma}{1-\rho} \sum_{k=1}^{N} \frac{\rho_k^2}{p_k} + \frac{\sigma}{1-\rho} \sum_{k=1}^{N} \frac{\rho_k}{p_k} -$$
$$\sum_{i=1}^{N} \rho_i s_i + \frac{\rho}{2\sigma} \sum_{i=1}^{N} p_i \, s_i^{(2)}.$$

Kleinrock and Levy, for the cases of exhaustive service and of gated service at all queues, give the individual mean waiting times as the solution of a set of $O(N^2)$ linear equations. For the completely symmetric case, they determine the mean waiting times (which now are all the same) explicitly for the exhaustive, gated and 1-limited strategies. It can easily be shown that (5.34) leads to the expression found by Kleinrock and Levy for this completely symmetric case.

REMARK 5.5
At this stage we'd like to point at the relative simplicity of the pseudoconservation law formulated in Theorem 5.1. In the righthand side of (5.31), only the first and second moments of the traffic

quantities and switchover times occur. In the general case the righthand side of (5.31) can be evaluated after N sets of N linear equations have been solved; in special cases such as purely random polling, the $E\mathbf{T}_{ki}$ can be determined explicitly in a straightforward manner. This should be contrasted with the fact that, apart from two-queue models and completely symmetric models, the *individual* mean waiting times are only known for purely random polling with exhaustive or gated service at all queues. For the more general Markovian polling, the individual mean waiting times might again be determined for these two service disciplines, following the approach of Kleinrock and Levy [1988]; but this seems to require the solution of a set of $O(N^3)$ linear equations. In the case of purely cyclic polling, several pseudoconservation-law based approximations for the individual mean waiting times have been investigated, see e.g. Boxma and Meister [1987] and Groenendijk [1988]; it might be worthwhile to try and obtain mean waiting time approximations for Markovian polling in a similar way, starting from (5.31).

REMARK 5.6
It is interesting to compare cyclic polling and random polling with equal visit probabilities ($p_{ij} \equiv 1/N$) for all queues. We restrict ourselves to the case that, for both models, all queues have exactly the same traffic characteristics, while all switchover times are independent, identically distributed s.v. with mean r. With an obvious notation, the difference between the mean workloads in both models is (cf. (5.1)):

$$E\mathbf{V}_{MP} - E\mathbf{V}_{cycl} = E\mathbf{Y}_{MP} - E\mathbf{Y}_{cycl}. \tag{5.35}$$

Comparison of the pseudoconservation laws for cyclic polling (Boxma and Groenendijk [1987]) and random polling (formula (5.34)) yields:

$$E\mathbf{V}_{MP} - E\mathbf{V}_{cycl} = \frac{N-1}{2}\frac{r\rho}{1-\rho}. \tag{5.36}$$

Not surprisingly, the random character of the server visits in random polling leads to a higher mean workload than for cyclic polling. Of course, if traffic is quite asymmetric, Markovian polling with a relatively high visit frequency of a relatively heavily loaded station may lead to $E\mathbf{V}_{MP} < E\mathbf{V}_{cycl}$.

When the service strategy at all queues is the same, then (5.36) leads to the following results of Kleinrock and Levy [1988]: for exhaustive and gated service, with an obvious notation,

$$E\mathbf{W}_{MP} - E\mathbf{W}_{cycl} = \frac{N-1}{2}\frac{r}{1-\rho}; \tag{5.37}$$

for 1-limited service,

$$E\mathbf{W}_{MP} - E\mathbf{W}_{cycl} = \frac{N-1}{2}\frac{r}{1-\rho-N\lambda_1 r}. \tag{5.38}$$

REMARK 5.7
We have also compared $\mathbf{V}_{MP}$ with $\mathbf{V}_{star}$, the amount of work in a single-server N-queue system ($N \geqslant 2$) with star polling, i.e. server visits according to the polling table [1,2,1,3,...,1,N]. We have assumed that

- Q_1 receives exhaustive service, whereas $Q_2, \ldots, Q_N$ receive 1-limited service (in both models);
- both models have the same traffic characteristics;
- all switchover times are equal to the constant r;
- $p_{ij} = p_j$, with $p_1 = \frac{1}{2}$, $p_2 = \cdots = p_N = 1/(2(N-1))$.

Comparison of $E\mathbf{V}_{MP}$ and $E\mathbf{V}_{star}$ amounts to comparison of $E\mathbf{Y}$ in both models. Using the evaluations of EY in the present paper and in Boxma et al. [1988], it can be shown that

$$EV_{MP} \geqslant EV_{star}; \tag{5.39}$$

in fact, if $\rho_2 = \cdots = \rho_N$, then

$$EV_{MP} - EV_{star} = \frac{1}{2} r\rho + r\frac{\rho-\rho_1}{1-\rho}[N-2+2\rho_1]. \tag{5.40}$$

It should be noted that this difference is
- roughly linearly increasing in N;
- tending to zero for $r \to 0$;
- dependent on λ_n and β_n only via their product ρ_n;
- equal to $\frac{1}{2} r\rho$ for $\rho = \rho_1$ ($Q_2, \ldots, Q_N$ receive no traffic), a result which is easily explained by noting that the comparison in this case amounts to a comparison of two M/G/1 queues with vacations.

Another special case of the Markovian polling model introduced in this paper is the cyclic service model with a Bernoulli schedule, as introduced by Keilson and Servi [1986]. This schedule operates as follows. If there are still customers present in Q_i after a service completion in this queue, the server decides with probability $1-p_i$ to serve the next customer at Q_i, and with probability p_i he switches to Q_{i+1}. He also takes the latter action when there are no more customers present in Q_i. Tedijanto [1988] has derived a pseudoconservation law for the cyclic service system with a Bernoulli schedule. Below we show how this pseudoconservation law follows as a special case of the pseudoconservation law in Theorem 5.1. Assume that all stations have a 1-limited service strategy, and take, for all $i \in I$,

$$\begin{aligned}
p_{ij} &= 1-p_i \quad \text{if } j=i, \\
p_{ij} &= p_i \qquad \text{if } j=i+1, \\
p_{ij} &= 0 \qquad \text{else}; \\
s_{ii} &= 0, \\
s_{i,i+1} &= s_i, \\
s_{i,i+1}^{(2)} &= s_i^{(2)}.
\end{aligned}$$

It should be noted that the server pays a geometrically distributed number of consecutive visits, with mean number $1/p_i$, to Q_i; even when Q_i has become empty, the server may still return a number of times, but this does not take time because $s_{ii}=0$. It follows that the server spends on the average EV_m/p_m at Q_m before he switches to Q_{m+1}. A work balancing argument now implies that

$$\frac{EV_m}{p_m} = \rho_m \frac{D}{1-\rho}, \quad m \in I, \tag{5.41}$$

with

$$D := \sum_{i=1}^{N} s_i.$$

For this special case we also have:

$$f(i) = p_i[s_{i-1} + \rho_i \frac{D}{1-\rho}], \quad i \in I, \tag{5.42}$$

$$q_i = \frac{1}{p_i} / (\sum_{m=1}^{N} \frac{1}{p_m}), \quad i \in I, \tag{5.43}$$

$$\sigma = D / (\sum_{m=1}^{N} \frac{1}{p_m}), \tag{5.44}$$

and (cf. (5.25))

$$\tilde{v}_{ij} = \sum_{k=j}^{i-1} \frac{1}{p_k}, \quad i,j \in I, \tag{5.45}$$

(note that this is a cyclic sum; for $j > i-1$, $\tilde{v}_{ij}$ is the sum of $1/p_j, \ldots, 1/p_N, 1/p_1, \ldots, 1/p_{i-1}$). Using the above formulas, (5.22) reduces to:

$$\mathbf{ET}_{ki} = \sum_{m=k+1}^{i} \frac{\mathbf{EV}_m}{p_m} + \sum_{m=k}^{i-1} s_m, \quad k,i \in I. \tag{5.46}$$

Indeed, in the cyclic service model with a Bernoulli schedule, $\mathbf{ET}_{ki}$ equals the mean amount of time the server spends at $Q_{k+1},...,Q_i$, plus the sum of the mean switchover times between Q_k and Q_i.

Using (5.41),...,(5.46), we obtain the pseudoconservation law for the cyclic service model with a Bernoulli schedule at all queues:

$$\sum_{k=1}^{N} \rho_k[1-\lambda_k p_k \frac{D}{1-\rho}]\mathbf{EW}_k = \tag{5.47}$$

$$\rho \frac{\sum_{i=1}^{N} \lambda_i \beta_i^{(2)}}{2(1-\rho)} + \frac{D}{1-\rho} \sum_{k=1}^{N} \rho_k^2 p_k + \rho \frac{D^{(2)}}{2D} + \frac{D}{2(1-\rho)}[\rho^2 - \sum_{i=1}^{N} \rho_i^2],$$

with $D^{(2)}$ the second moment of the sum of the N switchover times between Q_1 and Q_2, ..., Q_N and Q_1.

Expression (5.47) is the same as (3.6.6) in Tedijanto [1988]. Note that (5.47) reduces to the pseudoconservation law for cyclic polling with exhaustive (respectively 1-limited) service at all queues if all $p_k = 0$ (respectively all $p_k = 1$).

ACKNOWLEDGMENT
The authors are indebted to J.W. Cohen, F.A. van der Duyn Schouten, W.P. Groenendijk and H. Levy for useful comments and stimulating discussions.

REFERENCES

1. BOXMA, O.J. (1989). *Workloads and waiting times in single-server systems with multiple customer classes.* To appear in *Queueing Systems.*
2. BOXMA, O. J., GROENENDIJK, W.P. (1987). *Pseudo-conservation laws in cyclic-service systems.* J. Appl. Prob. **24**, 949-964.

3. BOXMA, O.J., GROENENDIJK, W.P., WESTSTRATE, J.A. (1988). *A pseudoconservation law for service systems with a polling table.* Report Centre for Mathematics and Computer Science, Amsterdam; to appear in *IEEE Trans. Commun.*
4. BOXMA, O.J., MEISTER, B. (1987). *Waiting-time approximations for cyclic-service systems with switchover times,* Performance Evaluation 7, 299-308.
5. CHUNG, K.L. (1967). *Markov Chains with Stationary Transition Probabilities (Springer, Berlin; 2nd ed.).*
6. CINLAR, E. (1975). *Introduction to Stochastic Processes* (Prentice Hall, Englewood Cliffs, NJ).
7. COHEN, J.W. (1982). *The Single Server Queue* (North-Holland, Amsterdam; 2nd ed.).
8. GROENENDIJK, W.P. (1988). *Waiting-time approximations for cyclic-service systems with mixed service strategies,* in: M. Bonatti (ed.), Proceedings ITC-12 (North-Holland, Amsterdam).
9. KEILSON, J., SERVI, L.D. (1986). *Oscillating random walk models for GI/G/1 vacation systems with Bernoulli schedules.* J. Appl. Prob. **23**, 790-802.
10. KELLY, F.P. (1979). *Reversibility and Stochastic Networks* (Wiley, New York).
11. KLEINROCK, L., LEVY, H. (1988). *The analysis of random polling systems.* Oper. Res. **36**, 716-732.
12. LEVY, H. (1984). *Non-Uniform Structures and Synchronization Patterns in Shared-Channel Communication Networks.* CSD-840049, Computer Science Department, University of California, Los Angeles, Ph.D. Dissertation.
13. LEVY, H., SIDI, M. (1988). *Correlated arrivals in polling systems.* Report Department of Computer Science, Tel Aviv University.
14. MITRANI, I., ADAMS, J.L., FALCONER, R.M. (1986). *A modelling study of the Orwell ring protocol.* In: Teletraffic Analysis and Computer Performance Evaluation, eds. O.J. Boxma, J.W. Cohen and H.C. Tijms (North-Holland, Amsterdam), pp. 429-438.
15. E. SENETA (1981). *Non-negative Matrices and Markov Chains* (Springer, New York; 2nd ed.).
16. TAKAGI, H. (1986). *Analysis of Polling Systems* (The MIT Press, Cambridge, MA).
17. TAKAGI, H. (1988). *Queuing analysis of polling models.* ACM Comput. Surveys **20**, 5-28.
18. TEDIJANTO (1988). *Exact results for the cyclic-service queue with a Bernoulli schedule.* Report Electrical Engineering Department and Systems Research Center, University of Maryland.

SIMPLE APPROXIMATIONS FOR SECOND MOMENT CHARACTERISTICS OF THE SOJOURN TIME IN THE M/G/1 PROCESSOR SHARING QUEUE

J.L. van den Berg
Centre for Mathematics and Computer Science
P.O. Box 4079, 1009 AB Amsterdam, The Netherlands

Abstract
In this paper simple approximations are derived for the second moment of the conditional and unconditional sojourn time in the M/G/1 processor sharing queue. The approximations are mainly based on general asymptotic results (e.g. heavy traffic) and on simple exact expressions for specific service time distributions. They depend on the service time distribution only through its first and second moment. Numerous numerical examples show that these simple approximations are accurate enough for many practical purposes. A refinement of the approximations is obtained by taking the third moment of the service time into account.

1. Introduction

The processor sharing (PS) service discipline is widely used to model time sharing in computer systems. During the last ten years considerable attention has been paid to the analysis of the M/G/1 PS queue, see e.g. the surveys by Jaiswal [4] and Yashkov [13]. The major problem in processor sharing queues is the problem of characterizing the sojourn time distribution. Only recently exact expressions for the Laplace-Stieltjes transform of the sojourn time distribution have been obtained by Yashkov [11], Ott [6] and Schassberger [7]. Due to their complexity these formulas are not attractive for practical applications. Only for the mean sojourn time a simple explicit expression exists; this expression is insensitive to the service time distribution apart from its first moment (Kleinrock [5]). Formulas for the second moment of the sojourn time (Ott [6], Yashkov [13]) require perfect information about the service time distribution, which is almost never available in practice. Moreover, these formulas contain a (double) integral which, in general, can only be evaluated numerically. As far as we know no attention has been paid to the derivation of approximations or asymptotic formulas which are useful for practical evaluation, apart from [1] and a paper by Yashkov [12]. Yashkov derives some asymptotic estimates for the conditional sojourn time variance for customers with small or large service times. An extension of these results is presented in [1].

The aim of the present study is to derive approximations for the second moment of the sojourn time distribution, which are quite simple and yet accurate enough for most practical purposes. We first show that a lower and upper bound for the second moment of the sojourn time can be expressed in terms of the first and second moment of the service time. Next some very simple approximation formulas based on the first and second moment of the service time are presented. The accuracy of the approximations is tested for a large number of different service time distributions and a wide range of traffic intensities. A refinement of the approximation is obtained by taking the third moment of the service time into account. This refinement yields remarkably accurate results with relative errors less than 1.5 percent in most cases.

The organization of the rest of the paper is as follows. In Section 2 we introduce the notations and give a summary of those known sojourn time results which are relevant for our study. We also present some new results and extensions. In particular an upper bound for the second moment of the sojourn time is derived. Section 3 is concerned with the second moment of the *conditional* sojourn time of a customer with service demand x. We propose an approximation formula which is based on an asymptotic result for $x \to 0$ derived in [1]. In Section 4 approximations are developed for the second moment of the *unconditional* sojourn time. We first propose an approximation which contains only information about the first and second moment of the service time distribution. A refinement of the approximation is derived for the case that the squared coefficient of variation of the service time distribution is smaller than one. Finally, for service time distributions with a squared coefficient of variation larger than one, we construct a very accurate approximation formula which is based on the first three moments of the service time distribution.

2. Notations and preliminary results

Customers arrive at a single server queue according to a Poisson process with rate λ. Their service requirements are i.i.d. non-negative random variables with a general distribution $B(\cdot)$ with first and second moment β and β_2. All customers are served simultaneously according to the processor sharing (PS) service discipline, i.e. whenever i customers are present, each customer receives service at a rate of $1/i$. We assume that $\rho := \lambda\beta < 1$ and that the system is in steady state. Let $S(x)$ be the sojourn time of a tagged customer who requires an amount x of service at his arrival. It is well-known that $E\,S(x)$ is linear in x: (see Kleinrock [5])

$$E\,S(x) = \frac{x}{1-\rho}\,. \tag{2.1}$$

The second moment of the distribution of $S(x)$ has been obtained by Yashkov [11] and Ott [6]:

$$E\,S^2(x) = \frac{x^2}{(1-\rho)^2} + \frac{2}{(1-\rho)^2} \int_{t=0}^{x} (x-t)(1-R(t))\,dt\,, \tag{2.2}$$

where $R(t)$ represents the waiting time distribution for the M/G/1 first come first served (FCFS) queue with service time distribution $B(\cdot)$,

$$R(t) = (1-\rho)\sum_{n=0}^{\infty} \rho^n F^{n^*}(t)\,, \tag{2.3}$$

$$F(t) = \frac{1}{\beta}\int_{u=0}^{t} (1-B(u))\,du\,.$$

Note, for the waiting time distribution $R(t)$ in (2.2), that $1-R(t) \leqslant 1-R(0) = \rho$, $t \geqslant 0$. Hence,

$$E\,S^2(x) \leqslant \frac{1+\rho}{(1-\rho)^2}x^2\,. \tag{2.4}$$

A lower bound for $E\,S^2(x)$ follows immediately from (2.1) and Schwartz' inequality (see also (2.2)),

$$E\ S^2(x) \geqslant \frac{x^2}{(1-\rho)^2}\ . \tag{2.5}$$

So,

$$\frac{x^2}{(1-\rho)^2} \leqslant E\ S^2(x) \leqslant \frac{1+\rho}{(1-\rho)^2}x^2\ . \tag{2.6}$$

Note that the upper bound is $100\rho\%$ higher than the lower bound and that these bounds depend only on the first moment of the service time distribution; this supports a certain robustness of $E\ S^2(x)$ for the service time distribution.

The heavy traffic behaviour of $E\ S^2(x)$ can be derived from (2.2) by noting that the heavy traffic behaviour of the waiting time distribution for the M/G/1 FCFS queue is, for $\rho \rightarrow 1$, negative exponential, i.e. (see Cohen [3])

$$R(\frac{t}{1-\rho}) \sim 1-e^{-t/d}\ , \qquad \text{for } \rho \rightarrow 1\ , \tag{2.7}$$

where $d = \frac{1}{2}\rho\beta_2/\beta$.

Substituting (2.7) into (2.2) yields

$$E\ S^2(x) \sim \frac{1+\rho}{(1-\rho)^2}x^2\ , \qquad \text{for } \rho \rightarrow 1\ . \tag{2.8}$$

Other asymptotic results apply to the case that the required service time x of the tagged customer becomes either very small or very large. In [1] it is shown that

$$var(S(x)) = E\ S^2(x) - (E\ S(x))^2 \sim \frac{\rho}{(1-\rho)^2}x^2 - \frac{\rho}{3\beta(1-\rho)}x^3\ , \qquad \text{for } x \rightarrow 0\ . \tag{2.9}$$

The x^2-term in this formula has been obtained before by Yashkov [12].
The quadratic behaviour of the sojourn time variance, $var(S(x))$, for $x \rightarrow 0$ contrasts with the linear behaviour for large x (see Yashkov [12], and [1]):

$$var(S(x)) \sim \frac{\rho\beta_2}{\beta(1-\rho)^2}x - \frac{1}{(1-\rho)^4}[\frac{1}{2}\rho^2\frac{\beta_2^2}{\beta^2} + \frac{1}{3}\rho\frac{\beta_3}{\beta}(1-\rho)]\ , \qquad \text{for } x \rightarrow \infty\ . \tag{2.10}$$

For exponential and deterministic service times, simple, explicit expressions for $E\ S^2(x)$ are known, see e.g. Ott [6]. For future use we state these expressions:

$$E\ S^2(x)_{EXP} = \frac{2\rho\beta}{(1-\rho)^3}x - \frac{2\rho\beta^2}{(1-\rho)^4}(1-e^{-x(1-\rho)/\beta})\ , \qquad x \geqslant 0\ , \tag{2.11}$$

$$E\,S^2(x)_{DET} = 2\frac{x^2}{(1-\rho)^2} - \frac{2\beta^2}{\rho^2(1-\rho)}(e^{\rho x/\beta}-1-\rho x/\beta)\,, \qquad 0\leqslant x\leqslant\beta\,. \tag{2.12}$$

Let S be the unconditional sojourn time, i.e.

$$Pr\{S\leqslant t\} = \int_{x=0}^{\infty} Pr\{S(x)\leqslant t\}dB(x)\,, \qquad t\geqslant 0\,. \tag{2.13}$$

From the above results for $S(x)$ it follows immediately that

$$E\,S = \frac{\beta}{1-\rho}\,, \tag{2.14}$$

$$\frac{\beta_2}{(1-\rho)^2} \leqslant E\,S^2 \leqslant \frac{1+\rho}{(1-\rho)^2}\beta_2\,, \tag{2.15}$$

$$E\,S^2 \sim \frac{1+\rho}{(1-\rho)^2}\beta_2\,, \qquad \text{for } \rho\to 1\,. \tag{2.16}$$

For exponential and deterministic service times,

$$E\,S^2{}_{EXP} = (1+\frac{2+\rho}{2-\rho})\frac{\beta^2}{(1-\rho)^2}\,, \tag{2.17}$$

$$E\,S^2{}_{DET} = \frac{2\beta^2}{(1-\rho)^2} - \frac{2\beta^2}{\rho^2(1-\rho)}(e^{\rho}-1-\rho)\,. \tag{2.18}$$

Remark. (2.15) implies that, for the M/G/1 PS queue, the dependence of $E\,S^2$ on the third moment of the service time distribution is limited. This should be contrasted with the behaviour of the second moment of the sojourn time distribution for the M/G/1 FCFS queue. For the FCFS discipline it depends linearly on the third moment of the service time distribution.

In the following sections the above results are exploited to develop simple approximations for $E\,S^2(x)$ and $E\,S^2$. We present extensive tables comparing the approximations with exact values. The service time distributions which we have chosen to test the approximations are:

- exponential distribution
- deterministic distribution
- k-stage Erlang distribution (E_k)
- two-stage hyper exponential distribution (H_2), in particular
  H_2 with balanced means (H_2^{BM}), and
  H_2 with gamma normalization (H_2^{GN})
- two-stage Coxian distribution (C_2)
- three-stage hyper exponential distribution (H_3)

These types of service time distributions are often used for practical applications in queueing theory, see Tijms [8] and Whitt [9,10].
In practice service times are often characterized by the mean, β, and the squared coefficient of variation, cv, defined by

$$cv = \frac{\sigma^2}{\beta^2},$$

where σ^2 denotes the service time variance, see Tijms [8]. In the rest of this paper we shall use cv rather than σ^2 to characterize the variability of the service times.
The H_2^{BM} and H_2^{GN} distributions have been introduced to reduce the number of parameters of the H_2 distribution, see Tijms [8]. The H_2^{BM} and H_2^{GN} distributions are uniquely determined by their first two moments. In particular, the H_2^{GN} distribution with mean β and $cv \geqslant 1$ has the same third moment as the gamma distribution with mean β and squared coefficient of variation cv. In Section 4 the class of H_2 distributions will be considered in more detail.

The tables presented at the end of the paper contain relative errors of the approximations for various service time distributions. The relative error is defined as

$$100\% \ \frac{\textit{approximation result} - \textit{exact result}}{\textit{exact result}}.$$

The exact values of $E\ S^2(x)$ and $E\ S^2$ have been obtained from the formulas derived in [1]. For H_k and C_k (and E_k) service time distributions these formulas require the roots of a polynomial of degree k and the solution of a set of k linear equations. Even for the case $k=2$, the resulting expressions are very large and complicated and do not give much insight into the influence of the parameters.

3. Approximation of $E\ S^2(x)$

In this section we show that the asymptotic result (2.9) yields a good approximation for $E\ S^2(x)$ for an important range of x-values.
We define (cf. (2.9)),

$$E\ S^2(x)_{APPX} = \frac{1+\rho}{(1-\rho)^2}x^2 - \frac{\rho}{3\beta(1-\rho)}x^3. \tag{3.1}$$

Note that $E\ S^2(x)_{APPX}$ satisfies the heavy traffic behaviour of $E\ S^2(x)$ (see (2.8)) and that $E\ S^2(x)_{APPX}$ is smaller than the upper bound of $E\ S^2(x)$ given by (2.4). Approximation $E\ S^2(x)_{APPX}$ is independent of the service time distribution apart from its first moment. Obviously it can not be applied for too large values of x because it becomes negative for $x > 3\beta(1+\rho)/(\rho(1-\rho))$. Moreover, assuming that the variance of $S(x)$ is a convex function of x (cf. (2.9) and (2.10)), we may not expect that $E\ S^2(x)_{APPX}$ is a good approximation for $x > x_1$, where $x_1 = \beta/(1-\rho) = E\ S$ is the point of inflection of (cf. (2.9))

$$f(x) = \frac{\rho}{(1-\rho)^2}x^2 - \frac{\rho}{3\beta(1-\rho)}x^3. \tag{3.2}$$

For $x<x_1$, $E\ S^2(x)_{APPX}$ is within the bounds of $E\ S^2(x)$ given by (2.6).

In Table 1 approximation results are compared with exact results for a number of different values of x ($x=\frac{1}{2}\beta, \beta, \frac{3}{2}\beta, 2\beta, \frac{\beta/2}{1-\rho}, \frac{\beta}{1-\rho}$) and different service time distributions. For each of these cases ρ varies from 0.1 to 0.9. For the sake of clarity only the relative approximation errors are given. It appears that for most cases the relative approximation errors are negative. As expected, the approximation becomes less accurate when x grows. For $0\leqslant x\leqslant\beta$ the relative errors are less than 2.34% in absolute value. For $0\leqslant x\leqslant 2\beta$ the maximum relative error is 6.56%. When x remains constant the maximum errors occur for $\rho\approx 0.3$. For $x=\beta/(2(1-\rho))$ and $x=\beta/(1-\rho)$ the relative errors tend to increase when ρ grows. For $x=\beta/(1-\rho)$ the maximum error is 11.29%. It is seen from the results for different service time distributions that the accuracy of the approximation tends to decrease when cv becomes larger.

Remark. In [2] we have derived an approximation for $E\ S^2(x)$ for the whole range of possible x-values ($x\geqslant 0$) by appropriately combining the two asymptotic formulas (2.9) and (2.10). The idea is as follows. Two values x_1 and x_2 are determined, such that for $x\leqslant x_1$ (2.9) yields a good approximation and for $x\geqslant x_2$ (2.10) yields a good approximation. For $x_1\leqslant x\leqslant x_2$, $E\ S^2(x)$ is approximated by the term $x^2/(1-\rho)^2$ plus a linear function of x, cf. (2.10). We took $x_1=\beta/(1-\rho)$. Details about the determination of x_2 are given in [2]. The approximation yields reasonably good results for service time distributions with cv not too large ($cv\leqslant 4$). For these cases we found relative errors which are typically less than 10%.

4. Approximation of $E\ S^2$

In this section we propose some approximations for the second moment of the unconditional sojourn time S. First we derive a very simple approximation which is based on the exact formula of $E\ S^2$ for exponentially distributed service times. This approximation uses only the first two moments of the service time distribution. Next it is shown how this simple approximation can be improved. For that purpose we distinguish between models with a service time squared coefficient of variation cv between zero and one, and models with cv larger than one. In the latter case also the third moment of the service time distribution is taken into account.

4.1. Simple approximation

It follows from (2.15) that an approximation $E\ S^2{}_{APP}$ of $E\ S^2$, which satisfies

$$\frac{\beta_2}{(1-\rho)^2} \leqslant E\ S^2{}_{APP} \leqslant \frac{1+\rho}{(1-\rho)^2}\beta_2\,, \tag{4.1}$$

yields relative errors which are bounded by $100\rho\%$ in absolute value. This observation and the relations for $E\ S^2$ given in Section 2 support the idea to derive an approximation for $E\ S^2$ which is based only on the first two moments of the service time distribution. We start with the exact formulas (2.17) and (2.18) for the case of exponential and deterministic service times. These formulas can be rewritten as follows:

$$E\ S^2{}_{EXP} = (1+\frac{\rho}{2-\rho})\frac{\beta_2}{(1-\rho)^2} = (1+\frac{1}{2}\rho+\frac{1}{4}\rho^2+\frac{1}{8}\rho^3+...)\frac{\beta_2}{(1-\rho)^2}\,, \tag{4.2}$$

$$E\,S^2{}_{DET} = 2(1-\frac{(1-\rho)}{\rho^2}(e^{\rho}-1-\rho))\frac{\beta_2}{(1-\rho)^2} \tag{4.3}$$

$$= (1+\frac{2}{3}\rho+\frac{1}{4}\rho^2+\frac{1}{15}\rho^3+...)\frac{\beta_2}{(1-\rho)^2}\,.$$

(4.2) and (4.3) suggest that $E\,S^2$ is almost linear in β_2. This observation and relation (2.15) lead to an approximation, $E\,S^2{}_{APP}$ for $E\,S^2$ which reads as follows:

$$E\,S^2{}_{APP} = \alpha\frac{1+\rho}{(1-\rho)^2}\beta_2 + (1-\alpha)\frac{\beta_2}{(1-\rho)^2}\,,$$

with $0 \leqslant \alpha \leqslant 1$.

To determine a suitable choice of the weight factor α we shall require that the approximation is exact for exponential service times. It is easily seen from (4.2) that this requirement yields $\alpha = 1/(2-\rho)$. So, we propose

$$E\,S^2{}_{APP} = (1+\frac{\rho}{2-\rho})\frac{\beta_2}{(1-\rho)^2}\,. \tag{4.4}$$

Note that this approximation has the following appealing properties.

Approximation properties
(1) The approximation is exact for exponentially distributed service times.
(2) The approximation yields values between the lower and upper bound of $E\,S^2$ given by (2.15).
(3) The approximation satisfies the heavy traffic behaviour of $E\,S^2$ (see (2.16)).
(4) The approximation yields the exact value of $E\,S^2$ for $\rho=0$:
$E\,S^2{}_{APP} = E\,S^2 = \beta_2\,, \quad$ for $\rho=0$.

The approximation results for the test set of service time distributions and traffic intensities are given in Table 2. It appears that the approximation yields reasonably good results. In all tested cases the relative approximation error is smaller than 9%. In particular for service time distributions with cv close to one ($0 \leqslant cv \leqslant 2$) the relative errors are less than 5.17%. Obviously this is due to the fact that the approximation is exact for exponential service times. For larger values of cv ($cv>2$) the approximation becomes worse. It is noticeable that the approximation is significantly better for the H_2 distribution with gamma normalization (H_2^{GN}) than for the H_2 distribution with balanced means (H_2^{BM}). In the next subsection we shall show that this is due to the influence of the third moment of the service time distribution on $E\,S^2$.

4.2 Detailed approximations

The simple approximation (4.4) tends to be less accurate if the squared coefficient of variation of the service time distribution becomes larger. In this subsection we shall develop two new approximations, one for the case that $0 \leqslant cv \leqslant 1$ ($APP\,1$) and one for the case that $cv \geqslant 1$ ($APP\,2$). $APP\,1$ is obtained by appropriately weighing the exact values of $E\,S^2$ for exponential and deterministic service times. $APP\,2$ is based on simple exact expressions for $E\,S^2$ for two classes of H_2 distributions.

The case $0 \leq cv \leq 1$

For service time distributions with $0 \leq cv \leq 1$ we propose a refinement of $E\,S^2{}_{APP}$, $E\,S^2{}_{APP1}$, which is based on the exact formulas (2.17) and (2.18) for exponential and deterministic service times. For $0 \leq cv \leq 1$ it is natural to approximate $E\,S^2$ by a linear interpolation between $E\,S^2{}_{EXP}$ and $E\,S^2{}_{DET}$:

$$E\,S^2{}_{APP1} = cvE\,S^2{}_{EXP} + (1-cv)E\,S^2{}_{DET}\,. \tag{4.5}$$

So,

$$E\,S^2{}_{APP1} = cv(1+\frac{2+\rho}{2-\rho})\frac{\beta^2}{(1-\rho)^2} + (1-cv)(\frac{2\beta^2}{(1-\rho)^2} - \frac{2\beta^2}{\rho^2(1-\rho)}(e^{\rho}-1-\rho))\,. \tag{4.6}$$

Additional to possessing the approximation properties of $ES^2{}_{APP}$ given in Subsection 4.1, $ES^2{}_{APP1}$ yields exact results for deterministic service times.

We tested $APP1$ for a number of service time distributions: E_4 ($cv=0.25$), E_3 ($cv=0.33$), E_2 ($cv=0.50$), C_2 ($cv=0.75$, 0.92). The results are given in Table 3. It appears, as expected, that $APP1$ is much more accurate than the simple approximation APP proposed in the previous subsection. The relative error of $APP1$ is less than 0.50% in all (test) cases.

The case $cv \geq 1$

For service time distributions with $cv \geq 1$ we shall develop an approximation, $APP2$, for $E\,S^2$ which is based on simple exact formulas for two classes of extreme H_2 distributions. This approximation contains the first three moments of the service time distribution.

We start with recalling some characteristics of the class of H_2 distributions. The H_2 distribution function is given by

$$B_{H_2}(t) = \alpha(1-e^{-t/\tilde{\beta}^{(1)}}) + (1-\alpha)(1-e^{-t/\tilde{\beta}^{(2)}})\,, \tag{4.7}$$

where $0 \leq \alpha \leq 1$, $0 \leq \tilde{\beta}^{(1)} \leq \tilde{\beta}^{(2)}$.

So, there are three parameters. Given the mean $\beta = \alpha\tilde{\beta}^{(1)} + (1-\alpha)\tilde{\beta}^{(2)}$ and $cv \geq 1$ there is thus one remaining degree of freedom, r, defined by

$$r = \frac{\alpha\tilde{\beta}^{(1)}}{\alpha\tilde{\beta}^{(1)} + (1-\alpha)\tilde{\beta}^{(2)}}\,.$$

$r = 1/2$ yields the class of H_2 distributions with balanced means (H_2^{BM}).

Obviously, if β and cv are given, r determines the third moment, β_3, of the H_2 distribution. For fixed β and β_2 (cv), the smallest possible value of β_3 is obtained for $r=0$. In that case $\beta_3 = \frac{3}{2}\beta_2^2/\beta$. For $r \to 1$, $\beta_3 \to \infty$, (see Whitt [9,10]).

Our numerical experience with respect to H_2 distributions indicates that $E\ S^2$ becomes smaller when β_3 grows (β and β_2 constant). So (cf. (2.15)), we expect that $E\ S^2{}_{H_2}$ has a limit for $\beta_3 \to \infty$, β and β_2 fixed. From the formulas for $E\ S^2$ given in [1] it is found that, for β and β_2 fixed,

$$E\ S^2{}_{H_2^{r=1}} = \lim_{\beta_3 \to \infty (r \to 1)} E\ S^2{}_{H_2} = \frac{\beta_2}{(1-\rho)^2} + \frac{2\rho}{2-\rho}\frac{\beta^2}{(1-\rho)^2}\,, \tag{4.8}$$

and, for $\beta_3 = \frac{3}{2}\beta_2^2/\beta$ $(r=0)$,

$$E\ S^2{}_{H_2^{r=0}} = (1+\frac{\rho}{2-\rho})\frac{\beta_2}{(1-\rho)^2}\,. \tag{4.9}$$

It is easily seen that, for $cv \geqslant 1$,

$$E\ S^2{}_{H_2^{r=1}} \leqslant E\ S^2{}_{H_2^{r=0}}\,, \qquad 0 \leqslant \rho \leqslant 1\,. \tag{4.10}$$

In (4.10), equality holds if $cv=1$ ($\beta_2 = 2\beta^2$), i.e. if the service times are exponentially distributed. Note that (cf. (4.4)),

$$E\ S^2{}_{H_2^{r=0}} = E\ S^2{}_{APP}\,.$$

This explains why approximation APP yields better results for H_2^{GN} service time distributions (with a relatively small third moment) than for H_2^{BM} service time distributions.

Now we introduce two approximation assumptions to extend the above results with respect to H_2 distributions to general service time distributions.

Assumption 1: $E\ S^2$ depends only on the first three moments (β, β_2, β_3) of the service time distribution.

Assumption 2: $E\ S^2$ decreases if β_3 grows (β and β_2 fixed).

Under these assumptions it follows from (4.8) and (4.9) that, for $cv \geqslant 1$, $\beta_3 \geqslant \frac{3}{2}\beta_2^2/\beta$,

$$\frac{\beta_2}{(1-\rho)^2} + \frac{2\rho}{2-\rho}\frac{\beta^2}{(1-\rho)^2} \leqslant E\ S^2 \leqslant (1+\frac{\rho}{2-\rho})\frac{\beta_2}{(1-\rho)^2}\,. \tag{4.11}$$

(4.8), (4.9) and (4.11) suggest an approximation, $APP2$, for $E\ S^2$ which reads as follows:

$$E\ S^2{}_{APP2} = \gamma(1+\frac{\rho}{2-\rho})\frac{\beta_2}{(1-\rho)^2} + (1-\gamma)(\frac{\beta_2}{(1-\rho)^2} + \frac{2\rho}{2-\rho}\frac{\beta^2}{(1-\rho)^2})\,, \tag{4.12}$$

where $\gamma := \gamma(\rho,\beta,\beta_2,\beta_3)$, $0 \leqslant \gamma \leqslant 1$.

The choice of the weight factor γ will be partially determined by the approximation properties listed below (4.4). Besides these four properties we require that

(5) for β, β_2 fixed,

$$\lim_{\beta_3 \to \infty} E\, S^2{}_{APP2} = \frac{\beta_2}{(1-\rho)^2} + \frac{2\rho}{2-\rho} \frac{\beta^2}{(1-\rho)^2},$$

(6) for $\beta_3 = \frac{3}{2}\beta_2^2/\beta$,

$$E\, S^2{}_{APP2} = (1+\frac{\rho}{2-\rho})\frac{\beta_2}{(1-\rho)^2}.$$

Note, that, without any further specification of γ, $APP2$ satisfies the approximation properties (1) and (4). Considering the other required properties ((2), (3), (5) and (6)) it is natural to choose γ as follows,

$$\gamma = \frac{1}{1+\gamma_1(\beta_3 - \frac{3}{2}\beta_2^2/\beta)(1-\rho)}, \tag{4.13}$$

where γ_1 represents the relative influence of β_3 on $E\, S^2$. γ_1 remains to be specified. We assume that γ_1 depends only on β and β_2. Note that γ_1 has to be chosen such that γ is dimensionless. The most obvious choices are $\gamma_1 = 1/\beta^3$ or $\gamma_1 = 1/(\beta\beta_2)$. For both cases we compared approximation results with exact results. Our test set consisted of H_2 service time distributions with cv ranging from 1 to 20. For each value of cv a large number of β_3 values was considered. It appeared that the choice $\gamma_1 = 1/(\beta\beta_2)$ yields much better results than $\gamma_1 = 1/\beta^3$. However, in most cases the choice $\gamma_1 = 1/(\beta\beta_2)$ underestimated $E\, S^2$. In particular for larger values of cv the approximation results became worse. Extensive tests of the approximation for some variants of $\gamma_1 = 1/(\beta\beta_2)$ led to a modification which yields remarkably accurate results:

$$\gamma_1 = \frac{1}{(cv-1)} \frac{1}{\beta\beta_2}.$$

So, the ultimate approximation formula is given by (4.12), with

$$\gamma = \frac{1}{1+(1-\rho)(\frac{\beta_3}{\beta\beta_2} - \frac{3}{2}\frac{\beta_2}{\beta^2})/(\frac{\beta_2}{\beta^2}-2)}. \tag{4.14}$$

It is seen from Table 4 that for H_2^{BM} and H_2^{GN} service time distributions (with $cv = 2, 4, 6$) $APP2$ yields very accurate results with relative errors less than 1%.
Table 5 illustrates the influence of β_3 on $E\, S^2$. This table shows exact values of $E\, S^2$ for a number of H_2 distributions with the same first and second moment but with a different third moment. The traffic intensity varies from 0.1 to 0.95. The relative approximation errors of $APP2$ are indicated below the exact values of $E\, S^2$. As we stated before $E\, S^2$ decreases when β_3 grows.

Note that even for large cv ($cv=10$) the relative approximation errors are less than 1.5%. It may be concluded from Table 5 that the influence of the third moment of the service time distribution on $E\ S^2$ increases when cv grows, cf. (2.15).
In Table 6 $APP2$ is tested for some arbitrarily chosen H_3 and C_2 service time distributions. The relative errors are in all cases less than 1.5%.

Originally, $APP2$ has been developed for service time distributions with $cv \geqslant 1$, $\beta_3 \geqslant \frac{3}{2}\ \beta_2^2/\beta$. For these cases $E\ S^2{}_{APP2}$ can be interpreted as an interpolation formula, see (4.12). Nevertheless, approximation formula (4.12) (together with (4.14)) can be applied to service time distributions with $cv<1$ or $\beta_3 < \frac{3}{2}\ \beta_2^2/\beta$ as well. In Table 7 some results are shown for deterministic, C_2 and E_k service time distributions with $cv<1$. It appears that the accuracy of $APP2$ for these cases is about the same as the accuracy of $APP1$.

5. Conclusions

In this paper we have studied the second moment of the conditional and unconditional sojourn time, $E\ S^2(x)$ and $E\ S^2$, for the M/G/1 processor sharing queue. An upper bound and some asymptotic properties (e.g. the heavy traffic behaviour) have been derived. Based on these properties and on exact expressions for specific service time distributions we developed some simple approximations. The approximations have been compared with exact results for a large number of different service time distributions and a wide range of traffic intensities. We conclude as follows.
- The influence of the third and higher moments of the service time distribution on $E\ S^2(x)$ and $E\ S^2$ is limited. An upper and a lower bound for $E\ S^2(x)$ can be expressed in terms of x (the service demand of a tagged customer) and the traffic intensity ρ, see (2.6). Upper and lower bounds for $E\ S^2$ contain only the second moment of the sojourn time distribution and ρ, see (2.15).
- Approximation $APPX$ for $E\ S^2(x)$, given by (3.1), is based on an asymptotic result for $x \rightarrow 0$ derived in [1]. It depends on the service time distribution only through its first moment. $APPX$ yields reasonably good results for not too large values of x, see Table 1. For $0 \leqslant x \leqslant \beta$ the relative error of the approximation is a few percent. The approximation becomes less accurate when x increases. For $x=2\beta$ the relative errors are typically less than 7%. $APPX$ satisfies the heavy traffic behaviour of $E\ S^2(x)$, see (2.8).
- The approximations for $E\ S^2$, APP, $APP1$ and $APP2$, given by (4.4), (4.6) and (4.12), have been constructed in such a way that they have the following appealing properties:
  * they are exact for exponential service times
  * they yield values between the lower and upper bound of $E\ S^2$
  * they satisfy the heavy traffic behaviour of $E\ S^2$
  * they yield the exact value of $E\ S^2$ for $\rho=0$.

In addition, $APP1$ yields exact results for deterministic service times; $APP2$ is exact for two classes of extreme H_2 distributions.
- Approximation APP is the most simple approximation. It depends on the first two moments of the service time distribution. For not too large values of cv ($cv \leqslant 6$) it yields fairly accurate results, see Table 2. In practical situations APP may be applied as a first order approximation for $E\ S^2$.
- $APP1$, a refinement of APP, has been constructed for service time distributions with $0 \leqslant cv \leqslant 1$. It depends also on the first two moments of the service time distribution. $APP1$ is very accurate.

The relative approximation error is less than 0.38% in all of our examples, see Table 3.
- $APP2$ depends on the first three moments of the service time distribution. It is based on exact formulas of $E\,S^2$ for two classes of extreme H_2 distributions. The details of the construction of $APP2$ are rather heuristic. Nevertheless, it yields remarkably accurate results. $APP2$ has been tested for a large number of different service time distributions with *cv* ranging from 0 to 10, see Tables 4 through 7. In all of these cases the relative error is less than 1.42%.

ACKNOWLEDGEMENT
The author is indebted to Prof. O.J. Boxma and Prof. J.W. Cohen for some useful comments.

REFERENCES

1 Van den Berg, J.L., Boxma, O.J. (1989). Sojourn times in feedback and processor sharing queues. In: *Teletraffic Science for new Cost-Effective Systems, Networks and Services,* ITC 12, ed. M. Bonatti (North-Holland Publ. Cy., Amsterdam).
2 Van den Berg, J.L. (1988). Unpublished work.
3 Cohen, J.W. (1982). The Single Server Queue, 2nd ed. (North-Holland Publ. Cy., Amsterdam).
4 Jaiswal, N.K. (1982). Performance evaluation studies for time sharing computer systems, *Perf. Eval.* **2**, 223-236.
5 Kleinrock, L. (1976). Queueing Systems, Vol. II (Wiley, New York).
6 Ott, T.J. (1984). The sojourn-time distribution in the M/G/1 queue with processor sharing, *J. Appl. Prob.* **21**, 360-378.
7 Schassberger, R. (1984). A new approach to the M/G/1 processor-sharing queue, *Adv. Appl. Prob.* **16**, 202-213.
8 Tijms, H.C. (1986). Stochastic Modeling and Analysis (Wiley, New York).
9 Whitt, W. (1982). Approximating a point process by a renewal process, I: Two basic methods, *Oper. Res.* **30**, 125-147.
10 Whitt, W. (1984). On approximations for queues, III: Mixtures of exponential distributions, *AT&T Bell Lab. Tech. J.* **63**, 163-175.
11 Yashkov, S.F. (1983). A derivation of response time distribution for an M/G/1 processor-sharing queue, *Problems Contr. & Info. Theory* **12**, *133-148.*
12 Yashkov, S.F. (1986). A note on asymptotic estimates of the sojourn time variance in the M/G/1 queue with processor-sharing, *Syst. Anal. Model. Simul.* **3**, 267-269.
13 Yashkov, S.F. (1987). Processor-sharing queues: some progress in analysis, *Queueing Systems* **2**, 1-17.

TABLE 1. Approximation of $E\ S^2(x)$. The table contains relative errors (%) of approximation APPX (given by (3.1)) for various service time distributions.

Service time distribution: H_2^{BM}, $cv=2$.

ρ	$x=\frac{1}{2}\beta$	$x=\beta$	$x=\frac{3}{2}\beta$	$x=2\beta$	$x=\frac{1}{2}\frac{\beta}{1-\rho}$	$x=\frac{\beta}{1-\rho}$
0.1	-0.19	-0.66	-1.35	-2.18	-0.23	-0.80
0.3	-0.32	-1.16	-2.40	-3.94	-0.62	-2.20
0.5	-0.27	-0.99	-2.08	-3.46	-0.99	-3.46
0.7	-0.15	-0.56	-1.19	-1.99	-1.44	-4.80
0.9	-0.04	-0.15	-0.30	-0.51	-2.31	-6.61

Service time distribution: H_2^{BM}, $cv=4$.

ρ	$x=\frac{1}{2}\beta$	$x=\beta$	$x=\frac{3}{2}\beta$	$x=2\beta$	$x=\frac{1}{2}\frac{\beta}{1-\rho}$	$x=\frac{\beta}{1-\rho}$
0.1	-0.22	-0.77	-1.55	-2.46	-0.27	-0.93
0.3	-0.39	-1.39	-2.83	-4.58	-0.76	-2.61
0.5	-0.35	-1.26	-2.58	-4.23	-1.26	-4.23
0.7	-0.21	-0.77	-1.60	-2.64	-1.93	-6.11
0.9	-0.06	-0.23	-0.48	-0.79	-3.39	-8.98

Service time distribution: H_2^{BM}, $cv=6$.

ρ	$x=\frac{1}{2}\beta$	$x=\beta$	$x=\frac{3}{2}\beta$	$x=2\beta$	$x=\frac{1}{2}\frac{\beta}{1-\rho}$	$x=\frac{\beta}{1-\rho}$
0.1	-0.24	-0.82	-1.63	-2.59	-0.29	-0.99
0.3	-0.42	-1.50	-3.02	-4.86	-0.81	-2.78
0.5	-0.38	-1.37	-2.80	-4.56	-1.37	-4.56
0.7	-0.24	-0.86	-1.78	-2.92	-2.14	-6.67
0.9	-0.07	-0.27	-0.55	-0.91	-3.85	-10.01

Service time distribution: exponential.

ρ	$x=\frac{1}{2}\beta$	$x=\beta$	$x=\frac{3}{2}\beta$	$x=2\beta$	$x=\frac{1}{2}\frac{\beta}{1-\rho}$	$x=\frac{\beta}{1-\rho}$
0.1	-0.14	-0.53	-1.11	-1.85	-0.17	-0.64
0.3	-0.23	-0.86	-1.86	-3.17	-0.45	-1.70
0.5	-0.17	-0.66	-1.46	-2.53	-0.66	-2.53
0.7	-0.08	-0.30	-0.67	-1.19	-0.83	-3.19
0.9	-0.01	-0.04	-0.09	-0.16	-0.96	-3.74

Service time distribution: H_2^{GN}, $cv=2$.

ρ	$x=\frac{1}{2}\beta$	$x=\beta$	$x=\frac{3}{2}\beta$	$x=2\beta$	$x=\frac{1}{2}\frac{\beta}{1-\rho}$	$x=\frac{\beta}{1-\rho}$
0.1	-0.25	-0.83	-1.58	-2.45	-0.30	-0.98
0.3	-0.45	-1.51	-2.91	-4.56	-0.85	-2.69
0.5	-0.41	-1.38	-2.68	-4.21	-1.38	-4.21
0.7	-0.26	-0.87	-1.69	-2.64	-1.99	-5.66
0.9	-0.08	-0.27	-0.52	-0.79	-2.82	-7.15

Service time distribution: H_2^{GN}, $cv=4$.

ρ	$x=\frac{1}{2}\beta$	$x=\beta$	$x=\frac{3}{2}\beta$	$x=2\beta$	$x=\frac{1}{2}\frac{\beta}{1-\rho}$	$x=\frac{\beta}{1-\rho}$
0.1	-0.34	-1.07	-1.98	-2.98	-0.41	-1.26
0.3	-0.64	-2.03	-3.79	-5.75	-1.17	-3.52
0.5	-0.61	-1.96	-3.68	-5.61	-1.96	-5.61
0.7	-0.41	-1.33	-2.50	-3.81	-2.93	-7.72
0.9	-0.14	-0.46	-0.86	-1.30	-4.33	-10.02

Service time distribution: H_2^{GN}, $cv=6$.

ρ	$x=\frac{1}{2}\beta$	$x=\beta$	$x=\frac{3}{2}\beta$	$x=2\beta$	$x=\frac{1}{2}\frac{\beta}{1-\rho}$	$x=\frac{\beta}{1-\rho}$
0.1	-0.38	-1.18	-2.16	-3.23	-0.45	-1.39
0.3	-0.72	-2.26	-4.17	-6.29	-1.31	-3.88
0.5	-0.69	-2.21	-4.11	-6.22	-2.21	-6.21
0.7	-0.48	-1.53	-2.85	-4.32	-3.33	-8.63
0.9	-0.17	-0.54	-1.00	-1.52	-4.98	-11.29

Service time distribution: E_2.

ρ	$x=\frac{1}{2}\beta$	$x=\beta$	$x=\frac{3}{2}\beta$	$x=2\beta$	$x=\frac{1}{2}\frac{\beta}{1-\rho}$	$x=\frac{\beta}{1-\rho}$
0.1	-0.03	-0.24	-0.68	-1.32	-0.05	-0.32
0.3	-0.00	-0.24	-0.86	-1.88	-0.06	-0.75
0.5	0.07	0.05	-0.26	-0.91	0.05	-0.91
0.7	0.11	0.27	0.33	0.24	0.32	-0.84
0.9	0.06	0.19	0.34	0.48	0.85	-0.53

TABLE 2. Approximation of $E\,S^2$. The table contains relative errors (%) of approximation APP (given by (4.4)) for various service time distributions.

Service time distribution: H_2^{BM}.

ρ	$cv=2$	$cv=4$	$cv=6$
0.10	0.82	1.48	1.77
0.30	2.30	4.22	5.07
0.50	3.38	6.31	7.63
0.70	3.63	6.87	8.37
0.90	2.08	4.00	4.91
0.95	1.19	2.30	2.83

Service time distribution: H_2^{GN}.

ρ	$cv=2$	$cv=4$	$cv=6$
0.10	0.40	0.73	0.86
0.30	1.08	1.97	2.36
0.50	1.52	2.78	3.34
0.70	1.53	2.83	3.40
0.90	0.81	1.51	1.82
0.95	0.45	0.84	1.02

Service time distributions with $cv<1$.

ρ	DET	E_2	C_2^*
0.10	-1.55	-0.40	-0.22
0.30	-3.92	-1.06	-0.60
0.50	-5.11	-1.46	-0.83
0.70	-4.79	-1.45	-0.84
0.90	-2.34	-0.75	-0.45
0.95	-1.29	-0.42	-0.25

$^*cv=0.75$

TABLE 3. Approximation of $E\,S^2$. The table contains relative errors (%) of approximation APP1 (given by (4.6)) for various service time distributions with $cv<1$.

ρ	E_4	E_3	E_2	$C_2^{(1)}$	$C_2^{(2)}$
0.10	0.18	0.14	0.13	0.00	0.00
0.30	0.35	0.35	0.28	0.00	-0.02
0.50	0.38	0.38	0.32	-0.07	-0.09
0.70	0.25	0.24	0.21	-0.13	-0.11
0.90	0.05	-0.01	0.04	-0.11	-0.07
0.95	0.02	-0.01	0.02	-0.08	-0.04

$C_2^{(1)}$: $cv=0.75$, $C_2^{(2)}$: $cv=0.92$.

TABLE 4. Approximation of $E\,S^2$. The table contains relative errors (%) of approximation APP2 (given by (4.12) and (4.14)) for various service time distributions with $cv\geq 1$, $\beta_3 \geq \frac{3}{2}\beta_2^2/\beta$.

Service time distribution: H_2^{BM}.

ρ	$cv=2$	$cv=4$	$cv=6$
0.10	-0.15	-0.27	-0.32
0.30	-0.32	-0.58	-0.69
0.50	-0.31	-0.53	-0.61
0.70	-0.12	-0.09	-0.04
0.90	0.08	0.34	0.51
0.95	0.07	0.26	0.39

Service time distribution: H_2^{GN}.

ρ	$cv=2$	$cv=4$	$cv=6$
0.10	-0.12	-0.21	-0.25
0.30	-0.23	-0.41	-0.48
0.50	-0.17	-0.30	-0.35
0.70	-0.01	0.01	0.03
0.90	0.09	0.20	0.26
0.95	0.07	0.14	0.18

TABLE 5. The influence of the third moment of the service time distribution (β_3) on $E\ S^2$. In the table the exact values of $E\ S^2$ are given. The relative approximation errors (%) of $APP2$ are indicated in parentheses below the exact values of $E\ S^2$.

H_2 service time distributions with $\beta = 1$, $cv = 4$.

β_3	$\rho=0.10$	$\rho=0.30$	$\rho=0.50$	$\rho=0.70$	$\rho=0.90$	$\rho=0.95$
37.500*	6.498 (0.00)	12.00 (0.00)	26.67 (0.00)	85.47 (0.00)	909.1 (0.00)	3810 (0.00)
49.084	6.434 (-0.25)	11.68 (-0.50)	25.65 (-0.40)	82.10 (-0.01)	889.2 (0.26)	3762 (0.19)
68.329	6.388 (-0.26)	11.43 (-0.58)	24.78 (-0.56)	78.74 (-0.15)	864.3 (0.34)	3698 (0.29)
105.42	6.353 (-0.19)	11.24 (-0.47)	24.02 (-0.54)	75.35 (-0.28)	830.6 (0.24)	3609 (0.26)
190.02	6.329 (-0.11)	11.09 (-0.29)	23.41 (-0.38)	72.17 (-0.31)	785.3 (0.01)	3441 (0.08)
310.86	6.318 (-0.07)	11.02 (-0.19)	23.12 (-0.26)	70.47 (-0.25)	751.6 (-0.13)	3295 (-0.09)
716.53	6.309 (-0.03)	10.97 (-0.08)	22.86 (-0.12)	68.85 (-0.14)	709.3 (-0.17)	3063 (-0.22)
1391.8	6.306 (-0.02)	10.95 (-0.04)	22.77 (-0.06)	68.21 (-0.08)	689.0 (-0.13)	2927 (-0.20)
2291.9	6.305 (-0.01)	10.94 (-0.03)	22.73 (-0.04)	67.94 (-0.05)	679.6 (-0.09)	2856 (-0.16)
4541.9	6.304 (-0.00)	10.93 (-0.01)	22.70 (-0.02)	67.74 (-0.03)	671.9 (-0.05)	2794 (-0.10)
∞	6.303 (0.00)	10.92 (0.00)	22.67 (0.00)	67.52 (0.00)	663.6 (0.00)	2724 (0.00)

* $\beta_3 = \frac{3}{2}\beta_2^2/\beta$

H_2 service time distributions with $\beta = 1$, $cv = 10$.

β_3	$\rho=0.10$	$\rho=0.30$	$\rho=0.50$	$\rho=0.70$	$\rho=0.90$	$\rho=0.95$
181.50*	14.29 (0.00)	26.41 (0.00)	58.67 (0.00)	188.0 (0.00)	2000 (0.00)	8381 (0.00)
223.21	14.17 (-0.29)	25.80 (-0.50)	56.78 (-0.35)	181.9 (0.05)	1965 (0.28)	8298 (0.19)
330.00	14.01 (-0.40)	24.95 (-0.80)	53.89 (-0.68)	171.3 (0.03)	1891 (0.70)	8110 (0.53)
502.34	13.90 (-0.31)	24.34 (-0.73)	51.62 (-0.72)	161.6 (-0.04)	1802 (1.00)	7859 (0.86)
710.16	13.84 (-0.24)	24.00 (-0.60)	50.26 (-0.65)	155.0 (-0.09)	1724 (1.13)	7612 (1.09)
1323.2	13.78 (-0.14)	23.61 (-0.37)	48.65 (-0.44)	146.4 (-0.11)	1588 (1.13)	7093 (1.37)
1845.6	13.76 (-0.10)	23.49 (-0.27)	48.11 (-0.34)	143.2 (-0.10)	1523 (1.04)	6797 (1.42)
4162.2	13.73 (-0.05)	23.31 (-0.13)	47.31 (-0.17)	138.4 (-0.06)	1401 (0.69)	6129 (1.26)
12263	13.72 (-0.02)	23.22 (-0.04)	46.89 (-0.06)	135.7 (-0.03)	1315 (0.30)	5543 (0.71)
30488	13.71 (-0.01)	23.19 (-0.02)	46.76 (-0.02)	134.8 (-0.01)	1285 (0.13)	5305 (0.34)
∞	13.71 (0.00)	23.17 (0.00)	46.67 (0.00)	134.2 (0.00)	1264 (0.00)	5124 (0.00)

* $\beta_3 = \frac{3}{2}\beta_2^2/\beta$

TABLE 6. Relative approximation errors (%) of $APP2$ for H_3 and C_2 service time distributions.

ρ	$H_3^{(1)}$	$H_3^{(2)}$	$C_2^{(1)}$	$C_2^{(2)}$	$C_2^{(3)}$
0.10	-0.20	-0.29	-0.15	-0.30	-0.32
0.30	-0.54	-0.71	-0.31	-0.63	-0.63
0.50	-0.65	-0.82	-0.25	-0.54	-0.48
0.70	-0.44	-0.48	-0.04	-0.04	0.05
0.90	-0.00	0.20	0.12	0.40	0.39
0.95	0.00	0.10	0.09	0.31	0.28

$H_3^{(1)}$: $cv=2.778$, $\beta_3=40.963$.
$H_3^{(2)}$: $cv=4.130$, $\beta_3=85.622$.
$C_2^{(1)}$: $cv=2.200$, $\beta_3=18.240$.
$C_2^{(2)}$: $cv=5.000$, $\beta_3=84.000$.
$C_2^{(3)}$: $cv=8.556$, $\beta_3=187.33$.
In all cases: $\beta=1$.

TABLE 7. Relative approximation errors (%) of $APP2$ for various service time distributions with $cv \leq 1$ and $\beta_3 \leq \frac{3}{2}\beta_2^2/\beta$.

ρ	DET	E_4	E_2	C_2^*
0.10	-0.02	0.16	0.12	0.06
0.30	-0.18	0.23	0.22	0.11
0.50	-0.36	0.16	0.18	0.10
0.70	-0.44	-0.02	0.05	0.04
0.90	-0.25	-0.10	-0.04	-0.02
0.95	-0.13	-0.08	-0.03	-0.02

* $cv=0.75$

SERIELLE DATENBUSSYSTEME IM KRAFTFAHRZEUG

Thomas Raith
Daimler-Benz AG
Forschung und Technik

7000 Stuttgart 60

Der Einsatz elektronischer Systeme mit zunächst eng abgegrenzten Einzelfunktionen hat zum Ziel, den Fahrer bei der Führung seines Fahrzeuges wirksam zu unterstützen, zu entlasten und störende bzw. unerwünschte Einflüsse von ihm fern zu halten. Das wachsendes Leistungspotential der Elektronik erfordert ein gesamtheitliches und durchgängiges Systemkonzept der Fahrzeugelektronik. Serielle Datenbussysteme ermöglichen durch Multiplextechniken den Informationsaustausch zwischen räumlich zentral und dezentral angeordneten Komponenten im Fahrzeug, verbunden mit hoher Flexibilität (Nachrüstung oder Weiterentwicklung einzelner Komponenten).
Der Beitrag beschreibt zunächst die Ausgangspunkte, die Ziele und die notwendigen Systemeigenschaften von seriellen Datenbussystemen im Fahrzeug. Auf wesentliche Konzepte der Systemarchitektur und der Standardisierung wird im mittleren Teil eingegangen, bevor im letzten Teil Methoden und Verfahren der Leistungsbeurteilung näher betrachtet werden. Ein kurzer Ausblick über Trends und Weiterentwicklungen bildet den Abschluß.

1. Ausgangspunkte, Ziele und Systemeigenschaften

Die Menge und die Komplexität der Elektronik im Fahrzeug führt zu einen zu Raum- und Verkabelungsproblemen und zum anderem zu Zuverlässigkeitsproblemen durch die steigende Anzahl der Steckverbindungen /FAZ87, MAR87/.
Soll das wachsendes Leistungspotential der Elektronik in Form von Funktionserweiterungen und -optimierungen genutzt werden, so ist ein gesamtheitliches und durchgängiges Systemkonzept der Fahrzeugelektronik Voraussetzung. Die primären Ziele sind dabei:

- Kosteneinsparung,
- Zuverlässigkeitssteigerung,
- und Funktionsoptimiertung.

Grundlegende Systemeigenschaften des Gesamtsystems Fahrzeugelektronik sind:

- die Nutzung latenter Redundanzen zur Steigerung der Verfügbarkeit und der Sicherheit des Gesamtsystems,
- die kostengünstige Realisierung durch Reduktion der Kabelsätze, Mehrfachnutzung von Sensoren und durch den Einsatz von standardisierten Systemkomponenten,
- die Möglichkeit der geräteübergreifenden Regelung und Diagnose durch die Kommunikation zwischen nebenläufigen Regelkreisen, die zum Teil in Echtzeit auf gleiche Aggregate eingreifen können,
- die Realisierung erweiterter Notlauffunktionen bei Ausfall von Teilkomponenten.

Eine Grundvoraussetzung des Systemkonzeptes ist die Vernetzung aller wesentlichen elektronischen Systemkomponenten (Steuergeräte, multifunktionale Sensoren und Stellglieder) durch einen seriellen Datenbus (Bild 1) und damit die Möglichkeit des Datenaustausches im Gesamtsystem /JOH85, BOT86, PHA86, STA86/.
Die Charakteristik des Datenverkehrs ist durch die Übertragung synchroner Daten, die durch die Zykluszeiten der Systemkomponenten bestimmt werden, und durch das Auftreten von seltenen asynchronen Ereignissen, die im allgemeinen durch einen Fahrerwunsch oder Umwelteinflüsse hervorgerufen werden, gekennzeichnet.

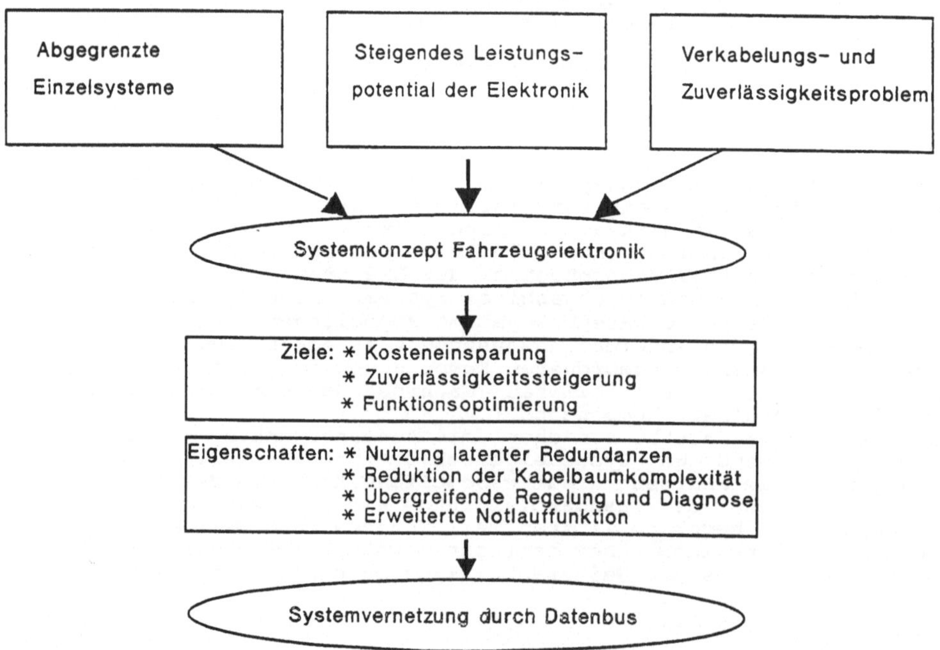

Bild 1: Ausgangspunkte, Ziele, Systemeigenschaften

2. Anforderungen an Datenbussysteme im Kraftfahrzeug

2.1 Allgemeine Netzwerkanforderungen

Durch die Vernetzung (physikalische Verbindung) der Steuergeräte entsteht zunächst das Problem der Verfügbarkeit des Datenbusses. Speziell Funktionen, die sich auf die Verfügbarkeit des Bussystems abstützen, müssen bei dessen Ausfall durch entsprechende (lokale) Notlauffunktionen begrenzt aufrechterhalten werden. Zum Beispiel sollte eine Beleuchtungseinheit als Datenbusstation bei einem Ausfall aus Sicherheitsgründen das Abblendlicht einschalten. Dies setzt eine entsprechende Fehlerlokalisierung und -behandlung auf Systemebene voraus, im einzelnen sollte dann zwischen lokalem Stationsausfall und globalem Ausfall des Datenbussystems unterschieden werden.

Lokale Fehlfunktionen oder Ausfall einer Station im Netzwerk sollten ebenfalls auf Systemebene erkannt werden, um gegebenenfalls erweiterte Notlauffunktionen zu initialisieren. Fällt zum Beispiel im Netzwerk das Geschwindigkeitssignal aus, dann kann im erweiterten Notlauf auf ein redundantes Signal mit geringerer Auflösung (z.B. E-Tacho) zurückgegriffen werden. Redundante oder diversitäre Schaltungsauslegungen zur Steigerung von Zuverlässigkeit und Verfügbarkeit werden aus Kostengründen im wesentlichen nur in Geräten zur Steuerung von sicherheitskritischen Fahrzeugfunktionen eingesetzt.
Um eine zuverlässige Funktion des Bussystems zu gewährleisten, müssen Maßnahmen zur Steigerung der Zuverlässigkeit (z.B. Burn-in der Halbleiterbauelemente, Höherintergration, ect.) und zur Erhöhung der Störsicherheit (z. B. HF-Filter, Datensicherung, Plausibilitätsprüfungen, ect.) getroffen werden.
Die Kostenfrage nimmt eine zentrale Stellung im Kraftfahrzeug ein. Steigende Kosten des Gesamtsystems, hervorgerufen durch den Einsatz eines Datenbusses, müssen durch höhere Funktionalität der Fahrzeugelektronik gerechtfertigt werden, bzw. durch Einsparungen an redundant vorhandener Sensorik und Aktuatorik oder durch Optimierung von Fertigungsabläufen (Reduktion der Kabelbaumkomplexität) kompensiert werden.
Weiterhin sollte das Bussystem einen hohen Grad an Modularität besitzen, um flexibel an die unterschiedlichen Ausbauanforderungen angepaßt werden zu können. Dadurch wird in der Fertigung die Typenvielfalt der Fahrzeuge unterstützt und ein Optimum an Montage- und Servicefreundlichkeit bereitgestellt.
Die Modularität und Flexibilität wird wesentlich unterstützt durch den Einsatz standardisierter integrierter Schaltkreise bei der Realisierung der Datenbusschnittstelle. Dies ermöglicht den Einsatz von Komponenten (Steuergeräten) unterschiedlicher Hersteller für den Aufbau der gesamten Fahrzeugelektronik.

2.2 Anforderungen an das Datenübertragungsprotokoll

Im folgenden werden spezifische Anforderungen an die Data Link Layer oder Sicherungsschicht (ISO-Referenzmodell Schicht 2) näher beschrieben.
Typische Informationen die im Fahrzeug anfallen und im Netzwerk übertragen werden, stammen aus Sensoren (z.B. Raddrehzahl, Lenkwinkel, Beschleunigung, ...) oder sind für Aktuatoren (z.B. Ventile, Relais, Motoren,...) bestimmt. Die Menge der zu übertragenden Daten beträgt deshalb oft nur wenige Byte (typischerweise 0..8), d.h. Anwendungen mit Filetransfer werden zur Zeit nicht diskutiert. Das Datenübertragungsprotokoll muß deshalb bezüglich kurzer Datenfeldlängen optimiert werden.
Eine weitere allgemeine Forderung besteht in der Broadcast- und Multicastfähigkeit des Übertragungsprotokolls. Das Protokoll muß dabei über Mechanismen zur netzwerkweiten Signalisierung lokal aufgetretener Störungen (eine Station hat die Broadcast-Nachricht fehlerhaft empfangen) verfügen.
Die Funktion des Buszugriffes kann entweder zentral durch einen Busmaster geregelt werden oder dezentral auf alle mit dem Bus verbundenen Stationen verteilt sein. Die dezentrale Organisation des Buszugriffs bietet ein hohes Maß an Ausfallsicherheit, da im System kein zentraler Master existiert. Bussysteme mit Multimaster-Fähigkeit führen wegen der dezentralen Lösung von Konfliktsituationen zu einem komplexeren Zugriffsprotokoll, bieten aber bezüglich Flexibilität und Ausfallsicherheit gegenüber zentralgesteuerten Zugriffsverfahren Vorteile.
Im Gegensatz zu den Zugriffsprotokollen bekannter LAN's, die eine faire Zuteilung der Übertragungskapazität anstreben, ist das Zutei-

lungskriterium im Fahrzeug die Wichtigkeit (Priorität) der zu übertragenden Daten. Dies ist insbesondere bei Echtzeitanwendungen von Bedeutung, in welchen sich die Transferzeiten der Nachrichten typischer Weise unter 1ms bewegen sollten. Darüber hinaus müssen Sicherheitskritische Informationen innerhalb weniger Mikrosekunden den Buszugriff erhalten. Die Realisierung dieser Anforderung erfolgt durch eine eindeutige Indentifikation der im Netzwerk definierten Nachrichten (Priorität) und nicht durch Adressierung der einzelnen Stationen. Die Priorität der Nachricht entscheidet im Konkurrenzfall über den Buszugriff. Nachrichten werden also netzwerkweit definiert und können damit von jeder Station selektiv empfangen werden. Die Realisierung von Multicast- und Broadcastfunktionen ist damit implizit gelöst.
Eine weitere Forderung besteht in der minimalen Belastung der CPU bzw. des Steuergerätecontrollers für Kommunikationszwecke. Daraus ergibt sich der Zwang, möglichst viele Kommunikationsaufgaben in Hardware zu lösen und den gesamten Verwaltungsaufwand der Datenübertragung zu optimieren. Der Netzwerkbetrieb im Fahrzeug setzt eine gewisse Robustheit gegenüber Störungen, speziell transienter elektromagnetischer Störungen, aus dem Fahrzeugumfeld voraus. Übertragungsstörungen oder allgemein der Ausfall ganzer Stationen müssen netzwerkweit sicher erkannt werden. Die sichere Erkennung defekter Stationen wird deshalb ebenso gefordert wie deren selbständige Isolation vom Bussystem.
Um speziell bei Echtzeitanwendungen eine möglichst schnelle Quittierung der Nachricht zu erhalten, erfolgt dies am Ende der Übertragung in einem dafür vorgesehenem Acknowledgeslot durch die Empfänger. Übertragungsfehler werden netzwerkweit signalisiert. Defekte Stationen stellen durch Eigendiagnose ihren Busverkehr ein.

2.3 Anforderungen an integrierte Schaltkreise

Intergrierte Schaltkreise für den Aufbau der Datenbusschnittstelle unterliegen bestimmten Anforderungen der Automobilindustrie.
Der hohe Kostendruck an integrierte Schaltkreise für den Einsatz im Kraftfahrzeug erfordert große Stückzahlen, die nur über eine Standardisierung der Schnittstellenbausteine erreicht werden. Bemühungen zur Bildung von internationalen Standards sind momentan im Gange. Weitere Forderungen der Automobilindustrie betreffen den möglichen Bezug der Halbleiterbauelemente über einen zweiten Lieferanten (Second Source) zur Vermeidung von Lieferengpässen, damit die Serienproduktion von Fahrzeugen in grossen Stückzahlen gewährleistet werden kann. Die integrierten Schaltkreise müssen beim Einsatz im Kraftfahrzeug fahrzeugspezifische Vorgaben erfüllen. Zum Beispiel bedeutet dies:

- die vollständige Funktion des Bausteins im Temperaturbereich von -40 bis +125 Grad,
- möglichst geringer Stromverbrauch (Einsatz von CMOS-Technologie, sowie die Implementierung von Sleep mode und Wake up Funktionen)
- gute elektromagnetische Verträglichkeit (EMV), geringe Störaussendung und hohe gestrahlte Störfestigkeit,
- Festigkeit gegenüber energiearmen, leitungsgebundenen Störimpulsen.

Zukünftig wird im Fahrzeug die Systemintegration eine steigende Rolle einnehmen. Ein erster Schritt der Systemintegration wird die Integration des Datenbusprotokolls (Data Link Layer) in den bereits vorhandenen Mikrokontroller des elektronischen Steuergerätes sein. Weitere Möglichkeiten der Systemintegration sind die Kombination von Leistungsteilen (z. B. Bustreiber oder Schaltendstufen) mit anderen integrierten Schaltungsteilen (z. B. Protokollbaustein). Dies kann zum einen durch Verbundtechnologien (BICMOS), die Bipolar-Technologie mit CMOS-Technologie auf einem Chip kombinieren oder zum anderen durch spezielle Verbindungstechniken, die auf einem Trägermaterial (Keramik oder Silizium) verschiedene Technologien verbinden, realisiert werden.

3. Systemarchitektur nach dem OSI-Referenzmodell

Die Forderung nach einem offenen System mit einer modularen und flexiblen Architektur setzt eine Systemstrukturierung voraus, die standardisierbar ist und damit letztlich die Kosten durch hohe Komponentenstückzahlen gering hält.
Das Basis-Referenzmodell für die Kommunikation offener Systeme (OSI-Referenzmodell) bildet auch im Fahrzeug die Grundlage für die Entwicklung einer Bussystemarchitektur sowie die Voraussetzung für mögliche Standardisierungen /ISO79/. Die sieben Schichten des Referenzmodells können allgemein in das Transportsystem (Schicht 1-4) und das Anwendungssystem (Schicht 5-7) unterteilt werden /FR086/, siehe Bild 2.

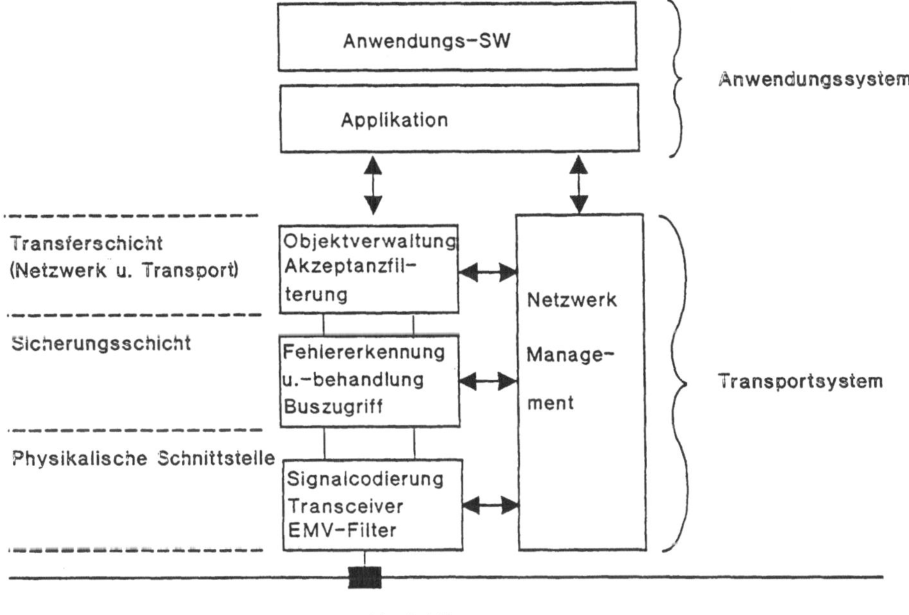

Bild 2: Systemarchitektur

Das Anwendungssystem kann prinzipiell unabhängig von der Netzwerktechnik betrachtet werden. Da im Fahrzeug Forderungen bezüglich Echtzeitfähigkeit und minimaler Kosten bestehen, werden Implementierungen dieser Schichten sehr stark anwendungsorientiert sein und sich auf ein Minimum beschränken. Eine wesentliche Rolle spielt dabei die Hardwareunterstützung der Schnittstelle zwischen Anwendungssystem und Transportsystem.

Das Transportsystem ermöglicht den sicheren Datenaustausch zwischen den Netzwerkteilnehmern. Nach dem ISO-Referenzmodell übernehmen die Schichten 3 und 4 Aufgaben, die speziell bei der Kopplung von Teilnetzen entstehen. Im Fahrzeug werden diese Schichten auch unter dem Namen Transferschicht zusammengefasst.

3.1 Übertragungsmedien

Allgemein werden Lichtwellenleiter (Plastikfaser) und Kupferleiter (Eindraht- bzw. Zweidrahtleitung) diskutiert. Im praktischen Einsatz besitzen Lichtwellenleiter noch wesentliche Nachteile:

- eine staubdichte Steckerkonstruktion ist nur sehr aufwendig möglich
- die Konfektionierung der Stecker in den Werkstätten ist schwierig, speziell, wenn nachträglich Arbeiten am Kabelsatz notwendig werden,
- die Biegeradien von Plastikfasern sind begrenzt,
- die Fasern altern nach größeren Zeiträumen, d.h. die Faser wird "milchig" und weist eine hohe Signaldämpfung auf,
- die optoelektrischen Wandler sind gegenüber elektromagnetischen Störungen sehr empfindlich, d.h. der durch die optische Übertragung gewonnene Störabstand geht u. U. an der optisch/elektrischen Schnittstelle wieder verloren,
- der im Fahrzeug erforderliche Temperaturbereich (-40...+125 C) wird nur mit erhöhtem Kostenaufwand erreicht,
- als Topologien sind zur Zeit nur stern- oder ringförmige Strukturen denkbar. Der Sternkoppler grenzt die Flexibilität des Netzwerkes bezüglich späterer Erweiterungen ein, die Ringstruktur stellt ein Zuverlässigkeitsproblem dar, das nur durch erhöhten Aufwand (Doppelringstruktur) gelöst werden kann.

Das Kupferkabel stellt im Fahrzeug zur Zeit sicher noch die beste Lösung dar. Das Hauptproblem des Kupferkabel besteht in der Antennenwirkung des Kabels, d.h. die Eigenschaft der elektromagnetischen Welle, sich bei bestimmten Frequenzen vom Kabel abzulösen bzw. in dieses einzukopplen. Möglichkeiten, diese unerwünschten Effekte zu verringern, bestehen zum einen durch den Einsatz von zwei miteinander verdrillten Leitungen, die zusätzlich durch ein Kupfergeflecht geschirmt werden und zum anderen dadurch, daß das Signalspektrum sendeseitig wie empfangsseitig durch einen Bandfilter begrenzt wird.
Konventionelle Koaxialkabel sind nur bedingt fahrzeugtauglich, da das im allgemeinen als Kunststoff ausgeführte Dielektrikum wie der Lichtwellenleiter Alterserscheinungen (schrumpfen) unterworfen ist und es im Laufe der Zeit durch Einfluß von Feuchtigkeit zu Kurzschlüssen kommen kann.
Klare Vorteile des Kupferkabels ergeben sich bezüglich der Korrosionsfestigkeit der Steckverbindung und der Kontaktierung der Stecker allgemein. Ein Problem ergibt sich bei der Kontaktierung des Kupfergeflechts bei geschirmten Kabeln. Die Busankopplung kann beim Kupferkabel entweder galvanisch gekoppelt oder galvanisch getrennt (induktiv, kapazitiv) erfolgen.

3.2 Schicht 1 - Bitübertragungschicht (Physical Layer)

Die Bitübertragungsschicht setzt zum einen den von der Schicht 2 gelieferten seriellen Datenstrom in eine geeignete physikalische Darstellung auf dem Übertragungsmedium um und schützt zum anderen die über das Übertragungsmedium kontaktierte Fahrzeugelektronik vor störenden und zerstörenden (elektromagnetischen) Einflüssen.

3.2.1 Signalcodierung

Im Fahrzeug werden zur Zeit drei Codierungsarten diskutiert:

- Pulse Width Modulation (PWM): Hier wird von einer Dreiteilung eines jeden Bits ausgegangen, die eigentliche Information liegt dabei im mittleren Teil. Der Vorteil dieses Verfahren besteht in der Rückgewinnung der Taktinformation aus dem Datensignal am Empfänger, z.B. durch kostengünstige RC-Resonatoren. Nachteilig wirkt sich die dreifache Grundfrequenz des Datensignals auf das Abstrahlverhalten des Datensignals aus.
- Manchester-Coding (MAN): Die Dateninformation wird durch eine steigende bzw. fallende Flanke in der Bitmitte dargestellt. Dies führt zu einer Verdopplung der Grundfrequenz des Datensignals. Die Manchester-Codierung besitzt damit die Vor- und Nachteile wie die PWM-Codierung, ermöglicht aber darüber hinaus die Realisierung eines gleichstromfreien Codes zur galvanisch getrennten Kopplung der physikalischen Schnittstelle mit dem Übertragungsmedium.
- Non Return to Zero (NRZ): Der NRZ-Code stellt schaltungstechnisch die einfachste Realisierung dar. Er benötigt allerdings die Implementierung einer Bit-Stuffinglogik und den Einsatz genauer Oszillatoren zur Taktrückgewinnung. Die Minimierung des Abstrahlverhalten (Störspektrum) läßt sich mit dem NRZ-Code am besten realisieren /KIE87/.

3.2.2 Sende- und Empfangsstufe (Transceiver)

Der Transceiver stellt die Verbindung von den hochintegrierten Bausteinen der Busschnittstelle (im allgemeinen in CMOS-Technologie) mit der Fahrzeugumwelt dar. Die Fahrzeugtauglichkeit des Transceivers setzt folgende Eigenschaften der Schaltungsauslegung voraus:

- Kurzschlußfestigkeit und Spannungsfestigkeit (Batteriespannung) bezüglich der Signalleitungen bzw. des Übertragungsmediums,
- maximaler Signal/Störabstand durch eine geeignete Dimensionierung des Signalpegels,
- eine Treiberleistung des Transceiver-Bausteins, die unter Einbeziehung der Schmutzwiderstände an allen Steckkontakten des Übertragungsmediums einen Signalpegel garantiert, der eine eindeutige Interpretation der Daten in allen Stationen des Bussystems ermöglicht,
- minimale Ruhestromaufnahme und damit geringe Leistungsaufnahme durch sleep und wake-up Funktion (über Busleitung gesteuert).

3.2.3 Signalfilterung

Die Sicherstellung guter elektromagnetischer Eigenschaften der Datenübertragung ist ebenfalls eine wesentliche Aufgabe der physikalischen Schicht. Der Einsatz von speziell für das Fahrzeug entwickelten Signalfiltern, die zum Teil in den Steckverbindungen integriert sind (Durchgangsfilter), führt zu folgenden Vorteilen:

- minimale Störaussendeung durch eine geeignete Bandbegrenzung des Datensignals,
- gute Störfestigkeit gegenüber externen elektromagnetischen Feldern, die durch kapazitive oder induktive Kopplung über das Übertragungsmedium leitungsgebundene energiearme Störimpulse erzeugen können,
- Schutz gegen die Entladung elektrostatischer Energie (ESD) im Fahrzeug über das Übertragungsmedium.

Zur Steigerung der elektromagnetischen Verträglichkeit werden in den Steckverbindungen der Busleitung an den Steuergeräten Durchgangsfilter diskutiert, die allerdings ein Kostenproblem darstellen.

3.3 Schicht 2 - Sicherungsschicht (Data Link Layer)

Die Sicherungsschicht wird, wie bei LAN's üblich, in die Schicht 2a (Media Access Control, MAC) und in die Schicht 2b (Logical Link Control, LLC) unterteilt.

3.3.1 Schicht 2a, Media Access Control

Schicht 2a beinhaltet das Zugriffsverfahren der Stationen auf das gemeinsame Medium, die Abbildung (Framing/Deframing) der Daten in einem vorgegebenen Rahmenformat und die Datensicherung. Im Fahrzeug werden zur Zeit nur Zugriffsverfahren mit unilateraler Konflikterkennung auf CSMA-CD Basis /SPA82/ diskutiert, d.h. es existiert ein dominanter logischer Signalpegel, der sich gegenüber dem rezessiven Pegel im Konfliktfall (gleichzeitige Busarbitrierung von zwei Stationen) durchsetzt. Unter der Voraussetzung einer geringen Netzwerkausdehnung, im Fahrzeug typischerweise wenige 10m, ist damit die Realisierung eines priorisierten Zugriffs auf das Medium möglich. Die Codierung der ersten Bits innerhalb der Nachrichtenübertragung bestimmt die Nachrichtenpriorität. Das Zugriffsverfahren entspricht damit einem System mit nicht unterbrechenden Prioritäten. Die priorisierte nachrichtenorientierte Zugriffssteuerung ermöglicht gleichzeitig durch die Implementierung eines Akzeptanzfilters in den Empfängern die Broadcast- und Multicastfunktion. Folgende Verfahren der Datensicherung und der Fehlererkennung werden in Schicht 2a angewendet:

- Vergleich von gesendetem Signalpegel mit dem Pegel, der sich auf der Busleitung einstellt (Bitfehler),
- Ein- oder Mehrfachabtastung eines Bits beim Empfänger,
- Cyclic Redundundcy Check (CRC), der auf die im Fahrzeug auftretenden kurzen Datenrahmen optimiert ist,

- Verfahren der Paritätsbildung (Längs-, Quer- oder Kreuzparität), wenn in bestimmten Anwendungen nicht erkannte Fehler der Übertragung bedingt toleriert werden können oder
- Plausibilitätskontrollen, die zusätzliche in der Systemebene durchgeführt werden,
- Überprüfung des Rahmenformats (Format-Error), der Signalcodierung (Codefehler) und der Bitstuffing-Regel.

Die Übertragung der Nachrichten erfolgt in einem definierten Format, das die Struktur der Nachrichtenübermittlung (zeichenorientiertes Protokoll, bitorientiertes Protokoll) sowie die Plazierung und Anzahl der Steuerinformation festlegt. Typische Steuerinformationen sind:

- Startbit (dient zur Synchronisation aller Busteilnehmer)
- Adressierung (Identifizierung bzw. Kennung oder Priorität der Nachricht)
- Prüfbits (Anzahl vom verwendeten Sicherungsverfahren abhängig)
- Quittierung (Bestätigung einer korrekten Übertragung durch einen oder mehrere Empfänger)
- Endekennung (Signalisiert das Ende einer Nachricht)

Zwischen zwei aufeinanderfolgend gesendeten Nachrichten muß von allen Busteilnehmern eine vom Übertragungsprotokoll festgelegte Zeit abgewartet werden (Inter-Frame-Space).

3.3.2 Schicht 2b, Logical Link Control

Schicht 2b ist unabhängig von dem eingesetzen Zugriffsverfahren und beschreibt die Verbindungart zwischen zwei Netzwerkteilnehmern. Prinzipiell wird zwischen folgenden Verbindungsarten unterschieden:

LLC Typ 1: Typ 1 beschreibt die verbindungslose Kommunikation zwischen den Netzwerkteilnehmern. Die Daten werden ohne den vorherigen Aufbau einer Verbindung gesendet (Datagrammprinzip). Fehlerhafte Übertragungen werden beim Empfänger erkannt und dort verworfen.

LLC Typ 2: Typ 2 beschreibt die verbindungsorientierte Kommunikation zwischen den Teilnehmern, d.h. es wird zunächst eine virtuelle Verbindung aufgebaut und anschließend numerierte Datenpakete ausgetauscht. Fehlerhafte Datenpakete werden beim Empfänger verworfen, die Wiederholung wird auf Grund eines Sequenzfehlers in der Paketreihenfolge veranlaßt (wie HDLC-Protokoll).

LLC Typ 3: Der quittierte Datagrammbetrieb (LLC Typ 3) beinhaltet eine Quittierung des Datagramms unmittelbar nach der Übertragung ohne das Übertragungsmedium erneut anzufordern oder durch einen separaten Quittierungsrahmen der erst nach einer erneuten Anforderung und Belegung des Übertragungsmediums durch den Empfänger übertragen wird.

Remote Transmission Request ist eine zusätzliche Kommunikationsform, die bei Kfz-Anwendungen gefordert wird. Die Übertragung der Daten wird durch den Empfänger angefordert bzw. initiiert. Dies erfolgt entweder durch einen speziellen Request-Frame mit anschließender separater Übertragung der Daten oder die angeforderten Daten werden in eine laufende Übertragung eingefügt.

3.4 Transferschicht und Netzwerkmanagement

Die Transferschicht übernimmt Aufgaben zur Verwaltung von Kommunikationsobjekten und der Akzeptanzfilterung empfangener Nachrichten. Die Objektverwaltung sollte den Mikrocontroller in der Applikation nur gering belasten. Dies wird zum einen durch Semaphore und zum anderen durch die Unterstützung von gängigen Adress/Datenbussen für Mikrocontroller realisiert.
Funktionen des Netzwerkmanagement werden zur Zeit für die Bitübertragungsschicht (z. B. Zustand der Busleitung etc. /VAL88/) diskutiert. Der Protokollvorschlag CAN /CAN89/ bietet durch eine interne Zustandssteuerung (ERROR-ACTIVE, ERROR-PASSIVE, OFF-BUS) die Möglichkeit, Zustandsinformationen der Sicherungsschicht dem Netzwerkmanagement zur Verfügung zu stellen.

4. Systemimplementierung

Bevor auf verschiedene Details der Systemimplementierung eingegangen wird, soll an dieser Stelle ein Vergleich bereits verfügbarer Systeme durchgeführt werden (Bild 3). Grundlage für den Vergleichs bildet die Klassifizierung der Bussysteme in

- Prozess-Busse (z. B. Feldbus, Bitbus, ...),
- Lokale Netzwerke mit Bustopologie (z. B. Ethernet, Token-Bus),
- Komponenten-Busse (z. B. I2C, VNP, ...) und
- Kraftfahrzeug-Busse (z. B. CAN, ABUS, J1850, ...)

	Multi-Master	Prioritäten	Datenrate	Kosten Schnittstelle	Datensicherung	Beispiele
Prozess-Bus	Nein					PDV-Bus Bitbus MilStd.1553B
LAN	Ja	Nein (Ja)	über 10MBd	hoch		Ethernet Token-Bus
Komponenten-Bus	Ja	Ja	0,1..10MBd	gering	mittel	I2C
Kraftfahrzeug-Bus	Ja	Ja	bis 1MBd	gering	mittel hoch	CAN J1850

Bild 3: Klassifizierung und Vergleich von Bussystemen

sowie eine Bewertung anhand der Beurteilungskriterien

- Multi-Masterfähigkeit,
- Nachrichtenpriorität,
- Datenrate,
- Kosten der Busschnittstelle und
- Datensicherung (Sicherheit und Zuverlässigkeit).

Die Prozeß-Busse besitzen im allgemeinen nicht Multi-Masterfähigkeit. Lokale Netzwerke mit Bustopologie scheiden im wesentlichen durch die hohen Kosten der Busschnittstelle aus. Die Komponenten-Busse sind nur bedingt einsetzbar, wegen ihrer im allgemeinen zu geringen Datenübertragungsrate oder zu geringen Datensicherung.

4.1 Informations-Bus für Multiplex Applikationen (IBUS)

Im folgenden werden nur die wesentlichen Protokolleigenschaften einer bei Daimler-Benz durchgeführten Implementierung eines Datenbusprotokolls für Multiplexanwendungen /DUD88/ näher beschrieben (Bild 4). Das IBUS-Protokoll ist ein zeichenorientiertes Protokoll, d.h. die gesamte Nachricht wird in 8 Bit breite Worte unterteilt, die jeweils mit Start-, Querparitäts- und Stoppbits verpackt, asynchron übertragen werden.
Die Übertragung erfolgt mit 19,2kbit/s, das Protokoll eignet sich daher nur für Multiplex-Anwendungen. Eine Nachricht setzt sich aus 6 Steuerworten (Startwort, Empfängerkennung, Senderkennung, Datenmenge, Längsparität, Endekennung) und einer variablen Anzahl von Datenworten zusammen (Bild 5). Der Verwaltungsaufwand für eine Nachricht beträgt

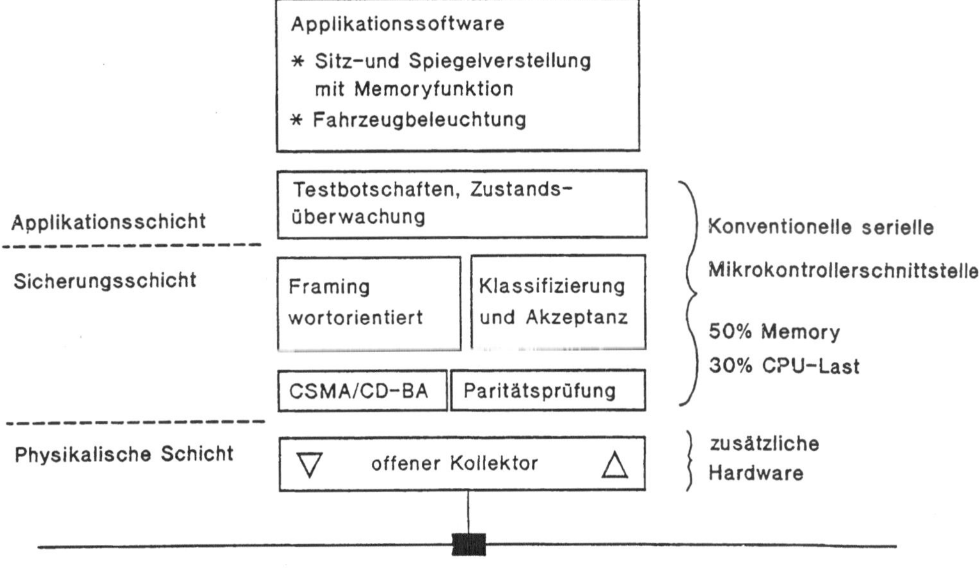

Bild 4: Informationsbus (IBUS)

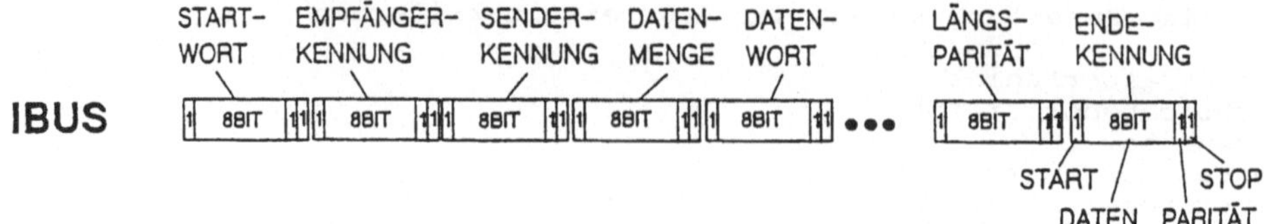

Bild 5: Nachrichtenformat IBUS

somit 66 Bit (Freizeiten zwischen den einzelnen Worten werden vernachlässigt). Die Freizeit zwischen zwei aufeinanderfolgenden Nachrichten (Inter-Frame-Space) beträgt 22 Bit.
Das IBUS-Protokoll wird für den Systemverbund von elektronischen Steuergeräten im Fahrzeug-Innenraum (Innenraum-Bus) und für ein Kabelbaummultiplex-System der Fahrzeugbeleuchtung eingesetzt. Das IBUS-Protokoll benützt die konventionelle, serielle Schnittstelle der gängigen Mikrocontroller und benötigt deshalb keine zusätzliche Hardware. Der Controller wird durch die Kommunikationssteuerung der seriellen Schnittstelle belastet.

4.2 Controller Area Network (CAN) für Echtzeitapplikationen

Das CAN-Protokoll (Controller Area Network) ist ein bitorientiertes Übertragungsprotokoll. Die Entwicklung des CAN-Protokolls wurde gemeinsam von den Firmen Bosch und Intel durchgeführt. Ausführliche Angaben sind in den angegebenen Literaturstellen zu entnehmen /CAN89, KIE86, LAW86, BEN87/. Bild 6 zeigt die Struktur der Datenbusschnittstelle mit CAN-Protokoll.
Um auch sicherheitsrelevante Daten übertragen zu können, wurde ein großer Aufwand für die Sicherung der zu Übertragenden Daten zugrundegelegt. Neben fest vorgesehener Formatbits, die eine ständige Überprüfung des Nachrichtenformats während der Übertragung ermöglichen, wird auch ein speziell für kurze Nachrichtenlängen ausgelegtes CRC-Verfahren zur Datensicherung verwendet. Weiterhin wurde durch eine Datenfeldlängenangabe die Flexibilisierung des Datenfeldes von 0 bis 8 Byte ermöglicht. Das Nachrichtenformat ist in Bild 7 dargestellt.
Durch ein spezielles Verfahren zur Fehlersignalisierung ist im Fehlerfall eine unmittelbare Wiederholung der Nachricht möglich, da die Übertragung bei einem erkannten Übertragungsfehler sofort abgebrochen werden kann. Der Inter-Frame-Space ist auf 3 Bit festgelegt. Zur Ausführung des CAN-Protokolls wird zusätzliche Hardware benötigt, das Steuergerät wird dann nur gering mit der Kommunikationssteuerung der seriellen Schnittstelle belastet. Die Schnittstelle zwischen Steuergeräte Mikrocontroller und dem CAN-Baustein wird entweder durch einen DPRAM (64 Byte) realisiert (FullCAN) oder durch spezielle Sende- und Empfangsregister (BasicCAN).
Die Datenrate ist programmierbar und auf 1Mbit/s begrenzt.
Die Realisierung BasicCAN benötigt weniger Chipfläche und eignet sich deshalb zur Integration in gängige Mikrocontroller. FullCAN ist ein standalone Baustein, der viele Kommunikationsaufgaben übernimmt und den Steuergerätecontroller stark entlastet.
FullCAN findet Anwendung im Motormanagement, BasicCAN wird bei Klimasteuerungen zum Einsatz kommen.
Die Implementierung der physikalischen Schnittstelle ist zur Zeit noch mit einigen ungelösten Problemen verbunden. Dies ist vor allem in den Forderungen nach einer kostengünstigen, fahrzeugtauglichen Implementierung mit geringem Platzbedarf begründet.

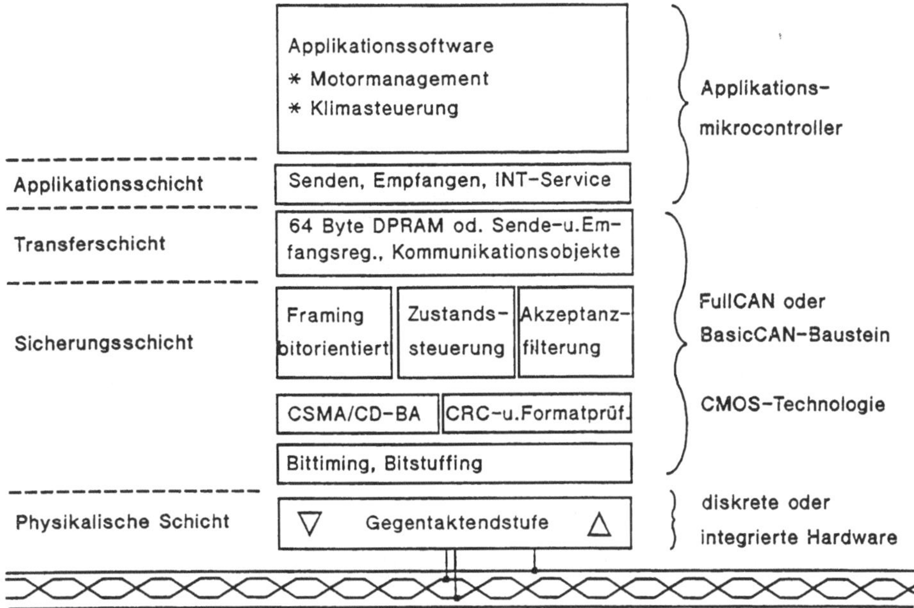

Bild 6: Controller Area Network (CAN)

Darüber hinaus prägt die physikalische Schnittstelle wesentlich die elektromagnetischen Eigenschaften der gesamten Datenbusschnittstelle und bildet die Grundlage für den Einsatz von Verfahren zur Fehlererkennung und -toleranz sowie zur Reduktion der Leistungsaufnahme. Die physikalische Schnittstelle wird deshalb auch einen starken fahrzeugspezifischen Charakter haben.
Im allgemeinen werden im Fahrzeug verdrillte Zweidrahtleitungen eingesetzt, die z.T. durch ein Kupfergeflecht geschirmt werden. Ein Problem besteht noch in der Anbindung der Schirmung an die Steuergerätemasse.

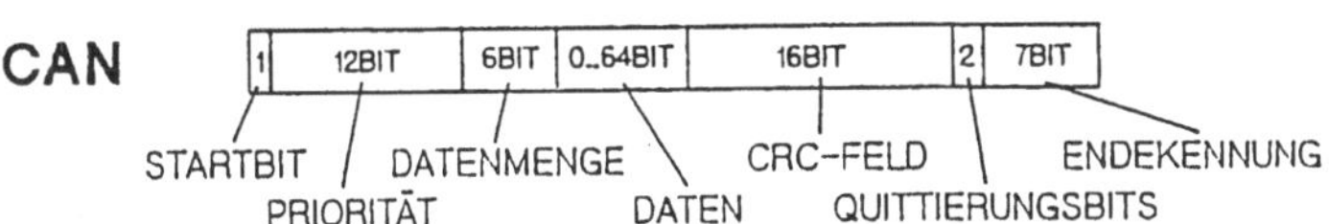

Bild 7: Nachrichtenformat CAN

Der Einsatz von Transceivern mit Gegentaktendstufen ermöglicht den Differenzbetrieb der Zweidrahtleitung und eine hohe Unterdrückung von Gleichtaktstörungen. Die Festlegung der Signalpegel wird immer einen Kompromiß zwischen der Abstrahlung des Datensignals (quadratische Abhängigkeit der Spannungsamplitude im spektralen Leistungsdichtespektrum, /KIE87/) und der gestrahlten Störfestigkeit sein. Da der CAN-Baustein intern eine Bitstuffinglogik besitzt, wird NRZ-Code verwendet bei Signalpegeln, die unterhalb 2V liegen. Eine kostengünstige Realisierung der physikalischen Schnittstelle kann zum einen entweder durch Integration aller geforderter Funktionen in einem speziell zu entwickelnden integrierten Baustein oder durch einen diskreten Aufbau mit minimalem Funktionsumfang erfolgen. Erste Lösungsvorschläge für die Integration der physikalische Schnittstelle von Bussystemen mit Datenraten bis 100kbit/s wurden von Valvo vorgestellt /VAL88/. Zum anderen wird zur Zeit die Integration von Treiberstufe und Protokollogik in einer Technologie (z. B. CMOS) oder in verschiedenen Technologien (BICMOS) untersucht.

5. Standardisierung

Im folgenden sollen Standardisierungsgremien, die sich mit Standards für Datennetze im Fahrzeug befassen, kurz vorgestellt werden.

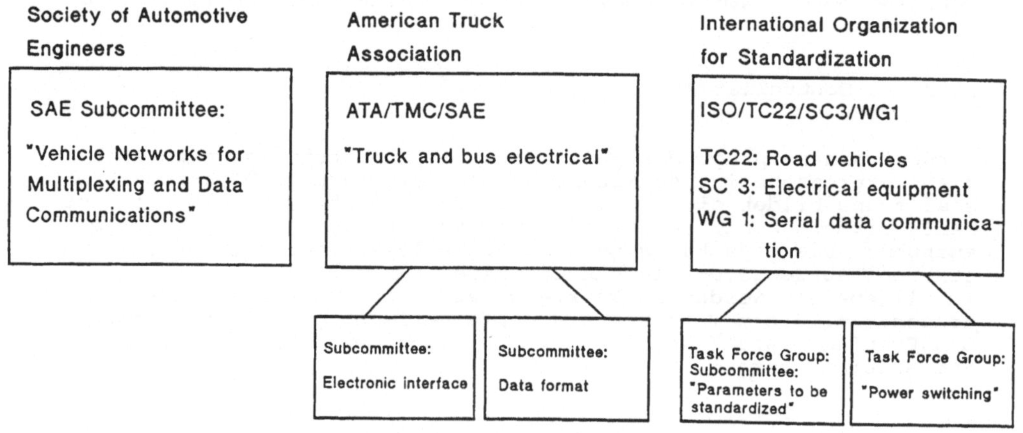

Bild 8: Standardisierungsgremien

5.1 Gremien für Kraftfahrzeugstandards

In den USA übernimmt die SAE (Society of Automotive Enginieers) und die ATA (American Truck Association) im Rahmen von verschiedenen Subcommittees Vorbereitungen zur Standardisierung. Die SAE ist etwa mit dem FAKRA (Normenausschuß Kraftfahrzeuge) der DIN vergleichbar. Zur Zeit exisiteren in USA das Subcommittee "Vehicle Networks for Multiplexing and Data Communications" sowie eine Task Force "Class C" der SAE und zwei Subcommittees der ATA, "Electronic Interface" und "Data Format" (Bild 8).

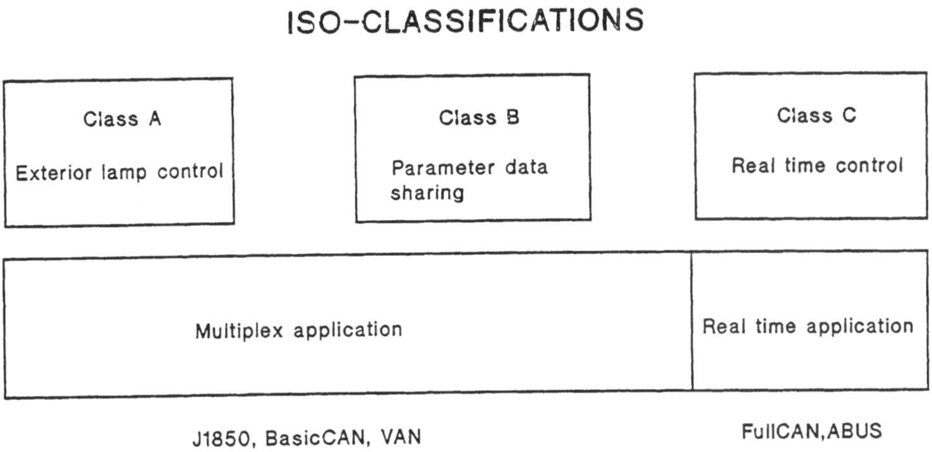

ISO-CLASSIFICATIONS

Class A	Class B	Class C
Exterior lamp control	Parameter data sharing	Real time control

Multiplex application	Real time application
J1850, BasicCAN, VAN	FullCAN,ABUS

Bild 9: Klassifizierung und Empfehlungen

Alle Committees und Task Forces erarbeiten Empfehlungen, die in der ISO (International Organization for Standardization) diskutiert und letztlich standardisiert werden.
In der ISO wurde eine Working Group ISO/TC22/SC3/WG1 eingerichtet, die sich mit serieller Datenkommunikation im Fahrzeug befasst.
Aus ihr sind zwei Task Forces hervorgegangen:

1) "Parameters to be standardized" - Festlegung und Definition von Datenbusschnittstellen für das Fahrzeug gemäß dem OSI-Referenzmodel.
2) "Power Switching" - Standardisierungen, die den Einsatz von intelligenten Leistungsschaltern im Fahrzeug betreffen.

Weiterhin werden in WG1 Themen wie Fahrzeugdiagnose, Erfassung von Abgaswerten CARB (California Air Research Board) und andere behandelt.

5.2 Europäische Standardisierungsbemühungen

Basierend auf unterschiedlichen Systemanwendungen bezüglich des Realzeitverhaltens, der Zuverlässigkeitsanforderungen und anderen Teilaspekten wurden von der WG1 der ISO Serial Data Communication drei Netzwerkklassen vorgeschlagen, die sich im wesentlichen durch die vorgegebene Datenrate unterscheiden (Bild 9):

Klasse A: Multiplexanwendungen mit niederer Datenrate (<1Kbit/s), z. B. zum Ein/Ausschalten von Lampen oder anderer Verbraucher im Fahrzeug.
Klasse B: Multiplexanwendungen mit mittlerer Datenrate (<125 Kbit/s) für den Austausch von Informationen und Daten zwischen elektronischen Steuergeräten.
Klasse C: Realzeitanwendungen mit Datenraten z.Z. bis 1Mbit/s für die Realisierung Steuergeräten übergeordneten Regelkreisen (Motormanagement).

Auf der Basis dieser Klassifizierung sind zur Zeit folgende Standardisierungsbemühungen im Gange:

- CAN, Controller Area Network (FullCAN, BasicCAN)
  Vorschlag von BOSCH und INTEL, /CAN89/
- ABUS, Automobile Universal Schnittstelle, /KRE86/
  Vorschlag von Volkswagen
- VAN, Vehicle Area Network
  französischer Vorschlag
- J1850, SAE-Empfehlung Klasse B
  gemeinsamer Vorschlag von GM, CRYSLER und FORD

Alle Vorschläge betreffen die Klassen B und C. Als Grenze zwischen diesen Klassen wird zur Zeit eine Datenrate von 125Kbit/s diskutiert. Bei CAN ist nur die Schicht 2 (Sicherungsschicht) Gegenstand der Standardisierungsbemühung.
VAN enthält im wesentlichen CAN als Subset und enthält erste Ansätze zur Realisierung von Netzwerkmanagementfunktionen sowie Vorschläge zur Realisierung der physikalischen Schnittstelle. J1850 ist letzlich ein Kompromiß zwischen den amerikanischen Protokollvorschlägen CCD von CHRYSLER /MIE86/ und VNP von FORD /JOH86/.

6. Systembewertung und Beurteilung

In Bild 10 sind die prinzipiellen Methoden der Leistungsbeurteilung dargestellt. Der Einsatz einer Methode hängt vom Entwicklungsstand des Systems ab. In den frühen Entwicklungsphasen existiert meistens noch keine konkrete Vorstellung über das System und über dessen Umwelt. Das System und seine Umwelt wird deshalb in einem Modell nachgebildet. Mit zunehmender Konkretisierung des zu entwickelnden Systems wird das Modell verfeinert und damit detaillierter. Die Leistungsbeurteilung kann in diesen Phasen der Systementwicklung zunächst durch eine mathematische Analyse und später mit einer zeittreuen Systemsimulation durchgeführt werden. Existiert ein Prototyp des Systems im Labor, dann muß die reale Umwelt simuliert werden, um den Prototyp testen zu können. Zur Bestimmung der Leistungsfähigkeit des Systems unter realen Bedingungen wird ein speziell entwickelter Meßmonitor eingesetzt, mit dessen Hilfe Zustandsänderungen und zeitliche Vorgänge während des Systembetriebs aufgezeichnet werden können /LAW87/.

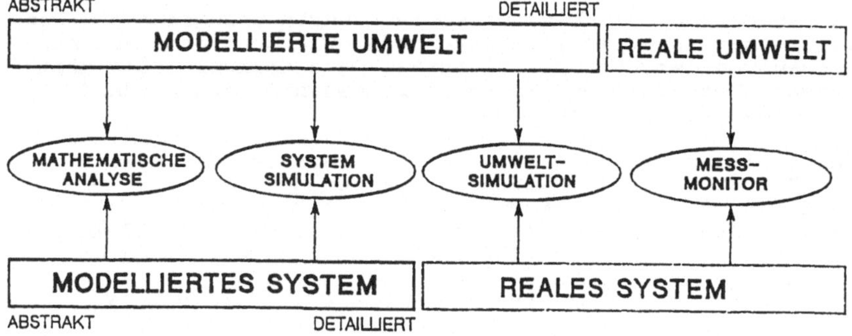

Bild 10: Methoden zur Leistungsbeurteilung

6.1 Approximative Analyse des Buszugriffsverfahren

Die Buszugriffsverfahren mit unilateraler Konflikterkennung können durch ein Warteschlangenmodell mit nichtunterbrechenden Prioritäten approximativ modelliert werden. In Bild 11 ist das Modell, welches der mathematischen Modellanalyse zugrundeliegt, dargestellt.
Das Bussystem wird durch die Anzahl der Nachrichten (n) und durch die Wartemöglichkeit der Nachrichten bis zum Buszugriff dargestellt.
Die Umwelt des Bussystems wird durch die Priorität i einer Nachricht (i = 1...n), durch den mittleren Ankunftsabstand der Nachricht (TAi) und durch die mittlere Busbelegungsdauer TBi modelliert.
Während die Busbelegungsdauer beliebig verteilt sein kann, indem erstes und zweites Moment der Verteilungsfunktion berücksichtigt werden, wird für die Verteilungsfunktion der Ankünfte eine negativ-exponentielle Verteilungsfunktion vorausgesetzt (Poisson Prozeß), die durch den mittleren Ankunftsabstand vorgegeben wird.
Das Verfahren der mathematischen Analyse geht auf eine Lösung, die 1953 von A. Cobham /COB54/ veröffentlicht wurde, zurück.
Die Anwendung der Cobham-Lösung auf Protokolle mit unilateraler Konflikterkennung für Kraftfahrzeuge wurde in /DUD88/ vorgestellt.

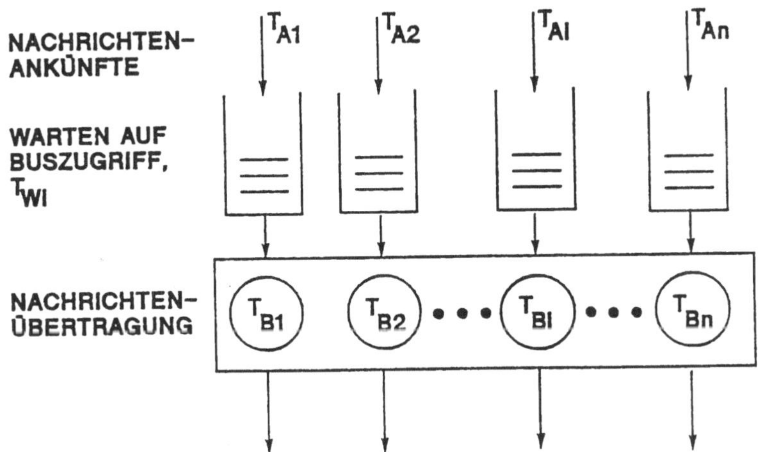

Bild 11: Modell der mathematischen Analyse

6.2 Zeittreue Simulation von Systemen mit CAN-Protokoll

In dem Simualtionsmodell nach Bild 12 ist das Steuergerät mit CAN-Schnittstelle, der allen Steuergeräten gemeinsame Datenbus und der Kanalstörer abgebildet.
Die Funktionen des Applikations-Prozessors werden in folgende Prozesse unterteilt:

- Sensordaten aufbereiten (abtasten, A/D-wandeln) und die digitale Information in den entsprechenden Nachrichtenspeicher eingetragen.
- Steuerung oder Regelung der Stellglieder. Durch die Aufteilung der Steuerphasen kann eine kontrollierte Unterbrechung des Applikations-Prozessors in seiner Steuer- und Regelphase ermöglicht werden.
- Transfer aufbereiteter Sensordaten zwischen Nachrichtenspeicher des Applikations-Prozessors und Sendespeicher des CAN-Bausteins
- Nachrichtentransfer korrekt empfangener Nachrichten in den RAM-Speicher des Applikations-Prozessors.

Die Reihenfolge, mit welcher die verschiedenen Prozeßphasen des Applikations-Prozessors durchlaufen werden, wird durch die Scheduling-Strategie bestimmt. Durch das Scheduling läßt sich auch eine priorisierte Behandlung bestimmter Phasen erreichen.
Das Teilmodell "CAN-Protokoll" bildet Protokollfunktionen wie stationsinterne Prioritätsauswahl, Akzeptanz-Filter, Fehlerstatistik und die stationsübergreifenden Funktionen wie Zugriffsregelung, Busbelegung usw. des CAN-Bausteins nach.

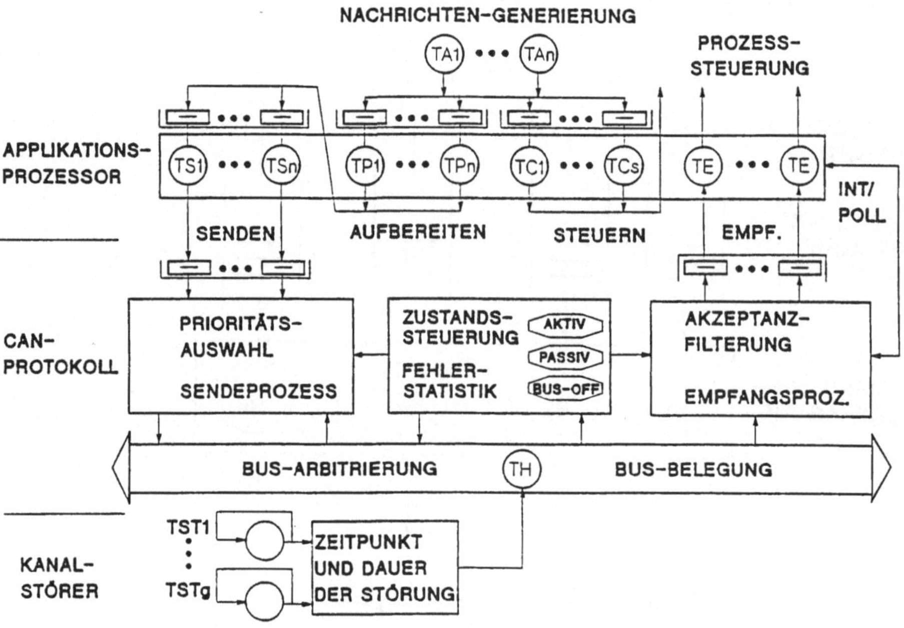

Bild 12: Simulationsmodell für ein Steuergerät mit CAN-Schnittstelle

Das Teilmodell "Kanalstörer" besteht aus beliebig vorgebbaren und voneinander unabhängigen Störgeneratoren. Jeder Generator kann bezüglich des zeitlichen Abstandes der Störungen (Mittelwert und Verteilung) sowie der Länge der Störung, d. h. Bit-Anzahl (Mittelwert und Verteilung) individuell eingestellt werden. Die Überlagerung aller (globaler) Störungen ergibt den Zeitpunkt und die Länge der Summenstörung. Lokale Störungen werden nicht von Kanalstörer erfaßt. Busbelegungen und Störungen sind voneinander unabhängig.

Für die Programmimplementierung wurde die genaue Einhaltung der CAN-Spezifikation, speziell der Fehlerbehandlung, gefordert. Darüber hinaus ermöglicht die flexible Modellierung der Software im Steuergerät eine möglichst detailierte Abbildung der zu simulierenden realen Implementierung.
Mit Hilfe des Kanal-Störers kann vom Einzelbitfehler bis zu Büschelstörungen beliebiger Form jede Störcharakteristik eingestellt werden. Die Anzahl der Stationen am Bus sowie die Anzahl der Prozesse im Applikations-Prozessor ist logisch unbegrenzt.
Die Produktion von Simulationsergebissen erfordert eine möglichst genaue Beschreibung der Steuergerätesoftware, d.h. detaillierte Angaben über das Prozeßkonzept, die Programmlaufzeiten und die Unterbrechungsbehandlung der Applikation. Zur Zeit besteht noch das Problem, daß Struktur und Laufzeiten der Applikationssoftware nur aufwendig zu bestimmen sind.
Durch den modularen und hierarchischen Programmaufbau ist eine Veränderung in den Blöcken Applikation, Protokoll und Störeinflüsse einfach zu realisieren. Einen Implementierung des BasicCAN-Protokolls um speziell auch Systeme mit FullCAN und BasicCAN zu untersuchen wird dadurch möglich.

6.3 Elektromagnetische Verträglichkeit (EMV)

Die Funktionssicherheit der Kfz-Elektronik muß auch in stark gestörter Umgebung gewährleistet sein. Transiente Störungen, wie sie vor allem beim Abschalten von Induktivitäten im Bordnetz aufteten, können aufgrund ihrer hohen Energie bzw. hoher Spannungsspitzen zu Ausfällen oder Störungen der Elektronik führen /FEU88/, da die räumliche Nähe der Datenbusleitung zu starken fahrzeuginternen Störquellen induktive oder kapazitive Einkoppelungsmechanismen begünstigt. Die für Pkw's typischen Busleitungslängen im ermöglichen andererseits Resonanzen mit externen elektromagnetischen Feldern, wie sie z.B. von Fernsehsendern ausgestrahlt werden. Meßmittel die zur Bestrahlung von elektronischen Komponenten oder gar ganzen Fahrzeugen dienen sind zum einen die TEM-Zelle (Transversal-Elektro-Magnetisch) oder auch Strip-Line genannt, zum anderen spezielle EMV-Hallen. Typische Testbedingungen sind z.B. Feldstärken von 200 V/m bei Frequenzen von 215 MHz.
Neben dem Schutz gegen Störungen die von außen auf die Elektronik einwirken (gestrahlte Störfestigkeit), ist speziell die Reduktion der Störaussendung für Datenbussysteme im Fahrzeug von besonderer Bedeutung.

5. Ausblick

Auf dem Gebiet der seriellen Datenbussysteme im Fahrzeug sind zur Zeit national und international verstärkt Entwicklungsaktivitäten in verschiedenen Automobilfirmen und in der Halbleiterindustrie zu beobachten. In der WG1 der ISO (Serial Data Communication) müssen in der nächsten Zeit bezüglich der Standardisierung der Data Link Layer Entscheidungen getroffen werden. Für die Klasse C (Real Time Control) hat das CAN-Protokoll zur Zeit gute Chancen standardisiert zu werden, vor allem weil einige Halbleiterhersteller (z. B. Intel, Motorola, Valvo) beabsichtigen, das CAN-Protokoll in Produkte mit zu integrieren oder eigensändige Bausteine zu entwickeln.

In der Klasse B bestehen zur Zeit starke amerikanische Bestrebungen, die SAE-Empfehlung J1850 als Standard für Multiplexanwendungen mit mittleren Datenraten festzusetzen. Dem steht die kostengünstige Version des CAN-Protokolls (BasicCAN) als Standardisierungsvorschlag gegenüber.
Der Einsatz von Netzwerken für die Rechner-Rechner-Kommunikation in Fahrzeugen wird auch in Demonstrator Fahrzeugen für PROMETHEUS im Vordergrund stehen, wobei zunehmende Anforderungen bzgl. Datenrate, Latenzzeit, Zuverlässigkeit an das Datenübertragungsprotokoll gestellt werden.
Der Einsatz von Lichtwellenleitern wird weiterhin erprobt, eine Serieneinführung wird weitgehend von der Entwicklung der elektrooptischen Wandler und von den Alterungseigenschaften der Plastikfaser abhängen.
Datenbussysteme werden zukünftig die Basis für eines hierarchisch gegliederten Gesamtsystems der Fahrzeugelektronik bilden.

Literaturverzeichnis

BEN87 Bendel, K., Juenger, G., "82C900 für serielle Kommunikation im Kfz", Elektronik im Kraftfahrzeug, Tagung, Essen, Juni 87.

BOT86 Botzenhardt, W., Litschel, M., Unruh, J., "Bussystem für Kfz-Steuergeräte", VDI Berichte Nr.612 1986, Seite 459-470.

CAN89 82526 - Serial Communication Controller, Architecture Overview, INTEL Corporation.

COB54 Cobham, A., "Priority Asignment in Waiting Line Problems", Operations Research 2, 1954, pp. 70-76.

DUD88 Dudeck, I., Hipp, U., Raith, T., "Investigations into the Electromagnetic Compatibility and Performance Evaluation of Databus Systems in Motor Vehicles", SAE Technical Paper Series 885083.

FAE87 Färber, G., "Bussysteme", 2. Auflage, Oldenburg Verlag München Wien 1987.

FAZ87 Spira, J.-C., "Bussystem statt Kabelsalat", FAZ, 8.9.1987.

FEU88 Feurer, K., "Ursachen und Auswirkungen Transienter Störungen im Kraftfahrzeug", EMV '88, Seite 65-78.

FRO86 Fromm, I., "Standardisierung Lokaler Netze (LAN)", Informationstechnik it 28. Jahrgang, Heft 1/1986.

ISO79 International Standard ISO 7498, "Information Processing Systems - Open System Interconnection, Basic Reference Model".

JOH85 Johnson, W.J., Kowalki, D.A., Volk, J.R., "System Considerations for Incorporating Vehicle Data Networks (Multiplex) into Automobiles", IMechE 1985, C218/85.

JOH86 Johnson, W.J., Volk, R., "A Proposal for a Vehicle Network Protocol Standard", SAE Technical Paper Series 860392.

KIE86 Kiencke, U., Dais, S., Litschel, M., "Automotive Serial Controller Area Network", SAE Technical Paper Series 860391.

KIE87 Kiencke, U., Cao, C.T., Litschel, M., "The impact of bit representation on noise emissions in automotive networks", Robert Bosch GmbH.

KRE86 Kreft, W., Stamm, K., "ABUS - die automobile, bit-serielle Universal-Schnittstelle", Impulse 7/1986, VW AG Wolfsburg.

LAW86 Lawrenz, W., Arnett, D.J., Zimmermann, P., "CAN - Controller Area Network for in Vehicle Network Applications", Proposed SAE Information Report J1583.

LAW87 Lawrenz, W., "Entwicklungswerkzeuge fuer Controller-Netzwerke", Elektronik 18/4.9.1987.

LUD84 Ludvigson, M.T., Milton, K.L., "A high speed bus for pave pillar applications", SAE Transactions 841456.

MAR87 "Saure Zeiten für die Kabelbäume", Markt und Technik, Nr.22 vom 29. Mai 1987.

MIE86 Miesterfield, F.O.R., "Chrysler Collision Detection (CCD) - A Revolutionary Vehicle Network", SAE Technical Paper Series 860389.

PHA86 Phail, F.H., Arnett, D.J., "In-Vehicle Networking - Serial Communication Requirements and Directions", SAE Technical Paper Series 860390.

SPA82 Spaniol, O., "Konzepte und Bewertungsmethoden für lokale Rechnernetze", Informatik Spektrum, 1982, 5: 152-170.

STA86 Stamm, K., "Möglichkeiten und Konsequenzen von Kfz-Netzwerken", VDI-Bericht,Wolfsburg.

VAL88 Philips Multiplex Transceiver (PMT), Preliminary Functional Description, Version 1.0, Valvo Applikationslabor Hamburg 1988.

Synchronisierte Software-Messungen zur Bewertung des dynamischen Verhaltens eines UNIX-Multiprozessor-Betriebssystems

Andreas Quick
Universität Erlangen-Nürnberg
IMMD VII
Martensstraße 3
D-8520 Erlangen
Federal Republic of Germany

Zusammenfassung

Bei der Entwicklung parallel ablaufender Programme in Multiprozessorsystemen benötigt man Informationen darüber, wie und wo das Programm im System bearbeitet wird. Da aber gerade das Zusammenspiel und die Wechselwirkung aller Systemkomponenten in den meisten Fällen nicht von außen sichtbar ist, muß dem Entwickler ein Einblick in das interne, dynamische Verhalten des Systems vermittelt werden. Um einen solchen Einblick in das dynamische Verhalten eines Multiprozessorsystems zu bekommen, ist das Beobachten des aktiven Systems (Monitoring) ein geeignetes Hilfsmittel. Im folgenden wird eine Methode vorgestellt, durch Definition relevanter Ereignisse und ihrer zeitgerechten Beobachtung das Verhalten eines Multiprozessorsystems nach außen hin sichtbar zu machen und in Ereignisspuren zu erfassen. Diese sind dann Grundlage für eine Analyse des Systemverhaltens. Anhand einiger Beispiele wird exemplarisch für ein UNIX-Multiprozessor-Betriebssystem die Auswertung von Ereignisspuren erläutert.

Schlüsselwörter

Ereignisgesteuerte Software-Messung, Leistungsbewertung, Ereignisspur, Instrumentierung, Quellbezug, Multiprozessorsystem, UNIX-Betriebssystem, Lastverteilung.

1. Einleitung

Die Anforderungen an die Leistungen von Rechenanlagen sind in letzter Zeit immer mehr angestiegen. Eine Möglichkeit, die Leistung zu erhöhen, besteht in der Verwendung von Multiprozessor- oder Multicomputer-Systemen (MP-/MC-Systeme). Gerade im Bereich der numerischen Simulation konnten durch die Verwendung solcher Systeme Problemgebiete erschlossen werden, deren Bearbeitung mit Monoprozessorsystemen nicht in vertretbarer Zeit möglich waren. Bei der Entwicklung neuer Hochleistungs-Rechenanlangen geht der Trend daher immer mehr von einer Monoprozessoranlage hin zu einem MP-/MC-System.

Diese Arbeit diente der Vorbereitung ähnlicher Messungen bei der Entwicklung des UNIX-MP-Betriebssystems für den Steuerrechner des SUPRENUM-Systems und wurde im Rahmen des SUPRENUM-Projekts gefördert.

Bei allen Vorteilen dieser Klasse von Rechenanlagen treten jedoch neue Probleme auf: Es müssen die Fragen beantwortet werden, wie Multiprozessorsysteme überhaupt programmiert werden können und wie die zu bearbeitende Aufgabe effizient auf die Prozessoren verteilt werden kann. Bei der Lastverteilung muß zwischen der Parallelisierung *einer* rechenintensiven Aufgabe durch Zerlegung in kleinere Aufgaben oder durch Aufteilung der Daten und der parallelen Bearbeitung mehrerer a priori unabhängiger Prozesse unterschieden werden.

Durchsatz und Antwortzeit eines MP-Systems hängen auch bei einer Last aus voneinander unabhängigen Prozessen davon ab, wie die Prozesse auf die einzelnen Prozessoren verteilt werden und wie groß der durch die Parallelisierung notwendigerweise entstehende Organisations- und Kommunikationsaufwand ist. Eine effiziente Lastverteilung und Systemorganisation setzt einen Einblick in die komplexen internen Abläufe des Systems voraus. In MP-Systemen ist es jedoch nicht möglich, die in Monoprozessorsystemen verfügbaren Werkzeuge zur Programmentwicklung und Fehlersuche (Debugging-Tools) zu verwenden, da diese es nicht erlauben, die Wechselwirkungen zwischen den Prozessoren zu beobachten. Aus diesem Grund muß für die Programmentwicklung und Fehlersuche in Multiprozessorsystemen ein Verfahren gefunden werden, das der Komplexität solcher Systeme gerecht wird. Der Programm- und Systementwickler benötigt Einsicht in das dynamische Verhalten des gesamten MP-Systems, um die Ursachen des nach außen hin sichtbaren Systemverhaltens zu ermitteln.

Das Monitoring (Beobachten der internen Abläufe in einem System) ist ein geeignetes Werkzeug, um die Abhängigkeiten und Wechselwirkungen aufzuzeigen und damit die Ursachen für das extern sichtbare Verhalten aufzudecken und die Möglichkeit der Analyse zu geben. Denn wenn die Ursachen des Systemverhaltens bekannt sind, weiß man oftmals schon, wie das Verhalten des Systems verbessert werden kann [Hofm87].

Die zentrale Frage bei MP-/MC-Systemen ist damit zur Zeit die Frage nach der geschickten Programmierung solcher Systeme. In verschiedenen Ansätzen werden verteilte Monitoring- und Debugging-Tools entwickelt, um dem Anwender die internen Abläufe transparent zu machen. So nehmen z.B. Burkhart und Millen in [Burk89] eine auf Meßdaten basierende Klassifikation von Verlustursachen bei Algorithmen in MP-Systemen vor. Zur Bewertung des Gesamtsystems werden hier gleichzeitig mehrere Monitortechniken miteinander kombiniert. In [LeBl85] wird durch die Visualisierung von Kommunikationsereignissen versucht, den Ablauf in lose gekoppelten Systemen zu verdeutlichen. Einen Überblick über Monitoring in MP-/MC-Systemen findet man in [Lutt89].

Die allgemeine Vorgehensweise beim Monitoring und die notwendigerweise zu treffenden Entscheidungen vor dem Einsatz eines Monitors werden in Kapitel 2 erläutert. In Kapitel 3 folgt die Darstellung eines Konzepts parallel arbeitender Software-Monitore. Dazu wird das Beispielobjekt und die Zielsetzungen bei der Entwicklung eines Betriebssystems vom Monoprozessor- hin zum Multiprozessor-Betriebssystem dargestellt. Daran schließt sich in Kapitel 4 die exemplarische Darstellung einiger Ergebnisse der Messungen an. Den Abschluß bildet in Kapitel 5 ein Ausblick.

2. Einsatz eines Monitors — Methoden und Grundlagen

Dem Einsatz eines Monitors geht zunächst einmal die Wahl der Monitoringmethode und des Monitors voraus. Beide Auswahlmöglichkeiten sind problemabhängig und sollen hier allgemein erörtert werden.

2.1 Wahl der Monitoringmethode

Die Fragen "*Was soll im Objektsystem beobachtet werden? Welche Schlüsse sollen aus den Meßdaten gezogen werden können?*" führen dazu, sich auf eine der beiden Monitoringmethoden, das **ereignisgesteuerte** oder das **zeitgesteuerte** Monitoring festzulegen.
Um einen Einblick in das dynamische Systemverhalten durch Monitoring bekommen und den Ablauf in einem System exakt rekonstruieren zu können, ist eine ereignisgesteuerte Messung (event driven

monitoring) notwendig. Nur mit dieser Monitoringmethode kann das — auf Ereignisse abstrahierte — Gesamtablaufgeschehen erfaßt und ein Quellbezug zum gemessenen Objekt, dem Programm, hergestellt werden. Die Meßdaten erlauben Rückschlüsse auf den Ablauf des gemessenen Programms; jedem Ereignis kann eindeutig ein Punkt im Programm zugeordnet werden. So kann die Auswertung auf einer für den Entwickler gewohnten Ebene (der problemorientierten Beschreibung des Algorithmus) durchgeführt werden. Die zweite gebräuchliche Monitoringmethode, in gleichmäßigen oder zufällig verteilten Zeitabständen den Systemzustand aufzuzeichnen, die sog. zeitgesteuerte Messung (sampling), ermöglicht nur statistische Aussagen über das Systemverhalten. Sie ist daher nicht geeignet, den Ablauf eines Programms oder das Verhaltens eines Systems so aufzuzeichnen, daß kausale Zusammenhänge erkennbar werden.
Bei der ereignisgesteuerten Messung mit einem Quellbezug zum Algorithmus (Software- und Hybrid-Monitoring) wird das zu beobachtende Programm mit Meß-Anweisungen versehen (Instrumentierung). Durch die Instrumentierung eines Programms produziert dieses während seines Ablaufs jeweils dann beobachtbare Ereigniskennungen, wenn eine der Meßanweisungen ausgeführt wird. So wird durch die Instrumentierung eines Programms dessen Ablauf nach außen hin sichtbar gemacht, durch die Instrumentierung eines Betriebssystems an für den Prozeßablauf relevanten Stellen wird durch diese Technik der Ablauf aller Prozesse und damit gleichzeitig das dynamische Systemverhalten durch eine Folge ausgezeichneter Ereignisse beobachtbar gemacht.

Der Instrumentierung sollte ein funktionales Modell des Programmverhaltens zugrunde liegen. Betrachtet man das Systemverhalten z.B. aus der Sicht von (Modell-)Zuständen und deren möglichen Übergängen, so werden in diesem Fall die Zustandsübergänge als (Modell-)Ereignis bezeichnet.
Mit Hilfe dieser Modellvorstellung kann das Objektprogramm an den für das Beobachtungsziel relevanten Zustandsübergängen mit Meß-Anweisungen versehen werden. Ein instrumentierter Zustandsübergang wird bei dessen Ausführung zum Meß-Ereignis. Durch die Instrumentierung wird es möglich, ein auf das Wesentliche abstrahiertes — aber sonst vollständiges — Ablaufgeschehen des Objektprogramms mit einem Monitor zu erfassen und in einer Ereignisspur aufzuzeichnen. Die Meß-Anweisungen sind abhängig von der gewählten Monitortechnik (Software- oder Hybrid-Monitoring).
Im folgenden wird mit **Ereignis** immer ein Meß-Ereignis bezeichnet. Da nur die Meß-Ereignisse extern sichtbar sind und nur diese in der Ereignisspur auftreten, braucht im weiteren zwischen Modell- und Meß-Ereignissen nicht mehr unterschieden zu werden.
Ein von zwei Ereignissen geklammerter Zustand wird als **Aktivität** bezeichnet. Dabei können zwischen den beiden begrenzenden Ereignissen beliebig viele weitere Ereignisse eintreten. Einer Aktivität ist eine Zeitdauer, nämlich die Differenz der Zeitmarken der beiden begrenzenden Ereignisse, zugeordnet.

2.2 Wahl des Monitors

Neben den beiden oben genannten Monitoringmethoden können drei verschiedene Monitorverfahren unterschieden werden: Software-, Hardware- und Hybrid-Monitor [Klar85].
Durch die Zielsetzung, Einblick in interne dynamische Abläufe mit einem Quellbezug zum beobachteten Programm auf der Algorithmenebene zu bekommen, entfällt die Wahl des **Hardware-Monitors**. Dieser meist für statistische Zwecke eingesetzte Monitor (z.B. Busauslastung) kann jedoch auch auf der Hardwareebene, ausgerüstet mit einem entsprechenden Ereignisdetektor, zur ereignisgesteuerten Messung eingesetzt werden. Bestimmte Hardwaresignale führen in diesem Fall zu einer Aufzeichnung eines Ereignisses. Da jedoch auf dieser Ebene nur schwer Aussagen über die Wirkungsweise von implementierten Algorithmen gemacht werden können, soll der Einsatz eines solchen Monitors nicht weiter verfolgt werden. Ein Beispiel für den Einsatz eines HW-Monitors findet man bei [Klar81].

Im Gegensatz zum Einsatz eines HW-Monitors entstehen durch den Einsatz eines **Software-Monitors** immer Rückwirkungen im Objektsystem: Der Monitor mißt sich mit. Diese Rückwirkungen sind von drei Parametern abhängig: Größe des Speicherplatzes zum Speichern der Ereignisse, Dauer zum Abspeichern eines Ereignisses, Anzahl der ausgeführten Meß-Ereignisse pro definierter Zeiteinheit. Eine wichtige Voraussetzung für den Einsatz eines SW-Monitors ist eine Uhr im Objektsystem, deren Auflösung ein

Bruchteil der Dauer der zu messenden Programmabschnitte ist. Eine detaillierte Messung ist nur mit einer hochauflösenden Uhr möglich (vgl. z.B. [Mill86, LeBl85]).
Ein weiteres Monitorverfahren, das die Vorteile von Hardware- und Software-Monitoring weitestgehend vereinigt, ist der **Hybrid-Monitor**. Die Ereignisse werden nicht intern im Speicher des Systems, sondern in einem externen HW-Monitor gespeichert. Dazu müssen die Ereignisse an einer beobachtbaren Systemschnittstelle ausgegeben werden. Dort werden sie vom HW-Monitor aufgenommen und dabei mit einem Zeitstempel von der Uhr des HW-Monitors versehen. Die Notwendigkeit einer hochauflösenden Uhr im Objektsystem entfällt hierbei. Ebenso wie beim SW-Monitoring muß auch beim Hybrid-Monitoring das Objektsystem bei ereignisgesteuerten Messungen instrumentiert werden. Beispiele für die Verwendung von Hybrid-Monitoren in MP-/MC-Systemen findet man z.B. bei [Burk89, Haba88, Zieh88, Bemm88, Mink87, Hofm87, Hofm88].

Die Wahl des Monitorsystems ist objektspezifisch und hat immer die Zielsetzung, die Abläufe im Objektsystem möglichst wenig zu stören. Sowohl beim SW-Monitoring als auch beim Hybrid-Monitoring muß nach Möglichkeiten gesucht werden, die Ereigniskennungen schnell abzuspeichern und die Zeitinformation zu lesen und zu speichern bzw. die Ereigniskennungen schnell an einer Systemschnittstelle auszugeben.

2.3 Aufbereitung der Meßdaten

Jedes Meßobjekt und jedes Monitorsystem hat seine spezifischen Besonderheiten. So ist es in der Regel nicht möglich, immer genormte oder genormt strukturierte Ereignisspuren zu bekommen. Um allgemein anwendbaren Auswertewerkzeugen beliebig strukturierte und formatierte Ereignisspuren zuführen zu können, wurden von Mohr im Rahmen des Projekts ZM4, wo ein universelles Hardware- und Hybrid-Monitorsystem entwickelt wird, die Werkzeuge **TDL** (**T**race **D**escription **L**anguage) und **POET** (**P**roblem **O**riented **E**vent **T**race interface) entwickelt [Mohr87, Mohr88]. Mit Hilfe dieser Werkzeuge kann auf Ereignisspuren unterschiedlicher Objektsysteme und unterschiedlicher Monitore einheitlich zugegriffen werden.
In der Beschreibung einer Ereignisspur (tdl-file) werden sowohl der Aufbau der einzelnen E-Records (Syntax) als auch die Bedeutung der einzelnen Komponenten eines E-Records und der Ereignis-Kennungen (event token) (Semantik) beschrieben. Aus der Beschreibungsdatei wird mit dem TDL-Compiler (tdlc) eine Schlüsseldatei (key-file) erzeugt, die als Eingabe für das Zugriffspaket POET dient. Zu jeder Ereignisspurklasse existiert eine solche Schlüsseldatei.
Durch die Verwendung dieser Werkzeuge wird die Ereignisspur als generische Datenstruktur betrachtet. Der Zugriff auf Ereignisspuren erfolgt über eine einheitlich definierte Schnittstelle. Durch diese Schnittstelle wird neben der objekt- und monitorunabhängigen Auswertung eine Unabhängigkeit zwischen der Messung und der Implementierung der Auswertesoftware erreicht. Die Auswertesoftware kann daher erstellt werden, ohne daß das Format der Ereignisspur oder die Bedeutungen der einzelnen Ereigniskennungen bekannt sind.
Die mit Hilfe der Ereignisspurbeschreibungssprache TDL und den Zugriffsfunktionen POET bereitgestellte Schnittstelle ist dann hilfreich, wenn das Monitorsystem nicht an ein bestimmtes Objektsystem adaptiert ist oder die Instrumentierung im Objektsystem nicht fest bestimmt ist. Letzteres ist während der Entwicklungsphase eines Programms immer der Fall, da nicht von Anfang an klar ist, was gemessen werden soll. Die hier gewählte Konzipierung des Monitorsystems als flexibles System zur Beobachtung beliebiger Objektsysteme wird in den meisten Fällen nicht durchgeführt (z.B. [Burk89]), oder es findet eine Beschränkung auf nur einige wenige Ereignisse (z.B. Kommunikationsereignisse) statt. Deshalb sind vergleichbare Ansätze, eine definierte Schnittstelle zwischen Messung und Auswertung zu definieren, nicht verbreitet.

2.4 Auswertung von Ereignisspuren

Bei der Auswertung von Ereignisspuren können zwei verschiedene Verfahren unterschieden werden:

- Bei der **Statistische Auswertung** werden aus der Ereignisspur Kenngrößen berechnet, die Aussagen über das Systemverhalten geben. Hierzu gehören z.B. Ereignishäufigkeit, Häufigkeit und Dauer der vom Anwender definierten Aktivitäten und Verteilungen dieser Größen, etc.

- Die **Ablauforientierte Auswertung** dient dazu, die kausalen Wechselwirkungen zwischen einzelnen Systemkomponenten oder Prozessen zu untersuchen. Dies kann entweder mit Hilfe eines Ablaufprotokolls (Listing der Ereignisspur mit problemorientierten Bezeichnern) oder durch Visualisierung des Systemverhaltens erfolgen. Die Visualisierung ist bei geeigneter Auswahl der Aktivitäten und/oder zeitlichen Ausschnitte der Ereignisspur bereits eine Auswertung und bildet die Grundlage für die Analyse des gemessenen Systemverhaltens. Bei der Visualisierung des Ablaufgeschehens können die folgenden zwei Methoden unterschieden werden:

  - Statische Visualisierung des Ablaufgeschehens (z.B. mit Gantt-Diagrammen)
  - Dynamische Visualisierung des Ablaufgeschehens (z.B. mit computeranimierter Grafik)

  Der Visualisierung muß eine Wahl der zu visualisierenden Aktivitäten und damit in den meisten Fällen auch eine Bearbeitung der Ereignisspur vorausgehen.

3. Einsatz paralleler Software-Monitore — ein Fallbeispiel

In diesem Kapitel wird ein Konzept parallel arbeitender Software-Monitore zur Beobachtung von Multiprozessoraktivitäten dargestellt. Die Tragfähigkeit des Konzepts wird mit einer Beispielimplementierung aufgezeigt, die das Ziel hat, die Entwicklung eines neuartigen parallelen UNIX-Multiprozessor-Betriebssystems zu fördern, das Parallelverarbeitung nicht nur von Anwenderprozessen, sondern auch von Betriebssystemfunktionen ermöglicht.
Um die durchgeführten Entwicklungsschritte bei der Betriebssystementwicklung bewerten zu können, wurden die Messungen gleichzeitig zur Entwicklung des Betriebssystems durchgeführt und ausgewertet, um dem Entwickler das Systemverhalten — und damit auch die Wirkungsweise der implementierten BS-Funktionen — offenzulegen. Die Messungen waren damit ein Bestandteil der Software-Entwicklung. Dazu war eine unmittelbare Zusammenarbeit der Betriebssystementwickler mit der Meßgruppe notwendig.
Wegen der verhältnismäßig leichten Implementierung, den vorhandenen Hardware-Voraussetzungen, der einfachen und detaillierten Durchführung von Messungen und den geringen Rückwirkungen beim Messen, wurde in diesem Beispiel ein parallel arbeitender Software-Monitor zur Beobachtung von Multiprozessor-Aktivitäten implementiert. Aufgrund vorgegebener Prozessoreigenschaften (nicht alle Prozessoren haben einen Zugriff auf eine E/A-Schnittstelle, siehe Abschnitt 3.1) im Objektsystem wäre eine Verwendung eines Hybrid-Monitors nur mit einigen Änderungen im Betriebssystem möglich gewesen.
Bevor die Implementierung der parallelen Monitore und die Durchführung der Messungen erläutert wird, werden in den folgenden Abschnitten das Objektsystem und die Zielsetzungen der Betriebssystementwicklung dargestellt.

3.1 Das Objektsystem

Die Betriebssystementwicklung wurde auf dem Multiprozessorsystem 3280 MPS der Firma Concurrent Computer Corporation (CCC) am Lehrstuhl für Betriebssysteme durchgeführt. Die Architektur des Objektsystems ist in Abb. 1 dargestellt. Bei diesem System handelt es sich um ein homogenes (alle Prozessoren sind technisch vom gleichen Typ), asymmetrisches (nicht alle Prozessoren dürfen alle Befehle ausführen) speichergekoppeltes Multiprozessorsystem (uniform memory access, UMA) [Rash87].

Die funktionale Asymmetrie des Systems ist durch die Einschränkung bedingt, daß nur einer der Prozessoren als Central Processing Unit (CPU) dadurch ausgezeichnet ist, daß nur er externe E/A-Befehle (Platte, Netz, etc.) ausführen kann. Dies wird durch einen unterschiedlichen Mikrocode realisiert. Die weiteren Prozessoren werden als Auxiliary Processing Unit (APU) bezeichnet. Alle Prozessoren sind über einen gemeinsamen synchronen Bus untereinander und mit dem Hauptspeicher verbunden. Das System besteht zur Zeit aus drei Prozessoren und kann um maximal drei weitere Auxiliary Processing Units (APU) erweitert werden, die Hauptspeichergröße beträgt 12 MB. Die Konsistenz der zu jedem Prozessor gehörenden Cache-Speicher wird durch einen Write-Through-Mechanismus in Verbindung mit einer Cache-Invalidierung gewährleistet [Conc86].

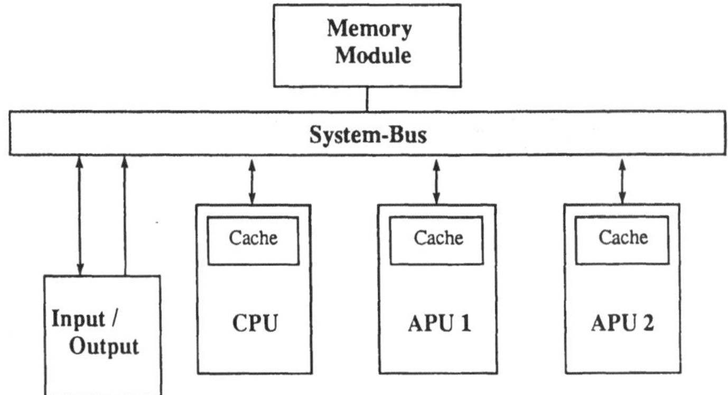

Abbildung 1. Architektur des Objektsystems CCC 3280 MPS

3.2 Zielsetzungen

Bei dem entwickelten Betriebssystem handelt es sich um das Betriebssystem XELOS, einer Portierung von UNIX System V.2 auf die Rechnerserie 3200 von Concurrent Computer. Das wichtigste bei einer Erweiterung vom Monoprozessor- zum Multiprozessor-Betriebssystem ist es, eine maximale Leistungssteigerung des Multiprozessorsystems durch eine optimale Lastverteilung auf die verfügbaren Prozessoren zu erreichen. Dabei soll hier die Verteilung von voneinander unabhängigen Anforderungen — wie sie üblicherweise in einem Mehrbenutzersystem auftreten — im Vordergrund stehen.
Der maximale Durchsatz eines speichergekoppelten Systems (siehe Abschnitt 3.1) kann bei einer nicht prozessorgebundenen Organisationsform des Betriebssystems erreicht werden. Dies ist jedoch bei der vorhandenen Architektur durch die funktionale Asymmetrie der Prozessoren nicht möglich. Ein Teil des Systemkerns wird daher immer prozessorgebunden sein.
Die Parallelisierung von unabhängigen Prozessen kann in drei Schritten durchgeführt und durch die auf den Auxiliary Processing Units (APU) ausführbaren Programmteile charakterisiert werden:

- Auf den APUs wird **nur Usercode** ausgeführt.
- **Unkritische Systemfunktionen** können auf den APUs ausgeführt werden.
- **Alle Systemfunktionen, die keine E/A-Befehle ausführen**, können auf den APUs ausgeführt werden.

Die Parallelisierung von BS-Funktionen bedeutet, daß diese Funktionen gleichzeitig von mehreren (unabhängigen) Prozessen auf verschiedenen Prozessoren ausgeführt werden können. Die Parallelisierung beschleunigt daher nicht die Ausführung einer Betriebssystemfunktion, sondern die Bearbeitung der im gesamten System anfallenden Last.

Durch die im letzten Schritt für das Objektsystem gegebene maximale Parallelisierung werden Ausschlußmechanismen notwendig, die die exklusive Ausführung der kritischen BS-Abschnitte (Zugriff auf

Datenbereiche im Systemkern, Zugriff auf Gerätetreiber) gewährleisten. Diese Ausschlußmechanismen wurden mit Monitoren[1] realisiert. Auf die Monitore wird im Abschnitt 4.2 näher eingegangen. Einzelheiten bezüglich der Erweiterung des Betriebssystems findet man bei [Hofm89].

Neben dem Tuning der bereits in der Monoprozessorversion verfügbaren Betriebssystemfunktionen lag der Schwerpunkt der Messungen auf den für das Multiprozessorsystem notwendigerweise neu implementierten Algorithmen zur Prozeßvergabe und Prozessorzuteilung (scheduling). Die von dem Betriebssystementwickler gestellten Fragen nach der Systemauslastung und der Prozessorbelegung (*"Welcher Prozeß wird auf welchem Prozessor wann bearbeitet?"*) sollten durch die Messungen beantwortet werden.

3.3 Monitoringmethode und Monitor

Die Messungen wurden als ereignisgesteuerte Messungen mit parallelen Software-Monitoren durchgeführt. Der Software-Monitor ist dabei als parallel ausführbare BS-Funktion konzipiert, so daß in allen Prozessoren gleichzeitig gemessen werden kann (Aufruf einer Funktion, der parallel erfolgen kann). Beim Aufruf dieser Funktion werden die aktuelle Zeit, die übergebene Ereigniskennung sowie — abhängig vom jeweiligen Ereignis — einige Parameter in einem reservierten Teil des Hauptspeichers, dem sog. *Meßdatenspeicher*, geschrieben. Ein solcher Datensatz wird als Ereignis-Record (kurz: E-Record) bezeichnet (siehe Abb. 4). Um die E-Records möglichst schnell abspeichern zu können und um Konflikte beim Speichern der E-Records zwischen den drei Prozessoren zu vermeiden, steht für jeden Prozessor ein eigener Meßdatenspeicher zur Verfügung. Jeder Prozessor benötigt also nur einen Zeiger auf die aktuelle Schreibposition *seines* Meßdatenspeichers. Dieser Speicher wird zyklisch beschrieben und bereits bei der Installation (booten) des Systems für die Messungen reserviert. Dieser Ansatz führt zwar zu einer Verkleinerung des freien Speichers, nicht jedoch zu einer erhöhten Auslagerungsaktivität (swapping), was die Messungen gegenüber dem "normalen" Betriebszustandes, ohne der Reservierung des Meßdatenspeichers, verfälschen würde. Daher wird dieser Zustand bereits bei der Installation des Systems hergestellt und als der "normale" Betriebszustand definiert.
Der Meßdatenspeicher hat eine Speichertiefe von 20000 Ereignisrecords je Prozessor und wird entweder nach der Messung oder bei längeren Messungen während der Messung periodisch ausgelesen.
Zu jedem E-Record gehört ein Zeitstempel, der den Zeitpunkt des Auftretens angibt. Die Zeitstempel werden einem Mikrosekundenzähler mit 32-Bit Wortlänge entnommen, der auf allen Prozessoren verfügbar ist und hardwaremäßig über den Systembus gestartet und synchronisiert werden kann. Auf diese Weise ist eine systemweit einheitliche Zeitbasis (globale Zeitbasis) vorhanden. Durch die sehr hohe Auflösung von einer Mikrosekunde der verwendeten Uhr wird — im Gegensatz zu den sonst unter UNIX verfügbaren Uhren mit einer Auflösung von mehreren Millisekunden — eine detaillierte Beobachtung des Systemverhaltens möglich.

Durch den Bezug auf eine globale Zeitbasis liefern die drei zunächst unabhängig voneinander, aber doch gleichzeitig durchgeführten Software-Messungen drei lokal erzeugte aber korrelierbare Ereignisspuren. Diese können zueinander in Beziehung gebracht werden, und es kann eine systemweit exakte Reihenfolge aller aufgenommenen Ereignisse hergestellt werden. Dies ist unbedingt notwendig, da Prozesse und sogar Systemfunktionen prozessorübergreifend ausgeführt werden können, d.h. während der Ausführung von Systemfunktionen können Prozessorwechsel (Der aktive Prozeß wird verdrängt und kann von einem anderen Prozessor wieder aufgenommen werden.) stattfinden. So läßt sich nicht nur das Verhalten eines Prozessors, sondern das Verhalten des gesamten MP-Systems beobachten und bewerten.

1. Der Begriff *Monitor* darf nicht mit den Monitoren zur Datenerfassung (Meßgeräten) verwechselt werden. Um Mißverständnisse zu vermeiden, wird das beim Monitoring benötigte Meßgerät zur Ereigniserfassung als **Software-** bzw. **Hardware-Monitor** bezeichnet. Mit **Monitor** sind die im Betriebssystem implementierten Ausschlußmechanismen gemeint.

Das dynamische globale Systemverhalten kann nur unter dieser Voraussetzung exakt rekonstruiert werden.

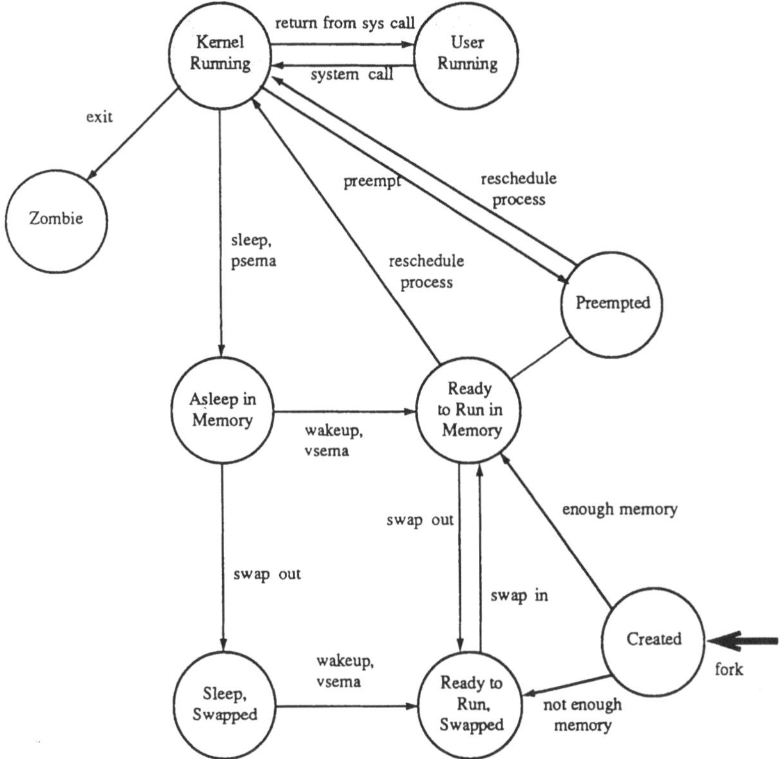

Abbildung 2. Das UNIX-Prozeßmodell: Basis für die Instrumentierung

Die für die Ausführung einer Meß-Anweisung benötigte Zeit ist abhängig von den zur jeweiligen Ereigniskennung gespeicherten Parametern und beträgt zwischen 7 µs (keine Parameter) und 15 µs (mehrere zu ermittelnde Parameter). Diese sehr kleine Zeit für die Aufnahme eines E-Records und die damit verbundene geringe Rückwirkung (overhead) beim Monitoring unterstreicht, daß das ereignisgesteuerte Monitoring ein fester Bestandteil bei der Softwareentwicklung sein kann. Die durch den SW-Monitor verursachte Rückwirkung ist abhängig vom Systemcode-Anteil der beobachteten Prozesse und beträgt zwischen 0,1% bei rechenintensiven Programmen und maximal 10% bei systemintensiven Programmen (z.B. *cat*). Abhängig von der Intention bei den einzelnen Messungen, ist es sinnvoll, nicht immer alle Ereignisse aufzuzeichnen. Deshalb können beim Start einer Messung die aufzuzeichnenden Ereignisse angegeben werden. Durch die Reduzierung auf die für die Auswertung notwendigen Ereignisse sinkt die Ereignisrate und der mit den Messungen verbundene Overhead. Der Programmablauf wird durch die Ausführung eines nicht ausgewählten Meß-Ereignisses lediglich um die Dauer einer IF-Abfrage verzögert.

Ein Leistungsvergleich verschiedener Monitorsysteme ist mit dem in der Literatur üblicherweise angegebenen Overhead nur schlecht möglich, da dieser Wert z.B. bei einer Instrumentierung des Betriebssystems sowohl lastabhängig (siehe oben) und auch vom Detaillierungsgrad der Instrumentierung bzw. von der Zielsetzung beim Messen, d.h. allgemein von der Ereignisrate abhängig ist. Ein weitaus besseres Maß für den Leistungsvergleich ist die Angabe der benötigten Zeit für die Ausgabe einer Ereigniskennung beim Hybrid-Monitoring bzw. Speicherung eines Ereignis-Records beim Software-

Monitoring, da das gemessene Programm beim Monitoring bei jedem Ereignis um diese Zeit unterbrochen wird. Neben der verwendeten Technik, das Ereignis auszugeben, ist die benötigte Dauer ebenfalls von der Geschwindigkeit des Objektsystems abhängig. Die einfachste Technik besteht beim Hybrid-Monitoring darin, die Ereigniskennung mit einem Befehl an eine Schnittstelle auszugeben. Hofmann et al. geben in [Hofm87] hierfür eine Dauer von 5 µs an. In [Haba88] wird der entstehende Overhead beim Hybrid-Monitoring bei einer bestimmten Ereignisrate angegeben. Aus diesen Werten kann eine Unterbrechungszeit von 1 µs bestimmt werden.

Das UNIX-Betriebssystem ist ein prozeßorientiertes Betriebssystem. Um das Systemverhalten beobachten zu können, muß das Verhalten aller Prozesse beobachtbar gemacht werden. Das UNIX-Prozeßmodell (Abb. 2) bildet daher als funktionales Modell die Basis für die Instrumentierung.
Es wurde eine Methode entwickelt, die sich bei der Instrumentierung auf alle Einsprung- und Aussprungstellen in den Betriebssystemkern sowie die für den Ablauf signifikanten Systemprozeduren (z.B. Dateisystem, Prozeßverwaltung, kritische Betriebssystemabschnitte) beschränkt. So kann das Systemverhalten aller Prozesse beobachtet werden.

Unter einem Prozeß unter UNIX wird die Ausführung eines Programms in der durch die Prozeßdatenstrukturen definierten Umgebung verstanden. Jedem Prozeß werden zwei Datenstrukturen (ein Element der Prozeßtabelle und die User-Struktur) zugeordnet. Diese beiden Datenstrukturen beschreiben die Umgebung, in der der Prozeß ausgeführt wird. Ein Prozeß führt ein Programm schrittweise aus; jeder Prozeß hat seine eigenen Daten und seinen eigenen Stack. Ein Zugriff auf Daten und Stack eines anderen Prozesses ist nicht möglich. Die Kommunikation zwischen zwei Prozessen wird mit Hilfe von Systemfunktionen realisiert. Das Prozessorzuteilungsverfahren ist in die Klasse "round robin with multilevel feedback" einzuordnen [Bach86]. Jeder Prozeß bekommt hierbei einen Prozessor für einen bestimmten Zeitbereich zugeteilt.
Der Ablauf eines Prozesses kann mit Hilfe des in Abb. 2 dargestellten Zustandsmodells beschrieben werden. Ein Prozeß nimmt während seiner Ausführung verschiedene dieser Zustände an. Für weitere Einzelheiten sei auf [Bach86] verwiesen. Dort ist auch eine typische Zustandsfolge eines UNIX-Prozesses beschrieben.

Um neben den Zuständen eines Prozesses weitere Informationen über den Prozeßablauf zu erhalten, werden neben der Ereigniskennung, die einen Punkt im Programmtext angibt (Quellbezug) noch weitere Parameter gespeichert. Diese zusätzlichen Parameter erlauben es, mit einer geringen Zahl von Ereigniskennungen, d.h. Meßanweisungen, auszukommen und trotzdem weitere Informationen über den Prozeßverlauf zwischen den Ereignissen zu erhalten.

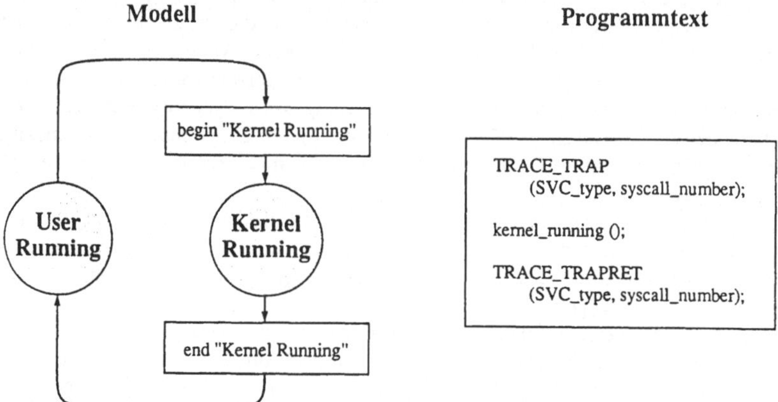

Abbildung 3. Instrumentierung des Zustands *Kernel running*

So reicht z.B. bei dieser Instrumentierungstechnik die Instrumentierung von Beginn und Ende des Zustands *Kernel running* mit der Speicherung der Parameter SVC-Typ und Sytemfunktionsnummer aus, um zu wissen, welcher SVC aufgetreten ist und falls eine Systemfunktion ausgeführt wird, um welche es sich handelt. Die Instrumentierung des Zustands *Kernel running* ist in Abb. 3 dargestellt.
Wie bereits oben ausgeführt, werden die E-Records für jeden Prozessor in einem eigenen Meßdatenspeicher gespeichert. Da jedoch Prozesse prozessorübergreifend ausgeführt werden können, muß zur Analyse des Systemzustands eine globale Ereignisspur gebildet werden (siehe Abb. 4): Die lokalen Ereignisspuren werden zu einer globalen Ereignisspur zusammengeführt. Das Ordnungskriterium der E-Records ist dabei die Zeit.

Um in der globalen Ereignisspur unterscheiden zu können, auf welchem Prozessor das Ereignis aufgetreten ist, wird der lokale E-Record beim Zusammenfügen der lokalen Ereignisspuren um eine Prozessorkennung erweitert.

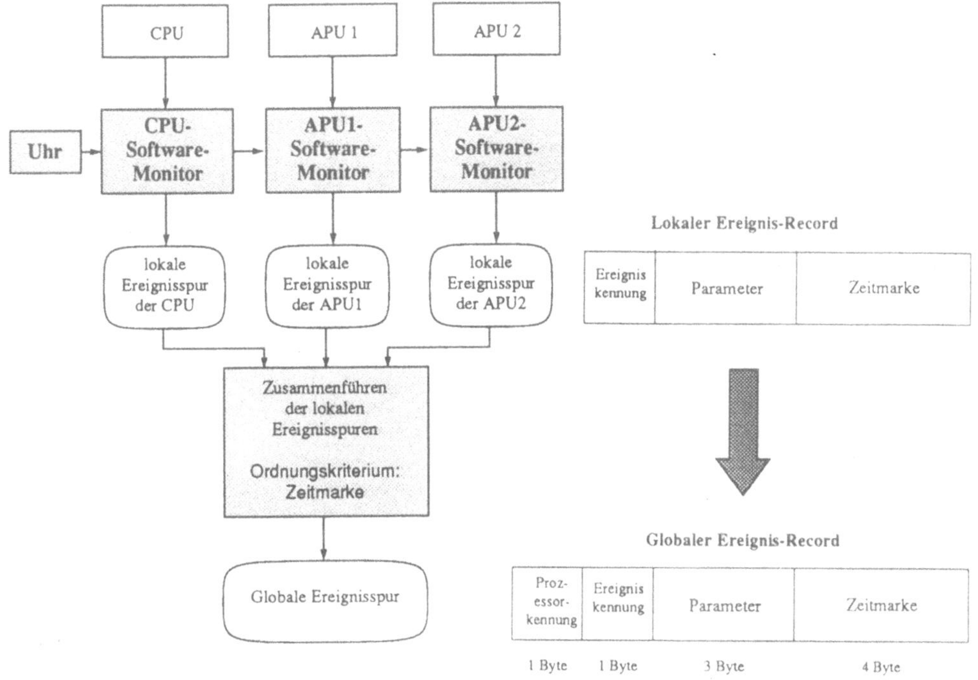

Abbildung 4. Erstellen der globalen Ereignisspur

4. Ergebnisse

4.1 Gewinn von Einsicht

Die durch das Monitoring gewonnene Einsicht soll anhand eines gemessenen Programmablaufs dargestellt werden. Als Musterlast wird das Programm cc (C-Compiler) genommen, da es ein allgemein verfügbares und häufig verwendetes Standardprogramm ist. Da von diesem Programm während der Bearbeitung — je nach Aufruf des Programms cc — noch drei bis fünf weitere Prozesse gestartet werden, eignet sich dieses Programm gut als Last zum Testen von Systemen. Es können zum einen mehrere Prozesse und zum anderen die Prozeßverzweigungen beobachtet werden. Der gemessene Ablauf dieses

Programms ist in Abb. 5 mit Hilfe von Gantt-Diagrammen visualisiert. Dabei ist für jeden Prozeß ein eigenes Gantt-Diagramm gezeichnet.
Mit Gantt-Diagrammen kann z.B. der Zusammenhang zwischen Prozessen sowie der Ablauf aller Prozesse dargestellt werden. Dabei werden auf der Abszisse die Zeit und auf der Ordinate die Aktivitäten aufgetragen. Ein Wechsel auf der Ordinate entspricht dem einleitenden Ereignis der Aktivität. Deshalb werden auch die Diagramme dadurch in Beziehung zueinander gebracht, daß sie in einem Bild mit einer gemeinsamen Zeitachse dargestellt werden.

Visualisierung der Prozeßaktivitäten Idle-, Preempted-, Kernel- und User-Mode

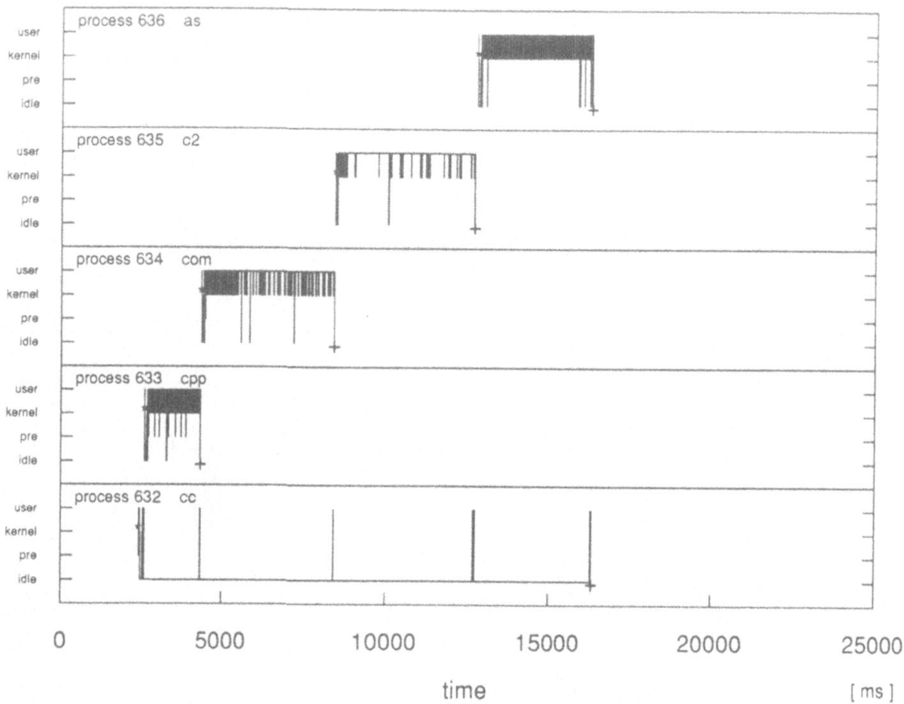

Abbildung 5. Gemessener Ablauf des C-Compilers (Programm cc)

Das in Abb. 5 dargestellte Beispiel zeigt, daß das Monitoring des Betriebssystems Einsicht sowohl in das Systemverhalten als auch in den Ablauf des bearbeiteten Programms gibt; und zwar qualitativ und quantitativ! Durch die Visualisierung aller aktiven Prozesse kann sofort die Auslastung des Systems (der erreichte Parallelitätsgrad) und die Prozeßvergabe analysiert werden. Weiterhin gibt die Visualisierung der Prozeßaktivitäten Auskunft über den Zustand eines Programms und dessen Bearbeitungszeit. Weitere in der Ereignisspur enthaltenen Informationen (zugeteilter Prozessor, ausgeführte Systemfunktion bei Wechsel in den Systemmodus, etc.) können interaktiv im Gantt-Diagramm angezeigt werden. Das Gantt-Diagramm in Abb. 5 nutzt nicht alle in der Ereignisspur enthaltene Information aus. So ist nicht ersichtlich, auf welchem der drei Prozessoren ein Prozeß bearbeitet wird. Dies kann aber entweder — für jeden Prozeß getrennt — in einem eigenen Gantt-Diagramm oder mit einer farblichen Unterscheidung bzw. mit einer Aufspaltung der Zustände *kernel* und *user* in die Zustände *kernel_CPU*, *kernel_APU_1* und *kernel_APU_2* (Zustand *user* analog) in dem in Abb. 5 dargestellten Gantt-Diagramm erfolgen. Das Problem bei der Visualisierung liegt darin, Zusammenhänge aufzuzeigen, ohne zu sehr auf Details einzugehen.

Der Visualisierung von komplexen Betriebssystemabläufen muß immer eine Wahl der zu visualisierenden Aktivitäten vorausgehen. So kann z.B. die Darstellung des Systemverhaltens durch eine Visualisierung der Prozeßaktivitäten, der Prozessoraktivitäten oder der Prozessorzuteilung erfolgen.
Als Hilfsmittel bei der Auswertung und der Visualisierung der gemessenen Abläufe wird das Statistikpaket S [Beck84] verwendet. Dieses Statistikpaket stellt neben verschiedenen statistischen Funktionen auch eine Datenbank-Abfragesprache zur Verfügung, die es erlaubt, nach bestimmten Zeit- oder Ereignisbedingungen in der Ereignisspur zu suchen. Daneben stehen verschiedene Treiber zur graphischen Ausgabe zur Verfügung. Durch eine von Mohr vorgenommene Erweiterung des Statistikpakets um eine S-POET-Schnittstelle (siehe [Quic88]) kann das Statistikpaket S unmittelbar zur problemorientierten Auswertung von Ereignisspuren eingesetzt werden.

Bei der Analyse der BS-Algorithmen lag der Schwerpunkt auf den für die Betriebssystementwicklung zum MP-Betriebssystem neu zu implementierenden Algorithmen zur Prozeßvergabe (scheduling) und zur Vergabe der Monitore. Exemplarisch werden daher im folgenden Abschnitt zwei durch die Messungen näher untersuchte BS-Algorithmen dargestellt.

4.2 Verbesserung von Betriebssystem-Algorithmen

Beispiel 1: Rechenzeitvergabe

Durch die Visualisierung des dynamischen Systemverhaltens kann der Ablauf der BS-Algorithmen untersucht werden. Im folgenden soll anhand der Beoabachtung mehrerer parallel ablaufender rechenintensiver Anwenderprogramme mit einem sehr geringen Systemanteil zu Beginn des Programms (Initialisierungsphase) die Prozessorvergabe an solche Prozesse analysiert werden. In Abb. 2 ist dargestellt, daß ein Wechsel vom Zustand *User running* in den Zustand *Preempted* nur über einen vorhergehenden Wechsel in den Zustand *Kernel running* realisiert werden kann.

Mit der Visualisierung der Anwenderprogrammabläufe konnte ermittelt werden, daß dieser Zustandswechsel falsch implementiert ist. In Abb. 6 ist der parallele Ablauf von sechs gleichen rechenintensiven Anwenderprogrammen (Programm *memcpu*), in dem in einer Iterationsschleife Speicherbereiche kopiert werden, um den Prozessor zu testen und die Geschwindigkeit des Speichers zu bestimmen, dargestellt. Auch hier werden wiederum die Prozeßaktivitäten visualisiert.
Auffallend bei den Gantt-Diagrammen ist die gleichmäßige alternierende Bearbeitung aller sechs Prozesse zu Beginn des Programmlaufs und die anschließende Bearbeitung von zunächst nur drei der sechs Prozesse (Prozesse 184, 185 und 188). Hier wäre aber auch eine gleichmäßige Bearbeitung aller sechs Prozesse wie in der Initialisierungsphase zu erwarten gewesen.
Durch die dynamische Prioritätenberechnung unter UNIX hätte die Priorität der bearbeiteten Prozesse sinken und die Priorität der wartenden Prozesse steigen müssen, sodaß nach einer gewissen Zeit die Priorität der wartenden Prozesse höher als die der bearbeiteten Prozesse wäre und diese verdrängt würden. Die Prozesse würden in den Zustand *Preempted* gelangen.

In der Bearbeitungsphase fallen bei zwei der drei bearbeiteten Prozesse die häufigen, im Sekundentakt auftretenden, Zustandswechsel zwischen Kernel- und User-Mode auf. Aufgrund der Tatsache, daß gleichartige Programme parallel ausgeführt werden und aufgrund des Programmtextes bekannt ist, daß nach einer kurzen Initialisierungsphase nur noch Usercode, also keine Systemfunktionen mehr ausgeführt werden, sind diese auftretenden Zustandswechsel nicht auf das Programm, sondern auf dessen Bearbeitung, also auf das Betriebssystem zurückzuführen.
Da die beobachtete Anomalie nur bei zweien der drei laufenden Prozesse auftritt (Prozesse 184 und 185), liegt die Vermutung nahe, daß die Zustandswechseln während der Bearbeitung auf die Asymmetrie des Multiprozessorsystems (CPU, 2 APUs) zurückzuführen ist und diese beiden Prozesse auf den APUs ausgeführt werden. Falls dieser häufige Wechsel nur bei einem der Prozesse aufgetreten wäre, wäre der Fehler bei der Bearbeitung auf der CPU zu suchen. Durch die Gleichmäßigkeit der Zustandswechsel (jede Sekunde) kann der Fehler auf die Bearbeitung der Clock-Interrupts eingegrenzt werden. Dafür spricht auch, daß die Zustandswechsel nur bei den auf den APUs ausgeführten Prozesse auftreten, da von

der CPU parallel zur Programmbearbeitung Interrupts an die APUs gesendet werden, während die momentan bearbeiteten Prozesse auf den APUs für die Bearbeitung des Interrupts in den Zustand *Kernel Running* gelangen. Durch diese Interrupts soll u.a. eine neue Prioritätenberechnung (für Details siehe bei [Bach86], Kap. 8 und bei [Klei87], Kap. 3.2) aller im System vorhandenen Prozesse und gegebenenfalls ein Prozeßwechsel ausgelöst werden (siehe oben), wenn höherpriore Prozesse auf die Bearbeitung warten. Da die anderen drei lauffähigen Prozesse jedoch nicht alternierend zu den laufenden Prozessen bearbeitet werden, kann auf einen Fehler bei der Interruptbearbeitung oder der Prioritätenberechnung (siehe Abschnitt 3.1) geschlossen werden.

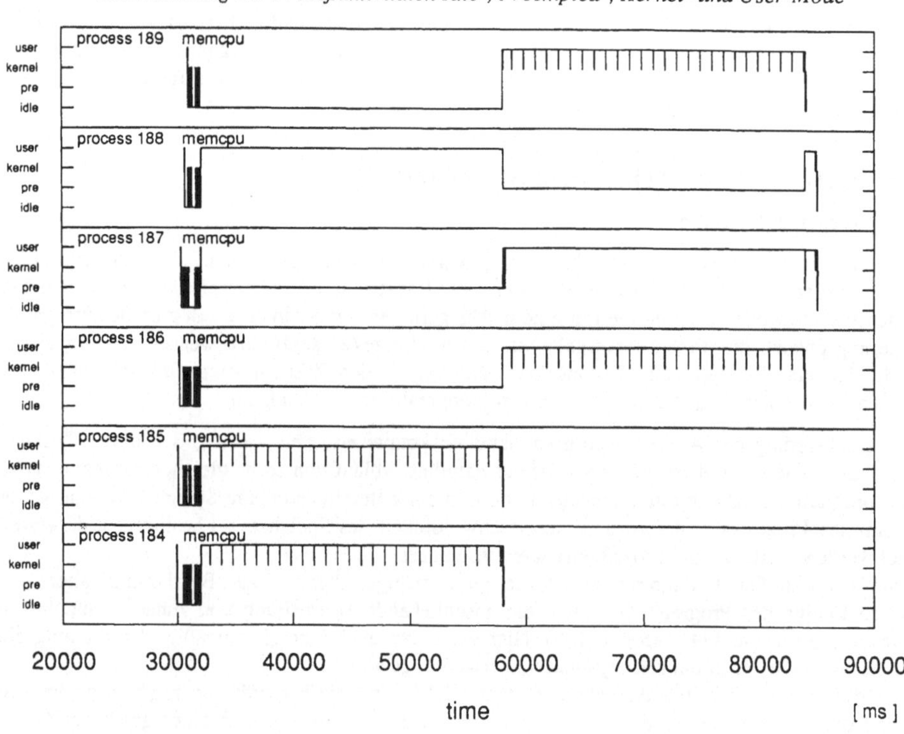

Abbildung 6. Gemessener Ablauf von sechs Anwenderprogrammen

Im dargestellten Beispiel führte die Visualisierung des Systemverhaltens zur Erkennung eines Fehlers. Die daraus abgeleitete Reimplementierung der Prioritätenberechnung und -zuordnung verschaffte allen Prozessen dieselben Bearbeitungschancen und führte zu einer alternierenden Bearbeitung der im System zu bearbeitenden Last.

An diesem Beispiel zeigt sich sehr deutlich, daß der Gewinn von Einsicht über das interne Systemverhalten sehr wichtig ist und daß eine geeignete Auswertung der Meßdaten sehr schnell zu einer Lokalisierung des Fehlers führt.

Beispiel 2: Verbesserung einer ineffizienten BS-Funktion

In einem weiteren Beispiel sei eine Verbesserung des Algorithmus zur Monitorvergabe durch eine kombinierte Auswertemethode aus Visualisierung der Prozeßaktivitäten und einer statistischen Auswertung dargestellt. Dieses Beispiel zeigt, wie wichtig die Untersuchung — gerade in MP-Systemen — des konkurrierenden Zugriffs mehrerer Prozesse auf Betriebsmittel ist.

Die Monitore dienen als Ausschlußmechanismus für die exklusive Ausführung der kritischen BS-Abschnitte. Sie müssen vor der Ausführung eines solchen Abschnittes angefordert werden und nach dessen Ausführung wieder freigegeben werden. Für jeden kritischen Bereich wird ein Monitor bereitgestellt. Da die Vergabestrategie der Monitore nach der Parallelisierung der BS-Funktionen die Systemleistung (bei einem hohen Anteil von Systemcode; und dies ist bei einem Entwicklungssystem wie dem Objektsystem der Fall) entscheidend beeinflußt, kommt der Analyse der Vergabestrategie eine große Bedeutung zu. Falls die Belegung der Monitore zu einem Engpaß wird, wird die Vergabe der Monitore zu einem zweiten Schedulingverfahren. Ein Monitor kann sich in den in Abb. 7 dargestellten Zuständen befinden. Ein Monitor ist *Idle*, wenn er von keinem Prozeß benötigt wird; nach einer Anforderung gelangt er in den Zustand *Belegt*. Ein weiterer Zustand, der Leerlauf eines angeforderten Monitors, wird dann angenommen, wenn der Monitor während der Belegtzeit von einem weiteren Prozeß angefordert wird. Die Zeit zwischen der Freigabe und der erneuten Belegung des auf den Monitor wartenden Prozesses wird als *Leerlaufzeit* bezeichnet. Die Leerlaufzeit ist also als die Zeit definiert, die ein Monitor nicht belegt ist, obwohl ein Prozeß auf die Belegung dieses Monitors wartet. Für die Berechnung der Leerlaufzeit mit dem Statistikpaket S sei auf [Quic88] verwiesen.
Die Monitorvergabe ist durch eine FCFS-Strategie realisiert. Dadurch wird eine faire Monitorvergabe an alle Prozesse, unabhängig vom Prozessor, auf dem sie ausgeführt werden, garantiert. Um die Qualität der Monitorvergabe bewerten zu können, muß jedoch neben der Belegtzeit auch die Leerlaufzeit eines Monitors berechnet werden.

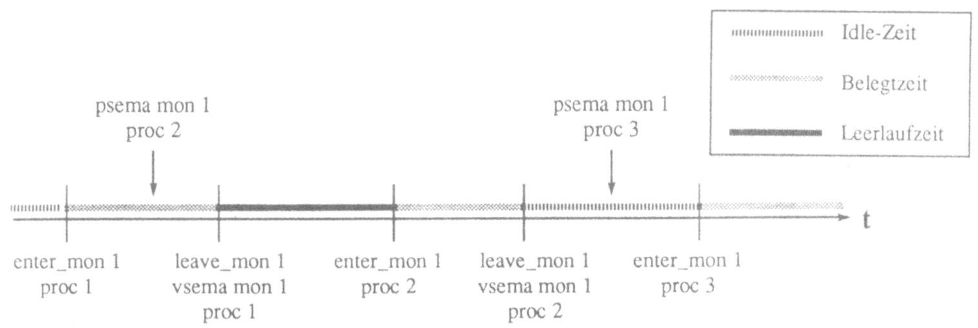

Abbildung 7. Idle-, Belegt- und Leerlaufzeit eines Monitors

Die für die Monitore entscheidenen Werte wie Belegtzeit, Belegungshäufigkeit und Leerlaufzeit können durch die Instrumentierung der entsprechenden Stellen im Betriebssystem ermittelt und mit Hilfe der vorhandenen Auswertewerkzeuge berechnet und graphisch dargestellt werden.

Da bei den betrachteten Programmen (mehrfach paralleler Ablauf des C-Compilers) auf das Filesystem — und damit auch auf den Filesystem-Monitor — am häufigsten zugriffen worden ist, sei anhand dieses Monitors exemplarisch erläutert, warum sich die implementierte Vergabestrategie ungunstig auf die Vergabe der Monitore und damit auch auf die Systemleistung auswirkt. Dabei sei die Vergabestrategie dann gut implementiert, wenn die Belegtzeit eines Monitors größer als dessen Leerlaufzeit ist.
Die Messungen haben für die Belegt- und Leerlaufzeit des Filesystem-Monitors die in Abb. 8 dargestellte Verteilungen ergeben. Daraus wird deutlich, daß die Leerlaufzeit im Mittel doppelt so groß wie die Belegtzeit des Monitors ist. Die relativ große Leerlaufzeit ist durch den benötigten Prozeßwechsel zu erklären, der erforderlich ist, um den auf den Monitor wartenden Prozeß zu aktivieren und weiter zu bearbeiten. Da dies die meiste Zeit zwischen der Freigabe des Monitors und der erneuten Belegung durch den wartenden Prozeß ausmacht, soll die Leerlaufzeit nicht weiter aufgeschlüsselt werden, sondern ein Verfahren beschrieben werden, das den Prozeßwechsel verhindert.
Die Verteilung der Belegtzeiten des Filesystem-Monitors (Abb. 8) zeigt, daß Monitorbelegungen häufig kürzer als 500 µs sind. Falls in dieser Zeit ein weiterer Prozeß den Monitor belegen will, wird dieser Prozeß blockiert (Zustand *Asleep in Memory*) und nach der Freigabe des Monitors wieder aufgenommen.

Der Prozeß geht über den Zustand *Ready to Run in Memory* in den Zustand *Kernel Running*. Gerade bei Belegtzeiten, die kürzer als die für einen Prozeßwechsel benötigte Zeit ist, kann die Monitorvergabe durch ein sog. *aktives Warten* auf den Monitor beschleunigt werden. Während dieser Wartezeit wird in bestimmten Intervallen vom wartenden Prozeß überprüft, ob der Monitor inzwischen freigegeben worden ist. Durch das aktive Warten wird zwar ein Prozessor belegt, bei einer geeigneten Wahl der aktiven Wartezeit kommt es aber wegen der nicht mehr notwendigen Prozeßwechsel (pro Prozeßwechsel ca. 200 µs) zu einer schnelleren Monitorbelegung. Daraus folgt eine geringeren Monitor-Leerlaufzeit und damit zu einer schnelleren Bearbeitung der Prozesse, was trotz Prozessorbelegung während der Wartezeit zu einer Erhöhung des Gesamtdurchsatzes des Multiprozessorsystems führt. Falls der Monitor nach der vorher bestimmten Wartezeit jedoch noch nicht freigegeben worden ist, wird die oben beschriebene Vergabestrategie mit Prozeßwechsel angewendet. Dies ist zum einen sinnvoll, um bei langen Monitorbelegungen, den Prozessor nicht unnötig lange mit einem auf einem Monitor wartenden Prozeß zu belasten und zum anderen notwendig, um Prozessen, die den Monitor in den aktiven Wartephasen nicht zugeteilt bekamen, weil ein anderer Prozeß den Monitor aufgriff, den Monitor nach der FCFS-Strategie zuzuteilen. Da es bei der momentanen Prozessorkonfiguration (drei Prozessoren) nur möglich ist, daß zwei Prozesse aktiv warten, während ein anderer Prozeß einen Monitor belegt, kommt der zuletzt genannte Fall allerdings nicht häufig vor. Bei mehreren Prozessoren muß die Monitorzuteilungsstrategie dahingehend überprüft werden, ob die Schedulingstrategie bei der Prozeßzuteilung nicht durch die Monitorzuteilungsstrategie aufgehoben wird. Hierauf soll jedoch im Rahmen dieser Arbeit nicht näher eingegangen werden.

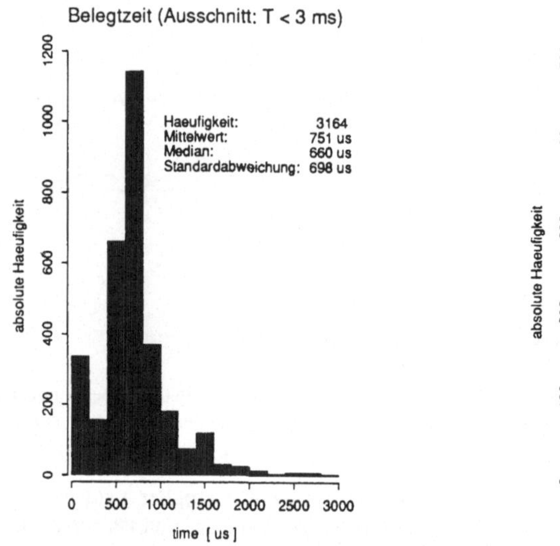

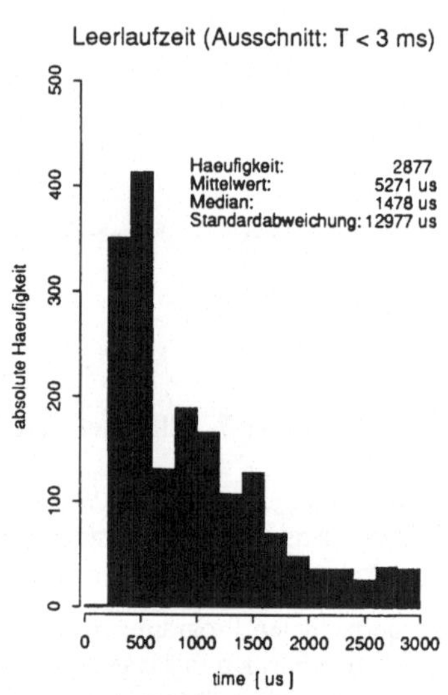

Abbildung 8. Verteilung der Belegt- und Leerlaufzeit des Filesystem-Monitors

Ein direkter Vergleich der Leerlauf- und Wartezeiten der Monitore zwischen beiden Implementierungen ist nicht möglich, da die Zeiten der beiden Implementierungen durch die Hinzunahme der aktiven Wartezeit nicht mehr unmittelbar miteinander vergleichbar sind. Es kann allerdings, da die Änderung der Monitorvergabestrategie den Durchsatz des Systems erhöhen sollte, ein Vergleich über dieses Leistungsmaß durchgeführt werden. Die nach der Reimplementierung erfolgten Messungen ergaben, daß der Durchsatz des MP-Systems bei monitorintensiven Programmen (z.B. C-Compiler cc) durch die neue, effizientere Monitorvergabe erhöht worden ist.

5. Ausblick

An zwei Beispielen haben wir gezeigt, wie notwendig und sinnvoll das Monitoring während der Programmentwicklung in MP-Systemen ist. Die Frage nach der Parallelisierung von bestehenden Programmen und nach der Programmierung solcher Systeme wird auch in Zukunft immer mehr in der Vordergrund rücken. Dazu reicht es nicht aus, nur die Leistung eines Systems (z.B. die erreichte Auslastung, Speed-up Werte, etc.) zu ermitteln, sondern es müssen auch die Ursachen für die Systemleistungen bekannt sein. Zur Erkennung dieser Probleme kann die Beobachtung des aktiven Systems mit Monitoren einen entscheidenden Beitrag über die Leistungsbewertung hinaus zum *"Gewinn von Einsicht"* leisten. Der kausale Zusammenhang zwischen Ursache und Wirkung muß in MP-/MC-Systemen unbedingt aufgezeigt werden.
Um aus den Ereignisspuren die für die Programmverbesserungen notwendigen Daten extrahieren zu können, müssen leistungsfähige Werkzeuge zur Auswertung und Visualisierung des dynamischen Systemverhaltens zur Verfügung stehen. Neben der quellbezogenen Auswertung muß eine systembezogene Auswertung, d.h. eine Auswertung nicht nur auf der Algorithmenebene, sondern auch auf der Systemebene (*"Wo wird welcher Teil des Algorithmus wann ausgeführt?"*) durchgeführt werden können. Gerade in verteilten Systemen ist eine Darstellung der Wechselwirkungen und des Zusammenspiels aller Komponenten notwendig.

Die Instrumentierung des Betriebssystems muß in MP-/MC-Systemen fester Bestandteil der Systementwicklung sein. Die angewendete Monitoring-Methode hat ebenso wie die Auswertung der Ereignisspuren die erhofften Ergebnisse gebracht und kann auf andere Objekte — z.B. bei der Entwicklung des UNIX-Multiprozessor-Betriebssystems für den SUPRENUM-Front-End-Rechner MPR 2300 [SUPR89]— übertragen werden.

Mein Dank gilt R. Hofmann und W. Hofmann, die diese Arbeit initiiert haben und für Fragen zum Monitoring und zur Betriebssystementwicklung immer ein offenes Ohr hatten. Stellvertretend für alle, die durch ihre Diskussionsbeiträge zu diesem Beitrag beigetragen haben, sei Dr. R. Klar gedankt.

Literatur

[Bach86] Bach, M.J.: **The Design of the UNIX Operating System.** Prentice-Hall, Inc., Englewood Cliffs, New Jersey 07623, 1986.

[Beck84] Becker, R.A., Chamber, J.M.: **S: An Interactive Environment for Data Analysis and Graphics.** The Wadsworth Statistics/Probability Series, Wadsworth Advanced Book Program, Belmont, California, A Division of Wadsworth, Inc., 1984.

[Bemm88] Bemmerl, T.: **Quellbezogenes Debugging von Multimikroprozessoren.** In: R. Valk (Hrsg), GI-18.Jahrestagung, Band I, Vernetzte und komplexe Informatik-Systeme, Hamburg, Oktober 1988, Springer Verlag, 1988.

[Burk89] Burkhart, H., Millen, R.: **Performance Measurement Tools in a Multiprocessor Environment.** In: IEEE Transactions on Computers, Vol. 38, No. 5, May 1989, Pp. 725-737.

[Conc86] Concurrent Computer Corporation: **3280 System Instruction Set and Reference Manual.** Tinton Falls, 1986.

[Haba88] Haban, D., Wybranietz, D.: **A Tool for Measuring and Monitoring Distributed Systems During Operation.** In: Architektur und Betrieb von Rechensystemen, 10. GI/ITG-Fachtagung, Paderborn, März 1988, Proceedings, Springer-Verlag, 1988.

[Hofm87] Hofmann, R., Klar, R., Luttenberger, N., Mohr, B.: **Zählmonitor 4: Ein Monitorsystem für das Hardware- und Hybrid-Monitoring von Multiprozessor- und Multicomputer-Systemen.** In: Messung, Modellierung und Bewertung von Rechensystemen, 4. GI/ITG-Fachtagung, Erlangen, September/Oktober 1987. Proceedings, Springer Verlag, 1987.

[Hofm88] Hofmann, R., Klar, R., Luttenberger, N., Mohr, B., Werner, G.: **An Approach to Monitoring and Modeling of Multiprocessor and Multicomputer Systems.** In: Proceedings of IFIP W.G. 7.3 International Seminar on Performance of Distributed and Parallel Systems, Kyoto, Japan, 7-9 December 1988, Edited by T. Hasegawa, H. Takagi and Y. Takahashi.

[Hofm89] Hofmann, W.: **Erweiterung des UNIX-Betriebssystems für Multiprozessoren.** Bericht Nr. 89/4 des Sonderforschungsbereichs 182 Multiprozessor- und Netzwerkkonfigurationen, Arbeitsberichte des IMMD, Band 22, Nummer 5, Universität Erlangen-Nürnberg, März 1989.

[Klar81] Klar, R.: **Hardware Measurement and Their Application on Performance Evaluation in a Processor-Array.** Computing, Suppl. 3, 65-88, 1981, Springer Verlag 1981.

[Klar85] Klar, R.: **Hardware/Software-Monitoring.** In: Das aktuelle Schlagwort, Informatik-Spektrum (1985) 8: 37-40, Springer Verlag, 1985.

[Klei87] Kleinöder, J.: **Prozessorzuteilungsstrategien in enggekoppelten Mehrprozessorsystemen unter UNIX.** Diplomarbeit am IMMD IV, Universität Erlangen-Nürnberg, April 1987.

[LeBl85] LeBlanc, R.J., Robbins, A.D.: **Event-Driven Monitoring of Distributed Programs.** In: Proceedings of the International Conference on Distributed Computing Systems, Denver 1985.

[Lutt89] Luttenberger, N.: **Monitoring von Multiprozessor- und Multicomputer-Systemen.** Dissertation, in: Arbeitsberichte des Instituts für mathematische Maschinen und Datenverarbeitung, Band 22, Nr. 7, Universität Erlangen-Nürnberg, März 1989.

[Mill86] Miller, B.P., MacRander, C., Sechrest, S.: **A Distributed Programs Monitor for Berkeley UNIX.** In: Software-Practice and Experience. Vol 16(2), 183-200, February 1986.

[Mink87] Mink, A., Draper, J.M., Roberts. J.W., Carpenter, R.J.: **Hardware- Assisted Multiprocessor Performance Measurement.** In: Proceedings of the 12th IFIP WG 7.3 International Symposium on Computer Performance Modeling, Measurement and Evaluation, Brussels, Belgium, 7-9 December 1987.

[Mohr87] Mohr, B.: **Entwurf und Implementierung eines Systems zur Entschlüsselung von Monitordaten.** Diplomarbeit am IMMD VII, Universität Erlangen-Nürnberg, April 1987.

[Mohr88] Mohr, B.: **Zählmonitor 4 — Quellbezogene Auswertung mit TDL und POET.** Interner Bericht 2/88 des IMMD VII, Universität Erlangen-Nürnberg.

[Quic88] Quick, A.: **Monitoring und Visualisierung des dynamischen Verhaltens eines UNIX-Mehrprozessor-Betriebssystems.** Diplomarbeit am IMMD VII, Universität Erlangen-Nürnberg, August 1988.

[Rash87] Rashid, R.: **Designs for parallel Architecture.** UNIX Review, April 1987, pp 36-43.

[SUPR89] SUPRENUM-Datenblatt: **Front-End-System: MPR 2300.** SUPRENUM GmbH, Bonn 03/89.

[Zieh88] Zieher, M., Zitterbart, M.: **NETMON - a Distributed Monitor System.** In: EFOC/LAN-88, The Sixth European Fibre Optic Communications and Local Area Networks Exposition, June 29-July 1, 1988, International Congresscentrum RAI, Amsterdam, The Netherlands.

Numerische Analyse zweier UNIX Multiprozessor-Betriebssysteme

Hermann Jung

Institut für
Mathematische Maschinen und Datenverarbeitung IV
Lehrstuhl für Betriebssysteme
Martensstraße 1
8520 Erlangen

Zusammenfassung

Der folgende Beitrag beschreibt die theoretische Untersuchung des Leistungsverhaltens zweier Betriebssystemvarianten für den Multiprozessor Concurrent Computer MPS 3280. Dazu werden die Prozesse des UNIX-Betriebssystems als Markovketten modelliert, deren Zustandsübergänge vom Zustand des Gesamtsystems abhängig sind. Die Gleichgewichtsverteilungen der so entstehenden Zustandsräume werden numerisch bestimmt und daraus die Leistungsgrößen berechnet. Die Systeme werden sowohl aus Betreibersicht wie auch aus Benutzersicht analysiert und bewertet, wobei die Modellparameter am realen System gemessen und die Ergebnisse dort validiert werden.

1. Motivation

Der Einsatz von Multiprozessoren kann mit zwei unterschiedlichen Zielrichtungen erfolgen. Zum einen soll durch Parallelverarbeitung eine *einzige* Aufgabe, zum anderen soll die Gesamtheit *aller* Aufgaben beschleunigt werden.

Für den Multiprozessor MPS3280 der Firma Concurrent Computer, der in einer interaktiven Programmentwicklungs-Umgebung eingesetzt wird, war und ist das zweite Ziel maßgebend; also die Beschleunigung *aller* Aufgaben. Dazu wurde das vorhandene UNIX-V.2 Betriebssystem für den Parallelbetrieb erweitert.

Durch die Unstrukturiertheit des Betriebssystemkerns ist eine Parallelisierung aufwendig und fehleranfällig. Ein erster Schritt war die Implementierung eines Systems, das nur Benutzercode parallel ausführen konnte. Intensive Tests ergaben eine Leistungssteigerung für rechenintensive Aufgaben, für die im interaktiven Betrieb aber vorherrschenden Aufgaben konnte keine wesentliche Leistungssteigerung festgestellt werden [Kleinöder87]. Durch eine Parallelisierung auch des Betriebssystemkerns wurde eine weitere Leistungssteigerung erwartet.

Aus Sicht der "theoretisch untermauerten Systementwicklung" war es das Ziel, die folgenden Fragen zu beantworten:

- Welches sind die kritischen, d. h. die leistungsbestimmenden Parameter des Systems ?

- Wie ändert sich die Leistung des Systems bei unterschiedlicher Struktur des Betriebssystems ?
- Wie ändert sich die Leistung des Systems bei unterschiedlicher Zahl von Prozessoren ?

Dazu wurde das System als eine zeitkontinuierliche Markovkette modelliert, für diese die Gleichgewichtsverteilung bestimmt und dann daraus die interessierenden Leistungsgrößen berechnet.

Da beide Betriebssystem-Varianten verfügbar sind, ist aus Sicht der Modellbildung auch die Beantwortung der folgenden Frage interessant:

- Wie genau sind die entworfenen Modelle ?

Die Arbeit ist dazu wie folgt aufgebaut: Nach einer Beschreibung der zugrunde liegenden Hardware werden die beiden Betriebssystemvarianten in ihren für die Analyse wichtigen Aspekten beschrieben. Anschließend wird ein Modell für das Systemverhalten entworfen und mit diesem die oben gestellten Fragen beantwortet.

2. Die zu untersuchenden Betriebssystemvarianten

2.1 Hardwarevoraussetzungen

Bei der zu untersuchenden Anlage handelt es sich um einen speichergekoppelten Multiprozessor MPS 3280 der Firma Concurrent Computer, der z. Zt. mit drei Prozessoren ausgebaut ist. Die Prozessoren sind über einen schnellen synchronen Bus (S-Bus) mit dem globalen Hauptspeicher verbunden und verfügen zu dessen Entlastung je über einen Cache-Speicher (Abb. 1).

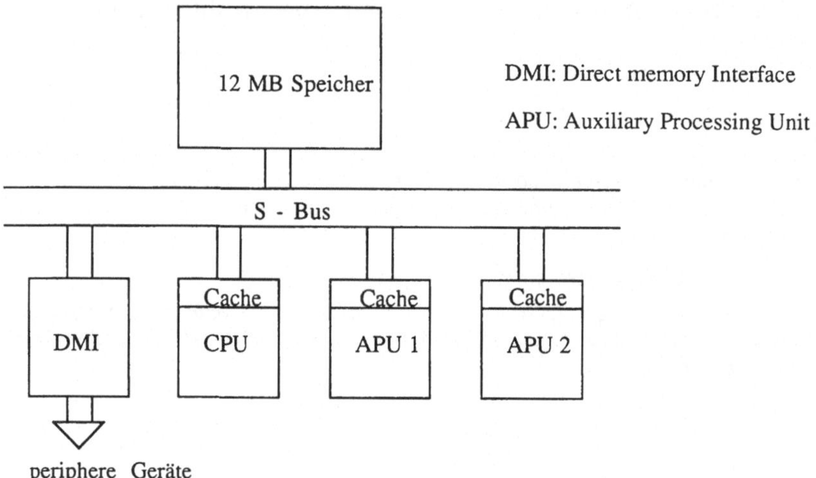

Abb. 1: Blockschaltbild des Rechners MPS 3280

Die Konsistenz des Cache wird durch einen *Write-Through* Mechanismus zusammen mit einer Cache-Invalidierung garantiert. Bei Schreiboperationen wird das zu schreibende Wort direkt in den Hauptspeicher geschrieben und gleichzeitig Einträge desselben Wortes in den anderen Cache-Speichern invalidiert. Über eine Anschaltung (*Direct Memory Interface* DMI) und den S-Bus sind die E/A-Bussysteme erreichbar, an denen die peripheren Geräte angeschlossen sind. Der Zugriff auf die peripheren Geräte ist somit prinzipiell von jedem Prozessor aus möglich, beim gegebenen Hardware-Ausbau aber auf einen Prozessor beschränkt.

Durch unterschiedlichen Mikrocode erhalten die hardwaremäßig identischen Prozessoren unterschiedliche Funktionalität. Ein zentraler Prozessor (CPU) besitzt die Kontrolle über die peripheren Geräte und alle anderen Prozessoren (*Auxiliary Processing Units* APUs) [Concurrent86].

2.2 Softwarevoraussetzungen

Als Basis der Betriebssystem-(Weiter-) Entwicklung wurde ein UNIX System-V.2 verwendet. Die kleinste Verwaltungseinheit des Betriebssystems ist ein Prozeß, der eine eigene Schutzumgebung besitzt und in UNIX typischerweise die Ausführung eines Kommandos (z. B. *cat, ls, cc*) oder eines Benutzerprogramms (z. B. eine numerische Berechnung) ist. Der UNIX-Systemkern besitzt eine prozedurale Schnittstelle, d. h. Prozesse rufen Systemdienste als Unterprogramme auf mit dem allerdings ein Wechsel der Schutzumgebung vom Benutzer- in den System-Kontext verbunden.

Das für Monoprozessoren konzipierte UNIX-System erlaubt keine Verdrängung von Prozessen im System-Kontext, ein Prozeßwechsel innerhalb des Systemkerns ist nur an natürlichen Wartestellen möglich. Der Zugriff auf kritische Datenstrukturen des Systems ist daher nicht gegen mehrfachen, gleichzeitigen Zugriff geschützt. Während Prozesse im Benutzerkontext völlig unabhängig voneinander ablaufen (aus Sicht des Zugriffs auf kritische Datenstrukturen des Betriebssystems), müssen gleichzeitige Aktivitäten im Systemkontext synchronisiert werden. Typische Aktivitäten im Systemmodus sind z. B. Zugriffe auf Dateien, die durch einen in der Software realisierten Cache-Mechanismus[1] optimiert werden. Ist das im Zugriff geforderte Datum schon oder noch im Cache vorhanden, so kann es sofort an den anfordernden Prozeß übergeben werden. Ist das Datum nicht vorhanden, so führt das zu einer E/A-Anforderung; das periphere Gerät wird beauftragt, der anfordernde Prozeß blockiert sich und wird erst wieder deblockiert, wenn seine E/A-Anforderung erfüllt ist.

Damit ergeben sich die folgenden Prozeßzustände:

Benutzerkontext: Der Prozeß arbeitet den vom Benutzer verfassten Code ab. Der den Prozeß bearbeitende Prozessor läuft im geschützten Modus, und der Prozeß kann an jeder beliebigen Stelle unterbrochen werden und später dort wieder aufgesetzt werden.

Systemkontext: Der Prozeß bearbeitet einen Systemaufruf. Der ihn bearbeitende Prozessor befindet sich im privilegierten Modus, der Prozeß kann nicht unterbrochen werden, sondern blockiert sich im Systemkern (und gibt somit den Prozessor frei) oder beendet den Systemaufruf und arbeitet wieder im Benutzerkontext (und kann dann dort unterbrochen werden).

Treiberkontext: Der Prozessor kommuniziert mit den peripheren Geräten, läuft im privilegierten Modus und kann nicht unterbrochen werden.

Blockiert sich ein Prozeß im privilegierten Modus, so haben evtl. wartende Anforderungen im Treibermodus die höchste Priorität. Anschließend werden wartende Prozesse im Systemmodus bearbeitet und dann diejenigen im Benutzermodus. Dabei werden Prozesse im Benutzerkontext nach einer *Multilevel-Feedback*-Strategie bearbeitet [Bach86].

1) Dieser logische Cache für Plattenblöcke ist nicht mit dem in der Hardware realisierten Cache für Speicherworte zu verwechseln.

3. Die beiden Betriebssystemvarianten

Da UNIX für Monoprozessoren konzipiert wurde, sind keinerlei Mechanismen für eine Parallelisierung des Betriebssystemkerns vorhanden und damit eine Erweiterung des Systems für Multiprozessoren zeitaufwendig. Relativ schnell erhält man ein lauffähiges Multiprozessor-System, wenn lediglich eine Parallelarbeit im Benutzerkontext zugelassen wird. Dabei muß jeder Prozeß, der einen Systemcall ausführen will, auf die CPU verlagert und dort wieder aufgegriffen werden. Dieses System soll als asymmetrische Variante bezeichnet werden [Kleinöder87].

Messungen ergeben, daß das UNIX-System ca. 50% seiner Zeit im privilegierten Modus arbeitet, eine Parallelisierung des Betriebssystemkerns also eine weitere Leistungssteigerung verspricht. Bei der gegebenen Hardware ist es möglich, bis auf die Kommunikation mit den externen Geräten (Treiber-Routinen) alle Systemfunktionen auf jedem Prozessor auszuführen. Hat ein laufender Prozeß eine E/A-Anforderung, so trägt er diese lediglich in eine Warteschlange (*Systemqueue*) ein und blockiert sich, bis diese Anforderung erfüllt ist. Die CPU überprüft bei jedem Prozeßwechsel und nach jedem Wechsel der Schutzumgebung die Systemqueue und arbeitet die dort abgelegten Anforderungen ab. Ein derartiges System soll als symmetrische Variante bezeichnet werden [Hofmann89].

Zur Beurteilung der Leistungsfähigkeit der beiden Systemvarianten kann auf Testprogramme zurückgegriffen werden, deren Laufzeiten zu messen und gegeneinander aufzurechen sind. Allerdings ist dieses Vorgehen erst nach Fertigstellung beider Systeme möglich; eine fundierte Aussage darüber, wie sich das Multiprozessorsystem z. B. bei Ausbau der Hardware um weitere Prozessoren verhält, ist aber selbst dann nicht möglich. Man ist also spätestens hier auf ein Systemmodell und daraus abgeleitete Prognosen angewiesen.

4. Rechensysteme als Markovsche Zustandsräume

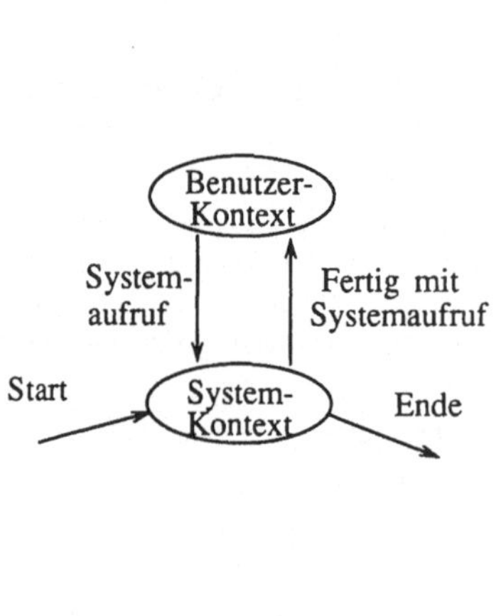

Abb. 2a: Lebenslauf eines UNIX-Prozesses

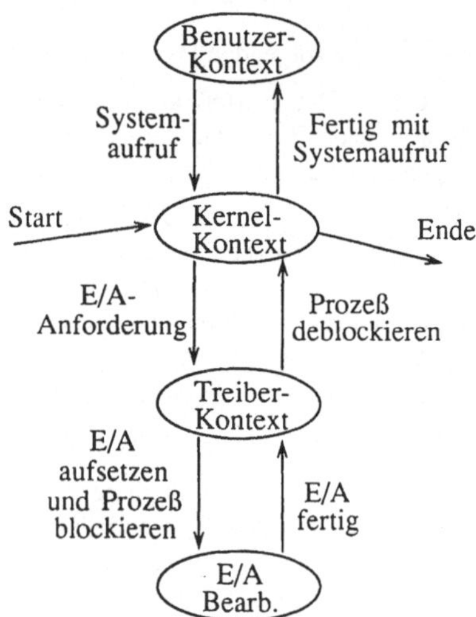

Abb. 2b: erweiterter Lebenslauf eines UNIX-Prozesses

Betrachtet man einen typischen UNIX-Prozeß, so stellt sich dessen Lebenslauf als eine (endliche) Folge von Benutzercodeabschnitten dar, die immer durch einen Systemaufruf getrennt werden. Sei weiter angenommen, daß der Lebenslauf im Betriebssystem beginnt und endet, so ergibt sich das in Abb. 2a angegebene Zustandsdiagramm.

Charakteristisch für das asymmetrische System ist, daß lediglich die obere Hälfte, d. h. der Benutzercode, auf jedem beliebigen Prozessor abgearbeitet werden kann, der Rest aber immer auf der CPU behandelt werden muß. Diese Einschränkung entfällt größtenteils beim symmetrischen Systemkern, lediglich die Kommunikation mit den externen Geräten muß auf der CPU erfolgen.

Der Zustand *Systemkontext* wird dazu aufgespalten in die beiden Zustände *Kernelkontext* und *Treiberkontext.* Ist ein Prozeß im Zustand *Kernelkontext,* so führt das mit einer gewissen Wahrscheinlichkeit zu einer E/A-Anforderung. Der Prozeß gelangt in den Zustand *Treiberkontext,* stößt die E/A-Anforderung an und wird blockiert, bis seine Anforderung erfüllt ist. Im Modell befindet sich der Prozeß auf dem externen Gerät und wird dort bedient (Abb. 2b).

Die bisherigen Ausführungen beziehen sich auf einen einzigen aktiven Prozeß. Im realen System ist nun gleichzeitig eine größere Zahl an Prozessen aktiv, von denen sich jeder in einem der vier beschriebenen Zustände befindet. Prozesse, die sich im Zustand E/A-Bearbeitung befinden, teilen sich die peripheren Geräte, die Prozesse in den anderen drei Zuständen den/die Prozessor/en.

Dabei gelten die folgenden Regeln:

- Prozesse im Treiberkontext haben die höchste Priorität.
- Prozesse im Kernelkontext haben die zweithöchste Priorität, werden von höherprioren Anforderungen aber nicht unterbrochen.

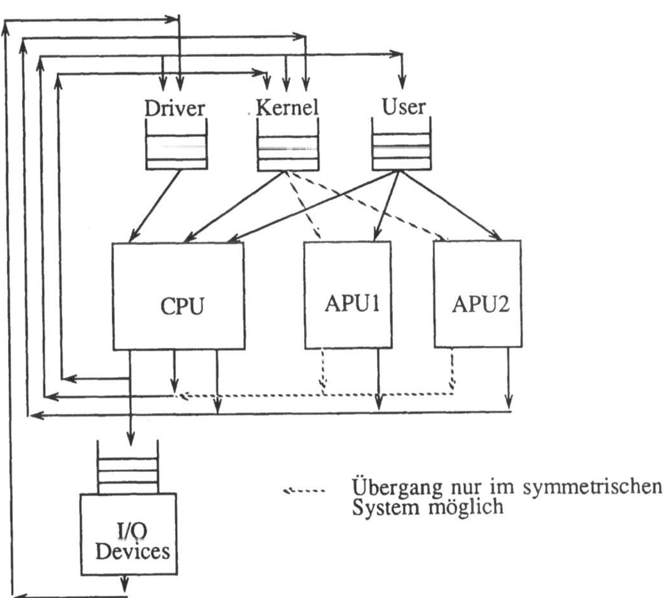

Abb. 3: Warteschlangenmodell des Multiprozessors MPS 3280

- Prozesse im Benutzerkontext haben die niedrigste Priorität und werden unterbrochen, sobald in einer der beiden anderen Prioritätsklassen eine Anforderung eintrifft.

Damit ergibt sich die in Abb. 3 gezeichnete Sicht des Multiprozessorsystems.

Ein konkreter Zustand des Systems ist charakterisiert durch den aktuellen Betriebsmodus der CPU (*user*, *kernel*, *driver*, *idle*), der beiden APUs (*user*, *kernel*, *idle*), die Zahl der Prozesse in den Warteräumen *Driver*, *Kernel*, *User* und die Zahl der Prozesse, die auf Fertigstellung einer E/A-Anforderung warten.

Um einen Eindruck der so entstehenden Zustandsräume zu vermitteln, ist in Abb. 4 der Zustandsraum eines Monoprozessors mit zwei aktiven Prozessen angegeben. Dabei wird zusätzlich angenommen, daß immer, wenn ein Prozeß fertig bedient ist, sofort eine neuer Prozeß gestartet wird, der sich identisch verhält.

Ziel der Analyse ist es, Gleichgewichtsverteilungen für Zustandsräume der obigen Art zu berechnen und dann daraus Leistungsgrößen wie z. B. Durchsatz, mittlere Antwortzeit oder Auslastungen von Prozessoren. Da die Klasse der analytisch leicht handhabbaren BCMP-Warteschlangennetze [Baskett75] durch die Prioritäten-Strategien und die unterschiedliche Funktionalität der Prozessoren verlassen wird und eine explizite Berechnung für größere Zustandsräume aussichtslos erscheint, werden die Gleichgewichtsverteilungen numerisch bestimmt [Stewart78], [Müller80], [Sauer81], [Marsan84].

Um den Zustandsraum als zeitkontinuierliche, homogene Markovkette interpretieren zu können, werden die folgenden Festlegungen getroffen:

- Die mittleren Bedienzeiten in den einzelnen Bedienphasen und an den E/A-Geräten sind voneinander unabhängig und jeweils exponentiell verteilt mit den Mittelwerten *usertime*, *kerneltime*, *drivertime*, *iotime*.

- Der Nachfolgezustand ist nur vom gerade aktuellen Zustand abhängig und nicht von dessen weiterer Vorgeschichte.

- Vom Zustand *kernel* wechselt ein Prozeß mit der Wahrscheinlichkeit *'pio'* in den Zustand *driver* und mit der Wahrscheinlichkeit *'1.0 - pio - pready'* in den Zustand *user*. Mit der Wahrscheinlichkeit *'pready'* ist ein Prozeß fertig bedient, verläßt das System und wird sofort durch einen neuen ersetzt, der sich identisch verhält (konstanter Multiprogramminglevel).

- Vom Zustand *driver* wechselt ein Prozeß mit der Wahrscheinlichkeit *'piodone'* in den Zustand *kernel* und mit der Wahrscheinlichkeit *'1.0 - piodone'* wird er am peripheren Gerät blockiert.

- Das System befindet sich im stationären Zustand.

- Prozessorwechselzeiten werden nicht berücksichtigt (asymmetrisches System).

- Alle Prozesse sind gleichzeitig im Hauptspeicher geladen, d. h. Ein- und Auslagerungen finden nicht statt.

- Alle Prozesse haben ein identisches Verhalten.

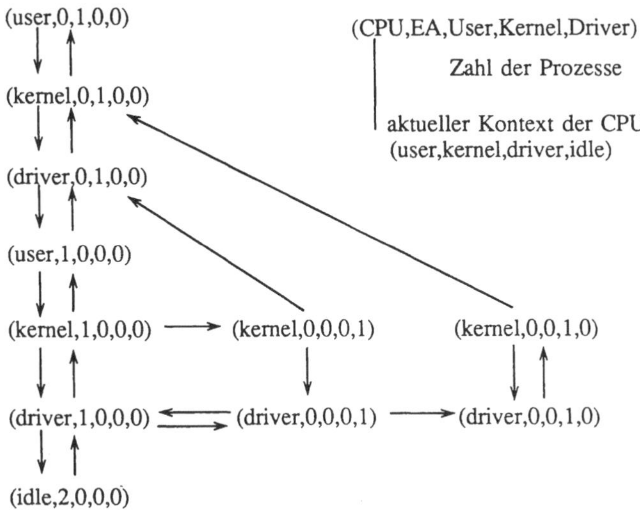

Abb. 4: Zustandsraum für einen Monoprozessor mit zwei Aufträgen

Messungen am realen System haben ergeben, daß der gemeinsame Bus bei drei Prozessoren (nur) zu etwa 50% ausgelastet ist und nicht zum Engpaß wird [Quick88]. Er kann somit in der Analyse vernachlässigt werden.

Durch Erweiterungen des Systemkerns um Meßsoftware konnten die Bedienzeitanforderungen direkt am System bestimmt werden. Ein Vergleich mit gemessenen Variationskoeffizienten zeigt, daß sowohl Treiberroutinen als auch Systemaufrufe sehr gut mit Exponentialverteilungen approximiert werden können, Bedienzeiten im Benutzerkontext allerdings noch einen größeren Variationskoeffizienten als eine Exponentialverteilung aufweisen.

Um die entstehenden Zustandsräume überhaupt kompakt beschreiben zu können, werden sie in einer formalen Weise spezifiziert, aus der automatisch die Generatormatrix für die Markovkette [Hennecke88] generiert wird. Mit Standardverfahren wird die Gleichgewichtsverteilung berechnet [Stewart78] und aus dieser die gesuchten Leistungsgrößen. Die detaillierte Beschreibung der verschiedenen Modelle kann im Anhang nachgelesen werden, ebenso wie die Definition der verwendeten Leistungsgrößen.

5. Modellberechnungen

Zur Untersuchung des Einflusses einzelner Parameter wird ein typischer Auftrag definiert, der für alle SVC-Raten die gleiche mittlere Bedienzeit anfordert. *Pready* wird also so bestimmt, daß die mittlere Bedienzeitanforderung im Benutzerkontext zuzüglich der mittleren Bedienzeit für SVCs konstant bleibt. Für die Modellparameter *kerneltime* und *drivertime* wurden 1.0 ms und 0.5 ms, für den Zugriff auf ein peripheres Gerät 15 ms gemessen. Das so eingeschränkte Modell wurde für eine Reihe von Parametersätzen *usertime*, *pio* und *multiprogramminglevel* durchgerechnet, deren Ergebnisse im folgenden diskutiert werden.

5.1 Betreibersicht

Aus der Sicht des Betreibers von Rechenanlagen ist das wichtigste Leistungsmaß der Durchsatz, der mit dieser Anlage erreicht werden kann. Das Antwortzeitverhalten spielt eine untergeordnete Rolle und wird nur indirekt über unzufriedene Benutzer an den Betreiber weitergegeben. Weitere Leistungsmaße sind die Auslastung von Teilkomponenten und die mittlere Zahl von Aufträgen im System.

In der Abb. 5 ist die Durchsatzverbesserung der beiden Systemvarianten gegenüber dem Monoprozessor in Abhängigkeit der mittleren Zeit zwischen zwei Systemaufrufen aufgetragen.

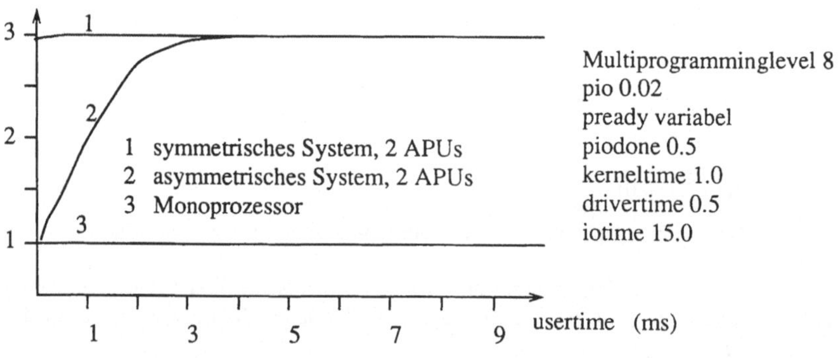

Abb. 5: relative Durchsatzverbesserung der beiden Systemvarianten

Die Grafik zeigt die Überlegenheit des symmetrischen Systems bei hoher SVC-Rate (d.h. die mittlere Zeit *usertime* zwischen zwei Systemaufrufen ist klein), während für rechenintensive Aufgaben kein Unterschied zwischen den beiden Varianten auftritt. Die Begründung dafür liegt im hohen Anteil von Systemkontext bei hoher SVC-Rate, der im asymmetrischen System nur auf der CPU abgearbeitet werden kann. Bei rechenintensiven Aufgaben fällt ein Systemcall weniger ins Gewicht, da immer hinreichend viele Prozesse im Benutzerkontext auf Prozessorzuteilung warten und somit die APUs seltener leerstehen.

Abb. 6 zeigt die Auslastung der Prozessoren für unterschiedliche SVC-Raten und einen Multiprogramminglevel von 8. Man sieht sehr deutlich die asymmetrische Auslastung der Prozessoren im asymmetrischen System für SVC-intensive Last. Im symmetrischen System sind die APUs sogar besser ausgelastet als die CPU, da Prozesse eine APU nur dann verlassen, wenn sie entweder verdrängt werden oder eine E/A-Anforderung absetzen und sich blockieren. Da der Durchsatz durch das System proportional zum Gesamtanteil an Kernelkontext aller drei Prozessoren ist, wird die CPU im asymmetrischen System zum Engpaß.

Bei hoher SVC-Rate ist auch die Wahrscheinlichkeit einer E/A-Operation höher, im symmetrischen System werden die peripheren Geräte zum durchsatzbeschränkenden Engpaß. Ein weiteres Verbessern des Systems für SVC-intensive Anwendungen ist nur dann möglich, wenn gleichzeitig die mittlere Bedienzeit an den peripheren Geräte verbessert wird. Von Seiten der Betriebssystementwicklung wäre es z. B. ein Ansatzpunkt, die Zugriffshäufigkeit auf die peripheren Geräte zu verringern, etwa durch Vergrößern des Caches für Plattenblöcke.

Nachdem beide Systemvarianten implementiert sind, konnten dem "analytischen Benchmark" real gemessene Benchmarks gegenübergestellt werden. Die im Modell berechnete Überlegenheit des symmetrischen Systems gegenüber der asymmetrischen Variante für SVC-intensive Aufgaben

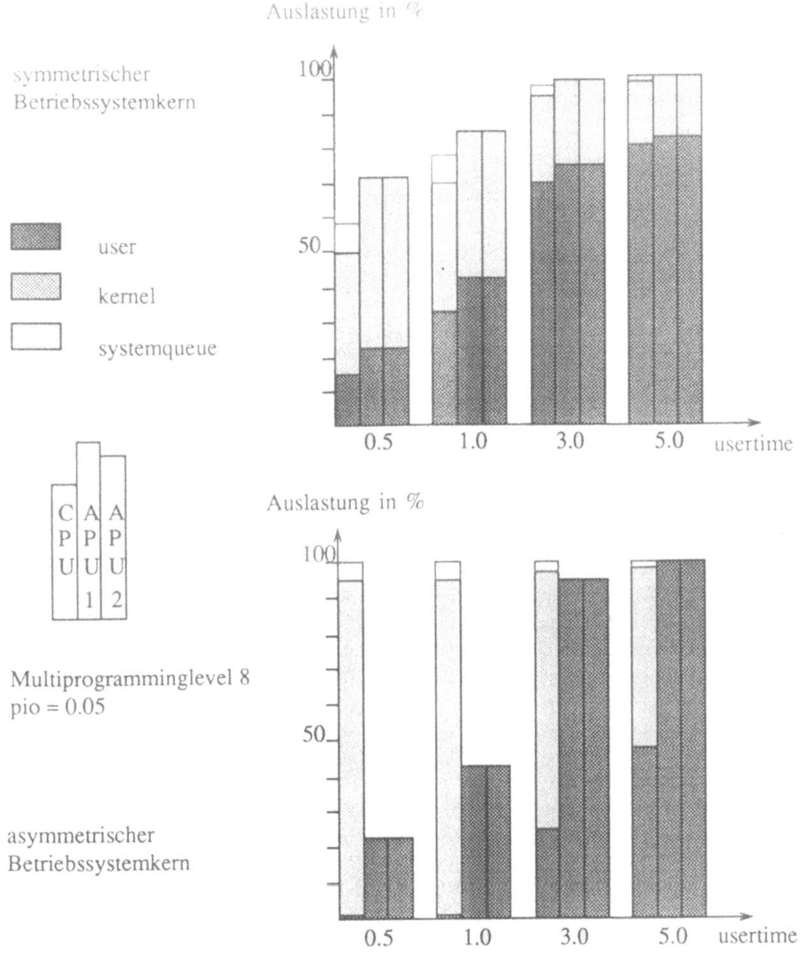

Abb. 6: Auslastung der Prozessoren in den beiden Systemvarianten
für unterschiedliche SVC Raten

(z. B. lesen von Dateien) wurde auch am realen System in dieser Deutlichkeit beobachtet, ebenso wie die gegenüber der CPU stärkere Auslastung der beiden APUs [Hofmann89]. Allerdings ist im realen Fall bei gleicher Last die Auslastung des symmetrischen Systems im Systemkontext höher als die des asymmetrischen; ein Effekt, der durch erhöhten Koordinationsaufwand und damit verbundenes aktives Warten verursacht wird.

5.2 Benutzersicht

Aus Sicht eines am Terminal sitzenden Benutzers ist der Durchsatz von untergeordneter Bedeutung, ihn interessiert im wesentlichen die Antwortzeit auf seine Aufträge. Um Aussagen über das

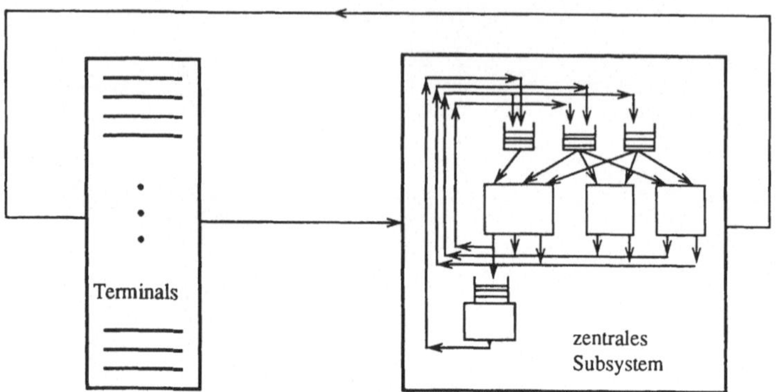

Abb. 7: Warteschlangenmodell des hierarchischen Systems

interaktive Systemverhalten gewinnen zu können, werden die Terminals explizit modelliert. Während bisher die Zahl der Aufträge im inneren System konstant war, ist jetzt lediglich die Obergrenze der Zahl der Aufträge durch die Zahl der Terminals bestimmt (Abb. 7)

Die Analyse des obigen Modells würde einen enormen Zustandsraum erfordern; ein Vorhaben, das nur für eine sehr geringe Zahl an interaktiven Benutzern Aussicht auf Erfolg verspricht. Ein Ausweg aus dieser Situation ist die hierarchische Analyse [Sauer81]; das zentrale Subsystem wird dazu als Bedienstation mit lastabhängiger Bedienrate modelliert und das entstehende Tandemnetz dann analytisch - das Warteschlangennetz ist separabel - oder wiederum numerisch bewertet. Die großen Unterschiede in der Denkzeit an den Terminals und den Bedienzeiten im Subsystem ergeben ein *fast vollständig zerlegbares* (*nearly completely decomposable*) Netz [Courtois77], das mit nur geringen Fehlern als hierarchisches Modell analysiert werden kann.

In Abb. 8 ist die Antwortzeit des zentralen Subsystems in Abhängigkeit der Zahl der aktiven Terminals und der Zahl der Prozessoren für die beiden Systeme aufgetragen. Die Leistungsunterschiede im Durchsatz werden für eine kleine Zahl an interaktiven Benutzern mehr oder weniger kompensiert und wirken sich erst für eine größere Zahl an interaktiven Benutzern aus. Interessant ist, daß das symmetrische System mit nur einer APU leistungsstärker ist als das asymmetrische System mit zwei APUs. Ein Effekt, der ebenfalls durch den Engpaß Systemkern verursacht wird. Ebenso wie bei der Analyse des inneren Systems wurde auch das Gesamtsystem mit Benchmarks getestet. Dabei konnte ebenfalls die Überlegenheit des symmetrischen Systems mit einer APU gegenüber dem asymmetrischen mit zwei APUs beobachtet werden.

6. Zusammenfassung und Ausblick

Mit Hilfe eines relativ einfachen Modells konnten zwei Betriebssystemvarianten analysiert werden. Die berechneten Effekte können auch in der Realität beobachtet werden. Im Rahmen der hier gemachten Untersuchung sind die absoluten Werte von untergeordnetem Interesse, da zur Beurteilung der beiden Entwurfsalternativen der relative Unterschied zwischen zwei Leistungsgrößen wichtig ist und nicht deren absoluter Wert.

Bezugnehmend auf die eingangs gestellten Fragen kann folgendes festgehalten werden:

- Die leistungsbestimmenden Parameter sind die SVC-Rate, die E/A-Wahrscheinlichkeit und die Reaktionszeit der peripheren Geräte. Während die E/A-Wahrscheinlichkeit durch

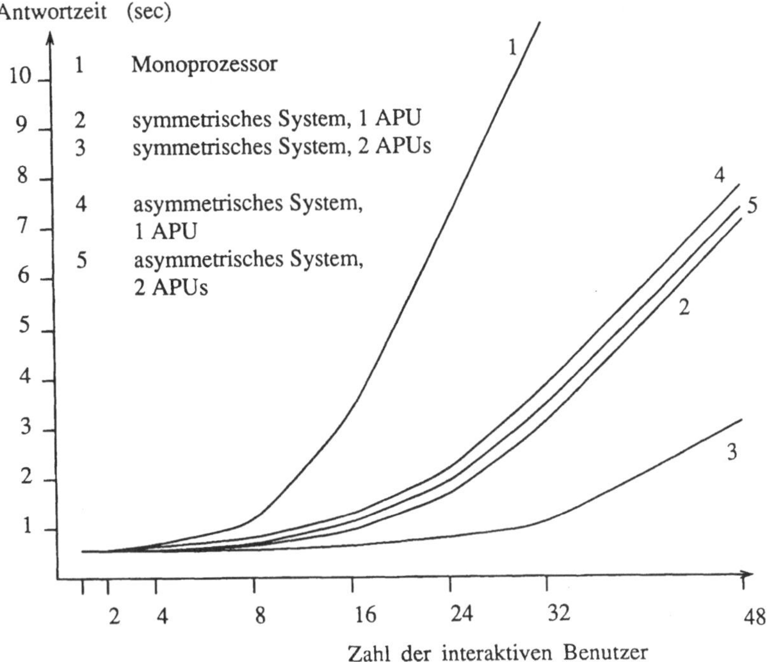

Abb. 7: Antwortzeitverhalten der beiden Systemvarianten bei steigender interaktiver Last

strategische Maßnahmen in gewissen Grenzen beeinflußt werden kann (z. B. Größe des Buffercaches), ist die SVC-Rate charakteristisch für die vorherrschende Systemlast und nicht beeinflußbar. Für hohe SVC-Raten und moderate E/A-Wahrscheinlichkeiten kann mit einer Parallelisierung des Betriebssystemkerns eine deutliche Leistungssteigerung erzielt werden; sind allerdings auch hohe E/A-Wahrscheinlichkeiten gegeben, so muß zusätzlich das Verhalten der peripheren Geräte verbessert werden.

- Die Untersuchung des hierarchischen Systems zeigt, daß bei Ausbau der Hardware um weitere Prozessoren ein symmetrischer Systemkern notwendig ist, um noch eine spürbare Leistungssteigerung zu erzielen. Allerdings verschiebt sich dann der Engpaß immer stärker in Richtung periphere Geräte.

Das Modell berücksichtigt nicht die Zeit, die im asymmetrischen System verstreicht, bis ein "SVC-williger" Prozeß auf der CPU laufbereit ist, ebensowenig Konflikte vor Monitoren im symmetrischen System. Beide Punkte sollen in Zukunft in das Modell integriert werden, wie auch die Annahme anderer Bedienzeitverteilungen in den einzelnen Prozeßphasen.

7. Literatur

[Bach86] Bach, M.J.: "The design of the UNIX operating system", Prentice Hall, Englewood Cliffs, New Jersey, 1986

[Baskett75] Baskett, F., Chandy, K. M., Muntz, R. R., Palacios, F. G.: "Open, closed and mixed networks of queues with different classes of customers", Journal of the ACM,Vol 22, No 2, pp. 248-260

[Concurrnet86] Concurrent Computer Corporation: "3280 System Instruction Set and Reference Manual", Tinton Falls, 1986

[Courtois77] Courtois, P. J.: "Decomposability: Queueing and Computer Systems Applications", Academic Press, New York 1977

[Hennecke88] Hennecke, Christian: "Parallele Implementierung der numerischen Analyse", Diplomarbeit, Institut für Mathematische Maschinen und Datenverarbeitung IV , Erlangen 1988

[Hofmann89] Hofmann, W.: "Erweiterung des UNIX-Betriebssystems für Multiprozessoren", Bericht Nr. 89/4 des Sonderforschungsbereichs 182 Multiprozessor- und Netzwerkonfigurationen, Teilbereich B2, Erlangen 1989

[Kleinöder87] Kleinöder, Jürgen: "Prozessorzuteilungsstrategien in enggekoppelten Mehrprozessorsystemen unter UNIX", Diplomarbeit, Institut für Mathematische Maschinen und Datenverarbeitung IV, Erlangen 1987

[Lavenberg83] Lavenberg, S. S.: "Computer Performance Modeling Handbook", Academic Press, New York1983

[Marsan84] Marsan, M. A., Conte, G., Balbo, G.: "A Class of Generalized Stochastic Petri Nets for the Performance Evaluation of Multiprozessor Systems", ACM Transactions on Computer, Vol 2, No 2, 1984, pp. 93-122

[Müller80] Müller, B.: "Zerlegungsorientierte, numerische Verfahren für Markovsche Rechensystemmodelle", Dissertation, Universität Dortmund, 1980

[Müller87] Müller-Clostermann, B.; Rosentreter, G.: "Synchronized Queueing Networks: Concepts, Examples and Evaluation Techniques", in Herzog, U., Paterok, M. (Hrsg): "Messung, Modellierung und Bewertung von Rechensystemen", 4. ITG/GI-Fachtagung, Springer, Heidelberg 1987

[Quick88] Quick, Andreas: "Monitoring und Visualisierung des dynamischen Verhaltens eines UNIX-Mehrprozessor-Betriebssystems", Diplomarbeit, Institut für Mathematische Maschinen und Datenverarbeitung VII, Erlangen 1988

[Sauer81] Sauer, C. H.; Chandy, K. M.: "Computer Systems Performance Modeling", Prentice Hall, Englewood Cliffs, New Jersey 1981

[Stewart78] Stewart, W. J.: "A Comparison of Numerical Techniques in Markov Modeling", Communications of the ACM, Vol. 21, No. 2, pp.144-152

Anhang

Im folgenden soll die Konstruktion des numerischen Modells näher beschrieben werden. Das in Abbildung 4 angegebene Warteschlangenmodell kann direkt in eine irreduzible, zeitkontinuierliche Markovkette umgesetzt werden.

Der Systemzustand läßt sich durch das folgende 8-Tupel charakterisieren:

(CPU,APUU,APUK,IO,USERQ,KERNELQ,DRIVERQ)

mit

CPU als aktuellen Kontext der CPU,
APUU als Zahl der APUs im Benutzerkontext,
APUK als Zahl der APUs im Kernelkontext
IO als Zahl von Prozessen, die auf Fertigstellung von EA-Anforderungen warten und
USERQ, *KERNELQ*, *DRIVERQ* als die Zahl von Prozessen in den Zuständen laufbereit Benutzerkontext, laufbereit Systemkontext und laufbereit Treiberkontext.

Der so spezifizierte Systemzustand ist auch für größere Konfigurationen zu verwenden, da nicht zwischen den einzelnen APUs unterschieden wird, sondern alle APUs quasi als *multiple server* mit zwei Auftragsklassen betrachtet werden. Um die Zustandsräume überhaupt kompakt beschreiben und automatisch in Generatormatrizen für Markovketten umsetzen zu können, werden sie formal spezifiziert und mit geeigneten Werkzeugen weiterverarbeitet [Hennecke88]. Nachstehend die Spezifikation des symmetrischen Systems und einige Erläuterungen zur Spezifikationssprache.

```
/*
Specification of a symmetric MPS system
*/
#define CPU      VAL[0]  /* state of CPU */
#define APUU     VAL[1]  /* # of APUs in usermode */
#define APUK     VAL[2]  /* # of APUs in kernelmode */
#define IO       VAL[3]  /* # jobs waiting io-completion */
#define USERQ    VAL[4]  /* # jobs waiting userkontext service */
#define KERNELQ  VAL[5]  /* # jobs waiting kernelkontext service */
#define DRIVERQ  VAL[6]  /* # jobs waiting driverkontext service */

/* possible states of Cpu  */

#define IDLE     0
#define USER     1
#define KERNEL   2
#define DRIVER   3

/* special service rate for immediate transitions */

#define imm              (1.0 - 2.0)

/* Modellparameters */

VAR int NO_APU;          /* # APUs */
VAR int N;               /* # of jobs circulating,level of multiprogramming*/
VAR double usertime;     /* mean time for service in userkontext */
VAR double kerneltime;   /* mean time for service in kernelkontext */
VAR double systime;      /* mean time for a systemqueue routine */
VAR double iotime;       /* mean time for an io operation */
VAR prob   pready;       /* prob. leaving system after finishing kernelkontext */
VAR prob   psyscall;     /* prob. doing a systemcall after
                                finishing userkontext */
VAR prob   pio;          /* prob. doing an io operation after
                                finishing kernelkontext */
VAR prob   piodone;      /* prob going to kernelkontext
                                after finishing systemqueue service */
```

```
/*  State space: (CPU,APUU,APUK,IO,USERQ,KERNELQ,DRIVERQ) */

DIM 7;

MIN  IDLE 0   0   0   0   0   0;
MAX  DRIVER  NO_APU (NO_APU-APUU)
                (N-APUU-APUK) (N-APUU-APUK-IO)
                (N-APUU-APUK-IO-USERQ) (N-APUU-APUK-IO-USERQ-KERNELQ);

/* feasable states, closed system, constant degree of multiprogramming */

NOT {CPU == IDLE}{APUU + APUK + IO + USERQ + KERNELQ + DRIVERQ != N };
NOT {CPU != IDLE}{APUU + APUK + IO + USERQ + KERNELQ + DRIVERQ != N-1 };

/* no more processes on apus than apus available */

NOT { APUK + APUU > NO_APU};

/* IO behaviour, M/M/1 Random */

RULE { IO > 0 }    -> IO -= 1 DRIVERQ += 1 : 1.0 / iotime : 1.0;

/* Cpu behaviour
                1. jobs finishing service */

RULE { CPU == USER } -> CPU = KERNEL : 1.0 /usertime : 1.0 ;

RULE { CPU == KERNEL } -> CPU = USER : 1.0 / kerneltime : 1.0-pio-pready;
RULE { CPU == KERNEL } -> CPU = DRIVER  : 1.0 / kerneltime : pio;

RULE { CPU == DRIVER } -> CPU = IDLE KERNELQ += 1 : 1.0/systime : piodone;
RULE { CPU == DRIVER } -> CPU = IDLE IO += 1 : 1.0/systime : 1.0 - piodone;

/*              2. preemptive scheme */

RULE { CPU == USER } { DRIVERQ + KERNELQ > 0 }
                -> CPU = IDLE USERQ += 1 : imm : 1.0;

/*              3. priority scheme for fetching jobs */

RULE { CPU == IDLE } { DRIVERQ > 0 }
                -> DRIVERQ -= 1 CPU = DRIVER : imm :1.0;
RULE { CPU == IDLE } { DRIVERQ == 0 } { KERNELQ > 0 }
                -> KERNELQ -= 1 CPU = KERNEL : imm : 1.0;
RULE { CPU == IDLE } { DRIVERQ == 0 } { KERNELQ == 0 } { USERQ > 0 }
                -> USERQ -= 1 CPU = USER : imm : 1.0;

/* APU behaviour */
/*              1. finishing jobs */

RULE { APUU > 0 } -> APUU -= 1 APUK += 1 : APUU/usertime : 1.0;

RULE { APUK > 0 } -> APUK -= 1 APUU += 1 : APUK/kerneltime : 1.0-pio-pready;
RULE { APUK > 0 } -> APUK -= 1 DRIVERQ += 1 : APUK/kerneltime : pio;

/*              2. preemptive scheme */

RULE { KERNELQ > 0 }  { APUU > 0 }
                -> APUU -= 1 USERQ += 1 APUK += 1 KERNELQ -= 1 : imm : 1.0;
```

```
/*              3. fetching new jobs */

RULE { APUU + APUK < NO_APU } { KERNELQ > 0 }
                  -> APUK += 1 KERNELQ -= 1 : imm : 1.0;
RULE { APUU + APUK < NO_APU } { KERNELQ == 0 } {USERQ > 0 }
                  -> APUU += 1 USERQ -= 1 : imm : 1.0;
```

Die Spezifikation ist wie folgt zu lesen:

- Der erste *define*-Block dient lediglich zur Erhöhung der Lesbarkeit; mit *VAL*[*num*] wird auf den Wert der *num*-ten Komponente des Zustandsvektors zugegriffen.
- Im *VAR*-Teil werden die Parameter des Modells spezifiziert; dieser Teil kann auch als die Lastbeschreibung interpretiert werden, während der Rest die Systembeschreibung ist.
- Mit *DIM* wird die Dimension des Zustandsvektors angegeben und mit *MIN* und *MAX* die Ober- und Untergrenzen der einzelnen Komponenten.
- Mit *NOT* werden die zulässigen Zustände spezifiziert, zwischen *{* und *}* steht ein Prädikat des Zustandsvektors, das erfüllt sein muß.
- Mit *RULE* werden Übergangsregeln zwischen Zuständen spezifiziert. Dabei steht links vom -> Zeichen eine Bedingung, die erfüllt sein muß, damit diese Übergangsregel anwendbar ist, und rechts vom -> Zeichen der Nachfolgezustand und mit *:* getrennt die Übergangsrate und die Verzweigungswahrscheinlichkeit.
- Mit *imm* ist ein Übergang bezeichnet, der zeitlos ist und lediglich der Strukturierung des Modells dient. Dieser Übergang ist vergleichbar mit den *immediate transitions* in stochastischen Petri-Netzen [Marsan84] oder den instabilen Zuständen in [Müller87].

Das aus der Spezifikation erzeugte C-Programm untersucht alle durch *MIN* und *MAX* eingeschränkten Zustände. Ist eine *NOT* Bedingung nicht erfüllt, so wird dieser Zustand weggeworfen, ansonsten werden alle spezifizierten Regeln untersucht und alle anwendbaren Übergänge aufgelistet. Man erhält auf diese Weise eine Matrix, die nach Entfernen der *instabilen* Zustände (d. s. die Zustände, bei denen mindestens eine wegführende Kante mit *imm* beschriftet ist) die Generatormatrix der Markovkette ist, und aus der dann mit Standardmethoden die Gleichgewichtsverteilung und die Leistungsgrößen bestimmt werden können [Stewart78].

Abschließend seien noch die Formeln angegeben, mit denen die weiter vorne zitierten Leistungsgrößen berechnet werden.

Um einen typischen Auftrag, der für jede SVC-Rate eine konstante Bedienanforderung stellt, zu definieren, muß *pready* jeweils in Abhängigkeit von *usertime* berechnet werden.

Die mittlere Anzahl *e* von Besuchen eines Prozesses in den einzelnen Zuständen und die mittlere Bedienzeitanforderung *s* berechnen sich wie folgt:

$$e_{user} = \frac{1.0\text{-}pio\text{-}pready}{pready} \qquad e_{driver} = \frac{1.0}{pready} * \frac{pio}{piodone}$$

$$e_{kernel} = \frac{1.0}{pready} \qquad e_{io} = \frac{(1.0 - piodone)}{pready} * \frac{pio}{piodone}$$

$$s = e_{user} * usertime + (e_{user} - 1) * kerneltime$$

Als primäre Leistungsgröße erhält man für jeden Zustand die Wahrscheinlichkeit seines Auftretens im statistischen Gleichgewicht. Durch Aufsummieren zusammengehöriger Zustände erhält man Randwahrscheinlichkeiten, Warteschlangenverteilungen und -mittelwerte, Durchsätze und mit dem Satz von Little Verweilzeiten. Im konkreten Fall sieht das wie folgt aus:

Sei Z die Menge aller Zustände, die das Modell annehmen darf und $z \varepsilon Z$ ein beliebiger Zustand aus Z. Dann gilt (Auswahl):

- $\bar{n}_{userq} = \sum_k k * P(\{z \in Z \mid userq = k\})$ mittlere Zahl laufbereiter Prozesse im Benutzerkontext

- $\rho_{cpu}^{user} = P(\{z \varepsilon Z \mid cpu = user\})$ Auslastung der CPU im Benutzerkontext

- $\nu_{apus}^{user} = \sum_k k * P\{(z \varepsilon Z \mid apuu = k\})$ mittlere Zahl an APUS im Benutzerkontext

- $\rho_{cpu}^{kernel} = P(\{z \varepsilon Z \mid cpu = kernel\})$ Auslastung der CPU im Systemkontext

- $\nu_{apus}^{kernel} = \sum_k k * P\{(z \varepsilon Z \mid apuk = k\})$ mittlere Zahl an APUS im Benutzerkontext

- $\lambda = \rho_{cpu}^{kernel} * \mu_{kernel} * p_{ready} + \nu_{apus}^{kernel} * \mu_{kernel} * p_{ready}$

  Systemdurchsatz

Erweiterung des Verfahrens MEDA zur analytischen Beschreibung empirischer Verteilungsfunktionen

Leonhard Schmickler

Lehrstuhl Datenfernverarbeitung
RWTH Aachen
Kopernikusstr. 16
D-5100 Aachen

Kurzfassung

Für die Approximation empirischer Verteilungsfunktionen haben sich die Eigenschaften der Erlang-Mischverteilung als vorteilhaft erwiesen. In einer früheren Arbeit [12] wurde das weitgehend heuristische Verfahren MEDA[1] zur Approximation der ersten drei gemessenen Momente und zur bestmöglichen Annäherung der Verteilungsfunktionsverläufe mit einer drei-zweigigen Erlang-Mischverteilung beschrieben. Im folgenden wird eine erweiterte und verbesserte Version des Approximationsverfahrens vorgestellt. Sie enthält das schnelle nichtlineare *flexible polyhedron search*-Optimierungsverfahren und bietet dadurch die Möglichkeit, bis zu sechs-zweigige Erlang-Mischverteilungen zu verwenden. Ferner können jetzt wesentliche Sprungstellen in den empirischen Verteilungsfunktionen berücksichtigt werden. Bei drei Zweigen ergibt sich ein erheblicher Geschwindigkeitsgewinn gegenüber dem heuristischen Verfahren. Die Hinzunahme weiterer Zweige erlaubt, durch Erhöhung der Komplexität die Genauigkeit der Approximation deutlich zu verbessern.

Auch die neue Version von MEDA ist als PASCAL-Quellcode verfügbar.

1 Einführung

Messungen der stochastischen Prozesse in Datenverarbeitungs- und Datentransporteinrichtungen stellen eine wichtige Voraussetzung für deren Modellierung und Leistungsbewertung dar. Die daraus gewonnenen empirischen Verteilungsfunktionen (Summenhäufigkeitsfunktionen) $F_E(x)$ können bei genügend großen Stichproben bis auf signifikante Sprünge mit einem linearen Verlauf zwischen den Stützstellen x_i dargestellt werden. Die empirischen Momente j-ter Ordnung $M_E(j)$ können ebenfalls aus den gemessenen Stichproben berechnet werden [13]. Für die Verwendung zur Leistungsbewertung läßt sich aus den Meßergebnissen z.B. ein Tabellenzufallsgenerator für ein Simulationsprogramm erstellen [2].

In vielen Fällen wird jedoch eine mathematische Berechnung eines Modells angestrebt und daher eine möglichst einfache, aber dennoch genügend detaillierte mathematische Beschreibung der empirischen Verteilung gesucht. Hierzu wurde in [12] das weitgehend heuristische Approximationsverfahren MEDA[1] vorgestellt, das die wichtigsten Voraussetzungen für Berechnungen und Simulationen in der Datenverkehrstheorie erfüllt:

- Verwendung einer Phasenverteilung mit wenigen Parametern (s. dazu [7,11,5]);
- Exakter Abgleich der ersten drei Momente $M_E(j)$ der empirischen Verteilung (s. dazu [4,8]);

[1] Mixed Erlang Distributions for Approximation

- Berücksichtigung der höheren Momente durch ausreichende Näherung der gemessenen Verteilungsfunktion in allen Stützstellen x_i (s. dazu [5]).

Die erste Version von MEDA verwendete eine zwei- bzw. drei-zweigige Erlang-Mischverteilung als Approximierende, deren erste drei Momente exakt mit den empirischen abgeglichen wurden. Bei der Näherung des empirischen Verteilungsfunktionsverlaufs berücksichtigte es alle (z.B. ≈ 1000) Stützstellen. Seine wichtigsten Elemente bildeten ein schneller Momentenabgleichalgorithmus, das neudefinierte Optimierungskriterium der Differenzfläche und eine sehr genaue Methode zur Ermittlung von Startwerten für die Parameter eines Zweiges aus dem Verlauf der empirischen Verteilungsfunktion.

Diese wesentlichen Eigenschaften der ersten Version des Verfahrens MEDA haben sich als zuverlässig und wirkungsvoll erwiesen und wurden bei der neuen Version beibehalten. Der Momentenabgleichalgorithmus (s. Abschnitt 2) mußte nur geringfügig erweitert werden. Einen Nachteil stellten jedoch die relativ lange Rechenzeit und die Beschränkung der Approximierenden auf maximal drei Erlangmischzweige dar.

Das in Abschnitt 3 beschriebene nichtlineare *flexible polyhedron search*-Optimierungsverfahren (FPS) konnte nach der Einführung der Gamma-Mischverteilung (s. Abschnitt 1.1) in MEDA einbezogen werden und den heuristischen Algorithmus GURU ersetzen. Es ermöglicht die schnelle gleichzeitige Optimierung nichtlinearer Funktionen mit mehreren Parametern und erlaubt die einfache Erweiterung der Approximierenden auf vier und mehr Zweige (s. Abschnitt 5). Am Beispiel einer Dateilängenverteilung werden in Abschnitt 6 Resultate der neuen Version von MEDA mit dem FPS-Verfahren gezeigt.

1.1 Die Gamma-Mischverteilung

Die Vorteile der Verwendung eines nichtlinearen Optimierungsalgorithmus zur gleichzeitigen Variation mehrerer Parameter lassen sich nur dann effizient nutzen, wenn alle Parameter kontinuierliche Größen sind. Für die Phasenzahlen k_i der Erlang-Mischverteilung gilt diese Voraussetzung nicht. Abhilfe schafft die Gamma-Mischverteilung, die sich nur in diesen Parametern von der Erlang-Mischverteilung unterscheidet und eine verallgemeinerte Erlang-Mischverteilung darstellt. Die resultierende approximierende Verteilungsfunktion soll jedoch weiterhin die Erlang-Mischverteilung sein, die als Phasenverteilung wichtige Eigenschaften für die Modellanalyse besitzt. Die Approximation wird also im ersten Teil mit der Gamma-Mischverteilung durchgeführt. Dann werden deren kontinuierliche Parameter r_i durch die bestmöglichen ganzzahligen Phasenzahlen k_i der Erlang-Mischverteilung ersetzt.
Die beiden Verteilungsdichtefunktionen für Z Zweige sind sehr ähnlich:

$$\text{Erlang-Mischverteilung:} \quad f_{A_E}(x) = \sum_{i=1}^{Z} p_i \, \frac{\lambda_i}{(k_i-1)!} \, (\lambda_i x)^{k_i-1} \, e^{-\lambda_i x} \qquad k_i \geq 1 \text{ ganzzahlig,} \tag{1}$$

$$\text{Gamma-Mischverteilung:} \quad f_{A_G}(x) = \sum_{i=1}^{Z} p_i \, \frac{\lambda_i}{\Gamma(r_i)} \, (\lambda_i x)^{r_i-1} \, e^{-\lambda_i x} \qquad r_i > 0 \text{ kontinuierlich.} \tag{2}$$

Für positive ganzzahlige $r_i = k_i$ geht die Gammafunktion in die Fakultät über

$$\Gamma(k_i) = (k_i - 1)! \tag{3}$$

und beide Verteilungsdichtefunktionen sind identisch.

In den Momentengleichungen der Gamma-Mischverteilung stehen lediglich die verallgemeinerten Parameter r_i anstelle der ganzzahligen k_i der Erlang-Mischverteilung. Dadurch kann der Momentenabgleichalgorithmus (Abschnitt 2) in der verallgemeinerten Form für beide Verteilungen gleich gehalten werden.

Da nach dem Momentenabgleich der Verteilungsfunktionsvergleich in allen Stützstellen der empirischen Verteilungsfunktion durchgeführt wird (s. Abb. 1), muß die Gamma-Mischverteilung an diesen Stellen berechnet werden. Die Verteilungsfunktion $F_{A_G}(x_i)$ der Gamma-Mischverteilung wird numerisch an den Stützstellen $x_i, i = 0,..,N$ nach der Simpsonschen Regel (in [1])

$$\int_a^{a+\Delta x} f(x)dx = \frac{\Delta x}{6}\left[f(a) + 4f\left(a + \frac{\Delta x}{2}\right) + f(a+\Delta x)\right] \tag{4}$$

durch Summation aus ihrer Verteilungsdichtefunktion $f_{A_G}(x_i)$ berechnet:

$$F_{A_G}(x_i) = \sum_{l=1}^{i} \int_{x_{l-1}}^{x_l} f_{A_G}(t)\, dt. \tag{5}$$

Die Werte der Gammafunktion $\Gamma(x)$ können einer Tabelle entnommen werden (z.B. in [3]).

2 Abgleich der ersten drei empirischen Momente

Der exakte Abgleich der ersten drei empirischen Momente ist eine grundlegene Voraussetzung für eine Verteilungsfunktionsapproximation. Da dieser Momentenabgleich für jeden Satz von Parametern der Gamma-Mischverteilung $F_{A_G}(x)$ bzw. Erlang-Mischverteilung $F_{A_E}(x)$ neu vorgenommen werden muß (s. Abb. 3), ist hier eine besondere Effizienz erforderlich. Der im folgenden beschriebene Abgleichalgorithmus war bis auf die Ersetzung der k_i durch r_i bereits Bestandteil der in [12] beschriebenen Version von MEDA, er konnte jedoch aus Platzgründen dort nicht dokumentiert werden. Seine wesentlichen Elemente, die ohne weiteren Entwicklungsaufwand eine Implementation ermöglichen, werden im folgenden erläutert.

2.1 Lösung des Gleichungssystems

Die Momente $M_{A_G}(j)$ einer Z-zweigigen Gamma-Mischverteilung lassen sich einfach aus ihren Parametern berechnen:

$$M_{A_G}(j) = \sum_{i=1}^{Z} p_i\, \lambda_i^{-j} \prod_{l=0}^{j-1} (r_i + l). \tag{6}$$

Für den Abgleich der aus der Messung stammenden ersten drei empirischen Momente $M_E(j)$ werden nur drei der $3Z - 1$ freien Parameter benötigt. Alle übrigen müssen zu Beginn möglichst gut geschätzt (Startvektoren) und anschließend mittels Variation optimiert werden. Allgemein lassen sich mit den Parametern p_1 ($p_2 = 1 - p_1$), λ_1 und λ_2 einer zwei-zweigigen Gamma-Mischverteilung relativ leicht numerisch-algebraisch deren erste drei Momente abgleichen (s. unten), wenn r_1 und r_2 bereits vorliegen. Die Momente einer Gamma-Mischverteilung mit mehr als zwei Zweigen können auf die einer zwei-zweigigen reduziert werden, wenn die Parameter der höheren Zweige, die deren Zweiganteile $M_{A_i}(j)$, $i > 2$ festlegen, ebenfalls bekannt sind. Der Abgleichalgorithmus bestimmt dann die Parameter p_1', λ_1 und λ_2 für eine Gamma-Mischverteilung mit zwei Zweigen und den reduzierten Momenten. Die nach dem Abgleich vorliegenden Zweigwahrscheinlichkeiten erfüllen $p_1' + p_2' = 1$ und müssen daher anschließend neu normiert werden (Gl. 16). Allgemein gilt für die zu approximierenden reduzierten empirischen Momente:

$$M_{APP}(j) = \left(M_E(j) - \sum_{i=3}^{Z} M_{A_i}(j)\right) / \left(1 - \sum_{i=3}^{Z} p_i\right). \tag{7}$$

Zur Berücksichtigung von Sprungstellen kann durch zusätzliche Subtraktion ihrer Momentenanteile der Abgleich von zwei Zweigen beibehalten werden (s. Abschnitt 4).

Die zwei-zweigige Gamma-Mischverteilung hat noch drei freie Parameter p_1' ($p_2' = 1 - p_1'$), λ_1, λ_2, wenn die Phasenzahlen r_1, r_2 vorgegeben werden. Sie hängen über die folgenden Gleichungen von den ersten drei reduzierten Momenten $M_{APP}(j)$ ab:

$$\begin{aligned} M_{APP}(1) &= p_1'\lambda_1^{-1}r_1 + p_2'\lambda_2^{-1}r_2; && (8)\\ M_{APP}(2) &= p_1'\lambda_1^{-2}r_1(r_1+1) + p_2'\lambda_2^{-2}r_2(r_2+1); && (9)\\ M_{APP}(3) &= p_1'\lambda_1^{-3}r_1(r_1+1)(r_1+2) + p_2'\lambda_2^{-3}r_2(r_2+1)(r_2+2). && (10) \end{aligned}$$

Diese drei Gleichungen lassen sich in ein Polynom sechsten Grades $p(y)$ überführen, das nur noch von der Intensität $\lambda_2 = 1/y$ des zweiten Zweiges abhängt:

$$\alpha_0 y^6 + \alpha_1 y^5 + \alpha_2 y^4 + \alpha_3 y^3 + \alpha_4 y^2 + \alpha_5 y + \alpha_6 = p(y) = 0. \tag{11}$$

Hierin gelten die Koeffizienten

$$\begin{aligned} \alpha_0 &= B_0C_0; & \alpha_3 &= B_0C_3 + B_3C_0 + B_1C_2 + B_2C_1 - A_1^2 - 2A_0A_2;\\ \alpha_1 &= -(B_0C_1 + B_1C_0 + A_0^2); & \alpha_4 &= -(B_1C_3 + B_3C_1 - B_2C_2 - 2A_1A_2);\\ \alpha_2 &= -(B_0C_2 + B_2C_0 - B_1C_1 - 2A_0A_1); & \alpha_5 &= -(B_2C_3 + B_3C_2 + A_2^2);\\ & & \alpha_6 &= B_3C_3 \end{aligned} \tag{12}$$

mit den Abkürzungen

$$\begin{aligned} A_0 &= a_{11}b_{22} - a_{21}b_{12}; & B_3 &= b_{11}c_{21} - b_{21}c_{11};\\ A_1 &= b_{11}b_{22} - b_{21}b_{12} + a_{11}c_{22} - a_{21}c_{12}; & C_0 &= a_{12}b_{22} - a_{22}b_{12};\\ A_2 &= b_{11}c_{22} - b_{21}c_{12}; & C_1 &= a_{12}c_{22} - a_{22}c_{12};\\ B_0 &= a_{11}a_{22} - a_{21}a_{12}; & C_2 &= c_{11}b_{22} - c_{22}b_{12};\\ B_1 &= b_{11}a_{22} - b_{21}a_{12}; & C_3 &= c_{11}c_{22} - c_{21}c_{12},\\ B_2 &= a_{11}c_{21} - a_{21}c_{11}; \end{aligned} \tag{13}$$

die sich ihrerseits auf die Parameter nach Gl. 8-10 zurückführen lassen:

$$\begin{aligned} a_{11} &= r_1r_2(r_1+1); & b_{22} &= \frac{M_{APP}(2)}{M_{APP}(1)}(r_2+1)(r_2+2);\\ a_{12} &= r_1r_2(r_2+1); & c_{11} &= M_{APP}(2)r_1;\\ a_{21} &= (r_1+1)(r_1+2)(r_2+1); & c_{12} &= M_{APP}(2)r_2;\\ a_{22} &= (r_1+1)(r_2+1)(r_2+2); & c_{21} &= \frac{M_{APP}(3)}{M_{APP}(1)}(r_1+1);\\ b_{11} &= M_{APP}(1)r_1(r_1+1); & c_{22} &= \frac{M_{APP}(3)}{M_{APP}(1)}(r_2+1).\\ b_{12} &= M_{APP}(1)(r_2+1)r_2; \\ b_{21} &= \frac{M_{APP}(2)}{M_{APP}(1)}(r_1+1)(r_1+2); \end{aligned} \tag{14}$$

Die Intensität λ_1 des ersten Zweiges kann nach der Bestimmung von $\lambda_2 = 1/y$ (s. Abschnitt 2.2) aus

$$1/\lambda_1 = \frac{(b_{12}y - c_{12})(a_{21}y - b_{21}) - (b_{22}y - c_{22})(a_{11}y - b_{11})}{(a_{12}y^2 - c_{11})(a_{21}y - b_{21}) - (a_{22}y^2 - c_{21})(a_{11}y - b_{11})} \cdot y \tag{15}$$

berechnet werden. Aus diesen beiden Intensitäten folgen die Zweigwahrscheinlichkeiten, die zur Einhaltung der Vollständigkeitsbedingung noch neu normiert werden müssen:

$$p_1 = (1 - p_2) = p_1' \cdot (1 - \sum_{i=3}^{Z} p_i) \quad \text{mit} \quad p_1' = \frac{M_{APP}(1) - 1/\lambda_2}{r_1/\lambda_1 - r_2/\lambda_2}. \tag{16}$$

2.2 Nullstellensuche

Die Nullstellensuche für das Polynom sechsten Grades in Gl. 11 zur Bestimmung des Parameters $\lambda_2 = 1/y$ der Erlangmischverteilung gliedert sich in sechs Schritte:

1. Bestimmung der ersten beiden Ableitungen $p'(y)$ und $p''(y)$ des Polynoms $p(y)$.
2. Ermittlung der Nullstellen von $p''(y)$ (vom Grad vier) rein algebraisch.
3. Numerische Bestimmung einer Nullstelle von $p'(y)$, wobei die Intervalle, in denen sich Nullstellen von $p'(y)$ befinden können, durch die Nullstellen von $p''(y)$ gegeben sind.
4. Ausklammern der numerisch bestimmten Nullstelle von $p'(y)$ sowie algebraische Berechnung der übrigen Nullstellen von $p(y)$.
5. Numerische Bestimmung zweier Nullstellen des Polynoms selbst (die Intervalle, in denen sich Nullstellen von $p(y)$ befinden können, sind durch die Nullstellen von $p'(y)$ gegeben).
6. Ausklammern der beiden numerisch bestimmten Nullstellen von $p(y)$ sowie algebraische Berechnung der übrigen Nullstellen von $p(y)$.

Die Ermittlung der Nullstellen eines Polynoms vierten Grades findet sich z.B. in [3]. Zu beachten ist, daß gilt:

$$y = 1/\lambda_2 \geq 0. \tag{17}$$

Die numerische Bestimmung der Nullstellen in Punkt 3. und 4. wird mit Hilfe des einfachen Bisektionsverfahrens vorgenommen. In beiden Fällen müssen jedoch noch eine möglichst genaue Ober- und Untergrenze für die äußeren Bisektionsintervalle bestimmt werden. Ein Satz zur Abschätzung der Nullstellen z_i eines Polynoms n-ten Grades wurde [10] entnommen:

$$\begin{aligned}
|z_i| &\leq 2 \cdot \max_{k=1,\ldots,n} \left(\sqrt[k]{\left|\frac{\alpha_k}{\alpha_0}\right|} \right); \\
|z_i| &\leq \sum_{j=0}^{n-1} \left| \frac{\alpha_j + 1}{\alpha_j} \right|; \\
|z_i| &\leq \max \left(\left|\frac{\alpha_n}{\alpha_{n-1}}\right|, \left|2\,\frac{\alpha_{n-1}}{\alpha_{n-2}}\right|, \ldots, \left|2\,\frac{\alpha_2}{\alpha_1}\right|, \left|2\,\frac{\alpha_1}{\alpha_0}\right| \right); \\
|z_i| &\leq \max \left(1, \sum_{j=1}^{n} \left|\frac{\alpha_j}{\alpha_0}\right| \right); \\
|z_i| &\leq \max \left(\left|\frac{\alpha_n}{\alpha_0}\right|, \left|1 + \frac{\alpha_{n-1}}{\alpha_0}\right|, \ldots, \left|1 + \frac{\alpha_1}{\alpha_0}\right| \right).
\end{aligned} \tag{18}$$

Der Satz gilt auch für komplexe Nullstellen z_i und für komplexe Koeffizienten α_j. Als Obergrenze der Nullstellen kann die kleinste Abschätzung verwendet werden, als Untergrenze die invertierte Obergrenze. Die nach diesem Verfahren bestimmten Grenzen für die Nullstellen liegen nahe an der größten bzw. kleinsten Nullstelle und sind an das jeweils zu untersuchende Polynom durch Einbeziehung seiner Koeffizienten angepaßt. In der Regel liefert die Nullstellensuche zwei gültige Lösungen für λ_2.

3 Der Optimierungsalgorithmus

Einen für das vorliegende komplexe nichtlineare Optimierungsproblem geeigneten Algorithmus stellt das ableitungsfreie *flexible polyhedron search*-Verfahren (FPS) von NELDER und MEAD (in [9]) dar. Ein Optimierungsverfahren, das mit Ableitungen arbeitet, konvergiert zwar schneller, führt jedoch bei der sehr hohen Anzahl von Stützstellen der empirischen Verteilungsfunktionen (z.B. 1000) zu sehr großen Rechenzeiten.

Das FPS-Verfahren erlaubt keine Berücksichtigung von Nebenbedingungen. Diese werden durch Hinzunahme einer *external penalty*-Funktion [6] einbezogen (s. Abschnitt 3.2).

Ein Verfahren zur Generierung eines Startvektors für die Parameter des dritten Zweiges einer Erlang-Mischverteilung und die als Zielfunktion (Gütekriterium) besonders geeignete Differenzfläche Δ (s. Abb. 1)

$$\Delta = \frac{1}{M_E(1)} \int_0^\infty |F_E(x) - F_A(x)| dx \tag{19}$$

wurden bereits in [12] beschrieben und werden weiterhin verwendet.

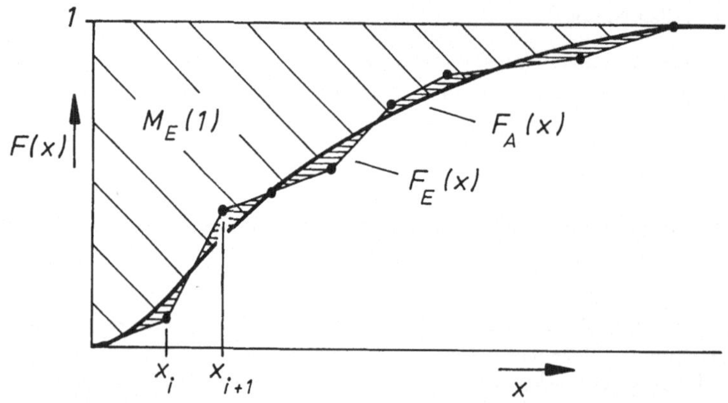

Abb. 1 Definition der Differenzfläche

Die Berechnung der Zielfunktion ist von der zeitweiligen Verwendung der Gamma-Mischverteilung $F_{A_G}(x)$ als Approximierender und der endgültigen Optimierung der Erlang-Mischverteilung $F_{A_E}(x)$ ebenso unabhängig wie der vorher beschriebene Momentenabgleichalgorithmus.

Die gleichzeitige Variation aller Parameter (fünf bis maximal vierzehn) kann auch vom FPS-Verfahren nicht in befriedigender Zeit und Güte durchgeführt werden. Das Verfahren wird daher immer nur auf einen Satz von bis zu sechs Parametern angewendet. Wird eine hohe Genauigkeit gefordert, können gar nur zwei Parameter gleichzeitig variiert werden (s. Abschnitt 5).

3.1 Das *flexible polyhedron search*-Verfahren

Das *flexible polyhedron search*-Verfahren ist ein Direktsuchverfahren, das eine nichtlineare Zielfunktion ableitungsfrei optimieren kann. Die Zielfunktion hängt von den n kontinuierlichen Parametern des Lösungsraums ab. Der Lösungsraum der Zielfunktion läßt sich als n-dimensionaler Vektorraum darstellen, in dem jeder Punkt durch einen Vektor $\mathbf{x}_j$ eindeutig beschrieben werden kann.

Das Verfahren benötigt zu Beginn einen Startvektor $\mathbf{x}_S = (x_1, x_2, ..., x_{n-1}, x_n)$ und einen Abweichungsvektor $\mathbf{d} = (d_1, d_2, ..., d_{n-1}, d_n)$. Aus Startvektor und Abweichungsvektor wird ein Startpolyeder $\mathbf{X}$ bestimmt, der als $n \times (n+1)$-Matrix darstellbar ist:

$$\begin{aligned} \mathbf{X} &= (\mathbf{x}_1, \mathbf{x}_2, ..., \mathbf{x}_n, \mathbf{x}_{n+1}) \\ &= \begin{pmatrix} x_1 & x_1 + d_1 & x_1 + d_1/2 & ... & x_1 + d_1/2 & x_1 + d_1/2 \\ x_2 & x_2 & x_2 + d_2 & ... & x_2 + d_2/2 & x_2 + d_2/2 \\ . & . & . & ... & . & . \\ . & . & . & ... & . & . \\ . & . & . & ... & . & . \\ x_{n-1} & x_{n-1} & x_{n-1} & ... & x_{n-1} + d_{n-1} & x_{n-1} + d_{n-1}/2 \\ x_n & x_n & x_n & ... & x_n & x_n + d_n \end{pmatrix}. \end{aligned} \tag{20}$$

Die Aufgabe des FPS-Verfahrens ist es, in diesem Lösungsraum den Parametervektor $\mathbf{x}_{j_{min}}$ zu finden, der den kleinsten Wert der Zielfunktion liefert. Zu diesem Zweck muß sich der Polyeder ‚fortbewegen' können und dabei seine ,Gestalt' der Umgebung, also der ,Gebirgigkeit' der Zielfunktion, anpassen. Die Größe des Polyeders kann zunehmen und abnehmen.

Ist ein Minimum $\mathbf{x}_{j_{min}}^{(k)}$ im k-ten Optimierungsschritt gefunden, muß der Polyeder auf diesen Punkt zusammenschrumpfen. Eine Bewegung des Polyeders wird dadurch erreicht, daß immer der Parametervektor $\mathbf{x}_{j_{max}}^{(k)}$, der den größten Wert der Zielfunktion $f(\mathbf{x}_j^{(k)})$, $j = 1, ..., n+1$ ergibt, durch einen Parametervektor mit einem kleineren Wert der Zielfunktion ersetzt wird. Dadurch konvergieren alle Parametervektoren des Polyeders zu einem Optimum.

Zur Generierung eines neuen Parametervektors $\mathbf{x}_j^{(k)}$ kennt das FPS-Verfahren die vier Grundoperationen Reflexion, Expansion, Kontraktion und Reduktion. Sie lassen sich an einem zweidimensionalen Beispiel verdeutlichen, s. Abb. 2.

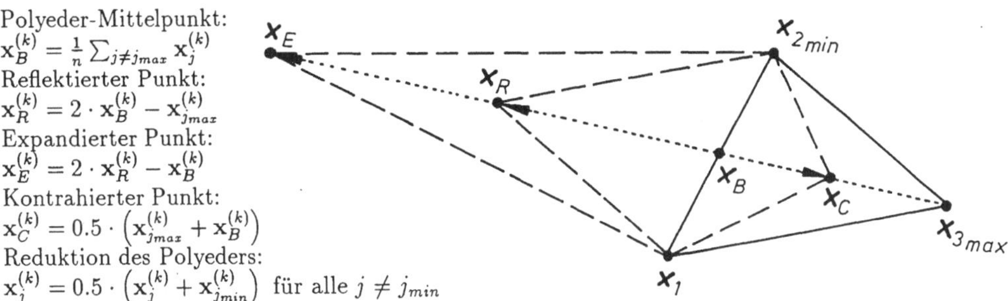

Abb. 2 Die vier Grundoperationen des *flexible polyhedron search*-Verfahrens mit einem zweidimensionalen Beispiel

Bei jedem Schritt des FPS-Verfahrens wird zunächst der Mittelpunkt $\mathbf{x}_B^{(k)}$ des Polyeders berechnet und der Punkt $\mathbf{x}_{j_{max}}^{(k)}$ mit dem momentanen Maximalwert der Zielfunktion an diesem gespiegelt. Liefert dieser sogenannte Reflektierte Punkt $\mathbf{x}_R^{(k)}$ eine Verbesserung gegenüber dem derzeitigen Minimalwert im Punkt $\mathbf{x}_{j_{min}}^{(k)}$, wird durch Expansion ein weiterer Punkt $\mathbf{x}_E^{(k)}$ in der gleichen Richtung untersucht. Der Punkt von beiden, der den besten Zielfunktionswert, der kleiner als das aktuelle Minimum sein muß, bringt, ersetzt den Punkt $\mathbf{x}_{j_{max}}^{(k)}$, s. Abb. 2. Bringen beide Punkte keine Verbesserung, wird erst durch Kontraktion ($\mathbf{x}_C^{(k)}$), dann durch Reduktion versucht, den bisherigen Minimalwert zu unterschreiten. Somit konvergieren die Punkte des Polyeders, wenn er nicht weiter in Richtung eines neuen Minimums verlagert werden kann. Der Suchalgorithmus bricht ab, wenn der quadratische Abstand der Zielfunktionswerte in den $(n+1)$-Punkten vom Zielfunktionswert im Mittelpunkt des Polyeders einen Grenzwert ϵ unterschreitet:

$$\sqrt{\sum_{j=1}^{n+1} \frac{(f(\mathbf{x}_j^{(k)}) - f(\mathbf{x}_B^{(k)}))^2}{n+1}} \leq \epsilon \tag{21}$$

Für die spezielle Anwendung in MEDA wurden der Abweichungsvektor und das Abbruchkriterium auf Werte festgelegt, die hinsichtlich der gewünschten Approximationsgüte und des CPU-Zeitbedarfs einen guten Kompromiß darstellen:

$$\mathbf{d} = 0.02\, \mathbf{x}_S; \tag{22}$$

$$\epsilon = 5 \cdot 10^{-5} \Delta_{opt}. \tag{23}$$

Δ_{opt} ist das Minimum der Zielfunktion, das vor dem Aufruf des FPS-Verfahrens mit dem vorherigen Variationssatz erzielt wurde (s. Abschnitt 5).

3.2 Die *external penalty*-Funktion

Die Parameter der Gamma-Mischverteilung haben folgende gültige Wertebereiche:

1. $0 \leq p_i \leq 1$ für alle $i = 1, ..., Z$. (24)
2. $0 \leq \sum_{i=3}^{Z} p_i < 1$. (25)
3. $1/\lambda_i > 0$ für alle $i = 1, ..., Z$. (26)
4. $r_i > 0$ für alle $i = 1, ..., Z$. (27)

Die daraus ableitbaren m Nebenbedingungen $g_i(\mathbf{x}) \geq 0,\ i = 1, .., m$ sind abhängig von der Wahl der bis zu sechs im jeweiligen Optimierungsschritt variierten Parameter (s. Abschnitt 5). Damit das FPS-Verfahren im gültigen Wertebereich der Parameter konvergiert, wird eine Lösung, die außerhalb des gültigen Wertebereichs liegt, mit Strafe (*penalty*) belegt. Die implementierte *external penalty*-Funktion ist in [6] beschrieben.

Aus einem allgemeinen nichtlinearen Optimierungsproblem P mit Nebenbedingungen $g_i(\mathbf{x}) \geq 0$ ergibt sich eine Folge von nichtlinearen Optimierungsproblemen H_k ohne Nebenbedingungen, die stattdessen gelöst werden muß:

$$\text{Minimiere: } H_k = f(\mathbf{x}) + \rho_k \cdot s(\mathbf{x}) \quad \text{mit} \quad s(\mathbf{x}) = \sum_{i=1}^{m} \max\{-g_i(\mathbf{x}), 0\}. \tag{28}$$

Für die Penaltyfunktion gilt $s(\mathbf{x}) = 0$ genau dann, wenn $g_i(\mathbf{x}) \geq 0$ für alle $i = 1, ..., m$ Nebenbedingungen erfüllt ist. Für die Folge des Penaltyfaktors ρ_k wird gefordert:

$$\rho_{k+1} > \rho_k > 0 \text{ für } k = 1, 2, 3, ... \text{ und} \tag{29}$$

$$\lim_{k \to \infty} \rho_k = \infty. \tag{30}$$

Die Zielfunktion $f(\mathbf{x})$ wird für die Hilfsfunktion H_k um die mit dem Penaltyfaktor gewichtete Penaltyfunktion erhöht, die die Nebenbedingungen enthält. Durch Vergrößerung von ρ_k kann somit die Hilfsfunktion an die Zielfunktion angepaßt werden. So konvergiert der Parametervektor der Hilfsfunktion H_k für $\rho_k \to \infty$ gegen einen optimalen Parametervektor, der alle Nebenbedingungen erfüllt.

In MEDA wird der Penaltyfaktor nach folgender Gleichung berechnet:

$$\rho_k = 2^k \cdot 1024. \tag{31}$$

Er wächst exponentiell, um eine schnelle Konvergenz zu erreichen.

Der Zielfunktion $f(\mathbf{x})$ entspricht in MEDA die Differenzfläche Δ. Vor der Berechnung der Zielfunktion wird geprüft, ob die Nebenbedingungen erfüllt sind. Werden während des Ablaufs des Algorithmus Parameter bestimmt, die außerhalb des Lösungsraums liegen, wird die Zielfunktion Δ entsprechend Gleichung (28) erhöht.

4 Die Einbeziehung von Sprungstellen

Empirische Verteilungsfunktionen weisen zwar oft einen erstaunlich glatten Funktionsverlauf auf, hin und wieder werden jedoch Sprungstellen (Dirac-Stöße in der Dichtefunktion) beobachtet, die streng genommen schon bei zwei gleichen Werten vorliegen. Diese Stufen können zwar leicht durch die Approximierende interpoliert werden, es kann sich aber um wichtige Erscheinungen handeln, die auch in die mathematische Beschreibung Eingang finden sollen.

Sprungstellen einer empirischen Verteilungsfunktion entsprechen mit den Zweigwahrscheinlichkeiten p_{0j} gewichten Dirac-Stößen in den Abszissenwerten $1/\lambda_{0j}$. Diese Erlang-∞-Verteilungen können als zusätzliche Zweige der Erlang-Mischverteilung dargestellt werden.

In MEDA werden Sprünge in der empirischen Verteilungsfunktion am Beginn eines Approximationslaufs untersucht und ab einer frei definierbaren Höhe als signifikant in die mathematische

Beschreibung aufgenommen (s. Abb. 3). Dazu müssen die Momentanteile der Sprungstellen wie die der höheren Zweige $M_{A_i}(j)$, $i > 2$ vor dem Momentenabgleich von den gemessenen Momenten $M_E(j)$ subtrahiert werden. Die endgültig an den in Abschnitt 2 beschriebenen Abgleichalgorithmus übergebenen reduzierten Momente $M_{APP}(j)$ lauten:

$$M_{APP}(j) = \left(M_E(j) - \sum_{i=3}^{Z} M_{A_i}(j) - \sum_{l=1}^{m} p_{0l}\frac{1}{\lambda_{0l}^{j}}\right) / \left(1 - \sum_{i=3}^{Z} p_i - \sum_{l=1}^{m} p_{0l}\right). \tag{32}$$

Der neue Nenner des Bruchs muß auch zur Normierung der Zweigwahrscheinlichkeiten (vgl. Gl. 16) verwendet werden.

Die Approximation der Verteilungsfunktion gemessener Dateilängen in Abschnitt 6 berücksichtigt eine Sprungstelle. Dies trägt wesentlich zur Güte der Näherungen bei.

5 Ablauf der Approximation

Die Approximation einer empirischen Verteilungsfunktion setzt sich aus mehreren Schritten zusammen (s. Abb. 3). Zu Beginn wird die empirische Verteilungsfunktion mit den daraus berechneten ersten drei Momenten eingelesen. Dann wird der Verteilungsfunktionsverlauf nach Sprungstellen durchsucht und deren Parameter ggf. festgehalten. Für die Zweige 3 bis Z müssen Startwerte gefunden werden, bevor durch das einfache Durchsuchen der k_1, k_2-Kombinationen (s. `K1K2`-Variation) und den Momentenabgleichalgorithmus die ersten beiden Zweige parametrisiert werden können.

In einer Eingabedatei des Programms MEDA sind für die Zweigzahlen 3 bis 6 jeweils Folgen von Variationssätzen (Kombinationen gleichzeitig zu variierender Parameter) abgelegt, die bei der Approximation optimiert werden. Die einzelnen Sätze werden analysiert. Es folgt entweder der entsprechende Prozeduraufruf oder der Variationssatz wird an den FPS-Algorithmus übergeben (s. Abb. 3).

Jeder dieser Sätze bewirkt die gleichzeitige Variation der in ihm genannten Parameter. Diese werden neu gewählt und die dadurch geänderten Momentanteile $M_{A_i}(j)$ der höheren Zweige von den empirischen Momenten abgezogen, während die übgrigen Parameter unverändert bleiben. Es folgt der Momentenabgleich mit den Parametern p_1, λ_1 und λ_2 und der Vergleich des Verteilungsfunktionsverlaufs der approximierenden mit der empirischen Verteilung in allen Stützstellen (Bestimmung des Gütekriteriums). Erst nach der Abarbeitung aller eingelesenen Variationssätze ist die Optimierung der Parameter vollständig abgeschlossen (s. Abb. 3).

Die speziellen Abfolgen der Variationssätze zur Optimierung der Erlang-Mischverteilung mit verschiedenen Zweigzahlen wurden heuristisch ermittelt und werden im folgenden beschrieben. Drei der verwendeten Variationssätze sind als eigene Prozeduren realisiert:

`K1K2` Optimierung der Phasenzahlen k_1 und k_2 der ersten beiden Zweige der Erlang-Mischverteilung. Beginnend mit $k_1 = k_2 = 1$ wird k_2 solange um eins erhöht, bis die Zielfunktion Δ nach einem lokalen Minimum Δ_{min_1} wieder wächst. Daraufhin wird k_1 um eins erhöht und wieder mit $k_1 = k_2$ beginnend k_2 bis zum Durchlaufen des nächsten lokalen Minimums Δ_{min_2} hochgezählt. Das Optimum ist gefunden, wenn gilt $\Delta_{min_{i+1}} > \Delta_{min_i}$.

`GE` Umwandlung der Gamma-Misch- in ein Erlang-Mischverteilung. Nacheinander werden die rationalen Werte r_i durch die nächst größeren bzw. nächst kleineren ganzen Zahlen k_i' ersetzt. Die vorherigen Zweigmittelwerte werden mit $1/\lambda_i' = (r_i/\lambda_i)/k_i'$ beibehalten. Die Wertepaare k_i', $1/\lambda_i'$, die die geringste Abweichung Δ der Verteilungsfunktionen liefern, werden als neue Parameter der Erlang-Mischverteilung festgehalten. Die Zweigwahrscheinlichkeiten p_i bleiben unberührt.

`K3L3` Ermittlung eines optimierten Wertes k_3' durch Hochzählen von $k_3' = 1$ an, bis die Zielfunktion Δ nach einem Minimum wieder wächst. Auch hierbei wird der vorherige Mittelwert des dritten Zweiges durch die Wahl $1/\lambda_3' = (k_3/\lambda_3)/k_3'$ beibehalten.

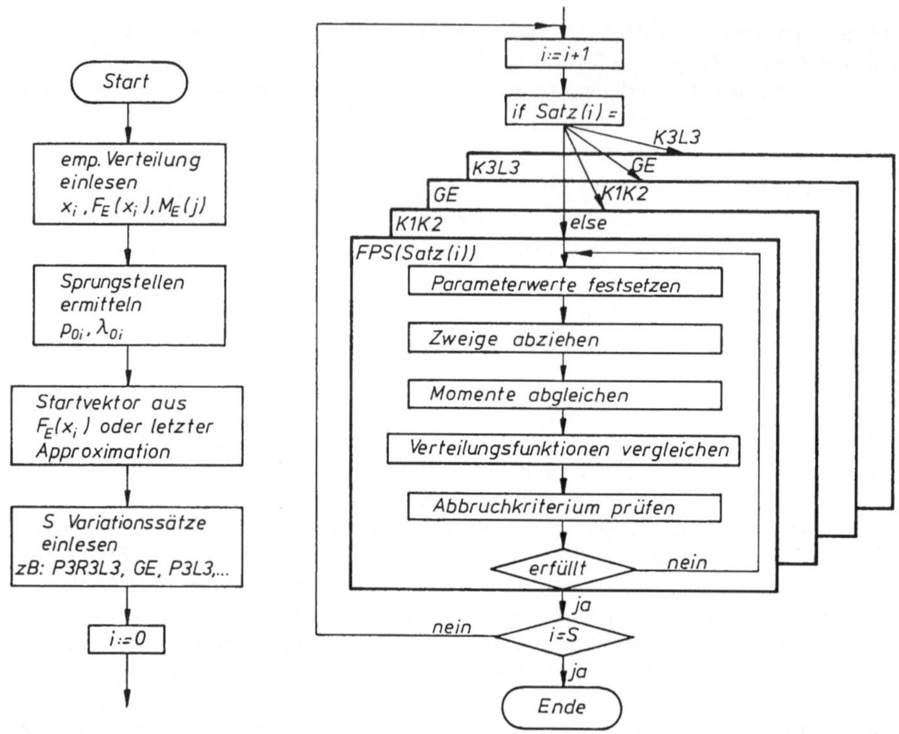

Abb. 3 Flußdiagramm zum Approximationsablauf

Zwei-zweigige Erlang-Mischverteilung: Verfahren Γ-FPS 2

Zur Optimierung der Parameter einer zwei-zweigigen Erlang-Mischverteilung wird nach der Ermittlung von Sprungstellen in der empirischen Verteilungsfunktion lediglich durch die im Variationssatz `K1K2` beschriebene Prozedur die beste k_1, k_2-Kombination gesucht. Der Momentenabgleich liefert die übrigen Parameter (s. Abb. 6a).

Drei-zweigige Erlang-Mischverteilung: Verfahren Γ-FPS 3

Vor der Optimierung der drei-zweigigen Erlang-Mischverteilung werden wieder die Sprungstellen ermittelt. Aus dem ersten steilen Anstieg der empirischen Verteilungsfunktion wird dann ein Startvektor für die Parameter p_3, λ_3, k_3 des dritten Zweiges bestimmt (s. [12] Abschnitt 2.4.1), der diesen repräsentieren soll. Die Folge der Variationssätze lautet:

`K1K2 P3R3L3 GE P3L3 K3L3 K1K2 P3L3.`

Zuerst wird also eine optimale k_1, k_2-Kombination von der `K1K2`-Prozedur ermittelt (wenn dort am Anfang $k_1 = k_2 = 1$ gesetzt wird, liegen damit zum ersten Mal alle Parameterwerte vor, die in die Gleichungen des Abgleichalgorithmus zur Bestimmung von p_1, λ_1 und λ_2 eingesetzt werden, vgl. Abschnitt 2). Anschließend werden die Parameter des dritten Zweiges p_3, r_3, λ_3 der Gamma-Mischverteilung an den FPS-Algorithmus übergeben und optimiert (`P3R3L3`). Daraufhin wird der kontinuierliche Parameter r_3 durch eine ganze Zahl k_3 ersetzt (`GE`) und damit die Gamma-Mischverteilung in eine Erlang-Mischverteilung überführt. Es folgt die gleichzeitige Optimierung von p_3, λ_3 (`P3L3`) und die anschließende Variation der Phasenzahl k_3 (`K3L3`). Hat sich durch die Variation von `K3L3` und `K1K2` der Wert der Zielfunktion Δ gegenüber dem Wert nach der ersten Variation von `P3L3` geändert, dann wird abschließend erneut der Versuch unternommen, die Parameter p_3, λ_3 des dritten Zweiges mit dem FPS-Algorithmus zu optimieren. Ein Ergebnis zeigt Abb. 6b.

Vier-zweigige Erlang-Mischverteilung
Vorher muß die Optimierung der Erlang-Mischverteilung mit drei Zweigen durchgeführt worden sein. Sowohl für die Parameter des dritten als auch des vierten Zweiges der approximierenden Erlang-Mischverteilung müssen zuerst Startwerte bestimmt werden. Dazu wird, ausgehend von den Parametern der eingelesenen Approximation mit drei Zweigen, der Anteil des dritten Zweiges gleichsam auf zwei Zweige aufgeteilt. Zwei Lösungen wurden entwickelt:
Erweiterung aus der drei-zweigigen Gamma-Mischverteilung: Verfahren Γ-FPS 4.1
Der dritte Zweig repräsentiert den ersten steilen Anstieg der empirischen Verteilungsfunktion. Die Ganzzahligkeit des Parameters k_3, der die Steigung des Verteilungsfunktionsanteils dieses Zweiges im wesentlichen bestimmt, schränkt die Anpassungsfähigkeit der Approximierenden ein. Die Abweichung der approximierenden drei-zweigigen Gamma-Mischverteilung (vor der GE-Umwandlung) ist daher meist besser als die der daraus abgeleiteten Erlang-Mischverteilung. Es liegt also nahe, auf die Gamma-Verteilung im dritten Zweig der Mischverteilung zurückzugreifen und sie durch eine Summe von zwei Erlang-k-Zweigen zu ersetzen. Ohne Beschränkung der Allgemeinheit nehmen wir $k_3' < k_4' = k_3' + 1$ für deren Phasenzahlen an und setzen k_3' und k_4' auf die beiden ganzen Zahlen links und rechts von r_3. Dieses Verfahren ist daher nur bei $r_3 > 1$ möglich. Beide Zweige sollen den gleichen Mittelwert und gemeinsam dieselbe Wichtung $p_3' + p_4' = p_3$ des alten dritten Zweiges haben. Somit treten sie zunächst möglichst exakt an dessen Stelle und können anschließend in der Optimierungsphase weiter an den empirischen Verteilungsfunktionsverlauf angenähert werden. Die Bestimmungsgleichungen lauten:

$$1/\lambda_3' = \frac{r_3}{\lambda_3 \cdot k_3'}, \qquad p_3' = p_3 \cdot (k_4' - r_3), \tag{33}$$

$$1/\lambda_4' = \frac{r_3}{\lambda_3 \cdot k_4'}, \qquad p_4' = p_3 \cdot (r_3 - k_3'). \tag{34}$$

Es ist offensichtlich, daß die Gamma-Mischverteilung nur zur Startvektorgenerierung verwendet werden kann, wenn gilt $r_3 > 1$, da sonst die ungültige Phasenzahl $k_3' = 0$ ermittelt werden würde.

Das Programm MEDA liest also die Optimalparameter der vorher optimierten drei-zweigigen Erlang-Mischverteilung ein und führt sie durch die Parametervariation P3R3L3 wieder in eine Gamma-Mischverteilung über. Anschließend erfolgt die oben beschriebene Ermittlung von Startparametern für den dritten und vierten Zweig. Die eigentliche Optimierung besteht aus der Folge:

```
P3L3P4L4 P3L3 P4L4 P3L3 P4L4.
```

Die einmal bestimmten Phasenzahlen k_3, k_4 werden also nicht mehr verändert. Lediglich die Zweigwahrscheinlichkeiten p_i und die Zweigintensitäten λ_i werden zur Optimierung an den FPS-Algorithmus übergeben, zuerst gemeinsam, dann zweigweise. Auch hierbei wird die zweite P3L3-Variation nur durchgeführt, wenn sich durch die P4L4-Variation eine Verbesserung der Approximation ergeben hat, usw.
Erweiterung aus der drei-zweigigen Erlang-Mischverteilung: Verfahren Γ-FPS 4.2
Der dritte Zweig repräsentiert den Anfangsbereich der empirischen Verteilungsfunktion und einer der ersten beiden Zweige den folgenden Abschnitt. Derjenige der ersten beiden Zweige mit dem kleineren Zweigmittelwert kann daher zur Fortsetzung des dritten Zweiges als Startvektor für den vierten Zweig genommen werden. Folgende heuristische Gleichungen führen zu den Startwerten für den dritten und vierten Zweig:

$$\begin{array}{rlll} & p_3' = 0.9 \cdot p_3, & k_3' = k_3, & 1/\lambda_3' = 1/\lambda_3; \\ \text{falls } k_1/\lambda_1 \le k_2/\lambda_2: & p_4' = 0.95 \cdot p_1, & k_4' = k_1, & 1/\lambda_4' = 1/\lambda_1; \\ \text{falls } k_1/\lambda_1 > k_2/\lambda_2: & p_4' = 0.95 \cdot p_2, & k_4' = k_2, & 1/\lambda_4' = 1/\lambda_2. \end{array} \tag{35}$$

Die Reduktion der Zweigwahrscheinlichkeiten um 10 bzw. 5 % wurde vorgenommen, um Spielraum für die weitere Optimierung zu schaffen. Die Folge zur Optimierung dieser Werte lautet:

```
K1K2 R2P3L3P4L4 K1K2 P3L3P4R4L4 GE P3L3P4L4 P3L3 P4L4 K1K2 P3L3P4L4 P3L3 P4L4.
```

Die wiederholte Optimierung einzelner Parameter hat sich bei den von uns approximierten Verteilungen als vorteilhaft erwiesen. In manchen Fällen ist sie jedoch nicht notwendig. Eine Anpassung sollte daher vom Anwender geprüft und in der Eingabedatei vorgenommen werden. Die Approximation der Beispielverteilung mit einer vier-zweigigen Erlang-Mischverteilung zeigt Abb. 6c. Der Startvektor für den vierten Zweig wurde vom ersten Zweig der drei-zweigigen Erlang-Mischverteilung abgeleitet, da der dritte ein Exponentialzweig ist ($k_3 = 1$).

Fünf- und mehr-zweigige Erlang-Mischverteilung: Verfahren Γ-FPS Z

Vorher muß die Optimierung der Erlang-Mischverteilung mit $Z - 1$ Zweigen durchgeführt worden sein. Die höheren Zweige $Z \geq 5$ sollen die Näherung der Verteilungsfunktion im oberen Bereich weiter verbessern. Dazu werden die Zweige der eingelesenen, nächst kleineren Approximierenden nach der Größe ihres Zweigmittelwertes sortiert und die Parameter des Zweiges mit dem zweitgrößten Zweigmittelwert als Startwerte k_Z', p_Z',und λ_Z' des zusätzlichen Zweiges gewählt. Die Zweigwahrscheinlichkeiten der nun dritten bis Z-ten Zweige werden um 5% bzw. 10% reduziert, um wieder Spielraum für die weitere Optimierung zu schaffen. Beim fünf-zweigigen Modell gilt speziell für die Wichtungen

$$p_3' = 0.9p_3 \quad p_4' = 0.9p_4 \quad p_5' = 0.95p_5 \tag{36}$$

und die Folge der Variationssätze

```
K1K2 P3L3P4L4P5L5 P3L3 P4L4P5L5 K1K2 P3L3P4L4P5L5 P3L3P4L4 P4L4P5L5 P3L3P5L5
K1K2 P3L3 P4L4.
```

Abb. 6d zeigt das Ergebnis für die Beispielverteilung. Bei sechs und mehr Zweigen werden die Zweigwahrscheinlichkeiten des 3-ten bis Z-ten Zweiges mit 0.95 gewichtet. Die Parameter einer sechs-zweigigen Erlang-Mischverteilung werden durch die Variationssätze

```
K1K2 P4L4P5L5P6L6 P3L3 P4L4 P5L5 P6L6 K1K2 P3L3 P4L4 P5L5 P6L6.
```

optimiert. Da die dem Autor vorliegenden empirischen Verteilungen durch die Hinzunahme eines sechsten Zweiges nur unwesentlich verbessert werden konnten, wurden keine Parameterfolgen für größere Modelle entworfen, obgleich der Algorithmus dies in einfacher Weise zuläßt.

6 Approximationsergebnisse

Die neue Version der Approximationsverfahrens MEDA wurde auf die am Lehrstuhl vorliegenden 21 empirischen Verteilungsfunktionen aus Verkehrsmessungen angewendet. Ein Vergleich der Approximation mit drei Zweigen mit den Ergebnissen des früheren heuristischen Verfahrens ergab keine merklichen Verbesserungen der Approximationsgüte, jedoch eine starke Verringerung der CPU-Zeit auf teilweise ein Fünftel. Dies erklärt sich durch die gleichzeitige Variation mehrerer Parameter, die mit dem FPS-Verfahren möglich ist.

Die vier Abbildungen 6 a-d zeigen am Beispiel einer empirischen Dateilängen-Verteilungsfunktion $F_E(x)$ die erheblichen Verbesserungen, die durch die Erweiterung des Verfahrens auf vier und mehr Zweige möglich sind. Die exakte Nachbildung der Sprungstelle (in $1/\lambda_{01}$ mit der Höhe p_{01}) ist dafür eine unabdingbare Voraussetzung. Die angegebenen CPU-Zeiten $T_{CPU}(Z)$ beziehen sich auf die SIEMENS 7.561 des Lehrstuhls. $N_{Comp}(Z)$ gibt die Anzahl der notwendigen Verteilungsfunktionsvergleiche an.

Am Übergang von zwei auf drei und vier Zweige läßt sich die Steigerung der Anpassungsfähigkeit der Approximierenden deutlich beobachten. Die Unterschiede zwischen den Approximationen mit vier und fünf Zweigen lassen sich jedoch nur noch am Verlauf der Differenz $\delta(x)$ ausmachen. Die Differenzfläche kann mit einem zusätzlichen sechsten Zweig noch einmal um 20 % verbessert werden. Dies geht jedoch zu Lasten der maximalen Abweichung δ_{max}, die dadurch um 50 % wächst.

Alle approximierten Verteilungsfunktionen konnten mit vier Zweigen hinsichtlich der Differenzfläche um 5−25% besser genähert werden als mit nur drei Zweigen. Der fünfte Zweig brachte in einigen Fällen eine Verbesserung um weitere 10%. Lediglich bei zwei empirischen Verteilungsfunktionen ergab die Approximation mit sechs Zweigen noch eine weitere, minimale Verbesserung.

$M_E(1) =$	22972	$N =$	4893	**Γ-FPS 2**	
$p_{01} =$	0.0241	$1/\lambda_{01} =$	4096		
$p_1 =$	0.8439	$1/\lambda_1 =$	3686	$k_1 =$	1
$p_2 =$	0.1320	$1/\lambda_2 =$	149717	$k_2 =$	1
$\Delta =$	0.2179	$N_{Comp.}(2) \approx$	20		
$\delta_{max} =$	0.1208	$T_{CPU}(2) =$	24.3 s		

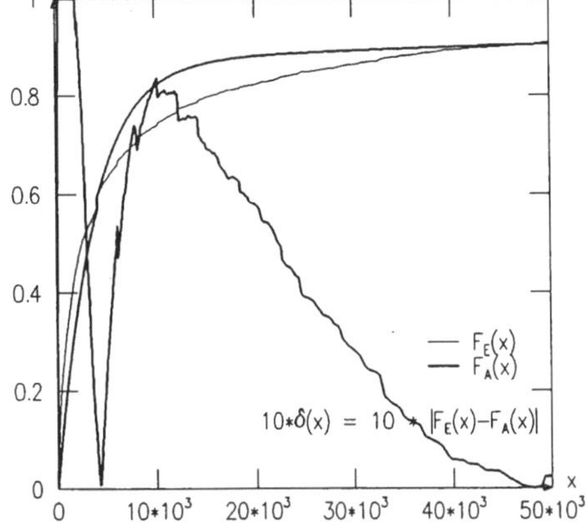

a) mit zwei Zweigen

$M_E(1) =$	22972	$N =$	4893	**Γ-FPS 3**	
$p_{01} =$	0.0241	$1/\lambda_{01} =$	4096		
$p_1 =$	0.2607	$1/\lambda_1 =$	29941	$k_1 =$	1
$p_2 =$	0.0430	$1/\lambda_2 =$	79746	$k_2 =$	4
$p_3 =$	0.6722	$1/\lambda_3 =$	2016	$k_3 =$	1
$\Delta =$	0.0804	$N_{Comp.}(3) \approx$	404		
$\delta_{max} =$	0.0743	$T_{CPU}(3) =$	855.9 s		

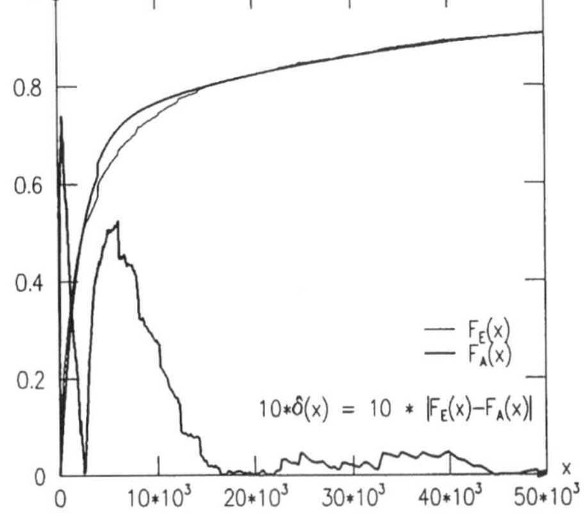

b) mit drei Zweigen

$M_E(1) =$	22972	$N =$	4893	**Γ-FPS 4.2**	
$p_{01} =$	0.0241	$1/\lambda_{01} =$	4096		
$p_1 =$	0.3598	$1/\lambda_1 =$	5490	$k_1 =$	1
$p_2 =$	0.0370	$1/\lambda_2 =$	69378	$k_2 =$	5
$p_3 =$	0.3658	$1/\lambda_3 =$	606	$k_3 =$	1
$p_4 =$	0.2133	$1/\lambda_4 =$	36776	$k_4 =$	1
$\Delta =$	0.0602	$N_{Comp.}(4) \approx$	$N_{Comp.}(3) + 862$		
$\delta_{max} =$	0.0450	$T_{CPU}(4) =$	$T_{CPU}(3) + 1102.8$ s		

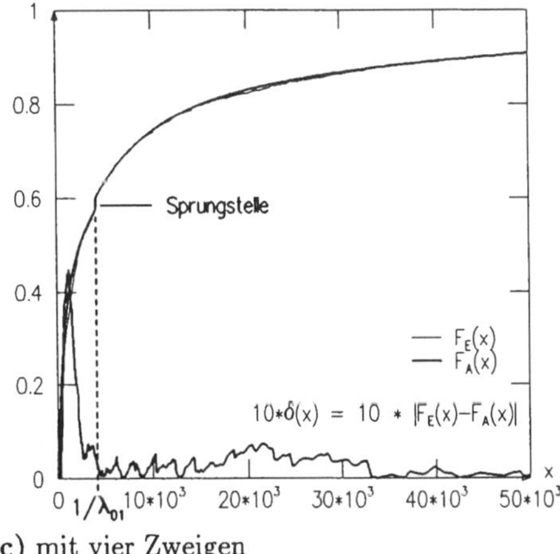

c) mit vier Zweigen

$M_E(1) =$	22972	$N =$	4893	**Γ-FPS 5**	
$p_{01} =$	0.0241	$1/\lambda_{01} =$	4096		
$p_1 =$	0.3022	$1/\lambda_1 =$	1228	$k_1 =$	1
$p_2 =$	0.0371	$1/\lambda_2 =$	69335	$k_2 =$	5
$p_3 =$	0.1189	$1/\lambda_3 =$	189	$k_3 =$	1
$p_4 =$	0.3046	$1/\lambda_4 =$	6002	$k_4 =$	1
$p_5 =$	0.2131	$1/\lambda_5 =$	36593	$k_5 =$	1
$\Delta =$	0.0598	$N_{Comp.}(5) \approx$	$N_{Comp.}(4) + 770$		
$\delta_{max} =$	0.0228	$T_{CPU}(5) =$	$T_{CPU}(4) + 958.2$ s		

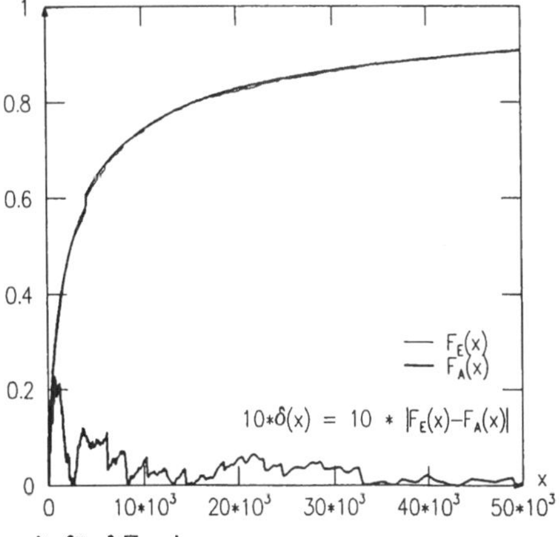

d) mit fünf Zweigen

Abb. 4 MEDA-Approximation $F_A(x)$ einer empirischen Dateilängen-Verteilungsfunktion $F_E(x)$ mit Fehlerkurve $\delta(x)$

Die Anzahl der notwendigen Stützstellen (Meßwerten) hängt ausschließlich von der Komplexität der zugrundeliegenden empirischen Verteilung ab. Eine Verteilung mit einem sehr glatten Verlauf kann bereits durch wenige Stützstellen ausreichend beschrieben werden, während eine Verteilung mit mehreren Stufen eine sehr hohe Anzahl von Meßwerten erfordert.

Bei einem zu geringen Stichprobenumfang, der durch die Meßmöglichkeiten bedingt sein kann, liefert MEDA zumindest eine gute Interpolation des empirischen Verteilungsfunktionsverlaufs. In ungünstigen Fällen kann es bei weniger als $N \approx 20$ Stützstellen vorkommen, daß die Startvektorgenerierung für den dritten Zweig nicht gelingt. Über die Eingabedatei läßt sich jedoch ein Startwert von außen vorgeben.

7 Zusammenfassung

Die beschriebenen Erweiterungen und Verbesserungen des bereits bewährten Approximationsverfahrens MEDA erhöhen dessen Einsatzmöglichkeiten weiter. Seine wesentlichen Merkmale wie Verwendung einer Phasenverteilung, exakter Abgleich der ersten drei empirischen Momente und bestmögliche Näherung der empirischen Verteilungsfunktion in allen Stützstellen wurden beibehalten. Das nichtlineare *flexible polyhedron search*-Verfahren (FPS) erlaubt die gleichzeitige Variation mehrerer Parameter. Durch die Einbeziehung der Gamma-Mischverteilung wird die Optimierung der Phasenzahlen mit dem FPS-Verfahren ermöglicht. Dies reduziert die Zahl der zeitintensiven Verteilungsfunktionsvergleiche gegenüber dem heuristischen Verfahren erheblich. Darüberhinaus erlaubt das FPS-Verfahren die einfache Erweiterung auf vier und mehr Zweige und damit eine weitere Steigerung der Genauigkeit. Signifikante Sprungstellen werden als gewichtete Dirac-Stöße in die mathematische Beschreibung aufgenommen. Eine Anzahl von vier bis fünf Zweigen der approximierenden Erlang-Mischverteilung scheint für die meisten Verteilungen völlig ausreichend. Da jede Approximation auf dem Ergebnis der vorhergehenden aufsetzt, kann der Anwender entscheiden, ob die erzielte Näherungsgüte ausreicht, oder ob durch eine Erhöhung der Komplexität der Approximierenden eine Verbesserung angestrebt werden soll.

Mit der neuen Version von MEDA steht nun ein flexibles und leistungsfähiges Approximationsverfahren zur Verfügung, das alle notwendigen Voraussetzungen für den Einsatz in der Datenverkehrstheorie erfüllt.

Schlußwort

Der Autor möchte sich bei Herrn Dipl. Inform. J.J. Kim bedanken, der mit seiner Diplomarbeit die Grundlagen der Erweiterungen und Verbesserungen von MEDA geschaffen hat. Dank gilt auch Herrn W.Y. Ding für die Lösung des Gleichungssystems und Herrn cand. ing. A. Basermann für die Entwicklung und Implementation der Nullstellensuche im Abgleichalgorithmus und die Pflege des MEDA-Programmcodes.

Literaturverzeichnis

[1] G. Björck, A. und Dahlquist. *Numerische Methoden.* Oldenbourg, München, Wien, 2. Auflage, 1979.

[2] P. Bratley, B. Fox, and L. Schrage. *A Guide to Simulation.* Springer-Verlag, New York, 1983.

[3] K.A. Bronstein, I.N. und Semendjajew. *Taschenbuch der Mathematik.* Verlag Harri Deutsch, Thun und Frankfurt(Main), 1986.

[4] W. Bux. *Single Server Queues with General Interarrival and Phase-Type Service Time Distributions - Computational Algorithms.* In *Proc. 9th Int. Teletraffic Congress (ITC)*, paper 413, Torremolinos, 1979.

[5] W. Bux and U. Herzog. *The Phase Concept: Approximation of Measured Data and Performance Analysis.* In *Proc. Int. Symp. on Computer Performance, Measurements, and Evaluation*, Elsevier North-Holland, New York, Amsterdam, 1977.

[6] S.G. Byron and W. Joel. *Introduction to Optimization Theory.* 1973.

[7] D.R. Cox. *A Use of Complex Probabilities in the Theory of Stochastic Processes.* Proc. Camb. Phil. Soc., Vol. 51, pp. 313–319, 1955.

[8] S. Halfin. *Delays in Queues, Properties and Approximations.* In *Proc. 11th Int. Teletraffic Congress (ITC)*, paper 1.4–3, Kyoto, 1985.

[9] D.M. Himmelblau. *Applied Nonlinear Programming.* MacGraw-Hill, New York, 1972.

[10] R. Jeltsch. *Numerische Mathematik für Ingenieure, Teil A.* Institut für Geometrie und Praktische Mathematik, RWTH Aachen, 4. Auflage, 1985.

[11] R. Schassberger. *Warteschlangen.* Springer-Verlag, Wien, New York, 1973.

[12] L. Schmickler. *Approximation von empirischen Verteilungsfunktionen mit Erlangmischverteilungen und Coxverteilungen.* In *Informatik Fachberichte: Messung, Modellierung und Bewertung von Rechensystemen, GI/NTG Fachtagung Erlangen*, pp. 118–133, Springer-Verlag, Berlin, 1987.

[13] B.L. van der Waerden. *Mathematische Statistik.* Springer-Verlag, Berlin, 3. Auflage, 1971.

Eine Modellwelt zur Integration von Warteschlangen- und Petri-Netz-Modellen

Falko Bause, Heinz Beilner
Informatik IV, Universität Dortmund
Postfach 50 05 00, 4600 Dortmund 50

Zusammenfassung:
Rechensysteme werden sowohl bzgl. funktionaler Aspekte (z.B. Betrachtung von Deadlocks) als auch hinsichtlich quantitativer Aspekte (z.B. Betrachtung des Durchsatzes) untersucht. Der Einsatz von Petri-Netzen bzw. Warteschlangennetzen bietet sich hierbei an, da für beide Modellwelten eine fundierte Theorie existiert. Um zu vermeiden, daß zwei Modelle für ein System erstellt werden müssen, existieren bereits einige Modellwelten zur Verbindung der Betrachtung funktionaler und quantitativer Gesichtspunkte eines Systems. Besonders stark ist dabei in den letzten Jahren die Linie der Time(d)-Petri-Netze hervorgetreten. Nachteilig wirkt sich bei ihnen allerdings aus, daß die Beschreibungsmittel stark an der Petri-Netz-Philosophie angelehnt sind und häufig für Warteschlangenmodellierer unpassend sind. Diese Arbeit stellt eine Modellwelt vor, welche Warteschlangennetz- und Petri-Netz-Beschreibungen vereint und so den Beschreibungsaufwand reduziert bzw. die Beschreibung vereinfacht. Ferner wird untersucht, inwieweit funktionale Ergebnisse Aussagen über das quantitative Modell zulassen.

1 Motivation und Einführung

Die bekanntesten Modelle zur Untersuchung von Rechensystemen sind Petri-Netze und Warteschlangennetze. Petri-Netze dienen dabei der Untersuchung funktionaler Aspekte (im Sinne "globaler" Funktionalität und Korrektheit), wogegen Warteschlangennetze zur Leistungsbewertung des modellierten Systems eingesetzt werden. In der Regel sollte ein System sowohl unter funktionalen Gesichtspunkten betrachtet als auch in seiner Leistungsfähigkeit bewertet werden.

Zur Leistungsbewertung (quantitativen Analyse) eines Systems eignen sich Petri-Netze nicht, da sie den Begriff der Zeit nicht beinhalten; andererseits sind Warteschlangenmodelle nur bedingt zur funktionalen Analyse einsetzbar, da Mechanismen wie Synchronisation von Prozessen, wechselseitiger Ausschluß und ähnliche Probleme zumindest standardmäßig nicht beschrieben werden können. Ist man an der Analyse eines Systems unter sowohl funktionalen als auch quantitativen Gesichtspunkten interessiert, so ist es häufig nötig, mehrere Modellbeschreibungen eines Systems zu erstellen.

Um diesen Mehraufwand zu vermeiden, sind bereits zahlreiche Modelle zur Verbindung der funktionalen und der quantitativen Analyse entwickelt worden. Schon Ende der 60er, Anfang der 70er Jahre wurden hierzu Modelle vorgestellt (GMC/GMB ([RAES77], [ESTR78]), E-Netze ([NUTT72a,b]), Pro-Netze ([NOE75])). Eine wesentliche Aufgabe dieser Modelle ist die Unterstützung des Benutzers bei der Modellierung und Bewertung von Rechensystemen. Oft erweitern diese Modelle die Petri-Netz-Struktur, um eine einfache Handhabung der rechnergestützten Modellierungswerkzeuge zu ermöglichen. Dies hat zur Folge, daß die Modelle fast ausschließlich simulativ bearbeitet werden können und eine funktionale Analyse nur für eingeschränkte Netzklassen praktikabel ist.

Ab Anfang der 70er Jahre enstanden sogenannte Time(d)-Petri-Netz-Modelle, die weitgehend auf eine Erweiterung der Petri-Netz-Struktur verzichten und so eine funktionale Analyse zulassen. Ferner wird in vielen Fällen die Zeit so in die Beschreibung integriert (Markov-Modelle), daß algebraisch-numerische Methoden, teils auch algebraisch-analytische Methoden zur Leistungsanalyse eingesetzt werden können ([MOLL81], [AMBC84], [AMBB85], [WTPN85]). Die Beschreibung der Time(d)-Petri-Netze ist allerdings (verständlicherweise) stark an der Beschreibung von Petri-Netzen angelehnt, was für Warteschlangenmodellierer häufig ungewohnt oder sogar unpassend ist.

In diesem Beitrag wird eine Modellwelt vorgestellt, welche versucht, Petri-Netze und Warteschlangennetze zu vereinen. Die Struktur eines Modells wird dabei durch ein Petri-Netz beschrieben und der Begriff der "Zeit" mit den Stellen des Netzes in Verbindung gebracht. Durch Einführung zeitbehafteter Vorgänge in einem Petri-Netz kann, außer für bestimmte Fälle (z.B. SPNs [MOLL81]), die Erreichbarkeitsmenge nicht mehr oder nur noch teilweise zur quantitativen Analyse genutzt werden, da der Zustandsraum des "zeitbehafteten" Modells i.a. andere Eigenschaften besitzt als der des "zeitlosen" Modells. Diese Probleme treten auch in der hier vorgestellten Modellwelt auf; sie sind allerdings auch bei bereits bekannten Modellwelten, z.B. der GSPN-Modellwelt ([AMBC84]), zu finden. Es wird untersucht, unter welchen Bedingungen sich gewisse funktionale Aussagen, wie z.B. Lebendigkeit, auf den quantitativen Fall übertragen lassen und ob ein einem funktional "korrekten" Modell zugeordneter Markov-Prozeß eine stationäre Zustandsverteilung besitzen kann. Damit wird das Ziel verfolgt, Aussagen der funktionalen Analyse als "Voranalyse" zur quantitativen Analyse zu nutzen.

2 Grundlegende Begriffe und Aussagen aus der Petri-Netz-Theorie

Ein Petri-Netz (auch Stellen/Transitions-Netz genannt) ist ein 5-Tupel $PN = (S, T, W^-, W^+, M_0)$

mit
- a) S ist eine endliche, nichtleere Menge von Stellen.
- b) T ist eine endliche, nichtleere Menge von Transitionen.
- c) $S \cap T = \emptyset$
- d) W^-, W^+ sind Funktionen aus $[S \times T \rightarrow \mathbb{N}_0]$.
  W^- heißt Rückwärtsinzidenzfunktion und W^+ Vorwärtsinzidenzfunktion.
- e) M_0 ist eine Funktion aus $[S \rightarrow \mathbb{N}_0]$, die Anfangsmarkierung von PN.

Wie üblich werden wir zur Bezeichnung der Ein- und Ausgabeelemente die Punktnotation verwenden, so ist z.B. $\bullet s = \{t \in T \mid W^+(s,t) > 0\}$ die Menge der Eingabetransitionen zu $s \in S$; analog ist $s\bullet$ die Menge der Ausgabetransitionen zu s, und $\bullet t$ bzw. $t\bullet$ die Menge der Ein- bzw. Ausgabestellen zu $t \subset T$. Eine Transition t heißt aktiviert in einer Markierung M genau dann, wenn $M(s) \geq W^-(s,t), \; \forall s \in \bullet t$. Eine aktivierte Transition $t \in T$ kann feuern; die Markierung M geht dabei über in die (Folge-)Markierung M' mit $M'(s) = M(s) + W^+(s,t) - W^-(s,t), \; \forall s \in S$ (Schreibweise: M -> M' oder M -t-> M'). Sei ->* der reflexive, transitive Abschluß der Relation ->. Dann bezeichnet $E(PN) = \{M \mid M_0 \text{ ->* } M\}$ die Erreichbarkeitsmenge von PN. Das Petri-Netz PN heißt beschränkt, falls E(PN) eine endliche Menge ist. PN heißt lebendig genau dann, wenn $\forall t \in T$, $\forall M \in E(PN): \exists M' \in E(PN)$: M ->* M': t ist aktiviert in M'.

Für allgemeine Petri-Netze sind Eigenschaften wie Beschränktheit und Lebendigkeit nur mit großem Aufwand nachprüfbar. Schränkt man die Netzstruktur dagegen ein, so existieren leicht nachprüfbare Bedingungen, insbesondere um die Lebendigkeit des Petri-Netzes zu sichern.

Definition (Netzklassen):

Sei PN = (S,T,W^-,W^+,M_0) ein Petri-Netz mit $W^-(s,t) \leq 1$ und $W^+(s,t) \leq 1$, $\forall$ $s \in S$, $t \in T$. PN heißt

a) marked graph-Netz :<==> $\forall$ $s \in S$: $|{\bullet}s| = |s{\bullet}| = 1$,

b) state machine-Netz :<==> $\forall$ $t \in T$: $|{\bullet}t| = |t{\bullet}| = 1$.

c) free choice-Netz :<==> $\forall$ $s \in S$: $|s{\bullet}| > 1$ ==> ${\bullet}(s{\bullet}) = \{s\}$. D.h. besitzen zwei Transitionen die gleiche Eingabestelle, so ist dies ihre einzige Eingabestelle.

d) simple-Netz :<==> $\forall$ $t \in T$: $|\{s \in {\bullet}t \mid |s{\bullet}| > 1\}| \leq 1$. D.h. jede Transition besitzt höchstens eine Eingabestelle, welche auch Eingabestelle einer anderen Transition ist.

Zwischen o.g. Netzklassen besteht folgende echte Teilmengenbeziehung:

state machine-Netze $\subset$ free choice-Netze $\subset$ simple-Netze

$\cup$

marked graph-Netze

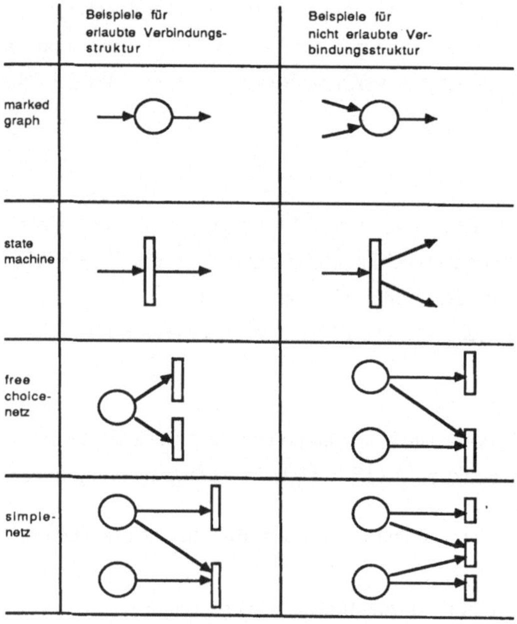

	Beispiele für erlaubte Verbindungsstruktur	Beispiele für nicht erlaubte Verbindungsstruktur
marked graph		
state machine		
free choice-netz		
simple-netz		

Abb. 1: Netzklassen

Um einfache Bedingungen zur Sicherung der Lebendigkeit obiger Netzklassen angeben zu können, spielen die Begriffe deadlock und trap eine wichtige Rolle. Ein Stellenmenge $S' \subseteq S$ heißt deadlock, falls ${\bullet}S' \subseteq S'{\bullet}$ bzw. trap, falls $S'{\bullet} \subseteq {\bullet}S'$, wo ${\bullet}S' := \cup_{s \in S'} {\bullet}s$ und $S'{\bullet} := \cup_{s \in S'} s{\bullet}$. Wenn ein deadlock unter einer Markierung M nicht markiert (leer; $M(s) = 0$, $\forall$ $s \in S'$) ist, so ist er auch unter allen Folgemarkierungen von M nicht markiert. Ist ein trap unter einer Markierung M markiert, so ist er auch unter allen Folgemarkierungen von M markiert.

Satz (Lebendigkeitscharakterisierung für free choice-Netze):
Sei PN ein free choice-Netz. Dann gilt:
PN ist lebendig <==> Jeder deadlock besitzt einen unter M_0 markierten trap.
Beweis: z.B. [REIS82] •

Satz (hinreichende Bedingung für die Lebendigkeit in simple-Netzen):
Sei PN ein simple-Netz. Dann gilt:
Jeder deadlock besitzt einen unter M_0 markierten trap ==> PN ist lebendig.
Beweis: z.B. [COMM72] •

Obige Bedingungen lassen sich mit geringem Aufwand nachprüfen, da sie sich nur auf die Struktur des Netzes und die Anfangsmarkierung stützen. Ebenso wie die Lebendigkeit läßt sich häufig auch die Beschränktheit eines Petri-Netzes mit einfachen Mitteln nachprüfen. So ist ein Petri-Netz beispielsweise beschränkt, wenn alle Stellen durch S-Invarianten überdeckt sind ([REIS82]).

3 Die WSPN-Modellwelt

Zur Bewertung der Leistung von Rechensystemen ("quantitative Analyse") werden verbreitet Warteschlangenmodelle eingesetzt. Für die Analyse solcher Modelle existiert ein großes Spektrum an Techniken, welche die Ermittlung von Leistungsgrößen gestatten. So ist neben dem Einsatz stochastischer, ereignisorientierter Simulation insbesondere die Abbildung der Modelle auf Markov-Prozesse üblich, welche einerseits einer direkten numerischen Analyse zugänglich sind, andererseits (bei geeigneter Einschränkung der Problemklasse und der zu ermittelnden Leistungscharakteristika) den Einsatz hocheffizienter Analyse-Algorithmen erlauben, wie etwa im Falle der sog. separablen Netze. Die Darstellung jener Eigenschaften eines Rechensystems, die für seine globale Korrektheit eine zentrale Rolle spielen (Synchronisation von Prozessen, Populationsbeschränkungen samt Blockaden u.a.m.) ist in Warteschlangenmodellen zwar möglich, geschieht aber üblicherweise mit Beschreibungsmitteln, die einer formalen, modellgestützten Beurteilung der Korrektheit nicht oder nur sehr schwer zugänglich sind. Es sei in diesem Kontext daran erinnert, daß zur leistungsorientierten Bewertung solcher Modelle im Prinzip nur "allgemeine" Techniken wie Simulation und numerische Markov-Analyse einsetzbar sind.

Zur Beurteilung der globalen Korrektheit von Rechensystemen ("funktionale Analyse") werden verbreitet Modelle des Petri-Netz-Typs eingesetzt. Für die Analyse solcher Modelle existieren sehr effiziente Algorithmen zur Feststellung gewisser Systemeigenschaften (Invariantenanalyse, Algorithmen für spezielle Netzklassen). Die Darstellung (der Länge) konkreter Zeitintervalle, die für die Leistung eines Rechensystems eine zentrale Rolle spielen, ist in reinen Petri-Netzen nicht vorgesehen. Zur Einbringung von Zeitaspekten in Petri-Netze bieten sich unmittelbar zwei alternative Möglichkeiten an, zum einen die Quantifizierung der Verweilzeit von Marken in Stellen, zum anderen die Quantifizierung der Schaltdauer von Transitionen, wobei die gegenwärtige Tendenz (mit nennenswertem Erfolg) fast ausschließlich der letztgenannten Alternative gilt. Es sei in diesem

Zusammenhang daran erinnert, daß das Verhalten des Modells (etwa in Form der möglichen Zustands-Trajektorien) in jedem Fall durch das Einbringen des Zeitaspekts potentiell verändert wird. Die im folgenden geschilderte "Warteschlangen-Petri-Netz-Modellwelt" (WSPN-Modellwelt) stellt einen weiteren Versuch dar, eine Spezifikation von Rechensystem-Modellen zu ermöglichen, die wesentliche Vorteile unter folgende Gesichtspunkten liefert:

- Sie vereint die Darstellungs-Annehmlichkeiten und -Gewohnheiten der Warteschlangennetze und der Petri-Netze.
- Sie ist zugänglich für die Analyse-Techniken der Warteschlangennetze (quantitative Analyse) und der Petri-Netze (funktionale Analyse).

Die Beschreibung der WSPN-Modellwelt geschieht hier verbal. Eine formale Charakterisierung findet sich in [BAUS86].

Die Struktur eines WSPN-Modells ist durch ein Stellen/Transitions-Netz gegeben. Die Stellen des Netzes dürfen durch Stationsbeschreibungen erweitert werden in der Form (s. Abb. 2), daß eine Strukturierung der Stelle in zwei Komponenten entsteht, wobei die "vordere" Komponente den Charakter einer Warteschlangen-"Station" annimmt, die "hintere" Komponente den einer "normalen" Petri-Netz-Stelle. Dynamisch gesehen, gelangen Marken ("Kunden"), die auf solch eine erweiterte Stelle ("zeitbehaftete Stelle") gefeuert werden, in einen (Stations-)Warteraum, aus dem heraus sie gemäß einer (Stations-)Bediendiziplin (z.B. FCFS, PS, ...) unmittelbar oder verzögert einer Bedieneinrichtung zugeführt werden; bei Bedienende werden Marken zeitlos im Abgabebereich der zeitbehafteten Stelle abgelegt, wo sie für Ausgabetransitionen dieser erweiterten Stelle verfügbar sind. Die Bedienzeit (der zeitliche Bedienbedarf) für Marken bezüglich des Stationsteils erweiterter Stellen wird durch kontinuierliche, auf [0,∞[konzentrierte Verteilungen beschrieben angenommen.

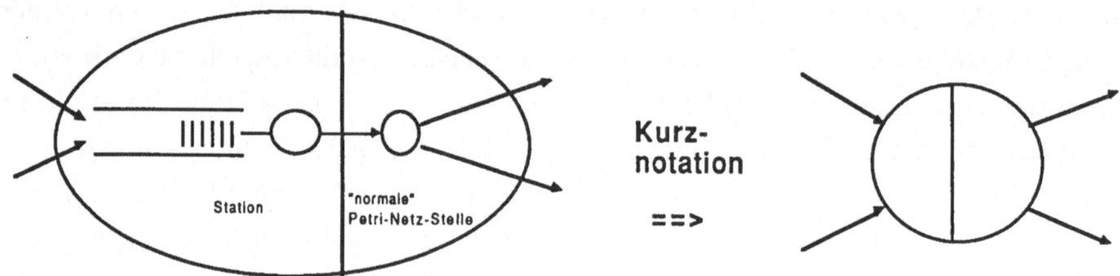

Abb. 2: Integration einer Station in eine Stelle des Petri-Netzes: zeitbehaftete Stelle

Alle Transitionen eines WSPN-Modells feuern im aktivierten Zustand zeitlos und unmittelbar nach Aktivierung. Sind mehrere Transitionen gleichzeitig aktiviert (Konfliktfall), so bestimmt eine diesbezügliche diskrete Verteilung die feuernde Transition. Der Anfangszustand eines WSPNs wird über die Anfangsmarkierung des zugrundeliegenden Petri-Netzes definiert. Marken, die zeitbehafteten Stellen zugeordnet sind, befinden sich dabei in der Stationskomponente, und die Station befindet sich in einem beliebigen zulässigen Stationszustand.

Die WSPN-Modellwelt gestattet damit beispielsweise die Beschreibung reiner Petri-Netze, wenn nämlich die Stellen nicht durch Stationsbeschreibungen erweitert werden, und die Beschreibung üblicher Warteschlangennetze, wenn als Netzstruktur ein state machine-Netz zugrundegelegt und in jede Stelle eine Stationsbeschreibung integriert wird. Beide Modellwelten werden also als Spezialfälle durch die WSPN-Modellwelt erfaßt. Der wesentliche Vorteil der WSPN-Modellwelt liegt darin, daß einerseits Mechanismen, wie Synchronisation, Populationsbeschränkungen etc. in der gewohnten Petri-Netz-Manier dargestellt werden, andererseits Scheduling-Strategien in der Form beschrieben werden, wie sie aus der Warteschlangentheorie bekannt sind. Die Beschreibung von Scheduling-Strategien mittels Time(d)-Petri-Netz-Modellen ist dagegen häufig recht umständlich. Durch die WSPN-Modellbeschreibung wird, analog zu Generalized Stochastic Petri Nets (GSPN)-Modellen ([AMBC84]), ein stochastischer Prozeß (SP) beschrieben. Der Zustandsraum des WSPN-Modells setzt sich zusammen aus den jeweiligen Stationszuständen und den Markierungen der Petri-Netz-Stellen bzw. der Petri-Netz-Komponenten zeitbehafteter Stellen.

Der SP verbringt nun eine positive Zeit in Zuständen, in denen keine Transition aktiviert ist, wohingegen er Zustände, in denen Transitionen aktiviert sind, in Nullzeit verläßt. Zustände des ersten Typs werden üblicherweise 'tangible' und die des letzteren 'vanishing' genannt (vgl. [AMBC84]). Werden die Bedienzeitverteilungen der Stationen durch Phasenverteilungen beschrieben, so kann der SP einen Markov-Prozeß beschreiben. Dieser Markov-Prozeß läßt sich mit üblichen Methoden analysieren, z.B. mit numerischen Techniken bei endlichem Zustandsraum oder auch mittels Simulation. Diese Analyse gilt quantitativen Eigenschaften des WSPN-Modells, wohingegen eine funktionale Analyse des WSPN-Modells auf Basis des zugrundeliegenden Petri-Netzes erfolgt.

4 Eigenschaften des WSPN-Modells

Durch Einführung von Stationen ändern sich gewisse Eigenschaften des zugrundeliegenden Petri-Netzes, da der Zustandsraum des WSPN-Modells i.a. nicht mehr die gleichen Eigenschaften aufweist wie die Erreichbarkeitsmenge des Petri-Netzes. Dies liegt hauptsächlich darin begründet, daß zum einen ein Zeitverbrauch an den Stellen des Netzes stattfindet, wodurch Marken für gewisse Zeiträume den Transitionen nicht zur Verfügung stehen und zum anderen aktivierte Transitionen in Nullzeit feuern müssen. So kann z.B. das zugrundeliegende Petri-Netz lebendig sein, wo diese Eigenschaft für das WSPN-Modell nicht erfüllt ist. Es treten ähnliche Probleme auf wie bei den Time(d)-Petri-Netzen. Eigenschaften werden an dieser Stelle informell beschrieben, und es werden auch nur "einfarbige" Netze behandelt. Aussagen über farbige WSP-Netze werden zum Abschluß des Kapitels angerissen. Eine formale Behandlung der angesprochenen Problemstellungen findet man in [BAUS86].

Anstelle des komplexen Zustandsraums des WSPN-Modells läßt sich ein reduzierter Zustandsraum betrachten, der dadurch entsteht, daß man bzgl. der Stationskomponente nur die Anzahl enthaltener Marken betrachtet und als gesichert annimmt, daß nach einer gewissen Zeitspanne eine (jede) Marke die Station verläßt. Viele Eigenschaften des WSPN-Modells lassen sich auf Basis dieses reduzierten Zustandsraumes analysieren, so beispielweise die "Lebendigkeit" des WSP-Netzes und Bedingungen für einen möglicherweise eingebetteten Markov-Prozeß.
Faßt man noch weitergehend im reduzierten Zustandsraum die Anzahl der Marken in der Stationskomponente und die Marken auf der Petri-Netz-Komponente zusammen, so ist der

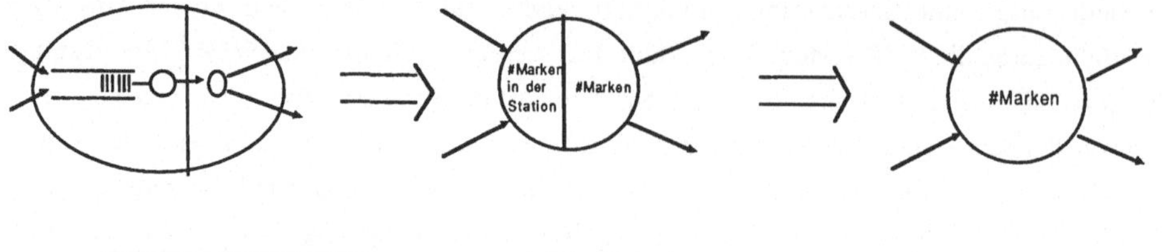

Abb. 3: Reduktion des Zustandsraumes

resultierende Zustandsraum eine (i.allg. echte) Untermenge der Erreichbarkeitsmenge des zugrundeliegenden Petri-Netzes. Es ist damit offensichtlich, daß der Zustandsraum des durch das WSPN beschriebenen SP endlich ist, sofern das zugrundeliegende Petri-Netz beschränkt ist und Stationen bei endlicher Population auch nur endlich viele Zustände annehmen können. Die Umkehrung gilt allerdings nicht, wie Abb. 4 zeigt.

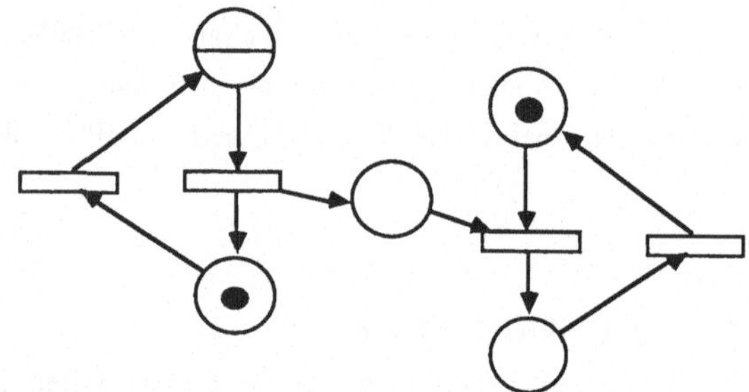

Abb. 4: WSPN dessen Zustandsraum endlich ist, obwohl das Petri-Netz nicht beschränkt ist

Obiges Netz stellt ein Produzenten/Konsumenten-System dar, wobei der Produzent zur Produktion eine gewisse Zeitspanne benötigt, der Konsument allerdings in der Lage ist, Produktionsgüter sofort zu verbrauchen. Im folgenden werden wir nur noch beschränkte Netze betrachten.

4.1 Lebendigkeit

Die Lebendigkeit des WSP-Netzes zu sichern, ist in vielen Fällen wünschenswert, da Transitionen zur Modellierung der Übergänge zwischen Stationen einzusetzen sind. Es ist daher wichtig sicherzustellen, daß jeder im WSP-Netz definierte Übergang bei Betrachtung des SP möglich ist. Wir wollen im folgenden annehmen, daß von jedem Zustand des SP aus mit Sicherheit ein 'tangible' Zustand erreichbar ist.

Definition (P-lebend):

Ein WSP-Netz heißt P-lebend genau dann, wenn das zugrundeliegende Petri-Netz (also das Netz ohne Beachtung der Stationsbeschreibungen) lebendig ist.

Definition (WP-lebend):
Ein WSP-Netz heißt WP-lebend genau dann, wenn es zu jeder Transition und zu jedem erreichbarem Zustand des SP einen Folgezustand gibt, in dem die Transition aktiviert ist.

Es läßt sich nun danach fragen, ob P-Lebendigkeit, welche für spezielle Netzklassen effizient nachprüfbar ist (vgl. Kap. 2), hinreichend für die WP-Lebendigkeit eines Netzes ist. Dies ist nicht allgemein der Fall, wie Abb. 5 zeigt.

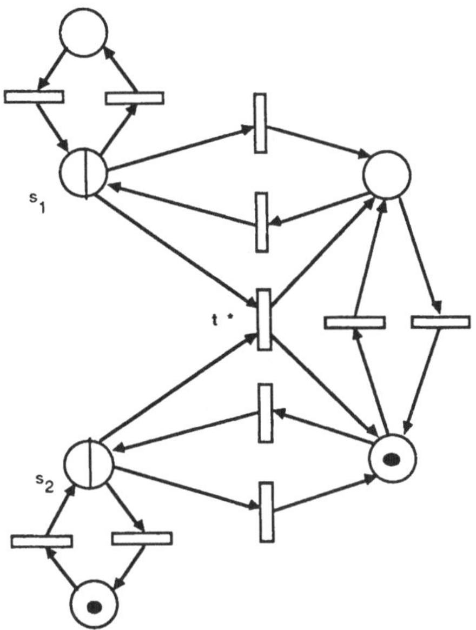

Abb. 5: WSP-Netz, welches P-lebend, aber nicht WP-lebend ist

Obiges Netz ist P-lebend, aber nicht WP-lebend, da eine der zu t* in Konflikt stehenden Transitionen jeweils sofort schaltet, wenn im Abgabebereich von s_1 (bzw. s_2) eine Marke abgelegt wird. Eine gleichzeitige Anwesenheit von Marken in beiden Abgabebereichen (die Aktivierungsbedingung für t*) tritt wegen der als kontinuierlich angenommenen Bedienzeitverteilungen nur mit Wahrscheinlichkeit 0 auf. Die Ursache der skizzierten Diskrepanz scheint in der Tatsache zu liegen, daß t* zwei vorwärtsverzweigte Eingabestellen besitzt. Es liegt somit nahe, die Struktur auf Netze einzuschränken, die maximal eine vorwärtsverzweigte Eingabestelle besitzen, also simple-Netze.

Satz:
Falls das dem WSP-Netz zugrundeliegende Petri-Netz ein beschränktes simple-netz ist, gilt:
WSP-Netz P-lebend ==> WSP-Netz WP-lebend.

Beweisskizze: Nimmt man an, daß das WSP-Netz nicht WP-lebend ist, so existiert eine Transition $t \in T$, welche ab einem bestimmten Zeitpunkt nicht mehr aktivierbar ist. Da das zugrundeliegende Petri-Netz ein simple-Netz ist, muß zumindest eine Stelle $s \in \bullet t$ existieren, welche nicht markiert und auch nicht mehr markierbar ist. Ist s nicht markierbar, sind auch alle $t \in \bullet s$ nicht mehr aktivierbar. Damit existiert eine Stellenmenge S_{leer} und eine Transitionenmenge T_{tot} mit $\bullet S_{leer} \subseteq T_{tot}$ und $S_{leer}\bullet = T_{tot}$ (wegen simple-Netz). Für die Stellenmenge S_{leer} gilt: $\bullet S_{leer} \subseteq S_{leer}\bullet$. S_{leer} wäre ein nicht markierter deadlock, im Widerspruch zur P-Lebendigkeit des WSP-Netzes. •

Die Umkehrung obiger Aussage gilt für simple-Netze nicht, wie Abb. 6 zeigt. Das angegebene simple-Netz ist nicht P-lebend, wie das Feuern von t_3 und nachfolgend t_1 zeigt. Es ist dagegen WP-lebend, da t_2 eher schalten muß als t_1.

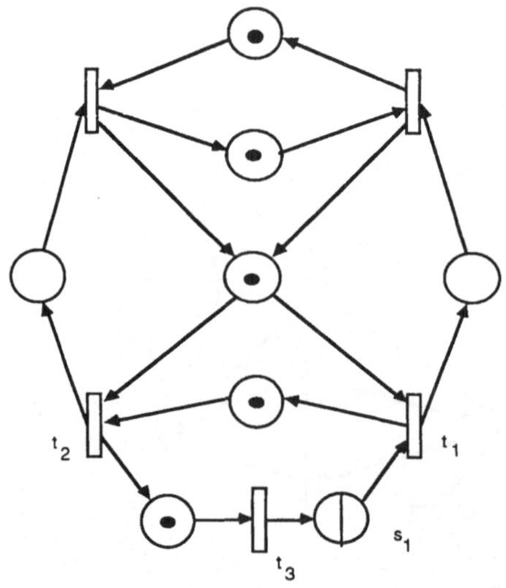

Abb. 6: simple-Netz, welches WP-lebend, aber nicht P-lebend ist

Schränkt man die Konfliktstruktur weiter ein, so ist folgende Aussage gültig.

Satz:

Falls das dem WSP-Netz zugrundeliegende Petri-Netz ein free choice-Netz ist, gilt:

WSP-Netz WP-lebend ==> WSP-Netz P-lebend.

Beweisskizze: Annahme: Das zugrundeliegende Petri-Netz sei nicht P-lebendig.

Dann gilt: $\exists$ deadlock S_d, welcher unter dem Anfangszustand keinen markierten trap enthält (vgl. charakterisierende Lebendigkeitsbedingung für free choice-Netze).

Es ist $\bullet S_d \subseteq S_d\bullet$. Definiere $S' = S_d$. Die Stellenmenge S' wird nun sukzessive verkleinert.

solange (S' kein trap) führe aus:

Wähle $t \in S'\bullet/\bullet S'$. Ein solches t existiert, falls S' kein trap ist.

Die Marken auf der Stellenmenge $\bullet t$ können bei Betrachtung des durch das WSP-Netzes beschriebenen SP mit positiver Wahrscheinlichkeit verfeuert werden. Die Marken auf diesen Stellen können somit als "verloren" für die Stellenmenge S_d gelten.

Setze $S' = S'/\{\bullet t\}$.

end(solange).

Da S_d eine endliche Menge ist, bricht o.a. Verfahren ab und es gilt abschließend: S' ist ein trap (evtl. S' = ∅). Damit existiert bzgl. des SP eine Trajektorie, die alle Marken aus der Stellenmenge S_d entfernt, da nach Annahme S_d keinen markierten trap enthält. Es ist also bzgl. des SP ein Zustand erreichbar, in dem S_d ein leerer deadlock ist. Widerspruch zur WP-Lebendigkeit des WSP-Netzes.•

Obige Sätze zeigen, daß sich für gewisse Netzstrukturen die Eigenschaft des Petri-Netzes, lebendig zu sein, auf das "zeitbehaftete" Modell überträgt und damit eine einfache Überprüfbarkeit der WP-Lebendigkeit gewährleistet ist. Für free-choice-Netze gilt sogar die Charakterisierung "P-lebendig <==> WP-lebendig". Hier kann also durch Einführung der Zeit eine Lebendigkeit des Netzes nicht erzwungen werden.

4.2 Existenz der stationären Verteilung des zugeordneten Markov-Prozesses

Von großer Bedeutung bei der Analyse von WSP-Netzen zugeordneten Markov-Prozessen ist die Existenz einer stationären Verteilung. Dabei ist es wünschenswert, effizient nachprüfbare Kriterien zur Sicherstellung der Existenz einer stationären Verteilung an der Hand zu haben, um unsinnige Analyseversuche vermeiden zu können. An dieser Stelle wird eine hinreichende Bedingung vorgestellt, um eine notwendige Bedingung für die Existenz der stationären Zustandsverteilung zu sichern. Zuvor muß jedoch sichergestellt werden, daß dem WSPN ein Markov-Prozeß zugeordnet werden kann. Eine Vorbedingung hierfür ist offensichtlich die Möglichkeit, einen Markov-Zustandsraum aus ausschließlich "tangible" Zuständen konstruieren zu können. Diese ist allerdings nicht immer gegeben, wie nachfolgende Betrachtung zeigt.

Nullzeit-Falle

Die Definition eines WSP-Netzes läßt Aktionsfolgen in Nullzeit zu, da auch zeitunabhängige Stellen erlaubt sind. Transitionen, die nur solche Stellen als Eingabestellen besitzen, feuern daher sofort nach Belegung dieser Stellen. Es ist durchaus möglich, daß von einem "tangible" Zustand aus nur noch Aktionen in Nullzeit ausgeführt werden, d.h. nicht mit Sicherheit ein neuer "tangible" Zustand erreicht wird. Das Netz befindet sich dann in einer "Nullzeit-Falle", wie im Bsp. der Abb. 7.

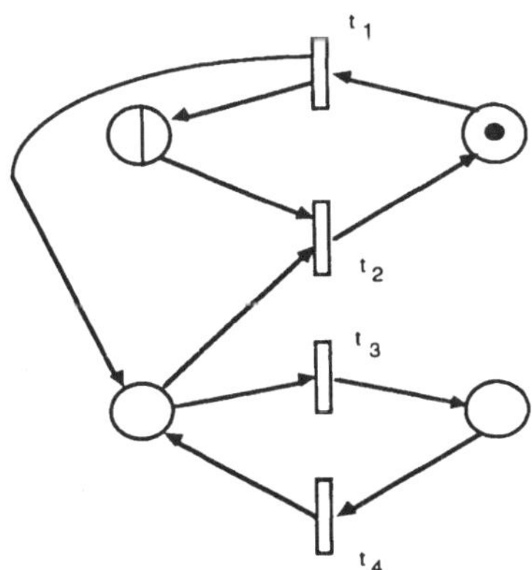

Abb. 7: WSP-Netz mit Nullzeitfalle

Nach Feuern von t_1 können t_3 und t_4 nacheinander beliebig oft in Nullzeit feuern. Ein Fortschritt des Netzverhaltens im Sinne seiner Zustandstrajektorie ist damit nicht mehr ohne weiteres beschrieben. Nullzeitfallen können unter folgender hinreichender Bedingung nicht auftreten.

Bedingung NZF:
Es sei $T' := \{t \in T \mid \bullet t$ enthält nur zeitunabhängige Stellen$\} \subseteq T$.
Es gelte: $\forall\, T_i \subseteq T', T_i \neq \emptyset$: $\bullet T_i \neq T_i \bullet$.

Satz: Ist Bedingung NZF erfüllt, so ist für beschränkte WSP-Netze eine Nullzeit-Falle ausgeschlossen.

Beweisskizze: Man betrachte hierzu für ein $T_i \subseteq T'$ die Stellenmenge $S' := (\bullet T_i / T_i \bullet) \cup (T_i \bullet / \bullet T_i) \neq \emptyset$. Falls S' eine Eingabestelle s von T_i enthält (welche dann nicht Ausgabestelle von T_i ist) werden, nach (evtl. mehrfachen) Feuerungen der Transitionen aus $s\bullet$, die Transitionen aus $s\bullet \cap T_i$ nicht mehr aktiviert. Daher können diese Transitionen nicht beliebig oft in Nullzeit feuern. Falls S' eine Ausgabestelle s von T_i enthält (welche dann nicht Eingabestelle von T_i ist) so wäre das Petri-Netz bei beliebig häufigen Feuerungen der Transitionen aus T_i unbeschränkt. Das beliebig häufige Feuern von Transitionen in Nullzeit ist daher nur für eine echte Teilmenge $T_j \subset T_i$ möglich. Dies wird durch Bedingung NZF ebenfalls verhindert.●

Besitzt das WSP-Netz keine Nullzeitfallen, so kann ihm ein Markov-Prozeß zugeordnet werden, wenn die Dynamik aller Stationskomponenten geeignet gewählt ist (z.B. im Falle der Wahl "üblicher" Stationstypen der separablen Netzklasse). Dieser Prozeß besitzt bei endlichem Zustandsraum eine stationäre Verteilung, falls er irreduzibel ist, d.h. genau eine abgeschlossene Zustandsmenge besitzt.

mehrere abgeschlossene Zustands-Teilmengen
Ist eine Nullzeit-Falle ausgeschlossen, so ist gesichert, daß der dem WSPN zugeordnete Markov-Prozeß rekurrente Zustände besitzt. Es kann aber der Fall eintreten, daß mehrere abgeschlossene Zustandsmengen existieren. In Abb. 8 ist ein solcher Fall dargestellt. Die jeweiligen abgeschlossenen Zustandsmengen zeichnen sich dadurch aus, daß auf den Stellenmengen $\{s_2,s_3\}$ und $\{s_4,s_5\}$ jeweils 0, 1 oder 2 Marken vorhanden sein können.

Für simple-Netze ist es möglich, die Existenz genau einer abgeschlossenen Zustandsmenge zu sichern.

Bedingung REK-SIMPLE:
$T_{RS} := \{t \in T \mid \forall\, t' \neq t: \bullet t' \cap \bullet t = \emptyset\}$. $\forall\, t \in T/T_{RS}$: $\bullet t$ enthält eine zeitbehaftete Stelle.

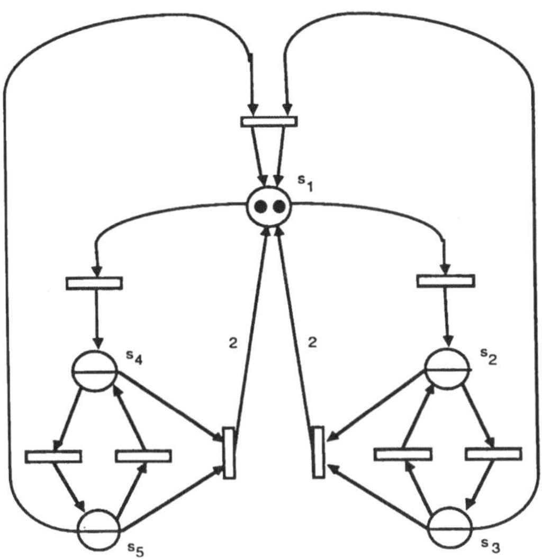

Abb. 8: WSPN mit mehreren abgeschlossenen Zustandsmengen

Satz:

Das dem WSP-Netz zugrundeliegende Petri-Netz sei ein beschränktes und lebendiges simple-Netz, dessen Erreichbarkeitsgraph genau eine abgeschlossene Zustandsmenge besitzt, und es gelten die Bedingungen REK-SIMPLE und NZF.

==> Der dem WSPN zugeordnete Markov-Prozeß besitzt genau eine abgeschlossene Zustandsmenge.

Beweisskizze: Die Beweisführung erfolgt über den reduzierten Zustandsraum (vgl. Abb. 3). Annnahme: Der reduzierte Zustandsraum besitzt mehrere abgeschlossene Zustandmengen.

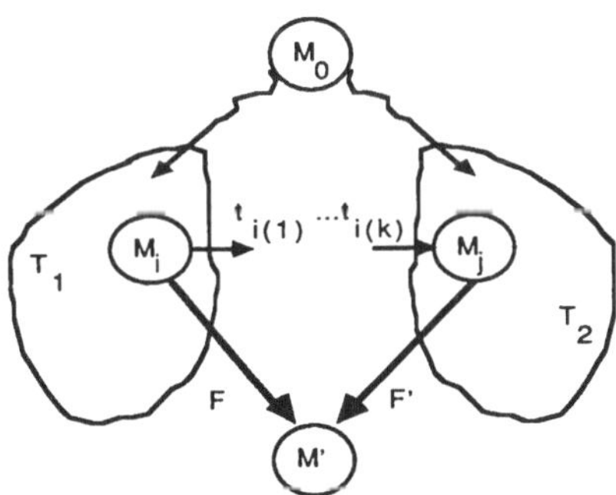

$H \subseteq T_{RS}$, $F' \equiv H$, $F \equiv t_{i(1)} \ldots t_{i(k)} + H$

Abb. 9

Betrachtet man die jeweiligen Markierungen (Zustände), so existieren zwei abgeschlossene Zustandsmengen T_1 und T_2 mit $T_1 \cap T_2 = \emptyset$ (vgl. Abb. 9). Seien $M_i \in T_1$ und $M_j \in T_2$ zwei beliebige Zustände in diesen Mengen. Da die Erreichbarkeitsmenge des zugrundeliegenden Petri-Netzes genau eine abgeschlossene Zustandsmenge besitzt, existiert eine Feuerungsfolge $t_{i(1)}...t_{i(k)}$, die durch sukzessives Feuern M_i in M_j überführt (oder umgekehrt; Œ gelte M_i ->* M_j). Gemäß obiger Annahme ist diese Feuerungsfolge im WSPN nicht mehr realisierbar. Von M_i ausgehend kann jetzt eine Feuerungsfolge F konstruiert werden, die alle Transitionen aus $t_{i(1)}...t_{i(k)}$ enthält sowie weitere Transitionen aus $H \subseteq T_{RS}$, und welche M_i in M' überführt. Da die Transitionen aus H nicht in Konflikt zu anderen Transitionen stehen können, existiert ausgehend von M_j eine Feuerungsfolge F' bestehend aus Transitionen aus H, welche M_j in M' überführt. Damit existiert eine Markierung $M' \in T_1 \cap T_2$. Widerspruch. ==> Behauptung. •

Vorbedingungen für die Existenz einer stationären Verteilung des angenommenen Markov-Prozesses lassen sich also über strukturelle Eigenschaften des zugrundeliegenden Petri-Netzes sichern. Man beachte, daß die oben geforderte Lebendigkeit des Netzes ebenfalls mit effizienten Mitteln nachprüfbar ist (vgl. Kap. 2).

4.3 Farbige WSP-Netze

Bisher wurden nur "einfarbige" Netze betrachtet, bei denen also nur eine Klasse von Marken (Kunden/Aufträge) existiert. Dies ist für die reale Anwendung von WSP-Netzen nur von eingeschränktem Interesse. Eine Erweiterung auf farbige Netze ist z.B. durch Einsatz farbiger Petri-Netze ([JENS81]) möglich.

Ein farbiges Petri-Netz ist ein 6-Tupel
CPN = (S, T, C, W^-, W^+, M_0) mit

a) S ist eine endliche, nichtleere Menge von Stellen.
b) T ist eine endliche, nichtleere Menge von Transitionen.
c) $S \cap T = \emptyset$.
d) C ist eine Farbenfunktion. Sie ist definiert auf $S \cup T$ in endliche, nichtleere Mengen.
e) W^-, W^+ sind (Inzidenz-)Funktionen definiert auf S×T, so daß $W^-(s,t)$ bzw. $W^+(s,t) \in [C(t) \rightarrow [C(s) \rightarrow \mathbb{N}_0]]$, $\forall (s,t) \in S \times T$.
f) M_0 ist eine Funktion definiert auf S, so daß $M(s) \in [C(s) \rightarrow \mathbb{N}_0]$, $\forall s \in S$. M_0 heißt Anfangsmarkierung des farbigen Petri-Netzes.

Ein farbiges Petri-Netz läßt sich eindeutig in ein Stellen/Transitions-Netz transformieren. Diese Transformation nennt man Entfaltung. Begriffe wie Aktivierung, Feuerung etc. lassen sich somit für farbige Petri-Netze über deren zugehörige entfalteten Stellen/Transitions-Netze definieren.

Definition (Entfaltung):
Sei CPN = (S, T, C, W^-, W^+, M_0) ein farbiges Petri-Netz. Die Entfaltung läßt sich wie folgt durchführen:

1. Erzeuge für jede Stelle $s \in S$ und für jede Farbe $c \in C(s)$ eine Stelle (s,c).
2. Erzeuge für jede Transition $t \in T$ und für jede Farbe $c' \in C(t)$ eine Transition (t,c').

3. Definiere die Funktionen W^-, W^+ und M_0 des entfalteten Petri-Netzes durch
W^- ((s,c), (t,c')) = W^- (s,t) (c') (c), W^+ ((s,c), (t,c')) = W^+ (s,t) (c') (c) und
M_0 ((s,c)) = M_0 (s) (c).
$PN = (\bigcup_{s \in S} \bigcup_{c \in C(s)} (s,c), \bigcup_{t \in T} \bigcup_{c' \in C(t)} (t,c'), W^-, W^+, M_0)$
ist dann das eindeutig bestimmte entfaltete Petri-Netz.

Farbige Petri-Netze sind damit, funktional betrachtet, nur kompaktere Beschreibungen für Stellen/Transitions-Netze. Alle angegebenen Sätze lassen sich somit auf entsprechende farbige Netze übertragen. Die Stationsbeschreibungen des WSP-Netzes können jetzt mehrere Kundenklassen behandeln, wobei die Stationsbeschreibungen den üblichen Beschreibungen aus der Warteschlangentheorie entsprechen. Allerdings treten dadurch zusätzliche Probleme auf, was zum Abschluß durch folgendes Beispiel illustriert werden soll.

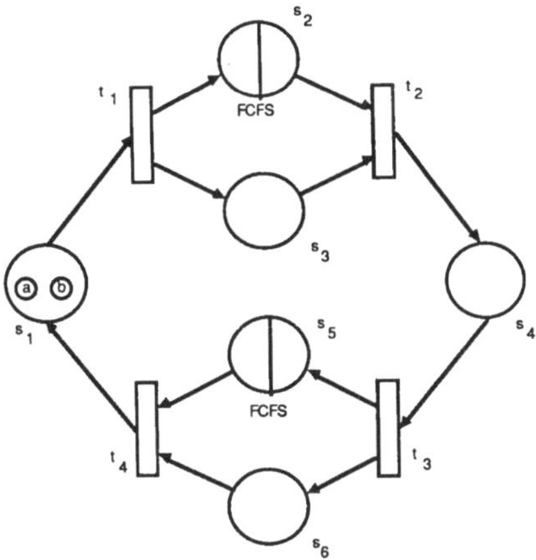

Abb. 10: Farbiges WSPN mit 2 abgeschlossenen Zustandsmengen

Das Verhalten des Netzes aus Abb. 10 ist wie folgt:
Die Stationen in den Stellen s_2 und s_5 sind beliebige FCFS-Stationen. Die Aktivierung und Feuerung der Transitionen ist folgendermaßen definiert: t_1, t_3 sind jeweils aktiviert, wenn eine Marke des Typs *a* oder *b* auf der Eingabestelle liegt. Feuert die Transition, so wird eine Marke des selben Typs auf s_2 bzw. s_5 abgelegt (gelangt in die Station). Auf s_3 bzw. s_6 wird jeweils eine Marke des "Komplementärtyps", also *{a,b}/Typfarbe*, abgelegt. Die Transitionen t_2, t_4 sind jeweils aktiviert, wenn eine Marke des Typs *a* oder *b* auf s_2 bzw. s_5 bedient wurde (sich also auf der "normalen" Petri-Netz-Stelle der Stelle befindet) und eine Marke der Komplementärfarbe auf s_3 bzw. s_6 vorhanden ist. Feuert die Transition, so werden die vorher genannten Marken entfernt und auf s_1 bzw. s_4 eine Marke des Typs *a* bzw. *b* (also des selben Typs wie zuvor auf s_2 bzw. s_5) abgelegt.
Aus der FCFS-Bedienstrategie der Stationen ergibt sich, daß eine einmal eingetretene Reihenfolge "*a* vor *b*" oder "*b* vor *a*" immer beibehalten wird. Der SP besitzt also zwei abgeschlossene Zustandsmengen.

Obiges Beispiel zeigt, daß Bedienstrategien, die alle Marken einer Farbe (Kunden einer Klasse) blockieren, also zwischenzeitlich von der Bedienung ausschließen, zusätzliche Probleme aufwerfen. Selbst für einfache Netzstrukturen (das Netz aus Abb. 10 ist ein marked-graph-Netz) sind wichtige Vorbedingungen für eine sinnvolle quantitative Analyse nicht mehr erfüllt.

5 Schlußwort

Die WSPN-Modellwelt ermöglicht die Integration von Warteschlangen- und Petri-Netzen in einer Modellbeschreibung und unterstützt so die Analyse des Modells unter sowohl funktionalen als auch quantitativen Gesichtspunkten. Die Modellbeschreibung ist dabei so gewählt, daß sie "Petri-Netz-Modellierern" und "Warteschlangen-Modellierern" entgegenkommt. Bestehende effiziente Analysetechniken aus der Petri-Netz-Theorie lassen sich dazu verwenden, Aussagen über wichtige quantitative Eigenschaften des Modells treffen zu können, sofern man sich auf gewisse Netzklassen beschränkt. Mit diesen Netzklassen lassen sich aber bereits wesentliche Merkmale eines Rechensystems beschreiben, die über sonst übliche Warteschlangen-Beschreibungen hinausgehen. Bei Einführung von mehreren Kundenklassen (farbige Netze) treten zusätzliche Probleme auf, die einer weiteren Untersuchung bedürfen.

6 Literatur

[AMBC84] M. Ajmone Marsan, M.G. Balbo, G. Conte, A Class of Generalized Stochastic Petri Nets for the Performance Evaluation of Multiprocessor Systems, ACM Trans. on Computer Systems, Vol. 2, No. 2, Mai 1984, S. 93-122.

[AMBB85] M. Ajmone Marsan, G. Balbo, A. Bobbio, G. Chiola, G. Conte, A. Cumani, On Petri Nets with Stochastic Timing, in [WTPN 85], S. 80-87.

[BAUS86] F. Bause, Funktional und quantitativ analysierbare Rechensystemmodelle; über Petrinetz- und Warteschlangen-Modellwelten, Diplomarbeit, Universität Dortmund, 1986.

[COMM72] F. Commoner, Deadlocks in Petri Nets, Report CA-7206-2311, Massachusetts Computer Associates, Wakefield, 1972.

[ESTR78] G. Estrin, A Methodology for Design of Digital Systems - supported by SARA at the Age of One, Proc. of the AFIPS National Computer Conference, Vol. 47, 1978, S. 313-324.

[JENS81] K. Jensen, Colored Petri Nets and the Invariant Method, Theoretical Computer Science 14, 1981, S. 317-336.

[MOLL81] M.K. Molloy, On the Integration of Delay and Throughput Measures in Distributed Processing Models, Ph. D. Diss., University of California, 1981.

[NOE75] J.D. Noe, Pro-nets: For Modeling Processes and Processors, Technical Report 75-07-15, Dept. of Computer Science, University of Washington, Seattle, 1975.

[NUTT 72a] G.J. Nutt, The Formulation and Application of Evaluation Nets, Ph. D. Diss., Computer Science Group, University of Washington, Seattle, Juli 1972.

[NUTT72b] G.J. Nutt, Evaluation Nets for Computer Systems Performance Analysis, Proc. of t he AFIPS Fall Joint Computer Conference, 1972, S. 279-286.

[PNPM87] Proceedings of the International Workshop on Petri Nets and Performance Models, Madison, Aug. 1987.

[RAES77] R.R. Razouk, G. Estrin, The Graph Model of Behaviour Simulator, Proc. of the Symposium on Design Automation and Microprocessors, Palo Alto, Feb. 1977.

[REIS82] W. Reisig, Petri-Netze. Eine Einführung, 1982.

[WTPN85] Proceedings of the International Workshop on Timed Petri Nets, Turin, Juli 1985.

Aspekte der Analyse und Simulation der Auftragslast eines wissenschaftlichen Rechenzentrums

W. Koch
TÜV Rheinland e. V.
5000 Köln 91

1. Einleitung

Viele Großrechenzentren sind unter anderem durch die folgenden Merkmale gekennzeichnet:

1. Sie müssen eine große Anzahl verschiedener Endbenutzer bedienen.

2. Die zu bearbeitenden Aufträge entstammen einer Vielzahl von unterschiedlichen Fachgebieten und Problemkreisen.

3. Die Betriebsmittelbeanspruchung durch die individuellen Aufträge ist im einzelnen unbekannt und kann global nur mit statistischen Methoden erfaßt und beschrieben werden.

Bei vielen wissenschaftlichen Rechenzentren kommen zu den drei oben genannten Charakteristika noch zwei weitere hinzu:

1. Der Spitzenbedarf an Betriebsmitteln, vor allem bezüglich der Rechenzeit, ist beliebig hoch und beliebig vermehrbar und wird nur durch die jeweils zur Verfügung stehende Kapazität beschränkt.

2. Die Kapazität der Rechenanlage kann nach wirtschaftlichen Gesichtspunkten aber nur in größeren Zeiträumen erweitert werden, wenn der gestiegene Grundbedarf und die zur Verfügung stehende Kapazität nicht mehr vereinbar sind.

Gerade in diesem Bereich ist deshalb eine Steuerung des Auftragsflusses nicht zu vermeiden. Dies geschieht in der Regel durch entsprechende Kontingentierungsmaßnahmen, die das Ziel verfolgen, möglichst alle vorhandenen Betriebsmittel optimal, das heißt gleichmäßig auszulasten.

Eine Kontingentierung der Rechnerbetriebsmittel zwecks Optimierung des Auftragsflusses ist aber nur dann sinnvoll, wenn der einzelne Benutzer bei der Formulierung seiner Aufträge überhaupt eine Wahlmöglichkeit bezüglich der benötigten Betriebsmittel besitzt, das heißt, wenn er ein überlastetes und deshalb kontingentiertes Betriebsmittel durch andere Betriebsmittel substituieren kann.

Eine Substitution von Betriebsmitteln liegt dann vor, wenn der ursprüngliche Auftrag in einen äquivalenten Auftrag, das heißt in einen Auftrag, der dasselbe Ergebnis erbringt, überführt werden kann, und dieser äquivalente Auftrag dann bestimmte Betriebsmittel weniger benutzt als der ursprüngliche, dafür aber andere Betriebsmittel um so mehr.

Der folgende Vortrag befaßt sich mit zwei Teilaspekten einer Untersuchung zur Substitution von Rechnerbetriebsmitteln, die in den Jahren 1982 bis 1987 bei der Gesellschaft für wissenschaftliche Datenverarbeitung mbH Göttingen (GWDG) durchgeführt wurde. Es werden die Methodik und Ergebnisse einer Untersuchung der Lastverteilung beschrieben und ein diskretes Simulationsmodell der Rechenanlage vorgestellt.

2. Untersuchung der Lastverteilung

Die Last einer Rechenanlage ist bestimmt durch die Gesamtheit aller individuellen Aufträge, Transaktionen und Daten, die während einer gegebenen Zeitperiode verarbeitet werden müssen, vgl. [TERP 77].

Im Zusammenhang mit dem Untersuchungsgegenstand ergaben sich die folgenden Fragestellungen:

1. Wie groß ist der Anteil der Programme, bei denen sich eine Umstellung lohnen würde?

2. Handelt es sich dabei um Benutzer- oder um Systemprogramme?

3. Können die entsprechenden Programme identifiziert werden?

Diese Fragen können nur beantwortet werden, wenn die Last des Rechenzentrums bekannt ist. Da bei dem untersuchten Rechenzentrum alle Aufträge von den Anwendern formuliert wurden und daher in der Regel nicht bekannt war, welche Art von Aufträgen zu erledigen ist, konnte die Kenntnis über die Zusammensetzung der Last nur im nachhinein durch empirische Untersuchungen der Accountingdaten gewonnen werden.

Dabei stand die Frage nach der Nutzung der Rechenprozessoren im Vordergrund, weil nur dieses Betriebsmittel einen Engpaß darstellte. Eine nähere Untersuchung des von den Programmen benötigten Hauptspeicherplatzes wurde nicht durchgeführt. Die Programmgröße wurde aber als ein Attribut der Lastelemente mit erfaßt. Die Nutzung des Ein-/Ausgabesystems, gemessen in I/O-Zeit, und der damit korrelierte Betriebssystemoverhead, gemessen in sogenannter CC/ER-Zeit, wurde dagegen näher untersucht.

Das Rechenzentrum der GWDG betrieb in dem Untersuchungszeitraum Rechenanlagen des Typs Sperry Univac 1100/80 unter dem Betriebssystem OS 1100.

Die empirischen Untersuchungen des Lastprofils wurden anhand von Monatslasten aus den Jahren 1984 und 1985 durchgeführt. Im Jahr 1984 war die Rechenanlage mit drei Rechenprozessoren, zwei Ein-/Ausgabeprozessoren, 2 MW Hauptspeicher- und ca. 5,2 GB Massenspeicherkapazität ausgestattet. In einem Drei-Schichtbetrieb wurden pro Monat etwa 50.000 Benutzeraufträge durchgeführt, von denen etwa 60 % auf den Dialogbetrieb und 40 % auf den Stapelbetrieb entfielen.

Als Beobachtungszeitraum wurden die Semestermonate April, Mai, Juni und Oktober, November, Dezember des Jahres 1984 ausgewählt, da in dieser Zeit ein normaler Benutzerbetrieb gewährleistet war.

Die für die Lastuntersuchungen benötigten Daten wurden dem System Logfile entnommen, das von dem Betriebssystem unter anderem für die Zwecke des Accountings und der nachträglichen Kontrolle der Betriebsabläufe unterhalten wird. Da alle Logfiles eines Betriebsmonats aus Gründen der Betriebsorganisation jeweils zu einem Monatsdatensatz zusammengefaßt wurden, lag es nahe, die Last eines Monats als Beobachtungsgegenstand zu wählen. Wegen des großen Umfangs des in einem Monatsdatensatz enthaltenen Datenvolumens erschien eine Monatslast auch das Maximum dessen zu sein, was in einem Programmlauf statistisch ausgewertet werden kann.

Für jeden beliebigen Betrachtungszeitraum können aus dem Logfile die folgenden auftragsspezifischen Lastparameter ermittelt werden:

Die Anzahl der durchgeführten Aufträge.

Für jeden Auftrag:

Der Name des Auftrags,
die Abrechnungsnummer des Auftraggebers,
der Verbrauch an CPU-Zeit,
der Verbrauch an CC/ER-Zeit,

der Verbrauch an I/O-Zeit,
die Anzahl der ausgeführten Programme.

Für jedes Programm:

Der Programmname,
der Verbrauch an CPU-Zeit,
der Verbrauch an CC/ER-Zeit,
der Verbrauch an I/O-Zeit,
die Programmgröße.

Eine erste Stichprobe ergab, daß pro Benutzerauftrag im Mittel etwa sechs Programme ausgeführt wurden. Da es offensichtlich keine eindeutige Zuordnung der Programme zu den Aufträgen gab, mußte eine Programmausführung als kleinste Beobachtungseinheit ausgewählt werden. Da ein Monatsdatensatz mehr als 300.000 Programmausführungen beinhaltete, ergab sich daraus die Folgerung, daß diese Rohdaten noch weiter zusammengefaßt werden mußten, bevor sie statistisch ausgewertet werden konnten. Deshalb wurden alle Ausführungen eines identifizierbaren Programms zu einem Datensatz zusammengefaßt. Durch diese Maßnahme, bei der keine für die Auswertungen wichtigen Informationen verloren gingen, konnte die Menge der auszuwertenden Datensätze von 300.000 auf ca. 2.500 reduziert werden. Die relativ hohe Anzahl von durchschnittlich mehr als 100 Programmausführungen pro Programm wurde durch die häufige Benutzung der Systemprogramme verursacht, von denen einige, wie zum Beispiel das File-Utility-Programm FURPUR, bis zu 60.000 mal pro Monat aufgerufen wurden.

Zur Identifizierung der einzelnen Programme wurde der im Logfile protokollierte Programmname benutzt. Da viele Benutzer des Rechenzentrums diesen zur Verfügung stehenden Namensraum nicht ausnutzten - beliebte Programmnamen sind zum Beispiel A, ABS, PROG, HAUPT -, mußte in diesen Fällen ein weiteres Kriterium zur Identifizierung der Programme herangezogen werden. Bei diesen häufig verwendeten Programmnamen wurde deshalb die Abrechnungsnummer des Auftraggebers als weiteres Unterscheidungsmerkmal verwendet.

Ein einzelnes für die Untersuchungen benötigtes Merkmal konnte nicht aus dem Logfile gewonnen werden, nämlich, ob es sich bei einem Programm um ein Benutzer- oder um ein Systemprogramm handelte. Deshalb wurde eine Tabelle mit den Namen aller Programme angefertigt, die sich während des Beobachtungszeitraums in den Systembibliotheken befanden. Anhand dieser Tabelle konnte dann entschieden werden, ob es sich bei einem speziellen Programm um ein Systemprogramm handelte.

Bei der anschließenden Auswertung und Verdichtung der Monatsdatensätze wurde für jedes in dem jeweiligen Monat gerechnete Programm ein Datensatz erzeugt, der die folgenden Informationen enthielt:

1. Die Anzahl der Programmausführungen,
2. einen Indikator für die Systemprogramme,
3. den Elementennamen des Programmnamens,
4. den Versionsnamen des Programmnamens,
5. die Abrechnungsnummer des Auftraggebers,
6. die Summe der CPU-Zeit aller Programmausführungen in Sekunden,
7. die Summe der CC/ER-Zeit aller Programmausführungen in Sekunden,
8. die Summe der I/O-Zeit aller Programmausführungen in Sekunden,
9. die Programmgröße in 512-Wort-Blöcken.

Die auf diese Weise gewonnenen Daten wurden anschließend mit einem Programm zur statistischen Datenanalyse ausgewertet. Für alle untersuchten Monate wurde eine deskriptive Statistik der Parameter aller Programme und zusätzlich der Parameter der Systemprogramme angefertigt. Dabei wurde auch bei allen Parametern, bei denen das sinnvoll erschien, der Anteil der Systemprogramme separat ausgewiesen.

Weiterhin wurden alle Datensätze entsprechend der Größe des anteiligen Verbrauchs der Programme an CPU-Zeit, CC/ER-Zeit und I/O-Zeit sortiert, und zu jedem der drei Lastparameter festgestellt, wieviel verschiedene Programme mindestens notwendig waren, um eine anteilige Last von 5, 10, 15, 20, ..., 95 % an der Gesamtlast zu erreichen. Diejenigen Programme, die benötigt wurden, um jeweilis 80 % der CPU-Last, der CC/ER-Last bzw. der I/O-Last zu erreichen, wurden in der Reihenfolge ihres anteiligen Verbrauchs namentlich aufgelistet, wobei die Anzahl der Programmaufrufe, der absolute und der relative Verbrauch mit aufgeführt wurden.

Die deskriptive Statistik aller Programme zeigte, daß alle untersuchten Lastparameter mit Ausnahme der Programmgröße eine erstaunlich hohe Schwankung zwischen den einzelnen Programmen aufwiesen, wobei die Standardabweichung den Mittelwert um mehr als das zehnfache überstieg. Im Monat April 1984 lag die maximale CPU-Zeit, die ein (Benutzer-)Programm verbrauchte, bei 281 Stunden 22 Minuten 18 Sekunden, das Minimum bei 0 Sekunden, der Mittelwert bei 2.444 Sekunden und der Medianwert bei 7 Sekunden. Ähnlich hohe Schwankungen wiesen auch der Verbrauch an CC/ER-Zeit und an I/O-Zeit sowie die Anzahl der Ausführungen pro Programm auf, während die Programmgröße mit einem Medianwert von 20 KW und einem Mittelwert von 32,5 KW bei einer Standardabweichung von 32 KB wesentlich gleichmäßiger verteilt war.

Die extreme Schiefe der Verteilung der CPU-Zeit, der CC/ER-Zeit und der I/O-Zeit deutete darauf hin, daß die überwiegende Menge aller Aufträge nur sehr geringe Betriebsmittelansprüche stellte, und daß die wesentliche Beanspruchung der Betriebsmittel durch einige wenige Programme verursacht wurde.

Die Auswertung der Programme in der Reihenfolge ihrer anteiligen Betriebsmittelnutzung führte zu dem Ergebnis, daß eine erstaunlich geringe Anzahl an Programmen den überwiegenden Lastanteil sowohl bei der CPU-Zeit, als auch bei der CC/ER-Zeit und bei der I/O-Zeit repräsentierte.

In den einzelnen Monaten des Jahres 1984 verbrauchten im Mittel nur 8 verschiedene Programme 50 % der CPU-Zeit. Die namentliche Auflistung dieser Programme ergab, daß es sich dabei ausschließlich um Benutzerprogramme handelte. Den Spitzenreiter bildete ein Programm namens EXE, das im Monat November 1984 allein 28 % der CPU-Zeit verbrauchte, in dem gesamten Beobachtungszeitraum zu den Großverbrauchern an CPU-Zeit zählte, und in den näher untersuchten neun Monaten insgesamt auf 153 % des CPU-Verbrauchs eines Monats kam.

Um einen Anteil von 50 % an der CC/ER-Zeit zu erreichen, wurden im Jahr 1984 im Monatsmittel etwa 7 Programme benötigt, und für 50 % der I/O-Last waren 11 verschiedene Programme notwendig. Bei den Großverbrauchern an CC/ER-Zeit handelte es sich ausschließlich um Systemprogramme, in der Reihenfolge Binder, Zeileneditor, File-Utility-Programm. Auch die Liste der Großverbraucher an I/O-Zeit wurde von den Systemprogrammen angeführt. Hier bildete das File-Utility-Programm den Spitzenreiter, gefolgt von dem Backup-Programm, dem Zeileneditor, dem FORTRAN-Compiler, dem Binder und dem Bildschirmeditor. Anschließend folgten erst die Benutzerprogramme. Das Benutzerprogramm mit dem größten Verbrauch an I/O-Zeit vereinigte etwa 2 % der monatlichen I/O-Last auf sich. Die Verteilung der Lastanteile auf die Programme wird exemplarisch für die CPU-Zeit für den Monat April 1984 im folgenden Bild dargestellt.

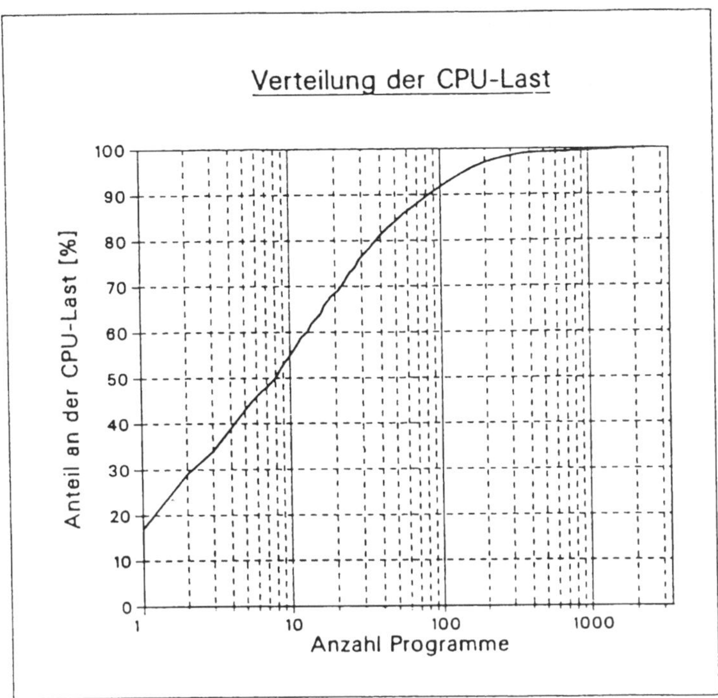

Die Untersuchung erbrachte das methodische Ergebnis, daß eine statistische Analyse der aufgezeichneten Logfiledaten auch bei an sich unbekannter Last eines Rechenzentrums ein genügend genaues Bild der Lastzusammensetzung ergibt. Es stellte sich heraus, daß dabei nicht ein einzelner Auftrag (Job), sondern eine Programmausführung als kleinste Beobachtungseinheit gewählt werden sollte. Der Stichprobenumfang in Form von einzelnen Monatslasten konnte aber erst dann von herkömmlichen Statistik-Analyse-Programmen verarbeitet werden, nachdem alle Ausführungen eines Programms innerhalb des jeweiligen Monats zu einem Datensatz zusammengefaßt wurden. Hierzu war es notwendig, die Programme namentlich zu identifizieren.

Als ein Ergebnis stellte sich heraus, daß die bekannte 80/20-Regel bei dem untersuchten Rechenzentrum bei weitem übertroffen wurde. Für 80 % der CPU-Last waren nur 1,5 % der Programme verantwortlich, 80 % der I/O-Last wurden durch 2 % der Programme verursacht.

3. Konstruktion eines Simulationsmodells

Bei der oben genannten Untersuchung im Rechenzentrum der GWDG wurden die Auswirkungen von Maßnahmen zur Substitution des Rechenprozessors auf den Gesamtdurchsatz der Rechenanlage mittels eines Simulationsmodells abgeschätzt.

Für die Modellierung des Durchsatzes von Rechenanlagen gibt es zwei prinzipiell verschiedene Ansätze (vgl. [BEIL 81]), die Entwicklung eines analytischen Modells oder die Konstruktion eines diskreten Simulationsmodells.

Analytische Modelle basieren auf den Erkenntnissen der Warteschlangentheorie. Die Arbeit eines Rechnersystems wird durch einen stationären, stochastischen Prozeß modelliert, wobei die verschiedenen Betriebsmittel durch sogenannte Bedienstationen repräsentiert sind. Die Last wird durch Lastelemente wie zum Beispiel Aufträge, Teilaufträge oder Transaktionen dargestellt, die die verschiedenen Bedienstationen in einer durch eine Zufallsvariable bestimmten Reihenfolge besuchen und für einen bestimmten Zeitraum, der in der Regel durch eine Exponentialverteilung bestimmt ist, in Anspruch nehmen. Die Übergänge der Lastelemente von einer Bedienstation zur nächsten werden durch eine Markov-Kette beschrieben (siehe auch [DENN 78]).

Diskrete Simulationsmodelle laufen im Gegensatz zu stetigen Simulationsmodellen nicht kontinuierlich, sondern ereignisgesteuert ab. Sie bilden den Zustandsraum und die Zustandsübergänge, die das zu modellierende Rechnersystem bei dem Auftreten von diskreten Systemereignissen (Events) durchführt, auf Zustandsvariablen und auf Prozeduren ab, die die verschiedenen Events bei deren Auftreten verarbeiten, vgl. [BAUK 76], [BEIL 81], [WEBE 83].

Nach Abwägung der Vor- und Nachteile der beiden Möglichkeiten der Modellierung wurde ein ereignisgesteuertes Simulationsmodell für die Simulation der auftragsinternen Substitution durch ein Rechnermodell ausgewählt. Dies geschah im wesentlichen aus zwei Gründen.

1. Die Übergangswahrscheinlichkeiten für den Wechsel zwischen den einzelnen Betriebsmitteln innerhalb der Programme der zu modellierenden Last waren zu wenig bekannt, und die im zweiten Abschnitt dargestellte Verteilung der Betriebsmittelanprüche der verschiedenen Programme wich von der üblicherweise verwendeten Exponentialverteilung zu sehr ab, als daß von einem analytischen Modell eine adäquate Wiedergabe der Wirklichkeit erwartet werden konnte.

2. Die individuelle Behandlung einzelner Programme, die für die Simulation der auftragsinternen Substitutionen notwendig war, konnte mit einem diskreten, deterministischen Simulationsmodell besser realisiert werden als mit einem analytischen Modell.

Aus diesem Grund wurde ein Simulationsmodell für eine Rechenanlage entwickelt, das mit den Accounting-Daten einer realen Rechenanlage getrieben werden kann, und das die Abarbeitung der Aufträge durch den Zentralkomplex der Rechenanlage, bestehend aus Rechenprozessoren, Ein-/Ausgabeprozessoren und Hauptspeicher simuliert. Dabei können die Eingabedaten einer spezifizierbaren Menge von Aufträgen so verändert werden, daß die gewünschte, auftragsinterne Substitution simuliert wird. Anhand der von dem Modell aufgezeichneten Auslastungszahlen und Durchsatzraten kann dann die globale Wirkung der Substitutionsmaßnahmen abgeschätzt werden.

Der Kern des Simulationsmodells geht auf einen bei [SERA 81] beschriebenen Algorithmus zurück, der auch in einem von Serazzi, Baiocchi, Capelo und Comincioli entwickelten dynamischen Modell eines Computersystems verwendet wurde. Dieses Modell gestattet die Beobachtung des zeitlichen Verhaltens von globalen Performance-Daten wie Response-Zeit, Multiprogramming-Level und Durchsatz, wobei für jeden simulierten Prozeß lediglich die Betriebsmittelansprüche pro Betriebsmittel und die Ankunftszeit des Prozesses bekannt sein müssen. Die Konstruktion dieses Modells geht von zwei wesentlichen Voraussetzungen aus.

1. Es wird angenommen, daß innerhalb eines Programms keine Überlappung zwischen der Inanspruchnahme des Rechenprozessors und des Ein-/Ausgabesystems stattfindet. Der hierdurch verursachte Fehler ist nach [DENN 78] in der Praxis sehr klein.

2. Es wird angenommen, daß alle Aufträge ein sogenanntes homogenes Verhalten bezüglich der Inanspruchnahme der Betriebsmittel zeigen. Damit ist folgendes gemeint. Ein Auftrag benötige während seiner Abarbeitung insgesamt n Betriebsmittel. Es sei C_j die Kapazität eines Betriebsmittels j, j = 1, ..., n und $t_j = \frac{s_j}{C_j}$, j = 1, ..., n die Zeit, während der Auftrag das Betriebsmittel j bei seiner Abarbeitung belegt. Dann bezeichnet $\tau_j = \frac{t_j}{\sum_{k=1}^{n} t_k}$, j = 1, ..., n den relativen Anteil des Betriebsmittels j an der Alleinlaufzeit des Auftrags, bzw. die relative zeitliche Inanspruchnahme des Betriebsmittels j durch den Auftrag. Ein homogenes Verhalten eines Auftrags liegt dann vor, wenn sich die relative zeitliche Inanspruchnahme aller Betriebsmittel während der gesamten Laufzeit des Auftrags nicht ändert. In diesem Fall werden alle Betriebsmittelansprüche des Auftrags kontinuierlich über die Laufzeit des Auftrags im Verhältnis τ_j zur Laufzeit aufgebaut und auch die relativen Anteile der noch nicht befriedigten Betriebsmittelansprüche können zu dem Zeitpunkt während der Abarbeitung des Auftrags durch die Werte τ_j beschrieben werden. Nach [SERA 81] kann zumindest für die Betriebsmittel Rechenprozessor und Ein-/Ausgabesystem ein solches homogenes Verhalten in praktisch allen Aufträgen gefunden werden.

Unter der zweiten Voraussetzung ist der interne Aufbau der zu modellierenden Lastelemente, speziell die Reihenfolge und die zeitliche Länge der Betriebsmittelbelegung für das Modell nicht mehr von Bedeutung, so daß sich sowohl die Modellierung der Last als auch die Modellierung der Betriebsmittel stark vereinfacht. Jeder Auftrag wird durch die Werte $\mathcal{T}_j$ und seine Alleinlaufzeit vollständig beschrieben und belegt während seiner gesamten Laufzeit alle von ihm benutzten Betriebsmittel kontinuierlich. Das Betriebsmittel mit der stärksten Gesamtbelastung, das sogenannte Bottleneck, bestimmt dabei die relative Fortschrittsgeschwindigkeit des Auftrags. Alle Aufträge, die dieses Bottleneck benutzen, haben die gleiche, relative Fortschrittsgeschwindigkeit und werden zu einem sogenannten Subsystem zusammengefaßt. Für alle anderen Aufträge, die dieses Bottleneck erster Stufe nicht benutzen, bildet ein anderes Betriebsmittel mit der nachfolgend stärksten Beanspruchung ein Bottleneck zweiter Stufe, das die relative Fortschrittsgeschwindigkeit aller Aufträge bestimmt, die dieses Betriebsmittel, nicht aber das erstere benötigen. Dieser iterative Prozeß wird fortgesetzt, bis alle Subsysteme bestimmt und damit die relative Fortschrittsgeschwindigkeit aller zu einem Zeitpunkt t aktiven Aufträge festgelegt ist.

Dieser Algorithmus wurde im wesentlichen auch bei der Konstruktion des im folgenden beschriebenen Rechnermodells verwendet, um die Abarbeitung der modellierten Last durch das Rechnermodell zu simulieren. Der bei Serazzi beschriebene Algorithmus wurde dabei so erweitert, daß von einem Betriebsmittel mehrere gleiche Exemplare vorhanden sein können, um zum Beispiel Multiprozessorkonfigurationen modellieren zu können, und daß Betriebsmittel eine von 1,0 verschiedene Grundgeschwindigkeit haben können, um sie mit einer einstellbaren Verarbeitungsgeschwindigkeit betreiben zu können.

Das Modell wurde in der Programmiersprache PASCAL implementiert, da unter den zur Verfügung stehenden Programmiersprachen PASCAL wegen seiner Datenstrukturen dynamische Variablen und Warteschlagen am besten nachbilden konnte.

3.1 Datenstrukturen

Das Modell verwaltet die folgenden Betriebsmittel:

a) CPU - modelliert ein oder mehrere Rechenprozessoren.
b) IO - modelliert ein oder mehrere Ein-/Ausgabesysteme.
c) THINK - modelliert die von den Prozessen verwendete Denkzeit oder freiwillige Wartezeit.
d) HSP - modelliert den Hauptspeicher.

Jedes Betriebsmittel besitzt die Attribute Kapazität, Anzahl, Nutzzahl, Belastung und Auslastung. Die Kapazität bestimmt die Grundgeschwindigkeit eines zeitartigen Betriebsmittels bzw. die Größe des Hauptspeichers. Die Anzahl gibt die Anzahl der vorhandenen Exemplare eines Betriebsmittels an, während Nutzzahl die davon nutzbare Anzahl Exemplare ausweist. Die Nutzzahl ist das Minimum von Anzahl und der Anzahl der gleichzeitig aktiven Aufträge. In der Variablen Belastung wird jeweils die augenblickliche Belastung des Hauptspeichers gespeichert, und Auslastung ist eine Beobachtungsvariable, die die Summe aller Bedienzeiten bzw. für den Hauptspeicher die Summe des Produkts von Bedienzeit und Hauptspeicherbelegung akkumuliert.

Jeder simulierte Auftrag wird durch einen Auftragsrecord beschrieben. Ein Auftragsrecord besteht aus den folgenden Variablen:

a) Folgenummer = Durchnumerierung aller Aufträge.
b) Runid = Name des Auftrags.
c) Typ = DEMAND oder BATCH.
d) Status = READIN, OPEN, WAITING, ACTIVE, FINNED.
e) Ankunftszeit = Modellzeit, zu der der Job in dem System eintrifft.
f) Startzeit = Modellzeit, zu der der Job aktiviert wurde.
g) Finzeit = Modellzeit, zu der der Job beendet wurde.
h) Verbrauch = Vektor der Betriebsmittelansprüche des Auftrags.
i) Tau = Vektor der relativen zeitlichen Betriebsmittelansprüche des Auftrags.
j) Rest = Vektor der noch nicht befriedigten Betriebsmittelansprüche.
k) ULaufzeit = Ursprüngliche Alleinlaufzeit des Auftrags.
l) RLaufzeit = Restliche Alleinlaufzeit des Auftrags.
m) Subsystem = Nummer des Subsystemens, zu dem der Auftrag gehört.
n) Quotazeit = Zeitpunkt der letzten Aktualisierung der restlichen Betriebsmittelansprüche.
o) Priorität = Priorität des Auftrags, A - Z.

Ein Subsystem wird durch den einen Record mit den Variabeln

a) Bottleneck = Name des Betriebsmittels mit der größten Auslastung.
b) Speed = Relative Geschwindigkeit des Subsystems, $0 < \text{Speed} \leq 1$.

beschrieben.

Das Modell verarbeitet die folgenden Systemereignisse (Events):

a) GETNEXT

Beim Eintreffen dieses Events wird die Beschreibung des nächsten Auftrags aus der Datei gelesen, die die zu modellierende Last enthält. Die Einführung dieses Ereignisses war aus implementatorischen Gründen notwendig, um die Anzahl der gleichzeitig von dem Modell zu verwaltenden Auftragsrecords steuern zu können, damit der für das Modell zur Verfügung stehende Datenspeicherplatz nicht überschritten wurde.

b) INIT

Dieser Event zeigt das Eintreffen eines neuen Auftrags im Rechnermodell an. Der Auftrag wird anschließend entweder aktiviert (siehe ACTIVATE) oder in die Hauptspeicherwarteschlange eingereiht.

c) ACTIVATE

Dieser Event aktiviert einen Auftrag. Der Auftrag bekommt den benötigten Hauptspeicherplatz zugewiesen und wird von dem Simulationsmodell "verarbeitet".

d) FINISH

Ein Event dieses Typs wird erzeugt, wenn ein Auftrag beendet werden soll. Dieser Event unterscheidet sich aus implementatorischen Gründen von allen anderen, da er nicht in die Event-Warteschlange eingetragen wird, sondern einen statischen Speicherplatz einnimmt, der bei dem erneuten Auftreten eines Events dieses Typs überschrieben wird. Der FINISH-Event beschreibt also immer den Zeitpunkt, zu dem der nächste Auftrag beendet werden soll.

e) RFIN

Dieser Event dient nicht zur Ablaufsteuerung des Modells, sondern um Beobachtungsvariablen bezüglich des realen Verhaltens der zu simulierenden Last zu aktualisieren. Er zeigt während der Kalibrierungsphase an, zu welchem Zeitpunkt ein simulierter Auftrag in Wirklichkeit beendet wurde. Hierdurch ist es möglich, den maximalen Multiprogramming-Grad und den maximalen Hauptspeicherbedarf der realen Last zu ermitteln.

Jeder Event besitzt die Attribute Typ (siehe oben), Timestamp (Modellzeit, zu der der Event bearbeitet werden soll), Job (Zeiger auf die Auftragsbeschreibung, auf die sich der Event bezieht) und Next (Zeiger auf den nächsten Event in der Event-Warteschlange). Die Event-Warteschlange besteht aus einem Anker, der den Zeiger auf den ersten Event, den Zeiger auf den letzten Event und die Anzahl der Events enthält und aus einer einfach verketteten Liste der Event-Records, die gemäß den Zeitstempeln der Events sortiert ist.

Außer der Event-Warteschlange verwaltet das Modell noch zwei weitere, für den Modellablauf notwendige Warteschlangen. Die Aktivliste enthält die Zeiger auf die Auftragsrecords aller aktiven Aufträge. Diese Liste wird benötigt, um nach dem Eröffnen eines neuen Auftrags oder nach der Beendigung eines Auftrags die dadurch veränderten, relativen Fortschrittsgeschwindigkeiten aller aktiven Aufträge berechnen zu können. Die Hauptspeicherwarteschlange enthält die Zeiger auf diejenigen Aufträge, die zwar eingelesen sind, aber wegen Überschreitung des eingestellten maximalen Multiprogramming-Grads für Stapelaufträge oder wegen fehlenden Hauptspeicherplatzes noch nicht aktiviert werden können. Die Haupspeicherwarteschlange enthält nur Stapelaufträge, da das Modell Dialogaufträge sofort bei Eintreffen in das System aktiviert. Sie wird jedesmal durchsucht, wenn ein Auftrag beendet wurde. Alle Aufträge, die danach in den Hauptspeicher passen, werden dann aus der Hauptspeicherwarteschlange entfernt und aktiviert.

Zur Datenstruktur des Modells gehören die folgenden Zustandsvariablen:

a) TIME - enthält jeweils die aktuelle Modellzeit seit dem Start der Simulation. Alle Zeiten werden in Millisekunden gemessen.

b) START - enthält die Startzeit der Simulation in Millisekunden seit dem 1. Januar des jeweiligen Jahres 0.00 Uhr. Die Startzeit der Simulation ist der Zeitpunkt, zu dem der erste zu simulierende Auftrag von dem realen Rechensystem eingelesen wurde.

c) STOPTIME - enthält die Modellzeit, zu der die Simulation beendet wurde.

d) ENDE - zeigt an, daß alle Aufträge eingelesen wurden und das Modell ausläuft.

e) MAXOPEN - gibt an, wieviel Stapelaufträge maximal gleichzeitig aktiviert werden sollen.

f) SGRAD - Substitutionsgrad für die Substitution des Rechenprozessors.

g) SAUFWAND - Substitutionsaufwand für die Substitution des Rechenprozessors.

h) SANZ - Anzahl Aufträge, bei denen die Substitution durchgefürht werden soll.

i) STABELLE - Tabelle der Aufträge, bei denen die Substitution durchgeführt werden soll.

Die folgenden Beobachtungsvariablen werden von dem Modell unterhalten.

a) Jeweils getrennt für Stapelaufträge und für Dialogaufträge:

ANZ(i)	-	Anzahl der simulierten Aufträge.
LAUFZEIT(i)	-	Summe der Laufzeiten aller simulierten Aufträge.
BACKLOGZEIT(i)	-	Summe der Wartezeiten aller simulierten Aufträge in der Hauptspeicherwarteschlange.
ALLEINZEIT(i)	-	Summe der Alleinlaufzeit aller Aufträge.
TLAUFZEIT(i)	-	Summe der Laufzeiten der realen Aufträge.
TBACKLOGZEIT(i)	-	Summe der Wartezeiten aller realen Aufträge in der Batch-Warteschlange.
JOPEN(i)	-	aktuelle Anzahl der aktiven simulierten Aufträge.
JOPENMAX(i)	-	maximal erreichte Anzahl gleichzeitig aktiver simulierter Aufträge.
TJOPEN(i)	-	aktuelle Anzahl der aktiven realen Aufträge.
TJOPENMAX(i)	-	maximal erreichte Anzahl gleichzeitig aktiver realer Aufträge.

b) HSPMAX - maximale Auslastung des simulierten Hauptspeichers.

c) THSP - aktueller Stand der realen Hauptspeicherauslastung.

d) THSPMAX - maximale reale Hauptspeicherauslastung.

e) TAUSLASTUNG - Vektor der Summe der realen Betriebsmittelansprüche.

f) TFIN - Beendigung des letzten Auftrags der realen Last.

g) SJOBS - Anzahl der Aufträge, bei denen die Substitution des Rechenprozessors durchgeführt wurde.

3.2 Prozeduraler Teil

Das Hauptprogramm des Simulationsmodells besteht im wesentlichen aus einer Schleife, in der jeweils der nächste Event aus der Event-Warteschlange entnommen wird und je nach dem Typ des Events in eine der Prozeduren job_input, job_start, job_activation, job_termination oder job_fin verzweigt wird.

Die Prozedur job_input verarbeitet Events des Typs GETNEXT. Sie liest jeweils den nächsten Auftrag aus der Auftragsdatei ein und erstellt den Auftragsrecord, der den Status des Auftrags während der gesamten Simulationsdauer beschreibt. Sie erzeugt für diesen Auftrag einen Event des Typs INIT und eventuell einen Event des Typs RFIN. Außerdem erzeugt diese Prozedur einen neuen Event vom Typ GETNEXT, damit der nächste Auftrag spätestens dann eingelesen wird, wenn der gerade eingelesene Auftrag eröffnet werden soll.

Die Prozedur job_start verarbeitet Events vom Typ INIT. Sie prüft, ob der Auftrag aktiviert werden kann. Falls dies der Fall ist, wird ein Event des Typs ACTIVATE erzeugt, ansonsten wird der Auftrag in die Hauptspeicherwarteschlange eingereiht.

Die Prozedur job_activation verarbeitet Events des Typs ACTIVATE. Vor der Aktivierung des Auftrags werden die restlichen Betriebsmittelansprüche aller aktiven Aufträge aktualisiert. Die Nutzzahl der Betriebsmittel CPU und I/O wird auf das Minimum der Anzahl der simulierten Exemplare des Betriebsmittels und des Multiprogramming-Grads (= Länge der Aktivliste) gesetzt. Von dem Betriebsmittel THINK sind so viele Exemplare nutzbar, wie Aufträge aktiv sind, so daß die Denkzeit niemals ein Bottleneck darstellen kann. Nach der Aktivierung des Auftrags wird für alle aktiven Aufträge die neue relative Fortschrittsgeschwindigkeit berechnet und der FINISH-Event für den nächsten zu terminierenden Auftrag erzeugt.

Die Prozedur job_termination verarbeitet Events des Typs FINISH. Sie führt die Beendigung eines Auftrags durch und prüft, ob danach weitere Aufträge, die sich in der Hauptspeicherwarteschlange befinden, aktiviert werden können. Für alle diese Aufträge wird ein Event des Typs ACTIVATE erzeugt. Außerdem aktualisiert diese Prozedur alle Beobachtungsvariablen, die sich auf die simulierten Aufträge beziehen, und gibt den Auftragsrecord frei.

Die Prozedur job_fin bearbeitet Events des Typs RFIN, die nur während der Test- und Kalibrierungsphase optional erzeugt werden. Sie aktualisiert Beobachtungsvariablen, die sich auf den Ablauf der realen Aufträge beziehen.

3.3 Beschreibung der simulierten Last

Das Simulationsmodell wurde mit den Accounting-Daten der realen Last des untersuchten Rechenzentrums betrieben. Dabei umfaßte jeder Simulationslauf jeweils die simulierte Last eines Betriebsmonats, wobei die gleiche Last verwendet wurde, die im Abschnitt 2 beschrieben wurde. Allerdings wurde für die Modellierung der Last ein einzelner Auftrag als kleinstes Lastelement verwendet. Da jeder Auftrag im Mittel etwa sechs Programmausführungen enthielt, konnte durch diese Maßnahme der Rechenzeitverbrauch des Simulationsmodells erheblich verringert werden. Für die beabsichtigte Simulation der Substitutionsmaßnahmen entstand dadurch kein wesentlicher Fehler, da die rechenintensivsten Aufträge im wesentlichen nur ein Programm ausführten, so daß bei diesen Aufträgen die wichtigsten Kennwerte für das Programm und den Auftrag übereinstimmten.

Die in den jeweiligen Monatsdatensätzen der System Logfiles enthaltenen Daten wurden für die Bearbeitung durch das Simulationsmodell in der Weise verdichtet, daß für jeden bearbeiteten realen Auftrag ein Datensatz erstellt wurde, der die folgenden Angaben enthielt:

1. Das reale Ankunftsdatum des Auftrags im Rechnersystem.
2. Der Name (RUNID) des Auftrags.
3. Der Programmname desjenigen Programms des Auftrags, das von allen Programmen, die der Auftrag ausführte, die höchsten Betriebsmittelansprüche stellte.
4. Die Betriebsart des Auftrags (Dialog- oder Stapelbetrieb).
5. Die von dem Auftrag verbrauchte CPU-Zeit in Millisekunden.
6. Die von dem Auftrag verbrauchte CC/ER-Zeit in Millisekunden.
7. Die von dem Auftrag verbrauchte I/O-Zeit in Millisekunden.
8. Die von dem Auftrag verbrauchte Denkzeit in Millisekunden.
9. Die durchschnittliche Hauptspeichergröße des Auftrags in K Wörtern.
10. Die externe Priorität des Auftrags (A - Z).
11. Die Abrechnungsnummer des Auftraggebers.
12. Der Zeitpunkt, zu dem der reale Auftrag aktiviert wurde.
13. Der Zeitpunkt, zu dem der reale Auftrag beendet wurde.
14. Die Wartezeit des realen Auftrags in der Batch-Warteschlange.
15. Die Laufzeit des realen Auftrags.
16. Der Zeitraum, in dem der reale Auftrag aktiv war.

Vor diesen Daten wurde die Ankunftszeit des Auftrags, der Verbrauch an CPU-Zeit, I/O-Zeit, Denkzeit und die mittlere Hauptspeichergröße des Auftrags für die Simulation der Auftragsabarbeitung verwendet. Der Name des "Hauptprogramms" und die Abrechnungsnummer des Auftraggebers wurden zusätzlich für die Simulation der auftragsinternen Substitution benötigt. Der Name des Auftrags wurde lediglich für die Identifizierung der Aufträge bei Kontrollausdrücken verwendet, und die Angaben über das Verhalten des realen Auftrags flossen in die Beobachtungsvariablen des realen Auftrags ein, die zur Kalibrierung des Modells verwendet wurden.

3.4 Vorgehensweise

Zunächst wurde die Funktionsweise des Simulationsmodells überprüft. Anhand von synthetischen Aufträgen, deren Abarbeitung durch das Modell vorausberechnet werden konnte, wurde verifiziert, daß die Algorithmen für die Verwaltung der Betriebsmittel und die Erzeugung und Verarbeitung der Events in der gewünschten Weise funktionierten. Daran anschließend wurden Belastungstests mit einer großen Anzahl von synthetischen Aufträgen durchgeführt. Dabei stellte sich heraus, daß die Rechenzeit, die das Modell benötigte, in einem günstigen Verhältnis zu der simulierten Rechenzeit stand, daß aber die zur Verfügung stehende Hauptspeichergröße überschritten wurde, wenn das Modell etwa 1.000 Aufträge gleichzeitig verwalten mußte. Dieses Ergebnis hatte zur Folge, daß das Einlesen der Auftragsrecords aus der Eingabedatei ebenfalls über den Event-Mechanismus gesteuert werden mußte, um auch mehrere tausend Aufträge in einem Simulationslauf verarbeiten zu können.

Anschließend wurde überprüft, ob das Modell bei der Abarbeitung der Aufträge durch den Rechnerkern die simulierte Wirklichkeit genügend genau nachbilden konnte. Diese Kalibrierung des Simulationsmodells wurde anhand mehrerer Monatslasten durchgeführt. Dabei handelt es sich um den Monat April 1984, bei dem insgesamt 50.615 Dialog- und Stapelaufträge gerechnet wurden, und um die Monate Oktober bis Dezember 1985, bei denen unter veränderten Betriebsbedingungen im Mittel etwa 3.800 Stapelaufträge bearbeitet wurden. Da bei diesen Tests die Verarbeitung der Aufträge durch den Rechnerkern angepaßt werden sollte, wurden die Auftragsrecords nach der realen Aktivierungszeit sortiert, und diese Zeit wurde als simulierte Ankunftszeit verwendet. Außerdem wurde der simulierte Hauptspeicher und die maximale Anzahl gleichzeitig zu eröffnender Stapelaufträge so groß gewählt, daß keine Wartezeiten wegen eines Hauptspeicherengpasses auftreten konnten, da bei dieser Abarbeitungsreihenfolge bei der realen Rechenanlage auch alle Aufträge zu dem angegebenen Zeitpunkt aktiviert werden konnten.

Bei der Kalibrierung des Modells stellte sich heraus, daß die Summe aus CPU-Zeit und CC/ER-Zeit nicht geeignet war, die tatsächliche zeitliche Beanspruchung der Rechenprozessoren durch die Benutzeraufträge zu approximieren. Dies lag zum einen daran, daß die CC/ER-Zeit nur pauschaliert abgerechnet wurde und daher keine gute Annäherung der von dem Betriebssystem tatsächlich benötigten CPU-Zeit darstellte, im wesentlichen aber daran, daß die Maschinenuhr, die die CPU-Zeit mißt, angehalten wird, wenn ein Rechenprozessor bei einem Hauptspeicherzugriff wegen eines Cache-Miss warten muß. Dieses Problem wurde dadurch gelöst, daß die CC/ER-Zeit für die Inanspruchnahme der simulierten Rechenprozessoren nicht mehr berücksichtigt wurde und stattdessen die Grundgeschwindigkeit der Rechenprozessoren auf das Verhältnis zwischen abgerechneter CPU-Zeit und Nutzungszeit, das bei der realen Abarbeitung der simulierten Last erreicht wurde, verringert wurde. Eine Feinabstimmung ergab, daß die so erhaltene Grundgeschwindigkeit bei Langläufern, das heißt bei Aufträgen, die mehr als eine Stunde CPU-Zeit verbrauchten, noch um etwa 5 % erhöht werden mußte, um eine gute Approximation der mittleren Aktivzeiten und des mittleren Multiprogramming-Grads zu erreichen. Mit dieser zusätzlichen Erhöhung der Grundgeschwindigkeit der Rechenprozessoren für Langläufer wurde der Effekt ausgeglichen, daß das Simulationsmodell die rechenintensiven Aufträge etwas zu langsam abarbeitete, weil es alle Aufträge gleich behandelte, während die rechenintensiven Programme in der Regel eine größere Lokalität besaßen als der Durchschnitt der Programme und deshalb in der Wirklichkeit auf weniger Cache-Misses und auf ein günstigeres Verhältnis von produzierter CPU-Zeit zur realen Laufzeit kamen.

Ein anderes Anpassungsproblem bestand darin, daß das Betriebssystem die Dialogaufträge gegenüber den Stapelaufträgen bei der Zuteilung der Rechenprozessoren bevorzugt, während das Simulationsmodell diese Möglichkeit nicht vorsah. Dieses Problem wurde dadurch gelöst, daß die Verarbeitungskapazität des Betriebsmittels Rechenprozessor auf zwei Betriebsmittel Dialog-CPU und Batch-CPU aufgeteilt wurde, wobei der Dialog-CPU eine in bezug auf die Betriebsmittelansprüche der Dialogaufträge relativ höhere Verarbeitungskapazität zugewiesen wurde als der Batch-CPU für die Bearbeitung der Stapelaufträge. Es stellte sich heraus, daß die Dialogaufträge um den Faktor 3 besser bedient werden mußten als die Stapelaufträge, um eine gute Approximation der mittleren Aktivzeiten zu erreichen.

Die Tatsache, daß das reale Rechnersystem bei drei Rechenprozessoren ein echtes Multiprocessing, das heißt die tatsächliche, parallele Abarbeitung mehrerer Prozesse eines Auftrags zuließ, während das Modell eine solche Möglichkeit nicht vorsah, führte außer bei einigen Systemprogrammen zu keinerlei Anpassungsproblemen, da diese Möglichkeit des Betriebssystems offensichtlich von den Benutzerprogrammen nicht ausgenutzt und von den Compilern auch nicht unterstützt wird.

Da diese Programme insgesamt nur sehr geringe Betriebsmittelansprüche aufwiesen, wurde dieses Problem dadurch gelöst, daß die drei Realtime-Programme aus der modellierten Last entfernt wurden.

Das Ergebnis der Anpassungen des internen Verarbeitungszyklus' ist in der folgenden Tabelle zu sehen. Tabelle 3.1 zeigt die Beobachtungswerte für die reale Last des Monats April 1984 und für die entsprechende Simulation mit der besten Approximation. Bei der Interpretation der Tabelle ist zu beachten, daß die mit einem Stern (*) gekennzeichneten Werte definitionsgemäß gleich groß sein müssen.

Beobachtungsvariable	Reale Last	Simulierte Last
Gesamtlaufzeit [hh:mm:ss]	724:50:11	723:12:37
Anzahl Stapelaufträge	22.199	22.199*
Anzahl Dialogaufträge	28.089	28.089*
Mittlere Laufzeit aller Aufträge [sec]	1.062,6	1.036,7
Mittlere Alleinlaufzeit aller Aufträge [sec]	776,0	776,0*
Mittlerer Servicefaktor aller Aufträge	1,369	1,336
Mittlere Laufzeit der Dialogaufträge [sec]	1.150,7	1.099,5
Mittlere Alleinlaufzeit der Dialogaufträge [sec]	1.058,0	1.058,0*
Mittlerer Servicefaktor der Dialogaufträge	1,087	1,039
Anzahl aktiver Dialogaufträge (im Mittel)	12,4	11,9
Anzahl aktiver Dialogaufträge (maximal)	78	80
Mittlere Laufzeit der Stapelaufträge [sec]	951,5	957,3
Mittlere Alleinlaufzeit der Stapelaufträge [sec]	419,0	419,0*
Mittlerer Servicefaktor der Stapelaufträge	2,271	2,284
Anzahl aktiver Stapelaufträge (im Mittel)	8,1	8,2
Anzahl aktiver Stapelaufträge (maximal)	21	40
CPU-Auslastung [%]	71,87	72,03
I/O-Auslastung [%]	12,69	12,72
Denkzeit-Auslastung [%]	1.148,9	1.151,5
Mittlere Hauptspeicherauslastung [KW]	864	793
Maximale Hauptspeicherauslastung [KW]	2.725	2.210

* Werte sind definitionsgemäß gleich

Tabelle 3.1

Kalibrierung des Simulationsmodells - Monat: April 1984

Das reale Verhältnis aus CPU-Zeit und der Differenz von Nutzzeit und Leerzeit betrug 73,72. Dementsprechend wurde die Grundgeschwindigkeit der Rechenprozessoren des Simulationsmodells auf 0,737 eingestellt und bei Aufträgen mit einer CPU-Zeit von mehr als einer Stunde noch um 5 % auf 0,774 erhöht. Der Dialogbetrieb wurde in dem Modell um den Faktor 3,0 besser bedient als der Stapelbetrieb. Das Modell simulierte entsprechend der Wirklichkeit drei Rechenprozessoren und fünf Ein-/Ausgabesysteme. Als Ein-/Ausgabesysteme wurden alle Ein-/Ausgabepfade gezählt, die die verschiedenen Aufträge gleichzeitig benutzen konnten. Die Alleinlaufzeit wurde nach der Formel (CPU-Zeit x Geschwindigkeitsfaktor + I/O-Zeit + Denkzeit) berechnet. Der Servicefaktor ergab sich als Quotient aus der Summe der Laufzeiten und der Summe der Alleinlaufzeiten.

Aus der Tabelle 3.1 ist zu ersehen, daß das Modell trotz des hohen Abstraktionsgrads, auf dem es arbeitete - für die Simulation einer gesamten Monatslast mit insgesamt 723 Stunden Laufzeit und einer CPU-Zeit von 1.562 Stunden, die von 50.288 Aufträgen verbraucht wurde, benötigte das Modell nur eine Rechenzeit von 29 Minuten - eine gute Approximation der Wirklichkeit erreichte. Lediglich bei der maximalen Anzahl gleichzeitig aktiver Stapelaufträge und bei der maximalen Hauptspeicherbelastung gab es signifikante Abweichungen zwischen Modell und Wirklichkeit, die nicht ausgeglichen werden konnten. Dies lag im wesentlichen daran, daß schon eine geringfügige Verzögerung im simulierten Ablauf einiger gleichzeitig aktiver Langläufer dazu führte, daß sich diese Maximalwerte durch die zusätzliche Abarbeitung vieler kleiner Aufträge verschoben. Eine genauere Untersuchung der Abweichung bei der maximalen Anzahl aktiver Stapelaufträge ergab, daß diese im wesentlichen durch eine Kette von 15 Stapelaufträgen verursacht wurden, die aus den genannten Gründen in dem Modell wegen der fest vorgegebenen Ankunftszeiten zu einem Zeitpunkt alle gleichzeitig aktiv waren, während sie in Wirklichkeit hintereinander abgearbeitet wurden. Diese Beobachtung führte zu dem Schluß, daß sich die Maximalwerte für den Multiprogramming-Grad und die Hauptspeicherbelegung generell nicht besonders gut für den Vergleich zwischen Simulation und Wirklichkeit eigneten.

Nachdem die Kalibrierung des Simulationsmodells abgeschlossen war, wurde das Modell benutzt, um die möglichen Auswirkungen von Maßnahmen zur auftragsinternen Substitution des Rechenprozessors auf den Durchsatz der Rechenanlage zu untersuchen. Dabei stellte sich heraus, daß bei gleichbleibender Anlagenkonfiguration die potentielle Durchsatzsteigerung nur etwa die Hälfte der potentiellen Einsparung an Rechenzeit betrug. Bei intensivem Einsatz von Methoden der Substitution der Rechenprozessoren würde sich der Engpaß der Rechenanlage von den Rechenprozessoren zum Hauptspeicher hin verschieben. Allerdings zeigten die Simulationsläufe auch, daß durch die auftragsinterne Substitution von jeweils 30 % der Rechenzeit in insgesamt nur sieben verschiedenen Benutzerprogrammen eine monatliche Durchsatzsteigerung der Rechenanlage von etwa 15 % erreicht werden könnte, ohne daß dabei andere Betriebsmittel überlastet würden.

4. Literaturhinweise

[BAUK 76] Bauknecht, K., Kohlas, J., Zehnder, C.A.: Simulationstechnik, Springer Verlag, Berlin - Heidelberg - New York (1976)

[BEIL 81] Beilner, Heinz: On the Construction of Computing System Simulators, in Experimental Computer Performance and Evaluation, North-Holland Publishing Company, Amsterdam - New York - Oxford (1981), S. 1 - 32

[DENN 78] Denning, P.I., Buzen, H.P.: The Operational Analysis of Queueing Network Models, in Computing Surveys, Vol. 10 (1978), S. 225 - 261

[SERA 81] Serazzi, Guiseppe: The Dynamic Behavior of Computer Systems, in Experimental Computer Performance and Evaluation, North-Holland Publishing Company, Amsterdam - New York - Oxford (1981), S. 127 - 163

[TERP 77] Terplan, Kornel: Leistungsfähigkeit der Datenverarbeitung, Verlag Dokumentation Saur AG, München (1977)

[WEBE 83] Weber, Karl, Trzebiner, Richard, Tempelmeier, Horst: Simulation mit GPSS, Verlag Paul Haupt, Bern - Stuuttgart (1983)

BEMO - Ein Lastsimulator

Ludger Meyer
Abteilung Leistungsanalyse
Nixdorf Computer AG
Pontanusstr. 55
D - 4790 Paderborn

Zusammenfassung: Im industriellen Alltag tritt oft die Aufgabenstellung auf, den Nachweis zu erbringen, daß auf einem bestimmten Rechensystem ein genau definiertes Lastkontingent bearbeitet werden kann. Da in der Regel die realen Anwendungen noch nicht oder nur teilweise für das System zur Verfügung stehen, sind synthetische Programme erforderlich, um die Last zu erzeugen. Gefragt ist hier ein flexibles Werkzeug, mit dem eine bestimmte Last wiederholbar erzeugt werden kann, ein Ändern des Lastaufkommens ohne größeren Aufwand möglicht ist, und welches zudem Aussagen zum Leistungsverhalten des Systems bereitstellt. Ein solches Werkzeug für UNIX® Betriebssysteme ist BEMO (BEnchmark-MOnitor), der im folgenden vorgestellt wird. Dabei wird sowohl auf den Aufbau und die Funktionalität von BEMO eingegangen, als auch an einem Beispiel die Mächtigkeit des Werkzeuges demonstriert.

1. Einleitung

Die Leistungsbewertung eines Rechensystems erfolgt in der Regel unter Verwendung verschiedener Lasten, mit denen das zu untersuchende System betrieben wird, wobei mit Hilfe von Meßwerkzeugen Leistungsdaten ermittelt werden. Die Last kann dabei aus realen Anwendungen, synthetischen Lasten oder einem Mix aus beiden bestehen. Zur Messung werden unterschiedliche Meßwerkzeuge wie Software- und/oder Hardwaremonitore verwendet, oder es wird nur die Systemuhr vor und nach einer Transaktion ausgelesen.

---

*) UNIX ist ein eingetragenes Warenzeichen von AT&T

Für UNIX* Betriebssysteme gibt es eine Vielzahl von Benchmarks. Diese bestehen in der Regel aus synthetischen Programmen, die meist einzelne Punkte des System untersuchen. So wird z.B. die Implementierung von Systemcalls, Dauer eines Funktionsaufrufes, usw., bewertet (vgl. z.B. [HINN84]). Andere untersuchen die Gleitpunktarithmetik (z.B. [KARP85]) oder Compiler-Implementation und Speicherarchitektur (z.B. [WEIC84]). Diese Liste läßt sich beliebig erweitern. Sie zeichnen sich aus durch leichte Installierbarkeit und weitgehende Systemunabhängigkeit.

Immer häufiger wird jedoch von Kunden der Nachweis der leistungsmäßigen Realisierbarkeit einer bestimmten Last auf dem ihm angebotenen System verlangt. Da die realen Anwendungen des Kunden in der Angebotsphase meist noch nicht auf dem System verfügbar sind, müssen synthetische Programme verwendet werden, die mit einem die Last repräsentierenden Mengengerüst parametrisiert werden können. Das Mengengerüst liefert der Kunde. Sind diverse reale Anwendungen schon auf dem System verfügbar, so sollen diese ohne weiteres mit zur Lasterzeugung verwendet werden können. Der Kunde ist an den Antwortzeiten und dem Durchsatz für "seine" Transaktionen interessiert und will wissen, wieviel verbleibende Reserven die verschiedenen Systemressourcen noch haben.

Ein Werkzeug, das die genannten Forderungen erfüllt, muß also folgende Eigenschaften in sich vereinigen:

- leichte Installierbarkeit
- weitgehende Systemunabhängigkeit
- Parametriesierbarkeit zur Umsetzung der Last von realen Anwendungen
- Reproduzierbarkeit des Lastaufkommens
- einfaches Ändern des Lastaufkommens
- Messung von Antwort- und CPU-Zeiten
- Messung von Systemdaten
- Unterstützung der Ergebnisauswertung und -aufbereitung

Diese Eigenschaften erfüllt der Benchmarkmonitor BEMO, welcher für die Beurteilung von UNIX* Betriebssystemen konzipiert und entwickelt wurde. Er bietet damit die Möglichkeiten einer gezielten und kontrollierten Lasterzeugung auf einem solchen System. So kann das System ohne großen Aufwand mit einer z.B. von einem Kunden spezifierten Last belastet und so eine ihn interessierende Bewertung vorgenommen werden. Außerdem ist in BEMO ein Konzept zur Durchsatzsteuerung und -kopplung aufgenommen worden, womit u.a. Überlastsituation behandelt werden können (vgl. hierzu [TOTZ87]). BEMO ist in einer ersten Version im Auftrag von Nixdorf von Dr. E. Janich (ehemals Uni Ulm, [JANI??]) geschrieben und bei Nixdorf weiterentwickelt worden.

In den folgenden Kapiteln werden zunächst die globale Struktur des BEMO und dann die für eine Simulation notwendigen drei Schritte, Beschreibung der Last, Durchführung der Simulation und Auswertung der Ergebnisse, beschrieben. Zum Abschluß wird dann ein kleiner Ausblick auf Weiterentwicklungen von BEMO gegeben.

2. Struktur von BEMO

Der Lastsimulator BEMO ist in C geschrieben und lauffähig unter UNIX* System V. Er beinhaltet eine Reihe von Elementarfunktionen, die zur Belastung verschiedener Sytemkomponenten (CPU, Platten, etc.) dienen, Pseudo-Zufallszahlengeneratoren zur Steuerung von Zeitabläufen, Funktionen zur statistischen Auswertung und grafischen Aufbereitung der Ergebnisse, sowie Kommandos, welche die allgemeine Handhabung von Testreihen unterstützen. Für die Zeitmessungen werden UNIX* Standardroutinen verwendet (z. B. times), und Systemleistungsdaten können während eines Simulationslaufes mit dem Softwaremonitor 'sar' erhoben werden.

Zur Durchführung einer Lastsimulation mit BEMO ist zunächst die Beschreibung der Last und deren zeitlichen Ablaufes erforderlich. Das dafür benötigte Mengengerüst zur Charakterisierung der Last ist vorab vom Benutzer zu ermitteln. Zum Start einer Simulation wertet das Verwaltungsprogramm von BEMO die angegebene Lastbeschreibung und (eventuelle) Umgebungsvariablen aus und erzeugt nach vorgegebenem Zeitraster die Lastprozesse. Zeitmeßergebnisse, sowie Warnungen und Fehlermeldungen werden in einer Ablaufverfolgungsdatei notiert und können nach Beendigung der Simulation ausgewertet werden. Die Programm-, Beschreibungs- und Ablaufverfolgungsdateien sind in den verschiedenen Katalogen des BEMO-Dateibaumes angesiedelt. Abbildung 2.1 zeigt den Datenfluß bei einer Simulation. Für sämtliche Kommandos und Bibliotheksfunktionen von BEMO ist auch eine online Dokumentation vorhanden.

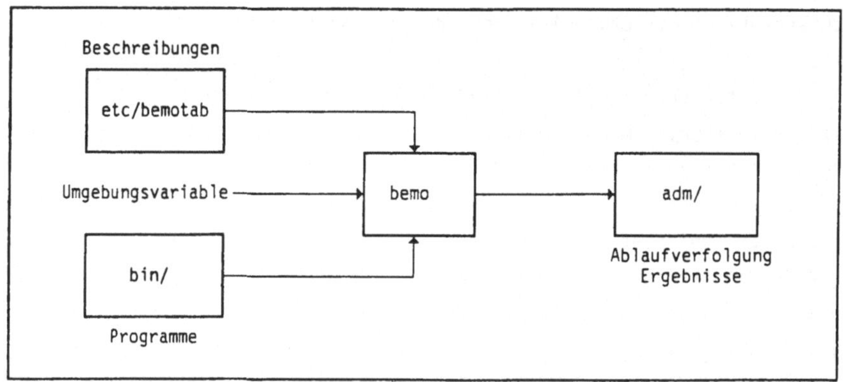

Abbildung 2.1: Datenfluß bei einer Lastsimulation mit BEMO

3. Lastbeschreibung

Die Beschreibung der Last besteht im wesentlichen aus zwei Teilen: Zum einen aus der Spezifikation der Kommandos, mit denen die Last erzeugt werden soll, zum anderen aus der Angabe eines Zeitrasters zur Definition des zeitlichen Ablaufes der Simulation. Diese Beschreibungen erfolgen in der Tabelle bemotab , die der 'inittab' nachempfunden ist. Zu jeder Zeile dieser Tabelle gehört genau ein Lastprogramm. Sie bestehen aus neun durch ":" voneinander getrennte Spalten, in denen nacheinander angegeben werden kann:

- Identifikation
- Level
- Priorität
- Attribute
- Erststartzeitpunkt und Neustartabstand
- Eingabedatei
- Ausgabedatei
- Fehlerausgabedatei
- Kommando

Abbildung 3.1 zeigt das Beispiel einer bemotab mit fünf Lastprogrammen. Die zugehörigen Kommandos sind durch die Makronamen $1-$4 und $f spezifiziert. Diese Definition der Kommandos über Makronamen erleichtert das mehrfache Verwenden desselben Kommandos in verschiedenen Zeilen der Tabelle und erhöht die Lesbarkeit der Datei. Ähnlich wie diese Makronamen können auch Umgebungsvariable zum Textersatz verwendet werden. Ist zur Durchführung der Simulation eine Vorabausführung von bestimmten Kommandos notwendig, so können auch diese durch das @-Zeichen gekennzeichnet in der Datei angegeben werden. Denkbar ist z.B. das Erzeugen einer Datei, aus der während der Simulation gelesen werden soll. Leer- und mit # beginnende Zeilen werden als Kommentarzeilen interpretiert.

Mit Identifikation wird der Anwendung ein Name gegeben, unter der sie in den Protokollen und Auswertungen eines Simulationslaufes wiederzufinden ist. Im Beispiel beginnen alle Namen mit "user". Es folgen Arbeitsplatz- und Kommandonummer. Die Priorität der Anwendung gegenüber anderen Lastprogrammen und insbesondere dem Verwaltungsprogramm von BEMO kann verändert werden, und ebenso können in der Tabelle die von der Anwendung zur Ein- und Ausgabe benötigten Dateien spezifiziert werden. Mit dem Makro $A ist im Beispiel der Katalog spezifiziert, in denen diese Daten abgelegt werden sollen. Die Dateien beginnen alle mit "e1" bzw. "e2" um diverse Auswertprogramme für diese Ausgaben einfacher verwenden zu können.

```
# bemotab beispiel

# Makrodefinitionen
$A=${BEMOHOME}/adm
$F=/tmp/ufile
$U=/dev/tty0200

#Makrodefinitionen der Kommandos
$1=synt -D195  -i10000 -F$F-1 -S1   -b128  -Q8 -msa -A2000 -B10  -C1 -T
$2=synt -D80   -i8000  -F$F-2 -S1   -b1024 -Q1 -msa -A1580 -B5   -C2 -T
$3=synt        -i6000  -F$F-3 -S5   -b1024 -Q3 -msa -A400  -B25  -C3 -T
$4=synt -D8640 -i7000  -F$F-4 -S21  -b512      -msw -A803  -B161 -C4 -T
$f=synt        -i1750  -F$F-f -S125 -b1024     -msw                 -T$U

# Zeitabelle
#ident :level:nice:attr:restart      :stdin:stdout :stderr    :command

# Arbeitsplatz 1
user1_1:ab   :09  :r1  :U(290,310) +0:     :$A/e1=u11:$A/e2=u11:$1/dev/tty0300
user1_2:ab   :09  :r1  :N(19,1) +  40:     :$A/e1=u12:$A/e2=u12:$2/dev/tty0300
user1_3:ab   :09  :r1  :U(290,310)+65:     :$A/e1=u13:$A/e2=u13:$3/dev/tty0300

# Arbeitsplatz 2
user2_1:a    :09  :r2  :L(4.0,1)+  11:     :$A/e1=u21:$A/e2=u21:$1/dev/tty0400
user2_2:a    :09  :r2  :N(150,5) + 30:     :$A/e1=u22:$A/e2=u22:$3/dev/tty0400
user2_3:a    :09  :r2  :N(150,5)   +45:    :$A/e1=u23:$A/e2=u23:$4/dev/tty0400

# Fixlast
fixjob-:a    :09  :rF  :     100   +20:    :$A/e1=fix:$A/e2=fix:$f
```

Abbildung 3.1: Beispiel einer bemotab

In der Spalte <u>Level</u> kann angegeben werden, in welchen Levels die Anwendung gestartet werden soll. Auf diese Weise ist es möglich, in einer Tabelle die unterschiedlichsten Lastsituationen zu beschreiben, so daß es z.B. für eine Testreihe mit unterschiedlichen Lastaufkommen pro Test nicht notwendig ist, verschiedenene Tabellen vorrätig zu halten. Jede Simulation wird mit einem bestimmten Level als Parameter aufgerufen, der dann die zu berücksichtigen Lastprogramme definiert. Im Beispiel werden mit Level a alle und mit Level b nur die ersten drei Zeilen der Tabelle berücksichtigt.

Mit Hilfe der <u>Attribute</u> kann zum einen die Behandlung bestimmter Situationen, in denen sich der Anwendungsprozeß während der Simulation befinden kann, festgelegt werden, zum anderen können hier Anwendungen verschiedenen Gruppen zugeordnet und zu Arbeitsplätzen zusammengefaßt werden. Zur ersten Gruppe der Attribute gehören solche, die Prozeßteile im Speicher fixieren, um ein mögliches Auslagern durch den Swapper zu verhindern, den sofortigen Neustart eines Prozesses veranlassen, wenn er abnormal beendet wurde (im Beispiel durch das Attribut 'r' veranlaßt) , oder den Neustart in Abhängigkeit von der Periode verzögern, wenn viele Prozesse aktiv sind, um eine Häufung von Aufträgen zu bestimmten Zeitpunkten abzuschwächen.

Die Attribute der zweiten Gruppe gestatten darüber hinaus eine Durchsatzsteuerung und -kopplung der Anwendungen. Dies bedeutet, daß Lastprogrammdurchsätze entweder absolut vorgegeben sind, oder daß sich die Durchsätze verschiedener Lastprogramme gemäß vorgegebenem Verhältnis einstellen sollen. Zunächst einmal können die Lastprozesse einer von drei Gruppen zugeordnet werden:

Fixgruppe — Die Prozesse dieser Gruppe stellen eine Grundlast dar, die auf jeden Fall realisiert werden soll. Deshalb bevorzugt BEMO die Prozesse der Fixgruppe gegenüber allen anderen Lastprozessen, d.h., wenn die Last der Fixgruppe nicht realisiert werden kann, erfolgt kein weiterer Start von sonstigen Anwendungen. Innerhalb der Gruppe werden die durch die Neustartabstände definierten Durchsatzverhältnisse eingehalten. Die Kennzeichnung der Gruppenmitglieder in der 'bemotab' erfolgt durch den Buchstaben 'F'.

Variable Gruppe — Auch die Durchsätze der Mitglieder dieser Gruppe sind gekoppelt. Die Anwendungen werden nur dann aktiviert, wenn eine vorhandene Fixgruppe realisiert werden kann. Ist die Last dieser Gruppe auf dem Testsytem nicht realisierbar, so wird der Neustart der zu dieser Gruppe gehörenden Anwendungen so lange verzögert, bis die dadurch verminderte Last realisiert werden kann. Dabei wird die Erhaltung des Mischverhältnisses der Lastprogramme gewährleistet. Bei einer erforderlichen Lastreduktion wird der Durchsatz von jedem Lastprogramm um denselben Faktor reduziert. Die Lastprogramme werden standardmäßig dieser Gruppe zugeordnet.

Unabhängige Gruppe — Hier erfolgt keine Kopplung der Durchsätze der Gruppenmitglieder. Eine Aktivierung der Lastprogramme hängt jedoch, wie bei der variablen Gruppe, von der Realisierbarkeit der Fixgruppe ab, falls eine solche definiert wurde. Die Mitglieder dieser Gruppe werden durch 'N' gekennzeichnet.

Neben der Einteilung der Anwendung in die obigen Gruppen, können sie zu Arbeitsplätzen zusammengefaßt werden. Innerhalb einer Arbeitsplatzgruppe ist zu jedem Zeitpunkt im Simulationszeitraum immer höchsten ein Prozeß aktiv. Im Beispiel sind zwei Arbeitsplätze mit je drei der variablen Gruppe zugeordneten Lastprogrammen angegeben (Identifikation user*) , sowie ein Lastprogramm in der Fixgruppe (fixjob).

Durch den Neustartabstand wird bestimmt, nach wieviel Sekunden die Anwendung nach dem letzten Start im Simulationszeitraum erneut gestartet werden soll. Der Neustart kann durch eine Konstante oder durch einen Pseudo-Zufallszahlengenerator bestimmt sein. Zur Verfügung stehen Generatoren zur Erzeugung von Poisson-, Binomial-, Gleich-, Exponential-, Normal- und Logarithmisch Normalverteilten Zufallszahlen. Neben dem Neustartzeitpunkt kann hier durch die Form +delta ein von Null verschiedener Erststartpunkt für die Anwendung angegeben werden. Im Beispiel ist der Erststartzeitpunkt von 'fixjob' im Simulationszeitraum 20 und der Neustardabstand wird durch die Konstante 100 bestimmt. Bei allen anderen Lastprogrammen ist der Neustartabstand durch eine Verteilung bestimmt: U(niform), N(ormal) oder L(ogaritmic normal).

Die Anwendung muß schließlich durch ein Kommando spezifiziert sein. Dafür gibt es drei Möglichkeiten:

- Erzeugung der Last mit Hilfe einer der von BEMO bereitgestellten Elementarfunktionen . Diese umfassen Rechenzeitverbrauch, Plattenaktivitäten, Hauptspeicherplatzbeanspruchung, Bildschirm- und Druckausgaben.

- Erzeugung der Last durch eigene C-Programme unter Verwendung von BEMO-Bibliotheksfunktionen im Programm.

- Erzeugung der Last durch beliebige auf dem Testsystem lauffähige Programme.

Die Elementarfunktionen entsprechen in ihrer Syntax den Kommandos des UNIX* Betriebssystems. Über Optionen werden an die Programme die Parameter übergeben, welche die zu erzeugende Last charakterisieren. Neben den Elementarfunktionen, die genau eine der genannten Aktivitäten initiieren, gibt es das Kommando synt , welches eine Zusammenfassung dieser Funktionen ermöglicht. Bei diesem Kommando ist die Angabe folgender Optionen erlaubt:

-F Datei	Dateiname.
-S Größe	Dateigröße.
-b Blockung	Blockungsfaktor für einen Dateizugriff.
-Q Anzahl	Anzahl der Dateizugriffe.
-m Modi	Angabe mit welchem Modus eine Datei behandelt werden soll, sowie die Art der Positionierung. Zum ersteren gehören Schrei-

ben, Lesen oder Anhängen, positioniert werden kann sequentiell, sägezahnartig oder gleichverteilt.

-T Terminal	Angabe des Terminals für Bildschirmausgaben.
-A Anzahl	Anzahl der Zeichen, die auf den durch -T spezifizierten Bildschirm ausgegeben werden sollen.
-B Blockung	Blockungsfaktor für die Bildschirmausgaben.
-C Zeichen	Zu verwendendes Ausgabezeichen.
-P Anzahl	Anzahl Zeichen für einen Spooldruckauftrag.
-D Anzahl	Anzahl Zeichen, die direkt gedruckt werden sollen.
-a KB	Zusätzlicher Bedarf an Hauptspeicherplatz in Kilobyte.
-i Anzahl	Anzahl von 1000 Maschineninstruktionen, um Rechenzeit zu verbrauchen. Die für den Verbrauch von einer Sekunde Rechenzeit benötigte Anzahl von Instruktionen ist Maschinenabhängig und muß vorab ermittelt werden.
-s Anzahl	Eine Pause von Anzahl Sekunden einlegen.
-n Anzahl	Alle Aufträge Anzahl-mal ausführen.

Die Vielfalt der Optionen bei 'synt' mag zunächst verwirrend erscheinen, ermöglicht aber eine einfache Umsetzung der durch ein Mengengerüst beschriebenen Last. Auf diese Weise kann mit dem synt-Programm die Lastsimulation von vielen realen Anwendungen erfolgen. Eine Lastsimulation besteht so möglicherweise nur aus einer Reihe von synt-Aufrufen, die sich nur in den an das Programm übergebenen Parametern unterscheiden. Genau diese Vorgehensweise ist im Beispiel angewendet worden: Hier bestehen alle Lastprogramme aus synt-Aufrufen.

Zur Umsetzung der durch die Parameter beschriebenen Last verwendet synt die BEMO-Bibliotheksfunktionsversionen der Elementarfunktionen. Dabei wird die Reihenfolge der Aufrufe der Funktionen durch einen Pseudo-Zufallszahlengenerator bestimmt, um so eine unrealistische Häufung von gleichen Aufträgen zu vermeiden.

Reichen die angebotenen Elementarfunktionen und das Programm synt nicht aus, um die Last in der gewünschten Form zu erzeugen, so steht es dem Benutzer frei, eigene C-Programme unter Benutzung der BEMO-Bibliotheksfunktionen zu schreiben. Schließlich erlaubt BEMO auch die Einbindung von beliebigen auf dem Testsystem lauffähigen Programmen bis hin zu realen Anwendungen. Dabei sollten jedoch keine Programme verwendet werden, die eine interaktive Eingabe erfordern, da dann eine Reproduzierbarkeit des Tests in Frage gestellt ist.

Um unnötigen Meßaufwand zu vermeiden, sollte bei der Lastbeschreibung darauf geachtet werden, die Last nicht in viele kleine, oft zu aktivierende Prozesse aufzusplitten, sondern sich soweit wie möglich auf einige wenige Prozesse zu beschränken. So reicht es z.B. bei der Simulation von Plattenlast aus, sich auf die Anzahl der im System parallel bedienbaren Plattenaufträge zu beschränken.

4. Durchführung einer Lastsimulation

Die Dauer eines Simulationslaufes wird durch drei Zeiten bestimmt:

Vorlaufzeit — die Zeit, während der sich das System einschwingen soll. Diese Zeit geht nicht in die spätere Auswertung ein. Sie wird vor allem bei der Angabe einer Fixgruppe dazu benutzt werden, die Realisierbarkeit bzw. Nichtrealisierbarkeit dieser Last festzustellen. Dazu werden zunächst alle zum Lastlevel gehörenden Lastprozesse zu ihrem Erststartpunkt aktiviert. Dann wird überprüft, ob die Prozesse der Fixlast einwandfrei durchlaufen. Falls nicht, so werden die restlichen Anwendungen verzögert, bis dies der Fall ist. Da diese Verzögerung nur den Neusstart der Anwendungen aus der variablen und unabhängigen Gruppe und nicht die schon gestarteten Prozesse aus diesen Gruppen betrifft, wird die Vorlaufzeit in Abhängigkeit von der Laufzeit dieser Prozesse gewählt.

Laufzeit — Die für die Auswertung der Ergebnisse maßgebende Zeit.

Nachlaufzeit — Prozesse, die am Ende der Laufzeit noch aktiv sind, werden nicht abrupt abgebrochen, sondern dürfen "auslaufen". Neue Prozesse werden nicht mehr gestartet.

Diese Zeiten können interaktiv beim Startkommando gesetzt bzw. verändert werden. Erfolgt keine Eingabe, so werden die Werte des letzten Simulationslaufes bzw. Voreinstellungen verwendet. Außerdem muß beim Startkommando angegeben werden, welches oder welche Level aus der bemotab gestartet werden sollen.

Gemäß dieser Spezifikation werden dann die Anwendungen im vorgegebenen Zeitraster aktiviert. Außerdem erfolgt eine Zeitmessung, die von jedem Prozeß die Antwort- und CPU-Zeiten ermittelt und in einer linearen Liste ablegt. Treten während des Simulationszeitraumes nicht gewünschte Effekte auf, so werden diese in einer Ablaufverfolgungsdatei protokolliert. Hierzu gehört das abnormale Beenden von Prozessen, Verzögerungen bezüglich des Neustarts, nicht Realisierbarkeit der Fixgruppe, etc.

Eine genaue Ablaufverfolgung mit allen Start und Endzeitpunkten aller Prozesse ist optional möglich.

Nach Ablauf der Simulationszeit werden alle noch aktiven und vom Benchmarkmonitor aktivierten Prozesse abgebrochen, sowie die für die einzelnen Lastprogramme ermittelten Zeiten unter dem Identifikationsnamen aus der bemotab in der Ablaufverfolgungsdatei abgelegt. Zudem wird der Rechenzeitverbrauch des Verwaltungsprogramms während des Simulationszeitraumes notiert, um so eine Kontrolle der durch BEMO selbst erzeugten Überlast zu haben.

Abbildung 4.1 zeigt das Beispiel eines Simulationsablaufes mit einem Prozess in der Fixgruppe (fixlast) und drei zu einem Arbeitsplatz zusammengefaßten Prozessen (proc*) in der variablen Gruppe. PREDUR, DURATION und EXTRADUR bezeichnen dabei Vorlaufzeit, Laufzeit und Nachlaufzeit. Die waagerechten Linien deuten die Aktivphasen der Prozesse und die kleinen senkrechten Striche die vorgesehenen Startpunkte im Simulationszeitraum an. Es zeigt sich, daß aufgrund der Zusammenfassung der 'proc'-Prozesse zu einem Arbeitsplatz ihre Startpunkte teilweise verschoben werden mußten, während der Fixgruppenprozess immer pünktlich aktiviert wurde.

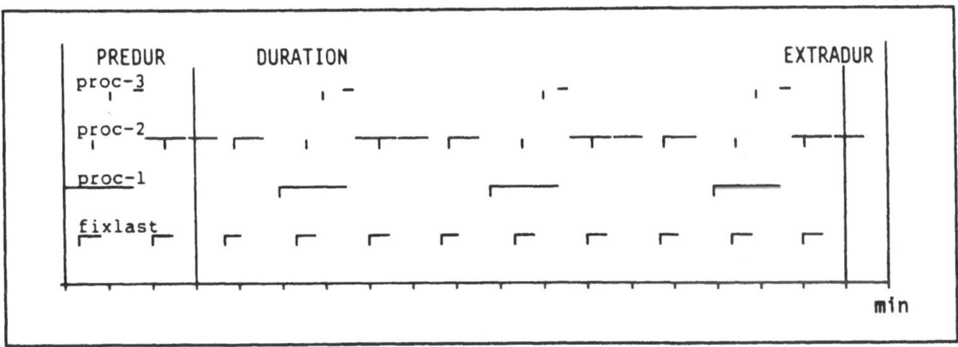

Abbildung 4.1: Beispiel eines Simulationsablaufes

5. Auswertung der Simulationsergebnisse

Mit Hilfe von Lastsimulationen soll zum einen das Verhalten des Systems unter der vorgegebenen Last beurteilt werden, zum anderen sind die Auswirkungen der Gesamtlast auf die einzelnen Lastprozesse von Interessse. Ersteres kann z. B. mit Hilfe des Softwaremonitors sar und einer Auswertung der von ihm gelieferten Daten erfolgen. Um den zweiten Punkt abzudecken, werden die Antwort- und CPU-Zeiten der einzelnen Prozesse ermittelt. Für diese Daten ist im Benchmarkmonitor eine umfangreiche Auswertung vorhanden, welche mehrere unterschiedliche Aspekte einer

Ergebnisauswertung ermöglicht:

- Ermittlung von statistischen Kenngrößen für die Antwort- und CPU-Zeiten. Hierzu gehören Mittelwert, Varianz, Standardabweichung, Minimum, Median und Maximum, sowie die Angabe verschiedener Quantile. Letztere haben den Zweck, mögliche Ausreißer in den Zeiten erkennen zu können.

- Vergleich der Soll und Ist-Durchsätze für die einzelnen Lastprogrammme. Außerdem werden die vorgegebenen und erreichten Relationen der Durchsätze in den drei möglichen Gruppen ermittelt.

- Auswertung in bezug auf Arbeitsplätze. Hierzu gehören mittlere Antwort- und CPU-Zeiten an den Arbeitsplätzen, Soll und Ist-Vergleich der Durchsätze und der vorgegebenen und erreichten Durchsatzrelationen der Arbeitsplätze untereinander.

- Graphische Darstellung der wichtigsten Kenngrößen.

In den meisten Fällem erstrecken sich die Leistungsbetrachtungen mit BEMO über eine Reihe von Simulationsläufen mit unterschiedlichen Lastaufkommen. So möge etwa unser Beispiel eine Spitzenlastsituation darstellen. Zunächst wird die Spitzenlast simuliert, um zu überprüfen, ob die Leistungsfähigkeit des Systems überhaupt ausreicht. Laut Abbildung 5.1 (s. Throughput) ist diese Last auf dem System lauffähig. Der Normalfall sei dagegen eine im Durchsatz halbierte Last bei den Arbeitsplatzprozessen. Für diesen Fall interessieren im Wesentlichen die Antwortzeiten. Um diese 'Normallast' zu beschreiben, sind in der bemotab nur die Parameter der Wahrscheinlichkeitsverteilungen anzupassen. Die Abbildungen 5.2 und 5.3 zeigen die Antwortzeiten für einen Simulationslauf mit der 'Spitzenlast'-bemotab (Abb. 3.1, Level a), während Abbildung 5.4 die Antwortzeiten für den 'Normallastfall' beinhalten.

In Abbildung 5.1 sind die Durchsatzstatistiken für die einzelnen Prozesse bzw. Arbeitsplätze aufgeführt. Die Angaben zu 'Relation' dienen zur Kontrolle der Durchsatzverhältnisse: Unter 'in' sind die vorgegebenen und unter 'out' die tatsächlich erreichten Verhältnisse angegeben. Für unser Beispiel sind diese eingehalten worden. Ebenso sind auch die vorgegebenen absoluten Durchsätze (s. Debit) erreicht worden (s. Credit). Der Prozentsatz der erfolgten Lastprogrammläufe erreicht zwar nicht in allen Fällen 100 Prozent (s. Throughput), doch da zur Berechnung der Durchsatzvorgaben die Mittelwerte der Verteilungen herangezogen wurden, ist hier mit kleinen Schwankungen zu rechnen. Zu berücksichtigen ist zudem, daß durch die Zusammenfassung der Prozesse aus der variablen Gruppe zu zwei Arbeitsplatzgruppen,

Load Statistics

Name	Grp	WS	Relation in	Relation out	Debit	Credit	Throughput	Distribution
fixjob-	Fix	-	1.00	1.00	180	180	100.00%	CONST
user2-5	Var	02	1.00	1.00	120	118	98.33%	N(150.0,5.0)
user2-3	Var	02	1.00	1.00	120	118	98.33%	N(150.0,5.0)
user2-1	Var	02	2.77	2.78	333	328	98.50%	L(4.0,0.1)
user1-3	Var	01	0.50	0.50	60	59	98.33%	U(290.0,310.0)
user1-2	Var	01	5.17	5.15	620	608	98.06%	N(29.0,1.0)
user1-1	Var	01	0.50	0.51	60	60	100.00%	U(290.0,310.0)
Global					1493	1471	98.53%	

Workstation Statistics

WS	Relation in	Relation out	Debit	Credit	Throughput	Cumulative Average Times real	user	sys
01	1.00	1.00	740	727	98.24%	43.01	22.91	1.01
02	0.77	0.78	573	564	98.43%	46.27	22.12	1.08
Global			1313	1291	98.32%	44.64	22.51	1.04

Abbildung 5.1: Durchsatzstatistik der einzelnen Lastprozesse und der Arbeitsplätze

zu keinem Zeitpunkt im Simulationszeitraum mehr als drei der Lastprozesse aktiv waren. Es ist dabei (vgl. Abb. 4.1) im Simulationsverlauf zu Startverzögerungen von einigen Prozessen gekommen, die zu einem späteren Simulationszeitpunkt wieder abgebaut waren. Hierzu sind auch entsprechende Einträge in der Ablaufverfolgungsdatei vorhanden. Da der vorgegebene Durchsatz schon im 'Spitzenlastfall' eingehalten wurde, ist dies beim 'Normallastfall' natürlich auch gegeben.

Die Ergebnisse in Abbildung 5.2 und 5.3 zeigen die größte Streuung der Antwortzeiten in den beiden Arbeitsplatzgruppen bei 'user1-1' bzw. 'user2-1'. Aus den Quantil-Angaben läßt sich darüberhinaus entnehmen, daß bei 'user1-1' diese Streuung hauptsächlich durch relativ wenige Ausreißer nach oben verursacht wurde, während bei 'user2-1' ein hoher Anteil von Prozessläufen mit längeren Antwortzeiten vorhanden ist. Dies ist auf den ersten Blick verwunderlich, da in beiden Fällen derselbe Lastprozeß (vgl. Abb. 3.1) zur Anwendung kommt. Eine Begründung ist jedoch darin zu finden, daß zum einen die Aufrufhäufigkeit von 'user1-1' erheblich geringer ist, und zum anderen 'user2-1' mit lastintensiveren Prozessen konkurrieren muß. Die Angaben in Abbildung 5.4 für den 'Normallastfall' bestätigen dies. Zwar sind wie erwartet die Antwortzeiten und deren Streuungen geringer geworden, doch die

Name	Grp	WS	Mean	Std.Dev	Min	Quantile 80%	90%	95%	100%
fixjob-	Fix	-	4.44	1.75	2.10	6.13	6.18	7.00	9.00
user2-5	Var	02	15.48	2.99	12.28	18.50	19.58	20.10	22.05
user2-3	Var	02	9.38	2.52	6.03	11.20	12.23	12.73	14.33
user2-1	Var	02	21.41	3.03	11.92	23.12	24.23	24.45	25.18
user1-3	Var	01	10.30	2.37	6.02	12.03	12.35	13.00	14.53
user1-2	Var	01	17.42	3.84	12.12	20.82	21.98	22.68	24.02
user1-1	Var	01	15.29	4.43	12.02	16.58	23.27	23.45	26.57
Average			13.38	3.10	8.92	15.48	17.11	17.63	19.38

Time Statistics for real times in seconds

Abbildung 5.2: Auswertung der Antwortzeiten des Beispiels

```
       Graph for real times
------------------------------------------------------------------------
sec          |0             7.7             15.5           23.2        31|
fixjob-  Fix |   -----a###*+----             .              .            |
user2-5  Var |              .        ------a#######**+---- .             |
user2-3  Var |           -------a####**+--- .               .            |
user2-1  Var |              .       --------------------a###**+-         |
user1-3  Var |           ---------a###*+----.               .            |
user1-2  Var |              .        ----------a#######***+---           |
user1-1  Var |              .       -------a###**************+------     |
------------------------------------------------------------------------
 (a: Mean, - left: Min, - right: Qnt 100%, #: Qnt 80%, *: Qnt 90%, +: Qnt 95%)
```

Abbildung 5.3: Grafische Darstellung von Minimum, Mittelwert und Quantilen der Antwortzeiten

Time Statistics for real times in seconds

Name	Grp	WS	Mean	Std.Dev	Min	Quantile 80%	90%	95%	100%
fixjob-	Fix	-	3.15	1.56	2.07	4.12	6.10	6.13	7.27
user2-5	Var	02	14.69	2.69	12.25	17.05	19.28	19.78	20.20
user2-3	Var	02	8.53	2.22	6.05	9.870	11.18	12.27	13.32
user2-1	Var	02	19.50	3.91	11.95	22.15	23.45	23.90	24.75
user1-3	Var	01	6.20	0.45	6.00	6.15	6.18	6.20	8.52
user1-2	Var	01	16.73	3.75	10.58	20.37	21.27	21.83	24.53
user1-1	Var	01	12.01	0.57	11.27	12.42	12.62	12.67	13.63
Average			11.54	2.52	8.59	13.16	14.29	14.68	16.03

Abbildung 5.4: Auswertung der Antwortzeiten des Beispiels mit halbierten Durchsatzvorgaben

Differenz zwischen den Mittelwerten der beiden Prozesse ist in etwa gleichgeblieben. Insgesamt gesehen ist die Verringerung der Antwortzeiten beim geringeren Lastaufkommen nicht sonderlich groß, so daß in allen Lastfällen bis hin zum 'Spitzenlastfall' mit nicht wesentlich verschiedenen Antwortzeiten gerechnet werden kann.

Zum Abschluß noch eine Bemerkung zu dem von BEMO erzeugten Overhead bei einer Simulation. Dieser Wert hängt selbstverständlich von der Anzahl der zu erzeugenden Prozesse und den zur Bestimmung der Neustartabstände verwendeten Verteilungen ab. Im allgemeinen benötigt BEMO jedoch zwischen 0.1% und 0.5% der CPU-Zeit. Für unsere Beispiele betrug der Verwaltungsoverhead von BEMO 0.29% bzw. 0.13% der CPU-Zeit.

6. Ausblick

Der Benchmarkmonitor BEMO ist zur Zeit lauffähig auf allen Targon-Systemen der Nixdorf Computer AG. Beim fehlertoleranten Multicomputer Targon/32 ist es zudem möglich, eine Verteilung der Last auf die vorhandenen Cluster zu veranlassen und die Lastprozesse verschiedenen Fehlertoleranzklassen zuzuordnen. Darüberhinaus ist BEMO auch schon erfolgreich auf den Systemen anderer Hersteller gelaufen.

Für den Monitor sind noch eine Reihe von Erweiterungen beabsichtigt. Dazu gehört z.B. das Bereitstellen von weiteren Elementarfunktionen, etwa zur Belastung von LAN's oder zur Nachbildung von Lokalitätseigenschaften. Weiterhin soll es für lange Simulationsläufe möglich sein, die Zeiten nicht für jeden Aufruf eines Lastprogrammes einzeln zu erhalten, sondern das arithmetische Mittel über jeweils n Aufrufe (Online-Analyse). Dadurch wird der Speicherbedarf von BEMO für die gemessenen Zeiten reduziert.

7. Literaturverzeichnis

[BEMO89] Meyer, L.:
BEMO: Ein Benchmarkmonitor für UNIX* Betriebssysteme
Nixdorf Dokument, Ausgabe 1.2, Januar 1989

[HINN84] Hinnant, D.:
Benchmarking UNIX Systems,
aus: Byte, August 1984

[JANI??] Janich, E:
Benchmark auf UNIX-Systemen,
wie mißt man die Leistung von UNIX-Systemen wirklich?
Interner Text der Universität Ulm, Sektion Informatik

[KARP85] Karpinski, R.:
Paranoia: A Floating-Point Benchmark,
aus: Byte, Februar 1985

[TOTZ87] Totzauer, G.:
Kopplung der Kettendurchsätze in geschlossenen
Warteschlangennetzwerken
Proceedings Messung, Modellierung und Bewertung
von Rechensystemen 1987
Inf. Fachbericht 154, Springer Verlag

[WEIC84] Weicker, R. P.:
Dhrystone: A synthetic system programming Benchmark,
aus: Comm. ACM, Vol. 27, No.10, pp 1013-1030, 1984 (10)

Lockperformance in Betriebssystemen

J. Cuellar, E. Schicker
Siemens AG
Otto-Hahn-Ring 6
8000 München 83

1 Einleitung

Im Multitasking-Betrieb eines Betriebssystems werden Betriebsmittel wie Prozessoren, Arbeitsspeicher oder auch Verwaltungsdaten nach Anforderung verteilt. Dies erfordert Koordinierung und Serialisierung in kritischen Pfaden. Die Realisierung erfolgt über Lockmechanismen.

Im Universalrechner-Betriebssystem BS2000 ist die Koordinierung der Prozessoren über Spin-Locks, die Koordinierung der Tasks über Suspend-Locks realisiert. Suspend-Locks werden auch dem Anwender, z.B. der Datenbankentwicklung, zur Verfügung gestellt.

Um den Einfluß der Locks auf das Systemverhalten besser zu verstehen und die Lockverluste quantitativ zu bestimmen, wird die Modellierung eingesetzt. Die neuen Erkenntnisse über das Lockverhalten wurden am BS2000 verifiziert und entsprechende Verbesserungen implementiert.

In Kapitel 2 werden Spin-Locks eingehend untersucht. Es wird gezeigt, daß die Varianz neben der Lockwahrscheinlichkeit einen deutlichen Einfluß auf die Lockverluste hat. Kapitel 3 beschäftigt sich mit Suspend-Locks, bei denen der overheadträchtige Taskwechsel, - als Folge eines Lockmisses -, eine wichtige Rolle spielt. Kapitel 4 enthält schließlich Verbesserungsvorschläge und die damit gewonnenen praktischen Erfahrungen.

In der Einleitung werden hier zunächst die verwendeten Modellierungstools vorgestellt und dann eine Übersicht über die benutzte Notation gebracht.

Für die Lockuntersuchungen wurde meist wie folgt vorgegangen: es wurden einfache Modelle überlegt und die Resultate mit realen Messungen verglichen. Manche Modelle ließen sich mathematisch lösen, meist mit nicht zu komplexen Markov-Ketten, andere Modelle wurden simulativ oder analytisch numerisch gelöst. Hier stand das Modellierungstool HIT zur Verfügung, das an der Universität Dortmund entwickelt wurde (siehe [BeSt], [BeMW]). HIT besitzt eine objektorientierte Struktur und läßt sich meist einfach und schnell anwenden. Einige simulative Modelle wurden auch mit BORIS erstellt, ein Simulationstool, das bei Siemens in München entwickelt wurde (siehe [MaSch], [MaSch1]). BORIS ist blockorientiert und bei Betriebssystemuntersuchungen oft schwerfälliger als HIT. Sind jedoch komplexe Koordinationen von Komponenten untereinander erforderlich, so zeigen sich die Stärken von BORIS. Beide Tools ergänzen sich daher gut.

Zuletzt sei noch die Notation der wichtigsten im Text verwendeten Symbole aufgelistet:

LT	mittl. Lockstrecke (Lock Time)
LF	mittl. Strecke ohne Lock (Lock Free Time)
p	Lockwahrscheinlichkeit (:= LT / (LF + LT))
s^2	normierte Varianz (Varianz / LT^2)
CS	Aufwand für Context Switch
n	Anzahl der Prozessoren
m	Anzahl der Tasks

2 Spinlocks

Im Nukleus eines Betriebsystems kommt es häufig vor, daß ein Prozessor bei Anforderung eines Betriebsmittels (z.B. eine zentrale Tabelle für Task- oder Speicherverwaltung) dieses auch unbedingt braucht, um weitere produktive Arbeit leisten zu können. Aus diesem Grunde, oder einfach weil die Kosten eines Kontextswitches teuerer als das Warten auf die Betriebsmittelfreigabe wären, sind in vielen Multiprozessorsystemen die meisten Nukleuslocks Spinlocks.

Ein Modell für Spinlocks ist üblicherweise das Maschinen-Interferenz-Modell (siehe [MaDo],[Gi],[HoSch],[BiVe],[Cu]). Dieses Modell ist eines der bekanntesten Warteschlangennetze und hat viele Anwendungen gefunden (siehe [Ta],[Ja],[Ke],[La],[Pr],usw.). Es besteht aus einem Zyklus mit einer Bedieneinheit vom Typ Delay (oder Unendlich-Server) und einem FCFS-Server (entspricht dem Lock). Die n Kunden können dann als n Prozessoren interpretiert werden, die zwischen einem lockfreien Zustand und einem Lockhaltezustand hin und her in einer Schleife kreisen. Der Zeitanteil, den ein Prozessor in der FCFS Queue verbringt, ist der Verlust des Systems.

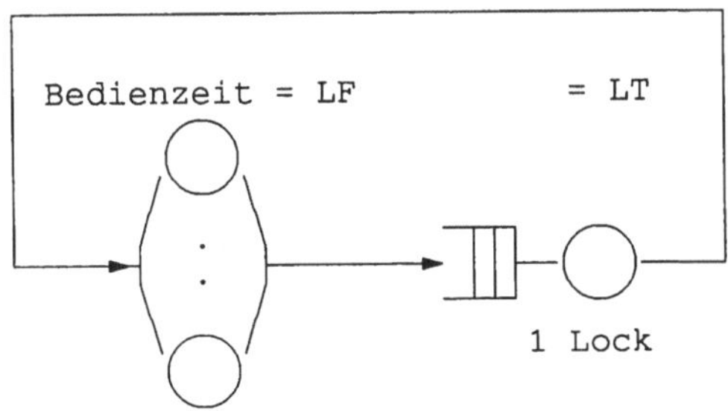

Die Formeln von Takacs-Jaiswal ermöglichen, den Verlust als eine Funktion von n, p und der Laplace-Transformation der Verteilung der Lockstrecken darzustellen ([Ta],[Ja]).Die Voraussetzungen dabei sind, daß einerseits alle Lockfrei- und Lockhaltezeiten untereinander unabhängig sind und anderseits zumindest eine dieser Bedienzeiten exponentiell verteilt ist. In praktischen Anwendungen ist diese Formel eine Näherung, da diese Bedingungen nicht gegeben sind: die einzelnen Bedienzeiten sind nicht unabhängig, deren Verteilung ist weit entfernt von exponentiell und ändert sich mit der Zeit, und die unterschiedlichen Locks in einem System beeinflussen sich gegenseitig. Auch wegen der geringen Transparenz dieser Formel ist es wünschenswert, einfachere Ausdrücke für den Verlust zu finden, die einen besseren Einblick in die relevanten Variablen und deren Effekt ermöglichen.

Takacs Formel ergibt als Grenzwert, wenn (fast) alle Lockhaltezeiten klein im Vergleich zur mittleren lockfreien Strecke sind, folgende Annäherung für den Verlust:

$$V \approx (n-1)\, p^2\, (S^2 + 1)/2 \tag{2.1}$$

Daß die Verluste proportional zu n-1 sind, ist unseres Wissens bislang noch unerkannt geblieben, obwohl es eine sehr einfache Erklärung dafür gibt. Betrachten wir dazu zwei aus den vorhandenen n Prozessoren und und interessieren wir uns für die Zeiten, in denen der erste Prozessor auf den zweiten wartet. Ist V' dieser Zeitanteil, so entspricht V' in etwa dem Verlust in einer entsprechenden Bi-Prozessor Anlage. Andererseits folgt aus der Symmetrie der ursprünglichen Konfiguration mit n Prozessoren, daß dieser Prozessor insgesamt (n-1)-mal diesen Zeitanteil V' gewartet hat, und damit, daß der Gesamtverlust V in etwa (n-1)V' ist.

Implizit wurde dabei angenommen, daß die Verluste klein sind. Das liegt daran, daß das Warten auf mehr als eine Lockfreigabe (weil mehrere Prozessoren in der Queue sind) nicht berücksichtigt ist. Diese zusätzlichen Verluste sind auch in der Formel (2.1) nicht enthalten. Vielmehr ist der Verlust in den für die Praxis relevanten Bereichen eine leicht konvexe Funktion von n, so daß er etwas stärker als linear mit (n-1) anwächst.

Der Faktor $(S^2 + 1)/2$ ist einleuchtend, wenn man sich den Beweis des Satzes von Pollaczek-Khinchine vergegenwärtigt: stößt ein Prozessor auf einen belegten Lock, so ist die mittlere Zeit, die er auf die Freigabe warten muß (die Restbedienzeit des Locks), gleich $LT \cdot (S^2 + 1)/2$. Einschränkend muß man sagen, daß die langen Lockstrecken die Varianz überproportional erhöhen, die Verluste jedoch nicht so stark zunehmen als dies Formel (2.1) vorgibt. Durch sehr lange, seltene Lockstrecken kann S beliebig groß anwachsen, aber der Verlust bleibt in jedem Zeitintervall durch (n-1)/n beschränkt, da höchstens n-1 Prozessoren auf den Lock warten. Das entspricht der bekannnten Tatsache, daß die Varianz einer Bedienzeit sich bei geschlossenen Netzen weniger als bei offenen auswirkt, wie es Buzen bemerkte ([La], [Pr]). Rührt allerdings diese Varianz von der Präsenz vieler kleiner Lockstrecken unterschiedlicher Länge her, so geht der Faktor $(S^2 + 1)/2$ voll in die Verluste ein.

Weiterhin haben wir durch zahlreiche Simulationen festgestellt, daß für alle uns bekannten realistischen Situationen die Verteilung der lockfreien Strecken ziemlich irrelevant ist, solange sie stätionär ist, d.h. wenn sie ihre Form mit der Zeit nicht ändert. In der Praxis sind leider die Verteilungen für LT und LF zeitabhängig. Dies hat zur Folge, daß sich die Lockwahrscheinlichkeit zeitlich verändert, wodurch sich Lockanforderungen häufen, sich die Konfliktsituationen vermehren und die Verluste wachsen. Es gibt sehr viele Gründe, warum solch eine zeitliche Verteilungsänderung der Lock- und der lockfreien Strecken in den Betriebssystemen zu beobachten ist. Es werden z.B. Betriebsmittel nach einem bestimmten Ereignis oder beim Erreichen eines Schwellwertes häufiger als sonst angefordert oder das Freigeben eines Task-Locks löst eine Lawine von Umkettungen aus.

Hinzu kommt, daß sich die Verluste der einzelnen überlappenden Locks im System gegenseitig beeinflussen: sie "verschieben" sich (zum Teil) von einem Lock zum anderen und nehmen linear mit der Überlappung ab, wie wir aus analytischen Modellen und Simulationen errechnet haben.

Wir haben einen guten Einblick in die performancerelevanten Effekte angeboten, aus denen man auf die möglichen Maßnahmen zur Begrenzung der Verluste zurückschließen kann. Weiterführend verweisen wir auf [Cu], wo ein Ansatz zur Methodologie zeigt, wie sich die nicht in der Formel (2.1) berücksichtigten Variablen quantitativ auswirken (Existenz sehr langen Lockstrecken, zeitliche Veränderung der Verteilungen, Überlappung von Locks).

Um eine Vorstellung über die Qualität der Formel (2.1) zu geben, geben wir in der folgenden Tabelle für zwei verschiedene Lasten die Summe der gemessenen und berechneten Spinlockverluste auf einem Quadroprozessor an:

	Gesamtverlust gemessen	Gesamtverlust nach (2.1) berechnet
LAST 1	4.49 %	3.15 %
LAST 2	4.32 %	4.79 %

Die Abweichungen sind - sowohl qualitativ als auch quantitativ - zum guten Teil auf die oben genannten Effekte zurückzuführen.

3 Suspendlocks

Während bei Spinlocks in einer Schleife auf Lockfreigabe gewartet wird, verliert bei Suspendlocks der Aufrufer die Kontrolle der CPU, sein gesamter Kontext wird gesichert. Der Aufrufer ist im BS2000 eine Task, im Falle eines Lockmisses findet ein Taskwechsel statt. Weitere den Lock anfordernde Tasks werden nach FCFS eingeordnet. Der Vorteil der Suspendlocks liegt darin, daß die CPU nicht blockiert wird, als Nachteil schlagen die direkten Kosten des Taskwechsels und eventuell indirekte, wie partieller Verlust des Working Sets während der Wartezeit, zu Buche.

Bei beiden Lockarten wachsen die Lockverluste quadratisch mit der Lockwahrscheinlichkeit p. Die durch Suspendlocks verursachten Verluste reagieren jedoch auf die Varianz der Lockstrecken kaum. Dies ist einsichtig, wenn man sich erinnert, daß die CPU sofort nach dem Contextswitch weitere Last aufnehmen kann. Negativ gehen nur die Kosten desselben ein.

Zur Veranschaulichung nehmen wir folgenden einfachen Fall an: es liege ein Monoprozessor vor, auf dem das Betriebssystem arbeitet. Schon hier treten Suspendlock-Verluste auf, da Tasks, die einen Lock halten, die CPU verlieren können, sei es wegen IO oder Paging oder einfach durch Verdrängung. Ferner existiere ein Lock, und es befänden sich im System m Tasks, die den Lock potentiell benutzen können, von denen respektive m1, m2 bzw. m3 in der CPU-Queue, auf einen Page-In bzw. auf eine IO-Termination warten (m1 + m2 + m3 = m).

Gehen wir zunächst davon aus, daß keine Staus vor Locks bestehen. Auf diesen Fall gehen wir später ein. Pro Lockanforderung ist dann der Lock proportional zu m1·p von einer Task in der CPU-Queue belegt. In diesem Fall ist ein Contextswitch erforderlich, der das System mit CS Instruktionen belastet. Der Verlust je Lockanforderung, verursacht durch eine lockhaltende Task in der CPU-Queue, beträgt somit durchschnittlich C1·m1·p·CS. In Analogie für das Warten auf Page-In und IO-Termination ist der Verlust pro Lockanforderung insgesamt

$$V_{LA} = K\ p\ CS$$

mit K = C1·m1 + C2·m2 + C3·m3 und C1,C2,C3 als Proportionalitätskonstanten, und der Gesamtverlust ergibt sich zu

$$V_{mono} = V_{LA} / (LF + LT + V_{LA}).$$

Ist nun die Lockwahrscheinlichkeit klein, etwa $(m-1)p < 0.1$, so gilt näherungsweise

$$V_{mono} = K\ CS\ p^2 / LT.$$

Betrachten wir nun den Multiprozessorfall. Als neuer Effekt kommt hinzu, daß eine Task einen Lock anfordert, während auf einer anderen CPU bereits eine Task unter diesem Lock läuft. Um die dadurch verursachten Verluste qualitativ zu erfassen, treffen wir folgende idealisierte Vereinbarungen:

- eine lockhaltende Task läuft ohne Unterbrechung bis zur Lockfreigabe,
- eine auf einen Lock wartende Task wird bei Freigabe des Locks sofort Task in Control an einer CPU, notfalls durch Verdrängung,
- dieses Verdrängen geschieht ohne Zeitverzögerung,
- LT und LF sind exponentiell verteilt.

Die ersten zwei Bedingungen gewährleisten, daß die auf Monoprozessoren auftretenden Verluste nicht nochmals mitgezählt werden. Die anderen gestatten es, ein einfaches Markov-Modell aufzustellen, das explizit lösbar ist.

Bezeichne (I,J) den Zustand:

> es gibt I Tasks, die den Lock zur Zeit weder halten noch wollen, und J (= 0 oder 1) Tasks, die den Lock halten. Ist J = 1, dann ist die lockhaltende Task an der CPU aktiv.

Dann bildet die Kette von Übergängen zwischen diesen Zuständen eine Markov-Kette:

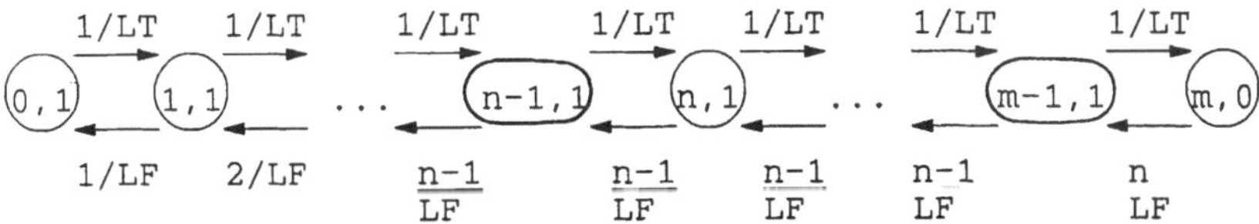

Die Transitionen nach rechts ergeben sich aus der Wahrscheinlichkeit, daß der Lock freigegeben wird. Diese ist umgekehrt proportional zu LT. Die Übergänge nach links verhalten sich umgekehrt proportional zu LF und proportional zu der Anzahl der Tasks in Control, die einen Lock anfordern können. Aus dieser Markov-Kette erhalten wir die Zustandsgleichungen

$$p_{i-1} / LT = i\, p_i / LF \quad , \quad i = 1 \ldots n-1,$$

$$p_{i-1} / LT = (n-1)\, p_i / LF \quad , \quad i = n \ldots m-1,$$

$$p_{m-1} / LT = n\, p_m / LF.$$

mit p_i als stationärer Wahrscheinlichkeit von (I,J).

Alle Transitionen nach links sind Lockanforderungen und bis auf den Übergang vom Zustand (m,0) nach (m-1,1) sind alles Lockmisses. Daraus berechnet sich der Anteil der Lockmisses an den Lockanforderungen zu

$$1 - \frac{n}{C} \frac{p_m}{1-p_m}$$

mit $C = LF/LT$.

Aus den obigen Zustandsgleichungen zusammen mit $\Sigma p_i = 1$ ergibt sich

$$\frac{1}{p_m} = 1 + \frac{n}{C-n+1} - \left[\frac{n-1}{C}\right]^{m-n} \cdot \left[\frac{n}{C-n+1} - \frac{n!}{C^n} \cdot \sum_{k=0}^{n} \frac{C^k}{k!}\right]$$

Daraus läßt sich der durch Lockmisses verursachte Verlust für große m berechnen:

$$V_{multi} \longrightarrow \frac{(n-1)\, CS}{C(C+1)LT + CS(n-1)} \qquad \text{für } m \rightarrow \infty .$$

Ist nun die Lockwahrscheinlichkeit p klein, so läßt sich obige Formel weiter durch

$$V_{multi} = (n-1)\, CS\, p^2 / LT \qquad (3.1)$$

approximieren. Völlig überraschend sind auch die multiprozessorspezifischen Lockverluste proportional zu p^2/LT.

Die diese theoretischen Überlegungen begleitenden Simulationen bestätigen diese Ergebnisse eindrucksvoll. Einen Überblick über das Verhalten von Tasklocks auf Quadro-Prozessoren ergibt folgendes Bild. Es zeigt den Leistungsfaktor in Abhängigkeit von der Länge der Lockstrecken LT, aufgetragen für mehrere Lockwahrscheinlichkeiten p. Ermittelt wurden diese Werte mit einem BORIS-Modell.

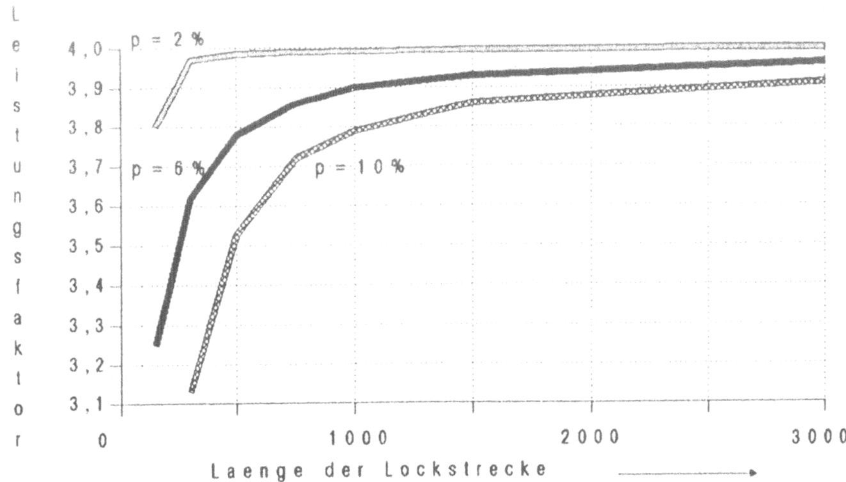

Fassen wir die beiden Teile zusammen, so ergibt sich als Gesamtverlust

$$V_{ges} = V_{mono} \text{ (IO-Rate, Paging-Rate, Verdrängung) } + V_{multi} (n) \, .$$

V_{mono} wächst mit der Anzahl der Tasks, steigt also mit der Leistung der Rechenanlagen, da auf schnellen Rechnern meist auch mehr Tasks ablaufen. V_{multi} hängt dagegen vor allem von der Anzahl der Prozessoren ab.

Es ist zwar schwer, bei Messungen die Anteile von V_{mono} und V_{multi} zu trennen, aber doch möglich. Wir hatten dazu folgende Meßanordnungen gewählt: zunächst wurde an einer Bi-Prozessor-Anlage mit je 15 MIPS Leistung und einer konfliktreichen Last von 400 Tasks die Anzahl der Lockmisses gemessen; es waren ca. 12000 pro Minute. An einem doppelt so schnellen vergleichbaren Mono-Prozessor wurden nur ca. 8000 gezählt. V_{multi} war damit bei dieser Meßanordnung zu etwa 1/3 am Gesamtverlust der Suspendlocks beteiligt, hochgerechnet auf Quadro ergeben sich wegen (3.1) bereits über 50%.

Messungen am realen System bestätigen außerdem, daß sich die Lockverluste proportional zu p^2/LT verhalten. Dies zeigen wir an einem Beispiel. Ein in bestimmten Extremsituationen konflikt-reicher Lock wurde wie folgt geändert:

Zwei Lockstrecken mit LT $\approx$ 500 Befehle wurden zusammengefaßt. Wegen der zwischen diesen Lockstrecken liegenden Strecke LF1 ≈ 600 Instruktionen ergibt sich daraus der neue Wert für LT zu ca. 1600 Befehle. Trotz einer damit verbundenen Erhöhung der Lockwahrscheinlichkeit konnte kein Ansteigen der Verluste beobachtet werden, was unsere Erwartung bestätigte. Im Gegenteil, da pro Zyklus ein Anfordern und Freigeben eines Locks wegfällt, konnte sogar eine CPU-Einsparung gemessen werden.

Auf die meisten Locks treffen die bisher festgestellten Effekte zu, einige Locks werden jedoch durch Konvoi-Effekte überlagert. Begünstigt werden Konvois durch folgenden Fall: Um die Lockwahrscheinlichkeit klein zu halten, wird der Lock nur über kritische Pfade gehalten, zwischendurch aber wenn möglich wieder freigegeben. Wir haben Programme gefunden, die einen Lock mehr als 10mal nacheinander holen und wieder freigeben. Ruft nun eine Task A ein solches Programm gelegentlich, aber regelmäßig auf, so ist die Lockwahrscheinlichkeit klein, LF ist aber äußerst ungleichmäßig verteilt. Behält Task A während des Programmablaufs die Kontrolle über die CPU, verhält sich das System gutartig, zumindest auf Monoprozessoren. Wird A aber verdrängt, und wird anschließend der belegte Lock von einer Task B angefordert, so muß B auf Lockfreigabe warten. Gibt später Task A den Lock zurück und benötigt ihn kurze Zeit später wieder, so muß A warten, da nun B Lockinhaber ist. Dieses gegenseitige Wegnehmen des Locks dauert so lange, bis eine der beiden Tasks den Lock für längere Zeit nicht mehr benötigt. Dieser dauernde Taskwechsel verursacht Systemoverhead, und die beiden Tasks selbst verweilen sehr lange im System. Dadurch steigt aber die Wahrscheinlichkeit, daß sich eine bisher unbeteiligte Task C diesem Spiel anschließt.

Dies zeigt, daß das Problem nicht so sehr darin liegt, daß Konvois entstehen, sondern vielmehr darin, **daß sich einmal gebildete Konvois nur sehr langsam auflösen.** Das wiederum führt zu einer dauerhaften Belastung des Systems.

Ein solches Konvoiproblem ist nicht zu unterschätzen. Konvois treten zwar oft erst ab Lockwahrscheinlichkeiten von einigen Prozent auf, was bei Suspendlocks selten ist. Kritisch kann es allerdings sein, wenn Locks über IO gehalten werden. Auch trügt die Angabe der aus einer längeren Messung ermittelten Lockwahrscheinlichkeit p. Liegt p beispielsweise bei unter 1%, so wiegt man sich in Sicherheit. Um so überraschter wird man sein, wenn gerade dieser Lock Ursache von Konvois ist. Dies liegt dann meist an einer zeitabhängige Verteilung der Lockwahrscheinlichkeit: der Lock wird nur selten geholt, zwischendurch aber häufig hintereinander. Dies ist in einem Multitaskingsystem keine Seltenheit. Dadurch entsteht in solchen konfliktreichen Momenten ein Konvoi, der sich aber erst auflöst, wenn die Lockanforderungsrate einen gewissen Minimalwert über einen längeren Zeitraum unterschreitet. Dieser Schwellwert kann unter Umständen so klein sein, daß sich einmal gebildete Konvois erst nach größeren Lastverschiebungen auflösen.

Unsere bisherigen Erfahrungen zeigen, daß die Empfehlung, nicht p sondern p^2/LT zu minimieren, für Tasklocks wichtig und meist auch optimal ist. Diese Empfehlung bedeutet, LT so groß wie möglich zu wählen, solange p nicht merkbar größer wird. Dies hat zur Folge, daß es pro Zeiteinheit weniger Lockanforderungen gibt. Die Empfehlung läßt sich daher wie folgt äquivalent formulieren:

Bei Suspendlocks ist das Produkt aus Lockwahrscheinlichkeit und Anzahl der Lockrequests pro Zeiteinheit zu minimieren.

Wegen der sinkenden Anzahl der Lockanforderungen bei etwas steigender Lockwahrscheinlichkeit lassen sich daher in Systemmodulen und Programmen Lockstrecken ohne Lockverluste zusammenlegen. Somit lassen sich sogar die oben beschriebenen Konvoiprobleme zumindest reduzieren. Weitergehende Vorschläge zu Tasklocks vor allem zur Konvoivermeidung finden wir im nächsten Kapitel.

4 Maßnahmen und Vorschläge

Über die bereits in den vorigen Kapiteln vermittelten Eindrücke hinaus, werden hier Maßnahmen zur Reduktion von Lockverlusten vorgeschlagen. Betrachten wir die häufig recht hohe Varianz bei Spinlocks. Sie läßt sich oft verringern, indem Systemtabellen oder Warteschlangen so umstrukturiert werden, daß alle Zugriffe darauf etwa die gleiche Pfadlänge besitzen. Routinen, die einen Lock viel länger als der Durchschnitt halten, besitzen nicht selten Konsistenzpunkte. An diesen sollte der Lock kurz freigegeben werden, zumindest dann wenn andere warten. Allein dadurch lassen sich meßbare Erfolge erzielen.

Weiter sollten wenn möglich Locks "zerschlagen" werden. Eine Vorgehensweise ist die Einführung von Super- und Sublocks. Normalerweise wird ein Lock benutzt, um eine Familie von Betriebsmitteln zu sperren, die häufig zusammen (z.B. paarweise) benötigt werden. Auch jede Tabelle oder Queue kann als eine (vielleicht "künstliche") Vereinigung von Subtabellen oder Subqueues angesehen werden. Ist man in der glücklichen Lage, daß jede Routine im Normalfall nur wenige Elemente dieser Betriebsmittelfamilie unbedingt braucht, kann man für jeden einzelnen einen Sublock einführen. Auf konventionelle Weise ist dies aber in der Regel imperformant. Eine Alternative sollte drei Bedingungen erfüllen: Jede Multi-Lock-Anforderung darf nur einmal einen komplexen Befehl wie Compare and Swap oder Lockmanageraufruf benutzen; zweitens soll der parallele Zugriff auf disjunkte Teilmengen der Familie gewährleistet sein; und drittens darf es nicht vorkommen, daß jemand ein Betriebsmittel sperrt und anschließend auf die Freigabe anderer benötigter Sublocks wartet.

Die Idee ist nun, ein Bit b für jedes Element BM der Familie einzuführen. b ist ein Sublock, auf den mit einfachen, billigen Befehlen unter dem Schutz eines Superlocks B direkt zugegriffen werden kann. Die Vorgehensweise ist die folgende: möchte z.B. eine Task auf die Mittel BM und BM' zugreifen, muß sie die dazugehörigen Bits b und b' unter dem B-Lock setzen. Etwas ausführlicher könnte das in Pseudocode so aussehen:

```
START: IF b oder b' belegt THEN spin-wait oder suspend;
                                GOTO START;
                                END IF;
set_lock(B);
IF b oder b' belegt THEN release_lock(B);
                         GOTO START;
                         END IF;
set_locks(b,b');
ändere_metainformation;
release_lock(B);
...(* Arbeite mit BM und BM' *)...;
release_locks(b,b');
```

Die skizzierte Lösung hat wesentliche Vorteile: die Parallelität wird erhöht, die Lockwahrscheinlichkeit und die Varianz von B und jedem Sublock wird klein, und durch das geeignete Wählen von Metainformationen wird ein langes Suchen von großen Tabellen oder Queues vermieden. Außerdem läßt sich diese Lösung sowohl für Spin- als auch Suspendlocks anwenden.

Zum Schluß zeigen wir noch, wie Suspendlock-Verluste durch gezielte Maßnahmen im System reduziert und Konvois weitgehend vermieden werden. Die Idee ist einfach: lockhaltende Tasks sollten solange wie möglich nicht von anderen Tasks verdrängt werden. Dazu wurde im Betriebssystem BS2000 für wichtige systeminterne Suspendlocks ein Anschluß zum Scheduler geschaffen. Jede Task, die einen entsprechenden Lock belegt, erhält einen Prioritätszuschlag, so daß die meisten anderen Tasks diese nun relativ hochpriore Task nicht mehr verdrängen können. Um Konvois zu verhindern, wird bei Lockfreigabe der Prioritätszuschlag zurückgenommen und an eine eventuell wartende Task weitergegeben. Sollte sich an diesem Lock wider Erwarten doch ein Konvoi gebildet haben, so wird damit verhindert, daß die erste Task den Lock nach kurzer Zeit gleich wieder anfordert. Statt dessen belegt nun die zweite Task die CPU und gibt schließlich den Lock frei: der Lock ist nicht mehr belegt. Jetzt kann die höherpriore der beiden Tasks auch bei neuen Lockanforderungen ohne Behinderung weiterarbeiten.

Dieser Algorithmus wird nachträglich für die wichtigsten Locks ab BS2000-Version V9.0 implementiert. Messungen an einem Prototypen auf einem BI-Prozessor mit einer konfliktreichen Last ergaben eindrucksvolle Ergebnisse: nur durch obige Maßnahme sank die Anzahl der Lock-Waits von 208 pro Sekunde auf 29 pro Sekunde! Der dadurch eingesparte Overhead verbesserte den Durchsatz meßbar, die Antwortzeiten wurden deutlich verringert.

5 Zusammenfassung

Es wurde gezeigt, daß es sowohl bei Spin- als auch Suspendlocks nicht ausreicht, nur auf die Minimierung der Lockwahrscheinlichkeit hinzuarbeiten. Vielmehr muß der Softwareentwickler bei Spinlocks den Einfluß der Varianz und bei Suspendlocks den Einfluß der Anzahl der Lockrequests berücksichtigen. Mehrere Beispiele veranschaulichten, daß sich diese aus der Modellierung gewonnenen Erkenntnisse auch bei realen Messungen am Betriebssystem BS2000 feststellen lassen. Zusätzliche Vorschläge wie die Einführung von Superlocks und Prioritätszuschläge für lockhaltende Tasks geben Hilfestellungen, wie auch in kritischen Pfaden Lockverluste auf ein akzeptables Minimum reduziert werden können. Wichtig ist, daß die hier vorgestellten Lockeffekte zwar nur an der Betriebssystemsoftware validiert wurden, sie sich jedoch auch auf Anwendersoftware, wahrscheinlich sogar auf Firm- und Hardware, übertragen lassen.

6 Literatur

[BeMW] Beilner,H; Mäter,J; Weißenberg,N: Towards a Performance Evaluation Enviroment: News on HIT. Proc "Modelling Techniques and Tools for Computer Performance Evaluation" Palma de Mallorca, 1988.

[BeSt] Beilner,H; Stewing,F.J: Concepts and Techniques of the Performance Modelling Tool HIT. Proc of the European Simulation Multiconference ESM '87, Wien,1987.

[BiVe] Bier,G.e; Vernon,M.K: Measurement and Prediction of Contention in Multiprocessor Operating Systems with Scientific Application Workloads. Proc of the International Conference on Supercomputing,1988,ACM Press.

[Bla] Blasgen,M; Gray,J; Mitoma,M; Price,T: The Convoy Phenomenon. ACM Oper Sys Rev 13(2),1979.

[Cu] Cuellar,J: Performance of Spin Locks, eingereicht.

[Gi] Gilbert,D.C: Modeling Spin Locks with Queueing Networks, ACM Oper Sys Rev 12(1),1978.

[HoSch] Hofmann,J; Schmutz,H: Performance Analysis of Locking in Operating Systems. IBM TR 81.03.003,1981.

[HoSch1] Hofmann,J; Schmutz,H: Performance Analysis of Suspend Locks in Operating Systems. IBM J. RES DEVELOP. 26(2), 1982.

[Ja] Jaiswal,N.K.: Priority Queues. Academic Press, NY 1968.

[Ke] Kelley,F.P: Reversibility and Stochastic Networks. John Wiley & Sons, N.Y.,1979.

[La] Lazowska,E.D: The Use of Percentiles in Modeling CPU Service Time Distributions. In: Chandy,K.M; Reiser,M (Eds.): Computer Performance. North Holland, 1977.

[MaDo] Madnick,S.e; Donovan,J.J: Operating Systems.McGraw-Hill, 1974.

[MaSch] Maierhofer,J.; Schmitt,H.: Modelling and Simulation with BORIS - a Block-Oriented Interactive Simulation System. Proc. of the International Conference on Modelling Techniques and Tools for Performance Analysis, Paris, 1984.

[MaSch1] Maierhofer,J.; Schmitt,H.: Modelling and Simulation on the Upper Levels of Computer System Design with BORIS. Proc. of the SCSC, 1984.

[Pr] Price,T.: A Note on the Effect of the Central Processor Service Time Distribution on Processor Utilization in Multiprogrammed Computer Systems. Journal of the ACM, 23(2),1976.

[Ta] Takacs,: Introduction to the Theory of Queues. Oxford University Press, 1962

Techniken zur Leistungsanalyse von Kommunikationssystemen

- Kurzfassung -

Wolfgang Hubig - Dietmar Weber
Siemens AG - München

1 Allgemeines

Der vorliegende Aufsatz beschreibt Techniken zur Leistungsanalyse von Kommunikationssystemen, wie sie bei der Hicom®-Systemfamilie erfolgreich eingesetzt werden. Ein wesentliches Kriterium für die Beurteilung der Leistungsfähigkeit eines derartigen Systems sind die sogenannten Reaktionszeiten. Hierbei handelt es sich um Zeiten, die das System braucht, um auf gewisse Anreize von außen in vorgegebener Art und Weise zu reagieren. Zur Bestimmung dieser Zeiten haben wir Simulationsmodelle für die wesentlichen Systemkomponenten (Teilnehmerverhalten, Kommunikationsrechner) entwikkelt. Diese Simulationsmodelle sind in der Simulationssprache GPSS (General Purpose Simulation System) geschrieben. Sie sind abgestimmt auf eine spezielle Meßmethodik zur Bestimmung von Programmlaufzeiten in Kommunikationsrechnern.

Ein Kommunikationssystem ist in hohem Maße Realzeitanforderungen ausgesetzt und arbeitet demzufolge nur dann ordnungsgemäß, wenn es auf Anreize von außen (Abheben, Wählen etc.) schnell genug reagiert. Dies bedeutet, daß für eine Reihe von Reaktionszeiten Maximalwerte eingehalten werden müssen.

Die Bestimmung von Reaktionszeiten und der Nachweis, daß bestimmte Maximalwerte eingehalten werden, ist eine wichtige Aufgabe der Hersteller solcher Systeme. Hierbei ist man mit folgenden konkreten Problemen konfrontiert:

- Es genügt nicht, Aussagen über Mittelwerte herzuleiten; es werden auch weitere detaillierte Informationen über die auftretenden Verteilungen gefordert (z.B. Quantile).

- Die Einfallprozesse, die in solchen Kommunikationssystemen bearbeitet werden, sind korreliert.

- Um quantitative Aussagen zum Echtzeitverhalten auch in Extremsituationen (z.B. unter Überlast) zu erhalten, muß man das System auch im instationären Fall analysieren.

Eine Berechnung der Reaktionszeiten ist aufgrund der Komplexität des Kommunikationssystems und des Verkehrsangebotes meist nicht durchführbar. Untersuchungen am realen System sind oft schon deshalb nicht möglich, weil es das reale System noch gar nicht gibt. Aber auch wenn das reale System verfügbar ist, sind diese Untersuchungen sehr aufwendig. Somit kann die Bestimmung der Reaktionszeiten meist nur mit Hilfe eines Simulationsmodells erfolgen.

Wir geben im folgenden einen Überblick über die Prinzipien des Simulationsmodells, das wir für die Hicom-Systemfamilie entwickelt haben. Das Simulationsmodell, das in der Simulationssprache GPSS geschrieben ist, umfaßt die Peripherie und die Steuerung des Kommunikationssystems.

2 Die Modellierung

2.1 Nachbildung der Peripherie

Bei der Nachbildung der Peripherie wird der Einfallprozeß so modelliert, daß die zeitliche Korrelation zwischen den Anreizen berücksichtigt wird und daß auch instationäre Phänomene - wie Überlastregelmechanismen - nachgebildet werden können. Der zeitliche Verlauf einer Belegung ist durch die Verteilung der Zwischenankunftsabstände der einzelnen Anreize (Abheben, Wählen etc.) gegeben. Wir beziehen diesen zeitlichen Verlauf in die Modellierung ein, da die Annahme, daß alle Anreize unabhängig voneinander einfallen, zu einer Unterschätzung der auftretenden Reaktionszeiten führt.

Im Simulationsmodell werden Belegungen unabhängig voneinander als GPSS-Transaktionen erzeugt. Von jeder dieser Belegungstransaktionen werden die zugehörigen Anreize als Transaktionen abgespalten und zur Systemsteuerung geschickt. Um Speicherplatzprobleme zu vermeiden, lassen wir die Belegungstransaktionen nur während der Verbindungsauf- bzw. -abbauphase bestehen. Hierbei wird von der Tatsache Gebrauch gemacht, daß man bei negativ exponentiell verteilten Verbindungsdauern das nächste Verbindungsende zu jedem Zeitpunkt neu berechnen kann. Diese Art der Modellierung wird durch GPSS-Sprachelemente unterstützt.

2.2 Nachbildung des Kommunikationsrechners

Bei der Modellierung des Kommunikationsrechners ist man mit der Tatsache konfrontiert, daß dort in der Regel Realzeit-Multitasking-Betriebssysteme eingesetzt werden. Um solch einen Kommunikationsrechner in angemessener Weise nachzubilden, ist es erforderlich

- über eine effiziente Modellierung des Task-Scheduling-Mechanismus zu verfügen und

- die Task-Kommunikation übersichtlich modellieren zu können.

Im Simulationsmodell werden diejenigen Tasks des Kommunikationsrechners, die die vermittlungstechnischen Anreize bearbeiten, mit Programmlaufzeiten und Meldungsverkehr nachgebildet.

Eine explizite Nachbildung des Task-Schedulers ist nicht erforderlich; diese wäre aufwendig in der Entwicklung und rechenzeitintensiv bei Durchführung der Simulationsläufe. Stattdessen lassen wir den in GPSS vorhandenen Verdrängungsmechanismus für Transaktionen auf die Transaktionen, die die Tasks im Modell repräsentieren, in der Weise wirken, wie es in der Realität der Task-Scheduler tut.

Bei einer Erweiterung eines bestehenden Modells um weitere Tasks bleiben die schon vorhandenen Tasks unverändert. Die neuen Tasks werden modular hinzugefügt. Auch die Aufteilung der Tasks auf verschiedene Rechner ist einfach zu realisieren.

Um das Modell zuverlässig und rasch mit dem Datenmaterial der Programmlaufzeiten der beteiligten Tasks zu versorgen, wurde eine spezielle Meßmethode entwickelt.

3 Simulationsergebnisse

Das beschriebene Simulationsmodell wurde benutzt, um die Leistungsfähigkeit verschiedener Hicom-Varianten zu untersuchen. Aus den Simulationsstudien haben wir als Beispiel die Bestimmung der Planungsbelastung eines Kommunikationsrechners und das Systemverhalten unter Überlast herausgegriffen. In der vorliegenden Kurzfassung beschränken wir uns auf die Bestimmung der Planungsbelastung. Unter Planungsbelastung verstehen wir die Rechnerbelastung, bis zu der vom CCITT vorgegebene Maximalwerte für Reaktionszeiten eingehalten werden.

In dem folgenden Bild ist das Verfahren beispielhaft für eine Reaktionszeit, nämlich den Wähltonverzug dargestellt; dies ist die Zeit, die zwischen dem Abheben des Handapparates und dem Anlegen des Wähltons verstreicht.

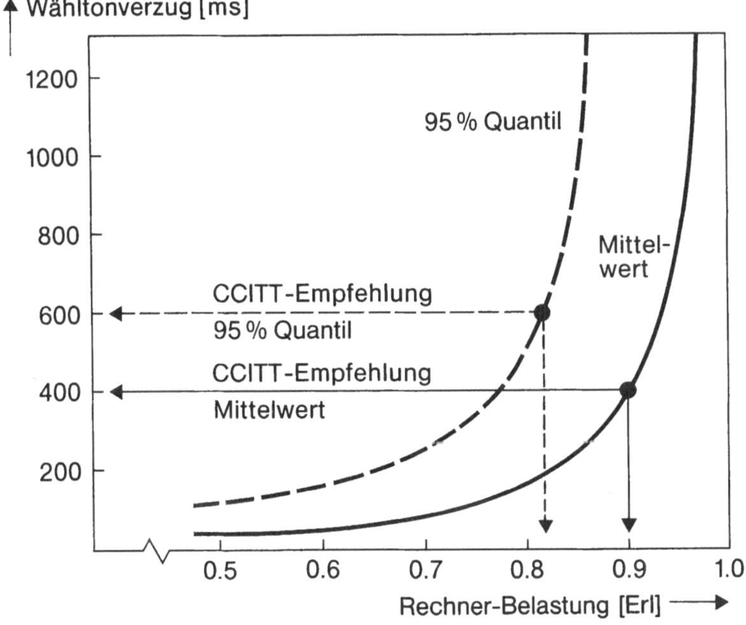

Bild Wähltonverzug in Abhängigkeit von der Rechnerbelastung

Wie man sieht, wird der CCITT-Wert (für Normallast) für das 95 %-Quantil bis zu einer Rechnerbelastung von 0,82 Erl eingehalten und der CCITT-Wert für den Mittelwert bis zu einer Rechnerbelastung von 0,90 Erl. Die Einhaltung des 95 %-Quantils ist somit die schärfere Forderung, und die Belastung, bis zu der diese CCITT-Werte eingehalten werden, beträgt 0,82 Erl. In gleicher Weise wird auch für alle anderen interessierenden Reaktionszeiten verfahren, und die Planungsbelastung wird auf das Minimum der sich bei den einzelnen Reaktionszeiten ergebenden Rechnerbelastungen gesetzt.

Stichprobenartig wurden aus den Simulationsläufen abgeleitete Aussagen an realen Anlagen überprüft. Hierbei wurde einerseits mit von Verkehrssimulatoren erzeugtem Verkehr gearbeitet, andererseits wurden Messungen an Feldversuchsanlagen durchgeführt. Es ergab sich sowohl in Bezug auf Reaktionszeiten als auch in Bezug auf Auslastung und Verhalten unter Überlast eine gute Übereinstimmung.

Charakteristische Eigenschaften von Transaktionslasten in DB/DC-Systemen

Volker Bohn
Universität Kaiserslautern, Fachbereich Informatik,
Postfach 3049, D-6750 Kaiserslautern

1. Einleitung

Informationen über charakteristische Eigenschaften von Transaktionssystemen (DB/DC-Systemen), d.h. über die Zusammensetzung der Transaktionslast, das Referenzierungsverhalten und Lokalitätsverhalten sowie über die Häufigkeit und Intensität von Lastschwankungen spielen eine entscheidende Rolle bei der Bewertung von Architekturvorschlägen zur Realisierung von Hochleistungs-Transaktionssystemen (normalerweise als Mehrrechner-Systeme konzipiert), bei der Entwicklung und Optimierung von leistungskritischen Algorithmen in transaktionsverarbeitenden Systemen und bei der Parametrisierung von Simulationsprogrammen. Wir haben deshalb eine allgemein anwendbare Methodik entwickelt, reale DB/DC-Anwendungen zu untersuchen. Die Vorgehensweise und die in langjährigen umfangreichen Analysen (Zeitraum von über 5 Jahren) ermittelten Ergebnisse über die Zusammensetzung der Transaktionslast, zeitlichen Änderungen in der Systembelastung und dem Referenzierungs- und Lokalitätsverhalten stellen wir in diesem Aufsatz vor.

2. Die Vorgehensweise

Wir halten unsere Vorgehensweise für eine allgemein einsetzbare Methodik zur Untersuchung DB-basierter Anwendungen. Ihre Vorteile liegen vor allem in der detaillierten Analyse von Datenbank- und Lastcharakteristika.

Wir beschäftigten uns mit DB/DC-Anwendungen, in denen die ausführbaren Transaktionen als vordefinierte Programme vorliegen und keine oder wenige Ad-Hoc-Anfragen mittels Anfragesprachen vorgenommen werden. Jedes vordefinierte Programm beschreibt einen Transaktionstyp, der vom (parametrischen) Benutzer aktiviert werden kann. Alle von uns untersuchten Anwendungen sind Anwendungen des CODASYL-Datenbanksystems UDS auf zentralen Verarbeitungsrechnern (keine Mehrrechner-Anwendungen) /UDSa, UDSb/.

Weil die umfangreichen Untersuchungen den Betrieb realer DB/DC-Anwendungen stark behindern würden und zudem sichergestellt werden muß, daß sich verschiedene Auswertungen auf denselben zeitlichen Ausschnitt der Systemlast der betrachteten Anwendungen beziehen, wurden diese Auswertungen nicht im laufenden Betrieb durchgeführt. Stattdessen wurden durch die Aktivierung verschiedener Meßpunkte im System anwendungsbezogene Protokolle erstellt, die eine Fülle von Detailinformationen beinhalten. Aus diesen, im allgemeinen aus mehreren Dateien bestehende Protokollinformationen wurde mit Hilfe eines Filterprogramms für jede der betrachteten Anwendungen ein sogenannter Seitenreferenzstring erzeugt. Die vorgestellten Auswertungen wurden dann anhand dieser Referenzstrings durchgeführt. Jeder Seitenreferenzstring enthält - unter anderem - folgende Informationen:

- Zeitpunkt des Beginns jeder Transaktion, den Namen des Transaktionstyps sowie eine eindeutige Kennzeichnung der Transaktion (TA-Nummer),
- Zeitpunkt und eindeutige Identifikation jeder durch die Transaktion (TA-Nummer) referenzierten Datenbankseite, die Art des Zugriffs (lesend, schreibend) und den
- Zeitpunkt des Transaktionsendes jeder während der Protokollphase ausgeführten Transaktion.

Damit ist jeder Seitenreferenzstring in Bezug auf den chronologischen Ablauf, den 'Transaktionsmix' und das Referenzierungsverhalten der Transaktionen ein genaues Abbild der protokollierten Systemlast und geeignet, umfangreiche Untersuchungen anwendungsbezogen durchzuführen. Die untersuchten Lasten unterscheiden sich hinsichtlich der Zahl der ausgeführten Transaktionen und der Zahl der Datenreferenzen sowie in der 'Länge' der Transaktionen (in Anzahl ausgeführter Referenzen) und der unterschiedlichen Anteile von Änderungs-Transaktionen sowie ändernden Zugriffen auf Datenbankseiten.

3. Analysen und Ergebnisse

Mit unseren Untersuchungen haben wir versucht, Antworten auf folgende Fragen zu geben:

- Zusammensetzung der Transaktionslast: Ist die Transaktionslast in Bezug auf die Länge der Transaktionen und dem Verhältnis von Lese-/Schreibzugriffen aus verschiedenartigen Transaktionstypen zusammengesetzt oder herrschen eher gleichartige Typen vor (Homogenität der Transaktionslast)?

- Zeitliche Änderung der Systembelastung: Ist die Systembelastung durch die zu bearbeitende Transaktionslast konstant oder sind starke Schwankungen erkennbar (Variation der Transaktions-Ankunftsraten)?
- Referenzierungsverhalten der Transaktionslast: Wie verteilen sich die Datenzugriffe über die Datenbank?
- Lokalitätsverhalten der einzelnen Transaktionstypen: Wie ist das 'Working-Set-Verhalten' der einzelnen Transaktionstypen und bei welchen 'Fenstergrößen' überwiegt der Zugriff auf disjunkte, dem einzelnen Transaktionstyp zuzuordnende Datenbereiche?

Aus den Antworten auf diese Fragen lassen sich Schlüsse ziehen über

- die Partitionierbarkeit realer Datenbanken im Hinblick auf Nutzung referentieller Lokalität, die, ggf. mit Unterstützung geeigneter Lastkontrollmechnismen, enormen Einfluß auf die Leistungsfähigkeit von Mehrrechner-Transaktionssystemen hat. Durch geschickte Zuordnung der Transaktionslast und der Datenpartitionen zu den einzelnen Verarbeitungsrechnern kann der Anteil rechnerübergreifender Kommunikationsoperationen (notwendig bei Zugriffen auf Daten, die von anderen Rechnern verwaltet werden als demjenigen, auf dem die Transaktion gestartet wurde), die einen enormen CPU-Overhead verursachen, drastisch reduziert werden.
- die Notwendigkeit von dynamischen Lastkontrollmaßnahmen in Mehrrechner-Transaktionssystemen, und außerdem
- das durch Lastkontrollmaßnahmen überhaupt nutzbare Optimierungspotential.

Zusätzlich stellen wir unsere Ergebnisse den Eigenschaften des sogenannten Debit/Credit-Benchmarks /An85, Gr85/, der oft als die Transaktionslast schlechthin bezeichnet wird, gegenüber und diskutieren dessen Eignung zur Bewertung der Leistungsfähigkeit von Hochleistungs-Transaktionssystemen.

3.1 Zusammensetzung der Transaktionslast

Die Zusammensetzung der Transaktionslast erwies sich im Hinblick auf die Zahl und Art der Referenzen der Transaktionstypen und die unterschiedliche Zahl von Ausführungen als sehr heterogen:

- Es gibt wenige dominante Transaktionstypen, die die deutliche Mehrzahl der ausgeführten Referenzen verursachen.
- Die Transaktionstypen unterscheiden sich erheblich in der Zahl der ausgeführten Referenzen und dem Anteil ändernder Zugriffe.
- Die Zahl der Transaktions-Aktivierungen ('Aufrufe') pro Transaktionstyp schwankt beträchtlich.
- Der am häufigsten ausgeführte Transaktionstyp ist nicht unbedingt derjenige mit den meisten Referenzen; vielmehr sind große Unterschiede der durchschnittlich pro Transaktion ausgeführten Referenzen zwischen den Transaktionstypen feststellbar. Die Transaktionstypen unterscheiden sich demnach in der 'Länge der Transaktion'.

3.2 Zeitliche Änderungen in der Systembelastung

Wir stellten fest, daß die Systemlast nicht nur heterogen zusammengesetzt ist, sondern zusätzlich starke Schwankungen aufweist, die so extrem sind, daß sie häufig einer 'Sprungfunktion' gleichkommen; dies gilt sowohl für die gesamte Last als auch für jeden einzelnen Transaktionstyp. Diese extremen Lastschwankungen sind zudem sehr häufig (etwa alle 1-5 Minuten). Dadurch, daß nicht alle Transaktionstypen zu jedem Zeitpunkt aktiv sind, ändert sich außerdem die Art der Systembelastung ('aktueller Transaktionsmix') laufend. Die Systembelastung ist also inhomogen und nicht konstant.

3.3 Referenzierungsverhalten und Lokalitätsverhalten der Transaktionstypen

Bis auf seltene Sonderfälle kann man sagen, daß es ideal partitionierbare Lasten für Mehrrechnersysteme nicht gibt - tatsächlich gilt dies für alle von uns untersuchten Anwendungen. Die Ursache liegt darin, daß auf die Datenbank sehr 'ungleichmäßig' zugegriffen wird:

- Es existieren dominante Referenzierungsbereiche, d.h., es gibt Datenbereiche, die die Majorität der Seitenreferenzen auf sich vereinen.
- Es existieren im allgemeinen mehrere Datenobjekte mit extremer Referenzierungs- und Konflikthäufigkeit ('Hot Spots', 'High Traffic Objects'); es handelt sich dabei um ein wohlbekanntes Phänomen, für dessen Entschärfung bereits Vorschläge existieren /Hä88/.

Bei diesen Aussagen spielt allerdings die Beobachtungsdauer eine große Rolle. Deshalb haben wir das 'Working-Set-Verhalten' der einzelnen Transaktionstypen, d.h. die referenzierten Seitenmengen und ihre

Beziehungen untereinander (disjunkt, ungleich oder überlappend) bei verschiedenen 'Fenstergrößen' (60s bis 900s) untersucht. Daraus ist ableitbar, ob für 'dynamische' Lastkontrollalgorithmen Optimierungspotential verfügbar ist und wie genau diese Algorithmen die aktuelle Systemlast inspizieren müssen. Das wichtigste und überraschende Resultat dieser Untersuchungen war der unerwartet hohe Anteil von Referenzen in disjunkten Datenbereichen von teilweise mehr als 50% und die Tatsache, daß diese Ergebnisse im Bereich von 1 bis 5 Minuten Fenstergröße liegen. Derlei Information kann von geeigneten Algorithmen duch geschickte Daten- und Lastzuordnung im Mehrrechnersystem genutzt werden, um den kommunikationsbedingten CPU-Overhead drastisch zu reduzieren.

4. Zusammenfassung und Folgerungen

Einfache Partitionierbarkeit realer Lasten im Hinblick auf Nutzung referentieller Lokalität ist - wie wir gesehen haben - im allgemeinen nicht gegeben. Um bei Mehrrechner-Architekturen eine hohe Leistungsfähigkeit zu erzielen, ist der Einsatz dynamischer Lastkontrollmechanismen erforderlich, die auf den aktuellen Systemzustand und die aktuelle Systemlast reagieren. Die Verfahren müssen schnell und wirksam auf Änderungen im Lokalitätsverhalten der Transaktionslast eingehen, denn Lastschwankungen sind häufig und heftig. Wie genau dynamische Lastkontrollverfahren Änderungen beobachten müssen, d.h., in welchen - aus Effizienzgründen möglichst langen - Zeitspannen Optimierungspotential vorhanden ist, wurde beantwortet: wir haben gezeigt, daß bei Fenstergrößen von wenigen Minuten überraschend gute Einsatzmöglichkeiten für derartige Algorithmen vorhanden sind.

Weil sich die untersuchten realen Anwendungen in den hier betrachteten Aspekten nicht unterscheiden, kann man annehmen, daß die vorgestellten Ergebnisse für ein sehr breites Anwendungsspektrum gültig sind. Die Ergebnisse können den Eigenschaften des sogenannten Debit/Credit-Benchmarks /An85, Gr85/, der heute oft zur Leistungsbewertung für transaktionsverarbeitende Systeme, insbesondere bei Systemvergleichen, herangezogen wird, gegenübergestellt werden. Die Eigenschaften des Debit/Credit-Benchmarks stehen in krassem Widerspruch zu unseren Ergebnissen, denn die Lastzusammensetzung des Benchmarks ist homogen: er kennt nur einen einzigen extrem einfachen Transaktionstyp, und Lastschwankungen kommen in der von uns festgestellten Form nicht vor. Das Referenzierungsverhalten ist bekannt und eignet sich hervorragend, den Datenbestand gleichmäßig unter den Verarbeitungsrechnern mit etwa gleichen Lastanteilen aufzuteilen. Deshalb stellt er eine ideal partitionierbare und damit realitätsfremde Anwendung dar. Reale Lasten zeichnen sich durch eine Vielzahl sehr unterschiedlicher Transaktionstypen aus, die mit verschiedener Häufigkeit die Datenbank 'ungleichmäßig' referenzieren. Hinzu kommen häufige starke Lastschwankungen. Hieraus resultierende Probleme bei der Nutzung des (trotzdem vorhandenen - siehe Kapitel 3.3) Lokalitätsverhaltens der Transaktionslast zur Minimierung externer Datenzugriffe treten im Debit/Credit-Benchmark nicht auf. Zukünftige Hochleistungs-Transaktionssysteme brauchen jedoch Komponenten, die auf die Beherrschung heterogener Lasten abgestimmt sind; derartige Probleme stellen sich jedoch mit Benchmarks wie Debit/Credit gar nicht. Dieser Benchmark ist demnach nur geeignet, Transaktionssysteme hinsichtlich eines sehr schmalen Spektrums leistungskritischer Algorithmen zu vergleichen, das die eigentlichen, durch reale Lasten gestellten Anforderungen nur zu einem geringen Teil abdeckt. Zieht man demnach diesen Benchmark oder ähnliche 'Monokulturen' zur Bewertung von Hochleistungs-Transaktionssystemen oder in der Entwicklung befindlicher Systeme heran, so werden mit hoher Wahrscheinlichkeit die unrealistischen Anwendungsfälle bewertet bzw. optimiert.

Eine detaillierte Darstellung der durchgeführten Untersuchungen und eine ausführliche Diskussion der Resultate findet sich in der Langfassung dieses Beitrags, der in einem Extraband ergänzend zum Tagungsband herausgegeben wird.

Literaturverzeichnis:

An85 Anon et. al.: A Measure of Transaction Processing Power, Datamation, February 1985, pp. 112-118.

Gr85 Gray, J.N., Good, B., Gawlick, D., Homan P., Sammer, H.: One Thousand Transactions per Second, Proc. IEEE Spring Compcon, San Francisco, Feb. 1985, pp. 96-101.

Hä88 Härder, T.: Handling Hot Spot Data in DB-Sharing Systems, Information Systems, Vol. 13, No. 2, 1988, pp. 155-156.

UDSa Aufbauen und Umstrukturieren, Benutzerhandbuch UDS V5.0C, Siemens AG, Best.-nr. U931-J-Z55-6, 1988.

UDSb Anwendungen programmieren, Benutzerhandbuch UDS V5.0A, Siemens AG, Best.-nr. U930-J-Z55-3, 1988.

Dieser Bericht wurde durch die Siemens-AG (Abteilung D-ST-DB-1) gefördert.

6 Jahre ETHERNET im Industrie-Einsatz
- Erfahrungen, Auslastungen und Grenzlastuntersuchungen -

Joachim Laue Versuchssysteme
VOLKSWAGEN AG Wolfsburg
Jung Sun Lie Institut für Betriebssysteme und Rechnerverbund
TU Braunschweig

Zusammenfassung

Im Jahr 1983 begann im Bereich der Forschung und Entwicklung (FE) der Volkswagen AG der Aufbau eines ETHERNET-Kommunikationssystems. Hiermit wurden zunächst 25 Prozeßrechner eines Herstellers miteinander verbunden. Heute umfaßt dieses Netz ca. 300 Knoten von mehr als 10 Herstellern, die sich auf das ganze Firmengelände (ca. 5 km^2) erstrecken.

Nachstehend wird über Netzaufbau, Einsatzerfahrungen, Auslastung und Untersuchung des Grenzbelastungsverhaltens berichtet.

1. Aufbau und Umfang des Firmen-LAN

1978 wurde in der FE ein Prozeßrechner-Verbund-System (PVS) zur Prüfstandsautomatisierung aufgebaut. Als Kommunikationswege für die 25 Rechner standen damals nur Punkt-zu-Punkt-Verbindungen zur Verfügung.

Mit Inbetriebnahme eines neuen Forschungskomplexes im Jahre 1983 wurde im Rahmen einer Erweiterung des PVS die bisherige Technik durch ein Ethernet-LAN ersetzt und ständig weiter ausgebaut.

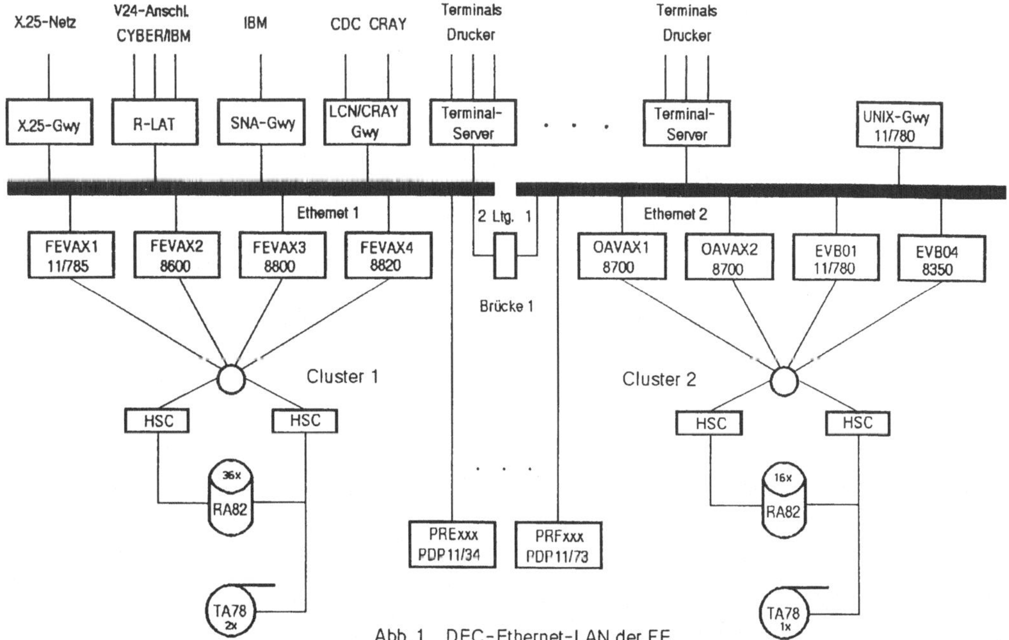

Abb. 1 DEC-Ethernet-LAN der FE

1988 wurden die in der Firma entstandenen Netzwerkinseln teils unterschiedlicher Hersteller zu einem gemeinsamen LAN verschaltet.

Das älteste, umfangreichste und am intensivsten benutzte Einzel-LAN ist das 1983 angefangene LAN für das damalige PVS. Dieser Umfang wird hier als PVS-LAN bezeichnet. In diesem LAN sind über 900 Benutzer registriert. Die Anwendungspalette reicht von Versuchsdurchführung und -auswertung, Bürokommunikation, CAD bis zur technisch-wissentschaftlichen Datenverarbeitung.

In Abb.1 sind die wesentlichen Komponenten des PVS-LAN dargestellt.

2. Einsatzerfahrungen

Das PVS-LAN hat sich in den 6 Jahren als äußerst betriebssicher erwiesen. Es gab in der Gesamtzeit keinen Kurzschluß durch Ausfall von Komponenten.

Die Fehlerrate auf dem PVS-LAN ist sehr gering, bei uns wird bei täglich etwa 18 Millionen übertragenen Paketen im Durchschnitt weniger als 1 fehlerhaftes Paket registriert. Bei gelegentlich höheren Werte ließ sich bei uns stets eine systematische Fehlerquelle finden.

Da die derzeitigen Spitzenwerte der Auslastung verschiedener Segmente schon 30 % erreichen, wurden zur weiteren Planung in Zusammenarbeit mit dem Institut für Betriebssysteme und Rechnerverbund der TU Braunschweig eine Analyse der derzeitigen Auslastung und Untersuchungen zur maximalen Auslastung durchgeführt.

3. Auslastungsuntersuchungen

In Abb.2 ist die prozentuale Auslastung des PVS-LAN über einen typischen Arbeitstag dargestellt. Die Messungen erfolgten mit einem LAN Protokoll Analysator vom Typ HP 4971 S.

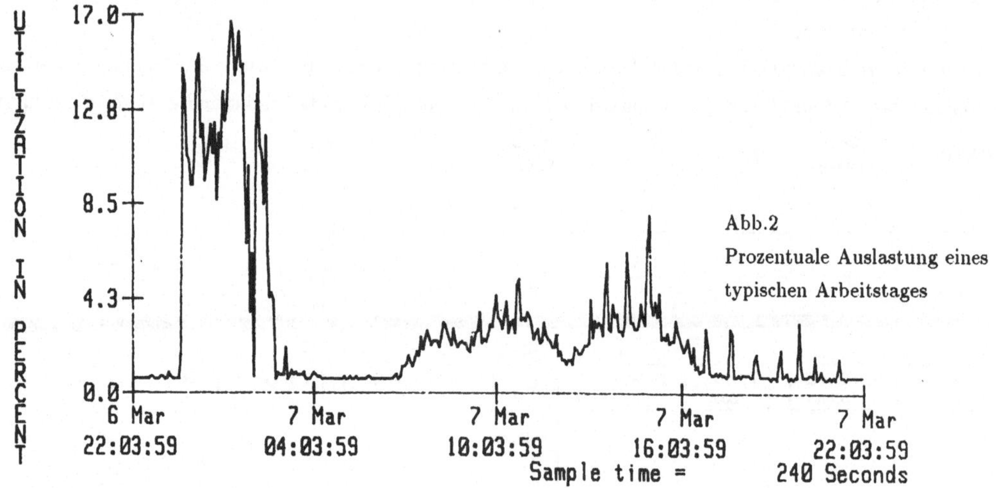

Abb.2
Prozentuale Auslastung eines typischen Arbeitstages

Die größte Last (17 %) wird in den Nachtstunden erreicht. Hierfür sind automatisch ablaufende Dateisicherungsmaßnahmen verantwortlich. Tagsüber bleibt die mittlere Belastung unter 10 %, der zeitliche Verlauf ist recht unregelmäßig.

Während die Zeitbasis in Abb.2 mit 240 Sekunden extrem hoch gewählt werden mußte, um 24 Stunden erfassen zu können, konnten mit einem anderen "LAN-Traffic-Monitor" die auftretenden Spitzenlasten gleichzeitig mit zwei unterschiedlichen Zeitbasen bestimmt werden. Bei Veränderung der Zeitbasis von 30 auf 1 Sekunde stieg z.B. die gemessene Spitzenlast von 21.3 auf 29.4 % (bzw. von 748 auf 911 Paketen/Sekunde).

Diese Erkenntnis ist für die Interpretation der Ergebnisse aus Abb.2 wichtig. Zu den mit dem HP 4971 S ermittelten Spitzenwerten sind noch erhebliche Zuschläge zu addieren. Für die Funktionalität von Ethernet sind die Spitzenwerte bei kleiner Zeitbasis maßgebend.

4. Grenzlastuntersuchungen

Zur Vorbereitung weiterer Planungen wurden Untersuchungen zur Grenzbelastung von Ethernet-Segmenten durchgeführt.

Als Lastgeneratoren dienten hierbei die vorhandenen Rechner selbst. Hierzu wurde ein Programm erstellt, daß in einer Schleife möglichst viel Datenpakete mit parametrierbarer Länge auf das zu untersuchende Segment senden kann. Als Protokolltyp für diese Pakete wurde das vom Hersteller für Anwender reversierte Protokoll DEC_C_USE (60-06) verwendet. Die Zieladressen dieser Belastungspakete hatten im zu untersuchenden Segment keine Empfänger, alle Brücken an dem Segment wurden für diese Adressen gesperrt, um andere Segmente von der Belastung freizuhalten.

Je nach Rechnerleistungsklasse und Paketlänge wurde hierbei pro System eine Ethernetauslastung von etwa 1 % (MicroVAX, 64 Bytes) bis zu 30 % (VAX 8800, 1514 Bytes) erreicht.

Diese Lastprogramm wurde in "betriebsfreien" Zeiten zeitplangesteuert auf bis zu 20 Rechnern gestartet. So konnte das PVS-LAN bis zu knapp 100 % belastet werden. Durch zeitlich verschobene Starts konnte die Last auch rampenförmig hochgefahren werden.

Parallel zu der künstlichen Belastung wurde zwischen (an der Erzeugung der Last unbeteiligten) Rechnern das Zeitverhalten von Anwendungen gemessen. Hierzu wurde ständig eine Datei transferiert und die jeweiligen Übertragungszeiten registriert.

Als Ergebnis bleibt festzuhalten, daß bei vorwiegend großen Paketen bis 70 %-Ethernet-Auslastung außer tragbarer Verminderung der Transferleistung keine Probleme auftraten. Bei 80 % war das Ethernet-LAN nicht mehr sinnvoll zu betreiben, da die beobachteten zeitweiligen Initialisierungen der Ethernet-Schnittstelleneinheiten in den Knoten unakzeptabel waren.

Im Gegensatz dazu wurde bei vorwiegend kleinsten Paketen schon bei 30 % Auslastung die unakzeptablen Initilisierungen beobachtet.

Als Ursache für das relativ frühe Versagen von Ethernet bei kleinsten Paketen war die hohe Zahl der Kollisionen (1200/Sek.) maßgebend, die bei der hier verwendeten Methode der Lasterzeugung auftrat.

5. Zusammenfassung

Ethernet-LAN's sind leistungsfähige, preiswerte und sehr zuverlässige Systeme zur Vernetzung von Rechnern und anderen Komponenten.

Die bisher erreichte Auslastung von bis zu 30 % bzw. 1000 Paketen/Sek. stellt noch nicht die Grenzbelastung dar, die Untersuchungen zeigen aber, daß an unserem Netz die Last ungefähr noch einmal verdoppelt werden kann, bis es zu merkbaren Beeinträchtigungen kommt.

Analysen und Bewertungen zur Realzeitfähigkeit des Token Rings

Sturm, Leonhard; Kröger, Karl-Josef
Universität-Gesamthochschule Paderborn
Fachbereich 14 - Datentechnik -
Pohlweg 47 - 49
4790 Paderborn

Bordewisch, Reinhard
Fa. Nixdorf Computer AG
Abteilung DF BS 4
Pontanusstraße 55
4790 Paderborn

1. Motivation und Zielsetzung der Untersuchungen

Eine zentrale Rolle bei Kommunikationssystemen in der automatisierten Fertigung spielt das zugrundeliegende Verbindungsnetzwerk, wobei vermehrt die Anwendung von lokalen Netzwerken zur Kopplung von Zellenrechnern diskutiert wird [1].
Diese Kommunikationssysteme müssen einigen speziellen Anforderungen genügen:

- Es wird Echtzeitverhalten gefordert,
- es sind mittlere bis hohe Übertragungsleistungen notwendig,
- das Kommunikationssystem muß "offen" sein für den Anschluß unterschiedlicher Rechner bzw. Prozessorsysteme,
- der Austausch von Informationen unterliegt speziellen Anforderungen bezüglich Störsicherheit und Zuverlässigkeit.

In dem hier vorgestellten Projekt wurde für die Vernetzung von flexiblen Fertigungszellen der Token Ring nach IEEE 802.5 ausgewählt, für den es das Controller-Chipset (TMS380) der Fa. Texas Instruments auf dem Markt gibt.
Für den Einsatz in der automatisierten Fertigung ist das Echtzeitverhalten die vorrangige Bewertungsgröße. Dieses ist nur gewährleistet, wenn die Kommunikation zwischen den Prozessen schnell genug ist, wobei die Zeitschranke für die Übermittlung einer Botschaft von Prozeß zu Prozeß allgemein mit 20 ms angegeben wird. Es wird angenommen, daß künftig viele Konzepte unter Auslassung der Protokollschichten 3 bis 6 direkt von der Anwendung auf den Data Link Layer (DLL) gemäß dem ISO-OSI-Referenzmodell [2] aufsetzen werden (vgl. dazu MINIMAP [3]).

2. Messungen der Laufzeiten

Die Messungen werden mittels Software-Monitoring gewonnen. Dazu wurde ein Meßprogramm erstellt, welches in der Lage ist,

- Zeitinformationen zu den gewünschten Zeitpunkten festzuhalten,
- die benötigten Stimuli für die Token-Ring-Adapter zu erzeugen,
- eine Echostation für die Ermittlung von Antwortzeiten zu realisieren,
- die Einstellung der Parameter vorzunehmen,
- durch automatische Wiederholung eine Vielzahl von Messungen einzuleiten,
- die system-meßspezifischen Voreinstellungen zu ermöglichen,
- die Ergebnisse statistisch auszuwerten,
- die Meßergebnisse für eine grafische Darstellung (Bildschirm und Drucker) aufzubereiten.

Es wurde ein Meßprogramm implementiert, welches auf dem DLL-Layer bzw. MAC-Layer aufsetzt. Als unabhängige Zeitbasis wurde eine Einsteckplatine mit drei 16-bit-Zählern aufgebaut, die mit den Frequenzen 1 MHz, 100 kHz und 10 kHz getaktet werden.

Vom Software-Monitor können Stimuli für ein Meßobjekt erzeugt werden, um die benötigten Betriebsbedingungen und Anstöße selbständig zu generieren sowie den zu messenden Prozeß zu installieren und zu aktivieren. Der Software-Monitor erlaubt die Einstellung sowohl der Parameter für den Meßgegenstand als auch der Parameter der Netzlast, die von anderen Stationen erzeugt werden. Parameter für den Meßgegenstand sind die Infofeldlänge und die Priorität. Parameter für die Netzlast können an jeder Station am Traffic-Generator eingestellt werden. Er erlaubt so eine vielfältige Variation der Netzlast. Zur Ermittlung von Antwortzeiten wird in der Partnerstation ein Echo-Programm gestartet, welches das empfangene Frame zurückschickt.
Der Traffic-Generator bietet eine Einstellung in 3 Parametergruppen: Framelänge, Priorität und

Verzögerung. Die Framelänge ist konstant bzw. gleich- oder normalverteilt variierbar; eine biodale Verteilung ist ebenfalls wählbar. Die Priorität kann konstant oder gleichverteilt vorgegeben werden. In der Gruppe Verzögerung wird die Zwischenankunftszeit (inter frame arrival time) festgelegt. Dabei kann ein Poisson-Prozeß als Standardlast nachgebildet werden, aber ebenso die intervallgesteuerte

Aussendung von Einzelframes. Weiterhin wird die Vollast erzeugt, wenn die Aufträge ohne Verzögerung in die Warteschlange gesetzt werden. Aus der Kombination dieser drei Gruppen lassen sich nahezu beliebige Lastprofile zusammenstellen.

Nach dem TG-Lauf steht eine Auswertung der Laufdaten zur Verfügung:

- Anzahl der gesendeten Frames,
- Anzahl der fehlerhaften Frames,
- benötigte Sendezeit,
  Anzahl der gesendeten Info-Bits
- und die Netto-Übertragungsrate (Info-Bits/Sendezeit).

Während der Messung werden die Meßwerte sowie deren Minimum und Maximum unter den Meßwerten festgehalten. Die Ergebnisse können auch in Kombination aus mehreren Meßläufen in Form von Grafiken angezeigt werden. Bis zu 99 Frames können mit diesem Netzmonitor vom Ring kopiert und am Bildschirm dargestellt werden. Dabei ermöglicht ein Filter eine Auswahl, so daß auch nur Anwenderpakte oder nur MAC-Frames erfaßt werden können. Bei MAC-Frames wird zusätzlich die inhaltliche Bedeutung der Protokollframes decodiert und explizit angezeigt.

3. Durchgeführte Experimente

Zunächst wurde die Meßsoftware geeicht: Es wurden die Eigenzeitverbräuche gemessen und für die Messungen am Objekt rechnerische eliminiert. Weiterhin wurde nachgewiesen, daß die Softwarepfade für die zu den Adapteraktivitäten prallelen Teile des Meßprogramms vor dem Eintreten der Ende-Bedingung abeschlossen sind, daß sie also durch die Ergebnisse nicht beeinflußt werden.
Zum Zwecke der Isolation soll der Einfluß der externen Parameter ausgeschaltet werden. Deshalb wurden Minimalkonfigurationen mit nur einer bzw. zwei Stationen am Ring gewählt.
Damit wurden u. a. Messungen zur Ermittlung von

- Übertragungsszeiten in Abhängigkeit der Pufferkonfigurierung,
- Übertragungszeiten bei verschiedenen Infolängen,
- Abhängigkeiten der Transferzeit bei unterschiedlichen Prioritäten und Netzlasten

durchgeführt.

4. Ein analytisches Modell für die Übertragungszeit (worst case)

Es soll eine Transferzeit T_{Trans} für den "worst case" berechnet werden. Dazu werden folgende Annahmen gemacht:

- Alle Frames der Interprozeß-Kommunikation sollen die gleiche Länge haben.
- Die Betrachtung wird für die Verhältnisse innerhalb einer Prioritätsebene gemacht.
- Jeweils eine "priority state machine" [4] befindet sich im aktiven Zustand.

Für die Übertragungszeit $T_ü$ gilt :

$$T_ü = T_{offs} + (m_ü + m_D)*L_I \quad \text{mit}$$

$T_ü$:	Übertragungszeit
T_{offs} :	Anlaufzeit der Adapter -Software (konstanter Anteil)
L_I:	Länge der Info-Felder in Bytes
$m_ü$ =	2.000 µs/Byte konstant (für 4 Mbit/s Übertragungsrate)
m_D =	0,3838 µs/Byte abhängig von DMA/Pufferung

Die Transferzeit : $(T_{Trans}) = T_{ü} + T_{w}$

T_w: Wartezeit auf ein freies Token

Mit den Parametern

"Bandbreite", besser Übertragungsrate: $B = 4$ Mbit/s = 500 kBytes/s

Signalausbreitungsgeschwindigkeit: $c_L = 200000$ km/s
Länge des" Frame-Overheads": $L_{FO} = 21$ Byte
Pfadlänge Anlauf des Ü-Kommandos (s.o.) $T_{offs} = 1100$ /us (gemessen)
Stationszahl: N
Leistungslänge: L_L

Steigungen (siehe oben): $m_{ü}$, m_D
Infofeldlänge: L_I

erhält man für Transferzeit:

$$T_{Trans} = (N - 1)*(L/c_L + ((N - 1)* 2{,}5 \text{ bit} + 36{,}5 \text{ bit}) / B + \\ + m_{ü}*(L_I + L_{FO}) + (m_{ü} + m_D)*L_I + T_{offs}$$

Damit wurden diverse Berechnungen durchgeführt: unterschiedliche Stationsanzahl, verschiedene Infofeldlängen, Bewertung der DMA-Parameter, Einfluß der Leistungslänge und Bandbreite.
Die größten Verbesserungen der Transferzeiten sind offensichtlich zu erwarten, wenn Token Ringe mit höheren Übertragungsraten eingesetzt werden.

5. Schlußfolgerung

Mit der bisher erhältlichen Technologie (Chipset TMS380) ist eine Eignung des Token Rings als Verbindungsnetzwerk für Fertigungsinseln festzustellen. Eine Interprozeß-Kommunikation unter Echtzeitbedingungen (20 ms) ist möglich, wenn bestimmte Grenzen für die Konfigurierung (Stationszahl) und die eingesetzten Anwendungsprotokolle (Paketlänge) eingehalten werden. Denkbar ist eine gezielte Belegung zeitkritischer Übertragungen mit hoher Zugangspriorität und kurzen Botschaften. Allerdings müssen die Pfadlängen der auf den DLL- Layer aufgesetzten Protokollelemente einkalkuliert werden. Der Schwerpunkt der Anstrengungen muß also weiterhin in der Entwicklung einer schnellen Protokollverarbeitung liegen, ob nun Protokollschichten explizit eingezogen oder in die Anwendung integriert werden. Dazu gehört eine auf diese Bedürfnisse abgestimmte Architektur eines Kommunikationssystems.

Literatur

1. Kröger, K.-J.: Monitoring an einem Tokenringnetz als LAN in der automatisierten Fertigung. Studienarbeit im Fachgebiet Datentechnik, Fachbereich 14, Universität-Gesamthochschule Paderborn, 1987.

2. IEEE Technical Report 8509, Information Processing Systems - Open Systems Interconnection - Service Conventions.

3. Suppan-Borowka, J., Simon, Th.: MAP, Datenkommunikation in der automatisierten Fertigung DATACOM Buchverlag, Pulheim 1986.

4. TMS380 Adapter Chipset User`s Guide, Texas Instruments Incorporated, 1986.

Leistungsbewertung von Hochgeschwindigkeitsnetzen

Peter Martini

Lehrstuhl für Informatik IV, RWTH Aachen
Ahornstr. 55; 5100 Aachen
☎: +241/80-4522; FAX: +241/80-6295; e-mail: ...unido!rwthinf!martini

Zusammenfassung

Fortschritte auf dem Gebiet der Glasfasertechnologie ermöglichen heute Netze mit Übertragungsraten im hohen Megabit/s-Bereich. Diese Netze sind darüber hinaus in der Lage, relativ große geographische Distanzen zu überdecken. Im Vergleich zum Design 'klassischer' Lokaler Netze wird eine weitgehende Neukonzeption der eingesetzten Protokolle erforderlich, von der insbesondere die unteren Ebenen des ISO-Referenzmodells betroffen sind. Der vorliegende Beitrag gibt einen Überblick über die derzeit wichtigsten Protokolle zur Regelung des Medienzugangs in Hochgeschwindigkeitsnetzen und diskutiert deren Leistungsfähigkeit unter besonderer Berücksichtigung von Fairness-Aspekten.

1. Einleitung und Überblick

Datenkommunikation, der Austausch von Informationen zwischen Maschinen, hat in den vergangenen Jahren mehr und mehr an Bedeutung gewonnen. Waren in der Pionierzeit der elektronischen Datenverarbeitung Rechner, Drucker und Kartenleser noch auf engstem Raum angeordnet, so suchte man doch von Anfang an nach Möglichkeiten, Terminals und andere Peripheriegeräte benutzernah, d.h. in immer größerer Entfernung vom Rechnerraum, zu installieren. Sinkende Preise machten es schließlich möglich, immer leistungsfähigere Rechenanlagen "vor Ort" einzusetzen. Aus dem bis dahin üblichen Master-Slave-Betrieb entwickelte sich die Kommunikation zwischen weitgehend voneinander unabhängigen, intelligenten Maschinen. Man erkannte schnell, daß Möglichkeiten und Grenzen der sich nun entwickelnden Rechnernetze wesentlich von der geographischen Ausdehnung derartiger Kommunikationssysteme geprägt sind. In Weitverkehrsnetzen, wie sie heute die Welt umspannen, sind andere Strukturen sinnvoll als in "Lokalen Netzen", die auf ein Gebäude oder auf das Gelände einer Fabrik oder einer Universität begrenzt sind. Werden im Weitverkehrsbereich vorwiegend unregelmäßige Netze eingesetzt, so haben sich im lokalen Bereich regelmäßige Topologien durchgesetzt: Ring, Stern und Bus. Für diese Lokalen Netze (engl. Local Area Network, LAN), bei denen in der Regel viele Stationen an ein gemeinsames Medium angeschlossen sind, müssen darüber hinaus Protokolle festgelegt werden, welche die zur Verfügung stehende Übertragungskapazität in geeigneter Weise auf die Benutzer verteilen. Bedingt durch unterschiedliche Anforderungen spezieller Einsatzgebiete einerseits und die Vielfalt denkbarer Lösungen andererseits wurde seit dem Ende der siebziger Jahre eine kaum noch überschaubare Fülle von Medienzugangsprotokollen vorgeschlagen. Sowohl bei Standardisierungsgremien als auch im kommerziellen Angebot haben sich heute zwei Konzepte durchgesetzt: "Random Access Protokolle", bei denen die angeschlossenen Stationen - von wenigen Einschränkungen abgesehen - zu jedem beliebigen Zeitpunkt Zugang zum gemeinsamen Medium haben, und "Token-Passing"-Systeme, bei denen das Senderecht durch eine spezielle Marke (engl. "Token") von Station zu Station weitergegeben wird.

Nun, da wir uns an die Benutzung von Lokalen Netzen mit Datenraten um 10 Mbit/s gewöhnt haben, liegt der nächste Schritt vor uns. Sich abzeichnende Standards, Prototypen und erste Produkte lassen Lokale Hochgeschwindigkeitsnetze (High Speed Local Area Networks, HSLANs) mit Datenraten um

100 Mbit/s in den Mittelpunkt des Interesses rücken. Wesentliche Leistungscharakteristika derartiger Systeme werden in den nachfolgenden Kapiteln diskutiert. Zuvor soll jedoch in Kapitel 2 durch eine Einordnung in das ISO-Referenzmodell, das in [ISO 84] als Rahmen für die Spezifikation von Kommunikationsprotokollen festgelegt wurde, deutlich gemacht werden, welche Probleme durch LAN-spezifische Protokolle gelöst werden. Den Schwerpunkt bildet dann eine Betrachtung der Modellierung und Bewertung mittels mathematischer Analyse und/oder Simulation. Abschließend werden der Einfluß höherer Protokollebenen und die Leistungsfähigkeit der heute kommerziell verfügbaren Kommunikations - Controller diskutiert.

2. Lokale Netze - Einordnung in das ISO-Referenzmodell

Das ISO-Referenzmodell [ISO 84] für Offene Rechnernetze (Open Systems Interconnection - Basic Reference Model) strukturiert die Protokolle zur Regelung der Datenkommunikation in sieben Schichten (Ebenen). Jede Schicht bietet für die über ihr liegende Schicht Dienste an und nutzt zur Erbringung dieser Dienste ihrerseits die Dienste der unter ihr liegenden Schicht, wobei die Übergabe von Daten und Kontrollinformationen jeweils an sog. "Dienstzugangspunkten" (engl. Service Access Points, SAPs) erfolgt. Auf jeder der so entstehenden Hierarchiestufen werden durch das zugehörige Protokoll spezielle Probleme der Datenkommunikation gelöst.

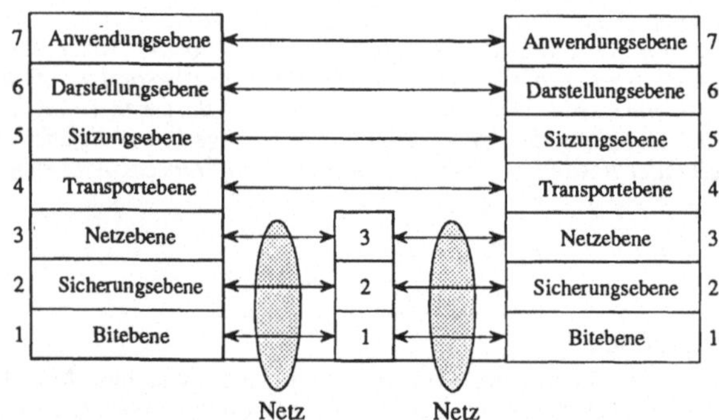

Abb. 2.1. Die Schichten des ISO-Referenzmodells

Die Abbildung 2.1. zeigt die Schichten des ISO-Referenzmodells, wobei Pfeile den Austausch von sog. "Protokolldateneinheiten" (engl. protocol data unit, PDU) symbolisieren, die jeweils eine spezielle Bedeutung (z.B. Verbindungsaufbauwunsch oder Daten) für die jeweilige Schicht haben. Netz-spezifisch sind die Schichten 1 bis 3, die sich mit physikalischer Signaldarstellung (Bitebene), mit Fehlererkennung und -korrektur (Sicherungsebene) sowie - neben einigen weiteren Aufgaben - mit Adressierung und Routing (Netzebene) befassen. Erfolgt eine Kommunikation über mehrere Teilnetze hinweg, dann können in verschiedenen Teilnetzen die Protokolle der Schichten 1 bis 3 völlig unterschiedlich sein. Dagegen ist der Informationsaustausch auf den Schichten 4 bis 7 durch End-to-End-Protokolle geregelt, die unabhängig von dem verwendeten Netz sind.

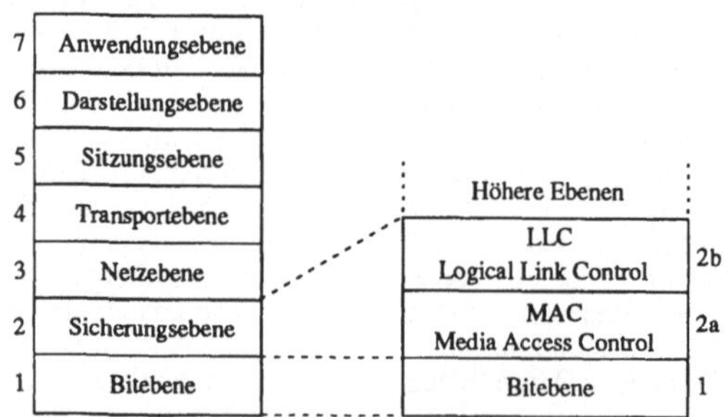

Abb. 2.2 Teilschichten der Schicht 2 für Lokale Netze

Befaßt sich in Weitverkehrsnetzen die Schicht 2 lediglich mit Fehlerbehandlung und Flußkontrolle, d.h. der Anpassung von Sendegeschwindigkeit und Empfangsgeschwindigkeit, so kommt in Lokalen Netzen eine weitere Aufgabe hinzu: da nun viele Stationen an ein gemeinsames Medium angeschlossen werden können, hat in Lokalen Netzen die Schicht 2 zusätzlich die Aufgabe der Regelung des Medienzugangs. Aus diesem Grunde wird die Schicht 2 in zwei Teilschichten zerlegt: Logical Link Control (LLC, Ebene 2b) - zur Abwicklung von Fehlerbehandlung und Flußkontrolle - und Media Access Control (MAC, Schicht 2a) - zur Regelung des Medienzugangs, vgl. Abb. 2.2.

Bekannte Standards für Lokale Netze - insbesondere sind hier [IEEE 85a], [IEEE 85b], [IEEE 85c] und [IEEE 85d] zu nennen - spezifizieren Protokolle für die Schichten 1, 2a und 2b. Diese Standards sind auf Datenraten im unteren Megabit/s-Bereich ausgerichtet, wobei 10 Mbit/s als typischer Wert zu betrachten ist. Technologische Fortschritte auf dem Gebiet der optischen Übertragung, d.h. auf der Bitebene, erlauben allerdings heute immer höhere Datenraten, was nicht ohne Folgen für die Protokolle der Sicherungsebene bleibt. Insbesondere haben sich beträchtliche Modifikationen oder sogar neu entwickelte Protokolle als erforderlich erwiesen, wenn die Regelung des Medienzugangs nicht zum Engpaß werden soll. Mit FDDI (vgl. [ANSI 87a], [ANSI 87b] und [ANSI 88]) liegen bereits wesentliche Teile eines Standards für Lokale Hochgeschwindigkeitsnetze vor und zahlreiche Hersteller bieten bereits Produkte an, die diesem Standard entsprechen.

3. Protokolle für lokale Hochgeschwindigkeitsnetze

Lokale Hochgeschwindigkeitsnetze stellen eine Weiterentwicklung der Lokalen Netze dar, die heute in großer Vielfalt kommerziell verfügbar sind. Es genügt allerdings nicht, auf Protokolle für LANs zurückzugreifen und einfach die Datenrate zu erhöhen. Dies zeigt z.B. eine kurze Betrachtung des im Bereich der Lokalen Netze bisher wohl erfolgreichsten Protokolls "CSMA/CD": Bei CSMA/CD wartet eine sendebereite Station zunächst so lange, bis der Kanal frei wird. Danach beginnt sie mit der Übertragung, wobei es allerdings möglich ist, daß mehrere Stationen quasi-simultan zu senden beginnen, denn die Signallaufzeiten sind in Lokalen Netzen zwar kurz aber nicht vernachlässigbar. Aus diesem Grunde kann der Fall eintreten, daß eine Station noch nicht erkennen kann, daß eine andere bereits eine Übertragung begonnen hat und daher ihrerseits zu senden beginnt, vgl. Abb. 3.1. Die so entstehende Kollision, die i.a. die Zerstörung aller beteiligten Nachrichten zur Folge hat, wird von allen Stationen erkannt. Dieser Konflikt bewirkt den Abbruch der Übertragung und die Wahl einer zufälligen Verzögerungszeit bis zum erneuten Sendeversuch, der aber erneut zum Konflikt führen kann.

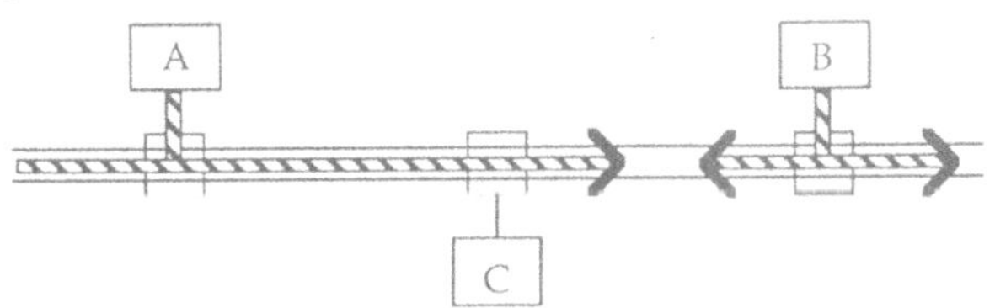

Abb. 3.1 Kollision bei CSMA/CD

Eine Stabilität des Kommunikationssystems versucht man z.B. dadurch zu erreichen, daß die Anzahl der aufeinanderfolgenden Konflikte gezählt und das Intervall, aus dem die zufällige Verzögerungszeit gemäß einer Gleichverteilung gewählt wird, mit jedem Konflikt vergrößert wird. Schließlich wird eine erfolgreiche Übertragung erzielt oder - bei zu vielen Konflikten - der Sendeversuch abgebrochen.

Details über die konkrete Festlegung des hier kurz skizzierten Verfahrens können dem Standard [IEEE 85b] entnommen werden. Im folgenden wird jedoch nicht näher auf dieses Protokoll eingegangen, da es für Hochgeschwindigkeitskommunikation ungeeignet ist. Der Grund liegt darin, daß die Signallaufzeit mit dem Übergang zu höheren Datenraten nicht kleiner sondern - bei Verwendung von Glasfaser statt metallischem Leiter - größer wird. Daher sind immer mehr Datenbits durch Kollisionen gefährdet, so daß CSMA/CD und ähnliche Random Access Verfahren für Hochgeschwindigkeitsnetze nicht oder nur bedingt in Frage kommen. Die ernsthaft diskutierten Vorschläge von MAC-Protokollen für HSLANs arbeiten aus diesem Grunde konfliktfrei, doch auch bei dieser Einschränkung bleibt eine kaum überschaubare Fülle denkbarer Ansätze erhalten. Einer Protokollspezifikation muß daher die Festlegung des angestrebten Anwendungsbereiches vorausgehen.

3.1. Einsatzgebiete für HSLANs

Synchronisationsprobleme bei Parallelübertragung und die damit verbundenen Längenbegrenzungen für parallele Bussysteme ließen zu Beginn der achtziger Jahre den Wunsch nach Kommunikationssystemen für schnelle serielle Übertragung aufkommen. Konkret stand dabei zunächst die Kopplung von Großrechnern und schnellen Speichern im Vordergrund.

Im Laufe der Zeit rückten andere Einsatzgebiete für HSLANs in den Mittelpunkt des Interesses. In diesem Zusammenhang sind "Backbone-Netze" zur Kopplung von LANs ebenso zu nennen wie die schnelle Kopplung neuartiger Endgeräte. Workstations mit zahlreichen Megabyte Hauptspeicher und hochauflösenden Bildschirmen werden in Zukunft die Art völlig verändern, in der wir Daten generieren, speichern, bearbeiten und übertragen.

Versucht man eine Klassifikation nach der geographischen Distanz, dann ergeben sich im wesentlichen drei Einsatzgebiete für HSLANs, die im folgenden näher betrachtet werden sollen:

a) Rechenzentrum

Das Anwendungsgebiet "Rechenzentrum" ist durch eine relativ kleine Anzahl von Geräten - im wesentlichen Großrechner und Peripherie - gekennzeichnet, wobei auch die Entfernungen zwischen den Geräten klein sind.

b) Kopplung von Workstations/Minicomputern

Wie bereits erwähnt, wird der Einsatz von Workstations mit hochauflösenden Bildschirmen in naher Zukunft leistungsfähigere Kommunikationssysteme als die herkömmlichen LANs erforderlich machen. Im Gegensatz zum Anwendungsgebiet "Rechenzentrum" wird nun eine ziemlich große Zahl von Geräten angeschlossen, die typischerweise häufig ausgeschaltet werden. Dies muß bei der Planung eines Netzes berücksichtigt werden, da für die hier skizzierte Anwendung nur der Anschluß mit geeigneten Konzentratoren, die stets eingeschaltet bleiben, die ggf. erforderliche Signalregenerierung sicherstellen kann.

c) Backbone-Netze

Als eines der wichtigsten Einsatzgebiete für HSLANs zeichnet sich die Verwendung als "Backbone-Netz" ab, d.h. als Kommunikationssystem zur Kopplung von Teilnetzen. Weil schon allein aus Kostengründen nicht jede Station unmittelbar an ein HSLAN angeschlossen werden kann, erscheint eine hierarchische Struktur sinnvoll: Teilnetze mit wesentlich geringeren Datenraten verbinden die Endgeräte und die so entstehenden Inseln werden durch ein HSLAN als Backbone-Netz gekoppelt. Backbone-Netze erlauben den Einsatz von Protokollen, die auf spezielle Lastprofile zugeschnitten sind, und ermöglichen dennoch die Kommunikation über die Grenzen der Subsysteme hinaus.

Natürlich sind auch Mischungen der hier genannten Einsatzgebiete möglich, was im Extremfall zu dem in Abb. 3.2 dargestellten Szenario führen kann.

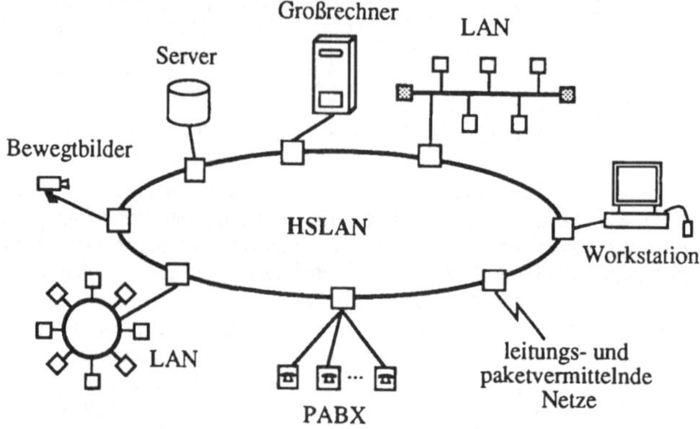

Abb. 3.2 Ein HSLAN für unterschiedlichste Anwendungen

3.2. Topologien

Lokale Hochgeschwindigkeitsnetze setzen als Übertragungsmedium Lichtwellenleiter ein. Da diese am einfachsten für Punkt-zu-Punkt Verbindungen zwischen aktiv angeschlossenen Stationen verwendet werden können, ergeben sich Ring und Stern als natürliche Topologien. Daneben werden Bussysteme diskutiert, die allerdings - im Gegensatz zu Bussystemen mit metallischem Leiter - aus technischen Gründen vorwiegend unidirektionale Übertragung einsetzen. Um dennoch Broadcast-Kommunikation ermöglichen zu können, wurden vor einigen Jahren "doppelter Bus" und "gefalteter Bus" als Topologien vorgeschlagen, vgl. Abb. 3.3. In diesem Zusammenhang ist zu erwähnen, daß der sich z.Z. für Metropolitan Area Networks (MANs) abzeichnende Standard [IEEE 89] den doppelten Bus als Topologie voraussetzt.

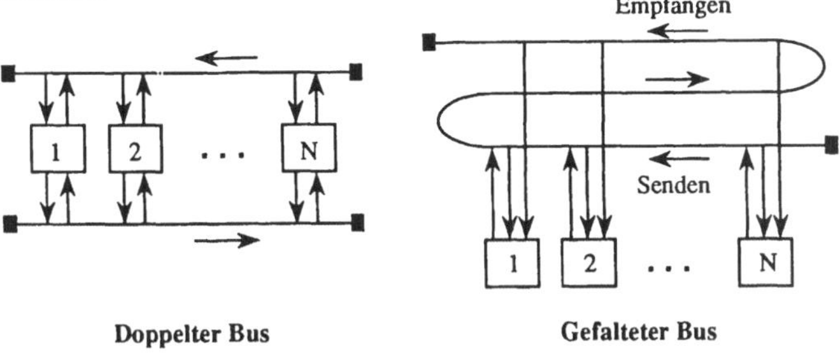

Abb. 3.3 Topologien für unidirektionale Bussysteme

3.3. Standards

In der Vergangenheit wurde eine Vielzahl von Protokollen für Lokale Hochgeschwindigkeitsnetze diskutiert, vgl. z.B. [SPA 87] oder [DAN 88]. Die nachfolgenden Betrachtungen im Rahmen diesen Beitrages beschränken sich auf die beiden Protokolle, die von Standardisierungsgremien vorgeschlagen worden sind: FDDI und DQDB.

3.3.1. FDDI

Das Fiber Distributed Data Interface (FDDI) ist der inzwischen sehr weit fortgeschrittene Entwurf eines ANSI-Standards (vgl. [ANSI 87a], [ANSI 87b] und [ANSI 88]) für HSLANs. Das auf einer optischen Ringstruktur basierende System, das mit einer nominellen Datenrate von 100 Mbit/s arbeitet, wurde speziell auf die Bedürfnisse der in Kapitel 3.1. dargestellten Einsatzgebiete zugeschnitten. Die Systemparameter wurden so gewählt, daß geographische Distanzen bis 200 km möglich sind; die maximale Zahl physikalischer Anschlüsse liegt bei 1000.

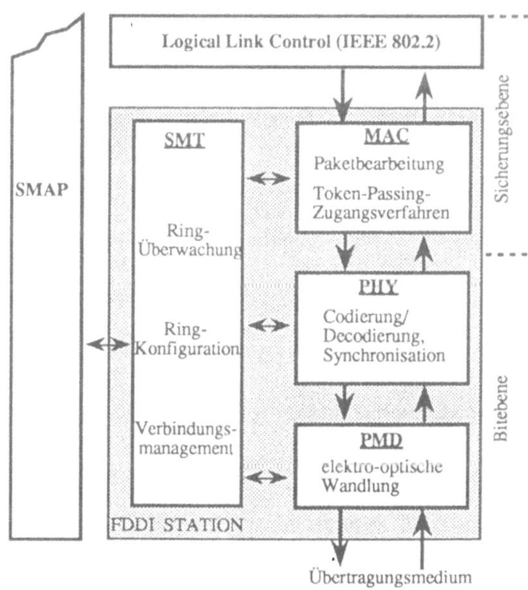

Abb. 3.4 FDDI-Blockdiagramm

Der FDDI-Standard umfaßt insgesamt 4 Teilstandards, deren Zusammenspiel in Abb. 3.4 dargestellt ist:
- Physical Layer Medium Dependent (PMD)
- Physical Layer Protocol (PHY)
- Medium Access Control (MAC)
- Station Management (SMT)

Der PMD-Standard befaßt sich mit Eigenschaften des optischen Übertragungssystems. Die Verbindung zwischen dieser Komponente und dem Medienzugangsprotokoll ist in PHY festgelegt, wobei PHY vor allem Synchronisation und Codierung/Decodierung (zur Erleichterung der Synchronisation werden 4 Bit auf 5 Bit codiert) spezifiziert. Überwacht und gesteuert wird der Protokollablauf vom Stationsmanagement SMT, doch ist der SMT-Standard z.Z. noch nicht verabschiedet worden. Bzgl. einer detaillierten Darstellung des FDDI-Protokolls sei auf den Standard selbst oder auf [MAR 89b] verwiesen. Da jedoch das MAC-Protokoll für die nachfolgenden Betrachtungen zur Leistungsfähigkeit des FDDI-Protokolls wesentlich ist, soll hier kurz darauf eingegangen werden.

Bei FDDI wird das Senderecht mittels einer speziellen Marke (engl. Token) von Station zu Station entlang des physikalischen Rings weitergegeben. Eine sendebereite Station entfernt das vorbeilaufende Token vom Ring und kann unmittelbar danach mit der Übertragung beginnen. Aus Synchronisationsgründen schreibt der FDDI-Standard vor, daß jedes Paket mit einer Präambel (PA) von 64 bit Länge entsprechend 0,64 µs Sendezeit beginnt. Dann folgen Anfangskennung (Starting Delimiter SD, 8 bit), Rahmenkontrolle (Frame Control FC, 8 bit), Ziel- und Quelladresse (Destination Address DA und Source Address SA, jeweils 16 bit oder 48 bit) sowie das Informationsfeld für Nutzdaten, vgl. Abb. 3.5. Das Rahmenende bilden Prüfsequenz zur Fehlererkennung (Frame Check Sequence FCS, 32 bit), Endekennung (End Delimiter ED, 4 bit) und ein Rahmen-Status-Feld (Frame Status FS, ≥ 12 bit), in dem erkannte Fehler und erfolgreicher Empfang vermerkt werden können. Insgesamt darf - wiederum aus Synchronisationsgründen - die Rahmenlänge nicht größer als 4500 byte werden, woraus sich eine Beschränkung für die Länge des Informationsfeldes ergibt.

PA	SD	FC	DA	SA	INFO	FCS	ED	FS

Abb. 3.5 Rahmenformat bei FDDI

FDDI unterstützt zwei Arten von Übertragung: "Synchrone Übertragung" (mit garantierter Sende- und Antwortzeit) und "asynchrone Übertragung" (mit dynamischer Sendezeitzuteilung). Synchrone Übertragung wird für Anwendungen mit vorhersagbarem Bedarf an Bandbreite und bekannten Antwortzeitschranken eingesetzt; die entsprechende Sendezeitvergabe erfolgt nicht durch das Medienzugangsprotokoll sondern durch das Management. Dagegen wird die Zuteilung der für asynchrone Übertragung verbleibenden Bandbreite durch ein Timer-gesteuertes Token-Verfahren geregelt: Jede Station verwaltet einen "Token-Rotations-Timer" (TRT), der zur Vergabe von asynchroner Sendezeit eingesetzt wird. Während der Ringinitialisierung wird eine angestrebte Rotationszeit (engl. "operative Target Token Rotation Time", T_OPR) zwischen den angeschlossenen Stationen ausgehandelt. Das MAC-Protokoll von FDDI garantiert eine mittlere Token-Rotationszeit, die nicht größer als T_OPR ist, und eine maximale Token-Rotationszeit, die nicht größer als 2 T_OPR ist.

Synchrone Übertragung

Eine FDDI-Station, der zuvor vom Management Sendezeit für synchrone Übertragung eingeräumt wurde, darf jedes empfangene Token für diese Art von Datentransfer nutzen. In diesem Zusammenhang ist zu beachten, daß die Unterstützung von synchroner Übertragung optional und für Interoperabilität nicht erforderlich ist.

Asynchrone Übertragung

Ein auf dem Ring festgestelltes Token darf nur dann für asynchrone Übertragung vom Ring entfernt und anschließend genutzt werden, wenn der TRT zeigt, daß die Zeit seit dem letzten Tokenempfang kleiner als T_OPR war. In diesem Fall wird die Differenz von TRT und T_OPR im Token-Holding Timer (THT) festgehalten und TRT zur Messung der nächsten Tokenrotationszeit auf T_OPR heraufgesetzt. Der THT mißt dann die Sendezeit für asynchrone Übertragung. Nachdem der THT abgelaufen ist, muß das Token baldmöglichst, d.h. nach dem Ende der aktuell laufenden Paketübertragung, weitergegeben werden.

Für nähere Informationen über das hier verkürzt dargestellte Timer-gesteuerte Token-Verfahren sei wiederum auf [MAR 89b] verwiesen. An dieser Stelle sei aber noch erwähnt, daß FDDI die Option zur Unterstützung von Prioritäten enthält: Jeder Prioritätsklasse wird ein "Schwellenwert" zugeordnet, und Token dürfen nur dann vom Ring entfernt und für Sendung von Daten einer gegebenen Priorität verwendet werden, wenn der TRT bei Empfang des Tokens den zugehörigen Schwellenwert noch nicht unterschritten hat. Entsprechend muß das Token weitergegeben werden, wenn der THT den Schwellenwert der entsprechenden Prioritätsklasse unterschreitet.

Von Anfang an wurde kritisiert, daß FDDI isochrone Dienste wie z.B. Sprachübertragung nur unzureichend unterstützt. Als Kompromiß wurde FDDI-II vorgeschlagen, ein Protokoll, das in gewissem Sinne eine Obermenge des bisher vorgestellten FDDI-I darstellt. Zusätzlich zu der in FDDI-I eingesetzten Paketvermittlung enthält FDDI-II durch Einsatz einer "Hybriden Ringkontrolle" die Möglichkeit der Leitungsvermittlung. Darüber hinaus wurden auch Modifikationen der Bitebene erforderlich, da FDDI-II auf einer zyklischen Rahmenstruktur basiert, die jeweils nach 125 µs wiederholt wird.

An dieser Stelle soll nicht näher auf FDDI-II eingegangen werden; auch hier enthält [MAR 89b] weitere Informationen. Erfolge bei den Standardisierungsbestrebungen im Bereich der "Metropolitan Area Networks" (vgl. Kapitel 3.3.2.) lassen erhebliche Zweifel daran gerechtfertigt erscheinen, daß FDDI-II in Zukunft eine größere Bedeutung erlangen wird.

3.3.2. DQDB

Als "Distributed Queue Dual Bus" (DQDB, vgl. [IEEE 89]) wird der Entwurf eines Standards bezeichnet, der von der Arbeitsgruppe IEEE 802.6 vorgelegt wurde. DQDB soll öffentliche Netzbetreiber in die Lage versetzen, bei geographischen Distanzen von 5 km bis zu über 50 km "Metropolitan Area Networks" (MANs) mit Datenraten im hohen Megabit/s-Bereich zu betreiben. Das DQDB-Protokoll, ursprünglich als "QPSX" (Queued Packet and Synchronous Exchange) bekannt, erlaubt sowohl Paketvermittlung mit mehreren Prioritätsklassen als auch Leitungsvermittlung. Im Gegensatz zum FDDI-Token Ring, der im wesentlichen eine Weiterentwicklung des für den Bereich der LANs standardisierten Protokolls [IEEE 85d] darstellt, ist DQDB in vieler Hinsicht ein völlig neuartiges Protokoll, das einem intuitiven Verständnis kaum zugänglich ist.

3.3.2.1. Netzarchitektur

Der DQDB-Standard sieht vor, daß ein Metropolitan Area Network durch Kopplung von mehreren DQDB-Teilnetzen entsteht, wobei jedes Teilnetz eine der in Abbildung 3.6 dargestellten Topologien verwendet. Die Stationen eines DQDB-Netzes sind an beide Busse mit je einem Lese- und einem Schreibkopf angeschlossen. Sie beobachten ständig die Aktivität auf dem Doppel-Bus, doch greifen sie nur dann aktiv in die Übertragung ein, wenn sie durch das Zugangsprotokoll (vgl. Kapitel 3.3.2.2) dazu die Berechtigung erhalten.

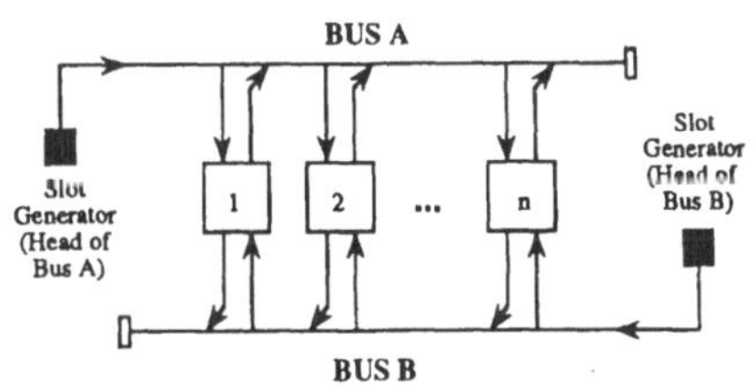

Offene Bus-Topologie

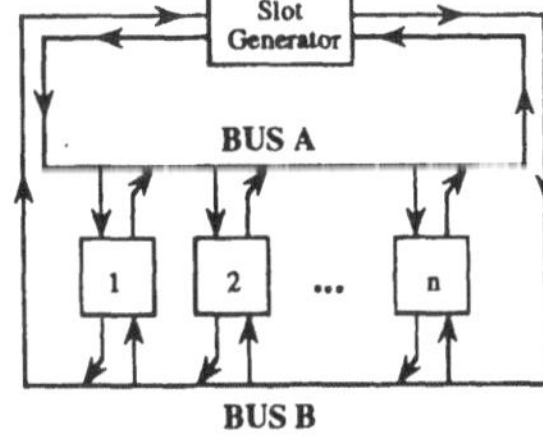

Geschlossene Bus-Topologie

Abb. 3.6 Bus-Topologien für DQDB

3.3.2.2. Medienzugangsprotokoll

Auf jedem Bus wird durch einen "Slot Generator" eine Slotstruktur generiert, wobei jeder Slot ein Feld der Länge 64 byte für den Transport von Nutzinformationen enthält.

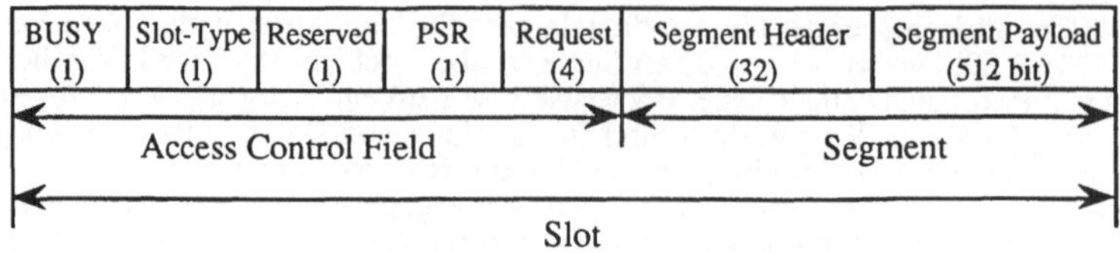

Abb. 3.7 Slotformat bei DQDB

Die Abbildung 3.7 zeigt das Slotformat von DQDB. Jeder Slot beginnt mit einem Feld für die Zugangskontrolle (Lese- und Schreiboperationen). In diesem Access Control Field (ACF) erfolgt die Kennzeichnung, ob der Slot Daten enthält (BUSY = 1) oder frei ist (BUSY = 0). Darüber hinaus enthält ACF die Information, ob der Slot für asynchronen Verkehr (Slot-Type = 0) oder für isochronen Verkehr (Slot-Type = 1) eingesetzt wird. Erstere Art von Slots wird "queued-arbitrated" genannt (Regelung des Zugangs durch verteilte Warteschlangen), letztere "pre-arbitrated" (Regelung des Zugangs durch das Management). Danach folgen zwei Bits, deren Verwendung derzeit noch nicht festgelegt ist und das Request-Feld, das für den Zugang zu queued-arbitrated Slots von Bedeutung ist (s.u.). Nach dem ACF folgt in jedem Slot ein "Segment Header", auf dessen Bedeutung hier nicht näher eingegangen werden soll, und ein "Segment Payload Field", das die Nutzinformationen aufnimmt.

Bzgl. einer Beschreibung des Zugangs zu pre-arbitrated Slots sei auf [IEEE 89] verwiesen. Wie bereits erwähnt, wird der Zugang zu queued-arbitrated Slots durch verteilte Warteschlangen (daher Distributed Queue Dual Bus) geregelt. Um diese verteilte Warteschlange wie eine zentrale Warteschlange arbeiten zu lassen, werden in jeder Station spezielle Zähler RC (Request Counter) und CD (Countdown Counter) für jeden der unidirektionalen Busse verwaltet, vgl. Abb. 3.8, wobei RC und CD durch die Busy- und Request-Bits in den ACFs gesteuert werden. Die Arbeitsweise dieser Zähler wird im folgenden am Beispiel des Zugangs zu Bus A erläutert (Zähler für Bus B arbeiten analog).

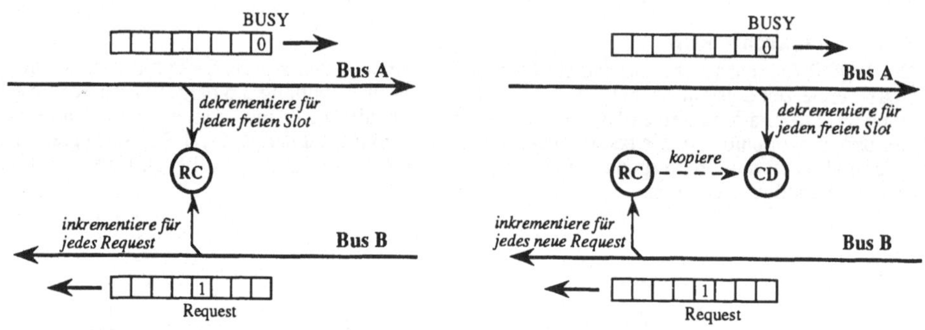

Abb. 3.8 Zähler für Bus A

Solange kein asynchrones Segment sendebereit ist, wird RC für jedes Request auf Bus B inkrementiert und für jeden freien Slot auf Bus A dekrementiert (Ausnahme: Bei RC = 0 wird nicht dekrementiert). Da Requests zur Mitteilung von Sendewünschen dienen, erhält so die hier betrachtete Station Informationen darüber, wieviele Stationen "stromabwärts" an Bus A auf einen freien Slot warten.

Sobald eine Station sendebereit für Bus A wird, sendet sie ein Request auf Bus B und kopiert RC in den Countdown Counter. Der aktuelle Wert von CD zeigt, wieviele freie Slots noch nicht für eigene Datenübertragung verwendet werden dürfen, weil "stromabwärts" liegende Stationen auf freie Slots warten und ihre Sendewünsche zuvor mitgeteilt hatten. CD wird also für freie Slots dekrementiert und erreicht schließlich den Wert 0, was die eigene Sendeberechtigung anzeigt. Soll nach der Übertragung eine weitere folgen, dann muß erneut ein Request gesendet und ausgehend von dem dann aktuellen RC der Sendezeitpunkt neu bestimmt werden. Abschließend sei noch darauf hingewiesen, daß die Darstellung des MAC-Protokolls von DQDB hier etwas vereinfacht wurde: DQDB sieht mehrere Prioritätsklassen vor und verwendet daher nicht ein sondern mehrere Request-Bits in jedem Slot. Der Protokollablauf entspricht jedoch in natürlicher Weise dem hier dargestellten.

4. Modellierung und Leistungsbewertung

Bei der Modellierung von technischen Systemen muß stets zunächst die Frage beantwortet werden, welches Ziel mit dieser Modellierung verfolgt wird. Bezogen auf Kommunikationssysteme ist hier zu unterscheiden zwischen Modellierung für die Protokollspezifikation und Modellierung für die Leistungsbewertung. Beide Zielsetzungen sind für die Praxis von außerordentlich großer Bedeutung: Nur formal korrekt spezifizierte Protokolle können sicherstellen, daß verschiedene Implementierungen interoperabel sind, nur im Rahmen einer formalen Spezifikation kann die funktionale Korrektheit von Protokollen überprüft werden. Darüber hinaus muß aber auch die Leistungsfähigkeit überprüft und ggf. durch entsprechende Protokollmodifikation sichergestellt werden. Es sind zwar einige Ansätze zur Modellierung für beide Zielsetzungen zu beobachten (vgl. z.B. [BAL 86]), doch beschränkt man sich nicht zuletzt aus Komplexitätsgründen in der Regel auf Modelle, die auf die speziellen Bedürfnisse einer Zielsetzung zugeschnitten sind. Um den Rahmen dieses Beitrages nicht zu sprengen, werden im folgenden nur Modelle mit dem Ziel der Leistungsbewertung betrachtet.

Vereinfacht kann ein Kommunikationssystem gemäß Abb. 4.1. modelliert werden. Protokolldatenheiten (PDUs) einer Schicht N+1 werden über entsprechende Warteschlangen an die Schicht N übergeben. Eine derartige "(N+1)-PDU" stellt für Schicht N eine sog. Dienstdateneinheit (Service Data Unit, SDU) dar, die auf Schicht N nicht interpretiert werden kann. Mit Kontrollinformationen (Protocol Control Information, PCI) der Schicht N wird die "(N)-SDU" zur Protokolldateneinheit der Schicht N, d.h. zur "(N)-PDU", die über einen Puffer an die Schicht N-1 übergeben wird. Diese und ggf. die darunter liegenden Schichten sind für den Transport zum Empfänger verantwortlich, wo auf Schicht N die von dieser Schicht eingefügte Kontrollinformation wieder entfernt und die Dienstdateneinheit unverändert "nach oben" weitergegeben wird.

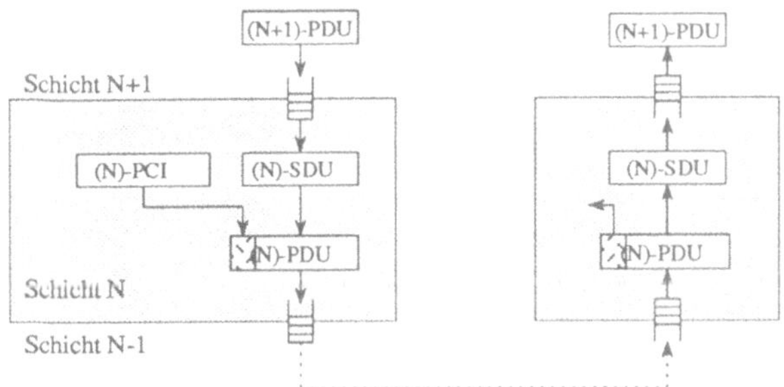

Abb. 4.1. Vereinfachtes Modell eines Kommunikationssystems

Dieses Modell liefert die Grundlage für das in Abb. 4.2 dargestellte Modell zur Leistungsbewertung der Schicht N. Die von den höheren Ebenen generierte Last wird in einem Lastmodell zusammengefaßt und der Transport durch die unteren Ebenen wird durch ein Modell des Transportkanals repräsentiert. Verarbeitungszeiten am Ziel können durch ein entsprechendes Modell des Empfängers berücksichtigt werden.

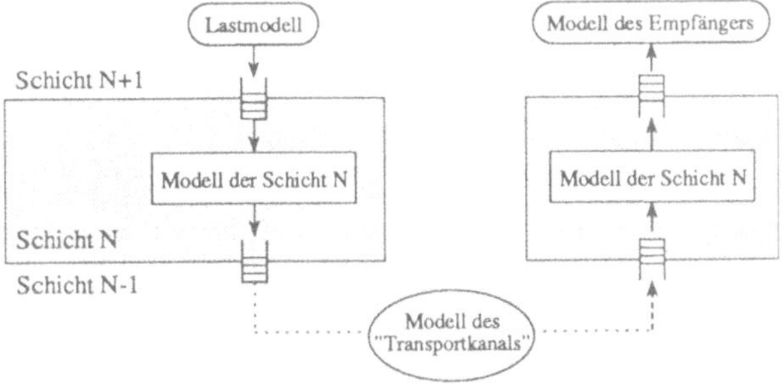

Abb. 4.2. Modell zur Leistungsbewertung der Schicht N

Wie bereits angedeutet, beinhaltet dieses Modell nicht die gesamte Komplexität eines realen Kommunikationssystems, denn es berücksichtigt z.B. in der hier dargestellten Form weder Segmentierung - aus einer (N-)-SDU werden mehrere (N)-PDUs - noch den Einsatz von Multiplex-Verfahren. Darüber hinaus geht das in Abb. 4.1. gezeigte Modell von einer gerichteten Kommunikation aus, wobei Quittungen des Empfängers nicht berücksichtigt werden. Dennoch können aufbauend auf diesem einfachen Modell schon viele wichtige Aussagen gewonnen werden. Verfeinerungen sind natürlich möglich und zur Untersuchung einiger wichtiger Fragestellungen zwingend erforderlich. Es ist jedoch zu berücksichtigen, daß mit wachsender Komplexität des Modells eine mathematische Analyse immer schwieriger, wenn nicht sogar unmöglich wird. Auch bei Simulationen von HSLANs zeigen sich sehr schnell die Grenzen von Speicherplatz und akzeptierbarer Rechenzeit. Der Grund ist vor allem darin zu sehen, daß bei Hochgeschwindigkeitsnetzen ggf. sehr viele relevante Ereignisse in sehr kurzer Zeit eintreten können - so hat z.B. bei DQDB mit 100 Mbit/s ein Slot eine zeitl. Dauer von 5,52 µs - andererseits aber häufig die Berücksichtigung von Lastschwankungen mit relativ großen zeitlichen Abständen - z.B. bei File-Transfer - wünschenswert ist.

Im Rahmen dieses Beitrages ist es nicht möglich, die allgemeine Diskussion von Systemmodellen im Rahmen des ISO-Referenzmodells zu vertiefen. Der interessierte Leser sei z.B. auf die Arbeiten [REI 86] oder [KÜHN 86] verwiesen.

4.1. Analysetechniken

Leistungsanalysen von existierenden oder geplanten Kommunikationssystemen gehen stets von Modellen aus, die als wesentlich erachtete Eigenschaften des technischen Systems widerspiegeln, das Gegenstand der Analyse ist. Die Art der Modellierung hat nicht nur einen wesentlichen Einfluß auf die Güte der Lösung, sondern sie bestimmt auch die Techniken, die bei der Analyse eingesetzt werden können. Je mehr das Modell des Kommunikationssystems und natürlich auch der Ankunftsprozesse den aus mathematischer Sicht "angenehmen" Systemen entgegenkommt, desto "exakter" kann die Analyse werden. Der Begriff "exakt" bezieht sich hier auf die Analyse des Modells, doch folgt daraus keineswegs schon eine wirklich genaue Lösung für den im Mittelpunkt des Interesses stehenden technischen Sachverhalt.

4.1.1. Exakte Analyse

Exakte Analysen zeichnen sich dadurch aus, daß die erzielten Ergebnisse dann exakt sind, wenn das zugrundeliegende Modell mit der Realität übereinstimmt. Die Vergangenheit hat gezeigt, daß es eine Vielzahl von Beispielen gibt, bei denen die Verkehrstheorie auch dann noch brauchbare Aussagen liefert, wenn sie auf Annahmen aufbaut, die nur näherungsweise erfüllt sind. Im Fall von Kleinrock's berühmt gewordener "Unabhängigkeitsannahme" handelt es sich sogar um eine Voraussetzung, die von Kleinrock selbst als *"clearly false"* [KLE 76, p. 322] bezeichnet wird. Obgleich also die Ergebnisse, die mit Hilfe der Warteschlangentheorie erzielt werden, in manchen Fällen erstaunlich robust sind, muß doch stets geprüft werden, ob diese Situation für das jeweils untersuchte Anwendungsgebiet gegeben ist. Gute Ergebnisse sind dann zu erwarten, wenn das der Analyse zugrundeliegende Lastmodell mit der Realität weitgehend überstimmt. Darüber hinaus darf sich das verwendete Modell nicht zu weit vom betrachteten technischen System entfernen. Gerade an diesem Punkt ergeben sich wesentliche Einschränkungen für die Einsetzbarkeit exakter Analysen: In vielen Fällen sind die technischen Systeme so komplex, daß genaue Modelle zwar vielleicht prinzipiell exakt analysiert werden könnten, aber weder Speicherplatz noch Rechenleistung ausreichen, um diese Lösung tatsächlich auch zu bestimmen.

4.1.2. Approximation

Neben der exakten Analyse gibt es noch eine andere mathematische Methode zur Leistungsbewertung: die Approximation. Techniken wie "Flußapproximation" oder "Diffusionsapproximation" beschreiben Ankunfts- und Abfertigungsprozeß nicht mit Verteilungsfunktionen sondern mit Mittelwerten und ggf. einigen höheren Momenten. Daneben gibt es eine Vielzahl von Heuristiken (vgl. z.B. [HAL 85] oder [JAG 82]), die unter bestimmten Randbedingungen brauchbare Resultate liefern. Da hier weniger Informationen in die Analyse eingehen als bei den exakten Analysen, bei denen in der Regel zumindest eine vollständige Beschreibung der Ankunftprozesse benötigt wird, kann i.a. nicht erwartet werden, daß eine derartige approximative Analyse stets gute Resultate liefert.

Ein Vergleich mit exakten Analysen kann schon einen gewissen Einblick in die Güte der Ergebnisse geben, doch bleibt ein derartiger Vergleich auf Sonderfälle beschränkt, in denen eine exakte Analyse möglich ist. Auch für diese exakt analysierbaren Sonderfälle kann eine Approximation sinnvoll sein, wenn sie deutlich weniger Speicherplatz und/oder deutlich weniger Rechenzeit benötigt und dennoch brauchbare Resultate liefert. Interessanter werden approximative Analysen aber meist erst dann, wenn sie zur Lösung von Problemen eingesetzt werden, die mit exakten Analysen nicht lösbar sind. Hier kann nur ein Vergleich mit Simulationsergebnissen zeigen, ob die Approximation für ein spezielles Einsatzgebiet geeignet ist.

4.1.3. Simulation

Wo mathematische Methoden zur Leistungsbewertung an die Grenzen der Einsetzbarkeit stoßen, da stellen Simulationen häufig das einzige Hilfsmittel dar, um die gewünschten Aussagen erzielen zu können. Simulationen sind nahezu universell einsetzbar und erlauben die Leistungsbewertung selbst hochkomplexer Modelle, wobei der Detaillierungsgrad für die Simulation unterschiedlich gewählt werden kann. Einschränkungen ergeben sich "nur" durch den Bedarf an Rechenzeit und Speicherplatz.

Muß bei mathematischen Analysen stets die Gültigkeit der zugrundeliegenden Annahmen überprüft werden, so müssen Simulationsprogramme kalibriert und validiert werden. Bei der Simulation von LANs kann man heute zu diesem Zweck auf Messungen zurückgreifen und auf diese Weise Simulationsergebnisse und Realität einander gegenüberstellen (vgl. z.B. [GRA 89]), doch ist diese Vorgehensweise bei den derzeit im wesentlichen erst geplanten HSLANs noch nicht möglich. Kalibrierung ist somit in der Regel nicht möglich, Validierung von Simulationsmodell und dem diesem Modell entsprechenden Programm kann nur (unter Berücksichtigung der entsprechenden Einschränkungen) durchVergleich mit mathematischer Analyse erfolgen. Dagegen ist ein formaler Korrektheitsbeweis nicht zuletzt aus Komplexitätsgründen meist undurchführbar.

Während bei Veröffentlichungen, die auf mathematischen Analysen (exakt oder approximativ) basieren, die zugrundeliegenden Annahmen und damit auch mögliche Schwächen meist klar genannt werden, gibt es leider eine Vielzahl von Veröffentlichungen, die Simulationsergebnisse präsentieren, ohne auf die Details der Modellierung einzugehen. Dies macht sowohl den Vergleich als auch die Überprüfung von Simulationsergebnissen schwierig, wenn nicht, wie z.B. in [GBU 89b], der Aufwand eigener Simulationen in Kauf genommen werden soll, um veröffentlichte Ergebnisse zu überprüfen.

4.2. Lastmodelle

Die Datenkommunikation im lokalen Bereich war in der Vergangenheit sehr stark von interaktivem Verkehr geprägt, der in guter Näherung mit Poisson-Ankünften modelliert werden konnte. Messungen in Lokalen Netzen zeigen jedoch, daß die Ankunftsprozesse in vielen Fällen nur noch unzureichend mit diesem Lastmodell charakterisiert werden können. Dies wird in neueren Veröffentlichungen wie z.B. [JAI 86] oder [GIHR 89] deutlich.

Niemand kann heute vorhersagen, was in einigen Jahren die "typische" Last für ein HSLAN wie das in Abb. 3.2 dargestellte sein wird. Analysen und Simulationen gehen vielfach noch von Poisson-Ankünften oder von "Burst-Poisson" (exponentiell verteilte Zwischenankunftszeiten von Blockankünften) aus, da diese für viele Anwendungen als Worst-Case-Annahmen erachtet werden können. Die nachfolgenden Betrachtungen werden darüber hinaus das in Abb. 4.3. dargestellte Modell von File Transfer und eine "gemischte Quelle" (10 % des Datenvolumens interaktiver Verkehr, 90 % File Transfer) berücksichtigen. Für File Transfer wird vorausgesetzt, daß je zwei File-Übertragungsphasen durch negativ exponentiell verteilte Idle-Zeiten t_{idle} getrennt sind. Sobald das erste Paket eines Files eingetroffen ist, folgen gemäß unserem Modell die weiteren Pakete in festem Abstand von 9,6 µs (minimaler Paketabstand nach [IEEE 85b]).

Details zu den zuletzt angedeuteten Quellen sind in [MAR 88a] und in [MAR 88b] zu finden.

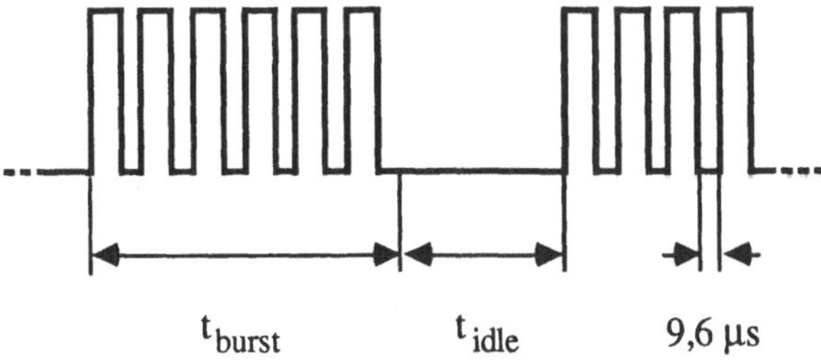

Abb. 4.3 Modellierung von File Transfer

Wenn auch heute nicht vorhergesagt werden kann, wie neue Dienste die "Workload" verändern werden, so kann doch auch in Zukunft gute Leistungsfähigkeit von den Protokollen erwartet werden, die in ihrem Leistungsverhalten weitgehend unabhängig von speziellen Lastmustern sind. Daher wird im folgenden stets kritisch die Frage geprüft, inwiefern die Performance der betrachteten Systeme von speziellen Lastmodellen abhängt.

4.3. Leistungsbewertung des MAC-Protokolls

Leistungsfähige Protokolle zur Regelung des Medienzugangs sind zwar nicht hinreichend aber doch notwendig für ein leistungsfähiges Gesamtsystem. MAC-Protokolle stehen daher seit Jahren im Mittelpunkt des Interesses, was zu einer großen Zahl von Veröffentlichungen in einschlägigen Fachzeitschriften und Tagungsbänden geführt hat. Dabei beschränkt sich die Mehrzahl der Untersuchungen auf das Lastmodell (am Übergang von Ebene 2b nach Ebene 2a) und ein Modell des Medienzugangsprotokolls, wobei der "Transportkanal", d.h. die physikalische Ebene, nur noch insofern in die Untersuchung eingeht, als Ebene 1 die Datenrate bestimmt. Von Details der physikalischen Ebene (z.B. Synchronisation, physikalische Signaldarstellung, Bypass-Mechanismen,...) wird i.d.R. abgesehen. Darüber hinaus wird in vielen Fällen auf die Modellierung des Empfängers verzichtet, d.h. die Bearbeitung von Datenpaketen wird in dem Moment als abgeschlossen betrachtet, in dem sie vollständig gesendet worden sind.

Die Vielfalt denkbarer Protokolle macht es unmöglich, ein universelles Modell für alle lokalen Kommunikationssysteme zu entwickeln und dieses dann mit einer einheitlichen Technik zu analysieren. Daher muß sich jede Analyse auf die Ermittlung von Aussagen für eine bestimmte Klasse von Protokollen beschränken. Da der vorliegende Beitrag nur die wichtigsten Standardisierungsvorschläge berücksichtigen kann, beschränken sich die nachfolgenden Untersuchungen auf die Leistungsfähigkeit von FDDI und DQDB, wobei ausschließlich asynchroner Verkehr betrachtet wird.

4.3.1. FDDI

FDDI kann als Token-Passing System durch das in Abbildung 4.4 dargestellte Warteschlangensystem modelliert werden (Ankunftsprozesse und Abfertigungsprozesse werden durch die entsprechenden Raten symbolisiert). In dem hier betrachteten System fragt der Server zyklisch alle Warteschlangen ab, ob ein "Kunde" zur Bedienung bereitsteht, wobei die Zeit für den Übergang von einer Schlange zur anderen i.a. nicht vernachlässigt werden darf. Findet der Server in einer Warteschlange mindestens einen wartenden Kunden vor, so sind als Extremfälle folgende Strategien denkbar:

- vollständiger Service (exhaustive Service)
  Der Server bleibt so lange bei dieser Schlange, bis sie leer ist.

- Einzelabfertigung (Single Service)
  Es wird pro Zyklus maximal ein Kunde pro Schlange abgefertigt.

Durch Zeitschranken, wie sie in FDDI Verwendung finden, kann jede beliebige Zwischenstufe erzielt werden, wobei die Wahl der angestrebten Tokenrotationszeit (T_OPR) jedoch starke Auswirkungen auf die Leistungsfähigkeit dieses Systems hat. Abb. 4.5 macht dies am Beispiel eines FDDI-Rings mit 10 Stationen und einer Rahmenlänge von 12660 bit deutlich. Variiert man die Ringlatenz, die sich als Summe von Stationslatenzen und Signallaufzeiten im Medium ergibt, so wird deutlich, wie stark der maximal erreichbare Durchsatz durch die Wahl von T_OPR geprägt ist. Würde man $T\_OPR = \infty$ wählen, so könnte eine Station ununterbrochen übertragen und der Durchsatz würde ausschließlich durch Datenrate und Länge des Informationsfeldes in jedem Rahmen bestimmt, vgl. Abb. 3.5. Bei $T\_OPR < \infty$ muß das Token zwar weitergegeben werden, doch steht bei großem T_OPR auch bei großer Ringlatenz - und die Ringlatenz ist ja fester Bestandteil einer jeden Tokenrotation - noch viel Zeit für Datenübertragung zur Verfügung. Schränkt man nun T_OPR mehr und mehr ein, dann wird der für Datenübertragung verwendbare Zeitanteil (und damit auch der maximale Durchsatz) immer kleiner. Wählt man gar T_OPR kleiner als die Ringlatenz, dann ist keine Datenübertragung mehr möglich. Es ist also stets abzuwägen zwischen dem Wunsch nach hohem Durchsatz, dem aufgrund der Overheadreduktion ein möglichst groß gewähltes T_OPR am nächsten kommt, und dem Wunsch nach gesicherten maximalen zeitlichen Abständen zwischen zwei aufeinanderfolgenden Zugriffsrechten ein und derselben Station.

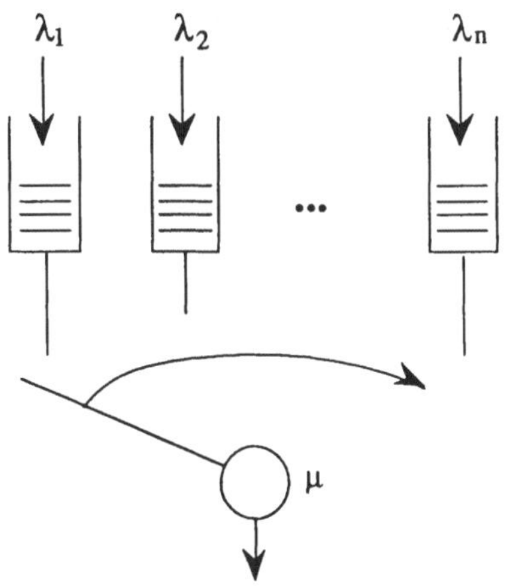

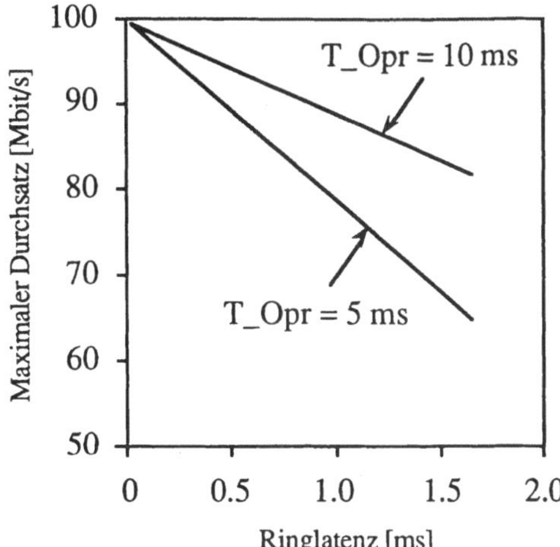

Abb. 4.4. Warteschlangensystem mit einem Server und zyklischer Abfertigung

Abb. 4.5. Maximaler Durchsatz vs. Ringlatenz (nach [DYK 87])

Bux und Dykeman haben in [DYK 87] neben der in Abb. 4.5 dargestellten Interdependenz von Durchsatz und T_OPR auch die Bedeutung der von FDDI als Option vorgesehenen Prioritäten untersucht. Abb. 4.6 zeigt Simulationsergebnisse für Durchsatz und Paketverzögerung bei 8 Stationen mit unterschiedlichen Prioritäten. Diesen Simulationen, bei denen negativ exponentiell verteilte Zwischenankunftszeiten verwendet wurden, lag eine Ringlatenz von 1,0236 ms und eine Rahmenlänge von 12660 bit zugrunde.

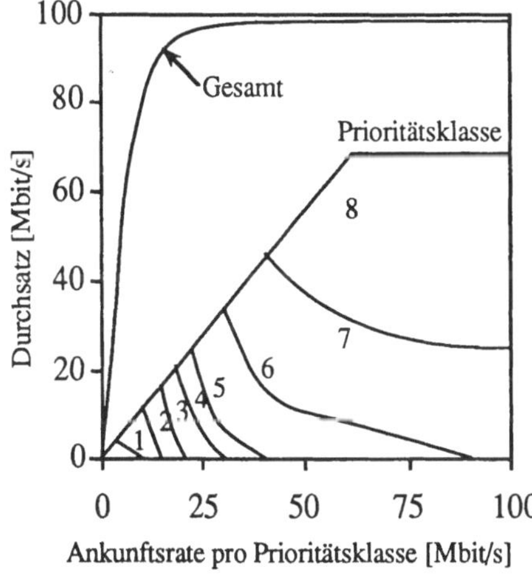

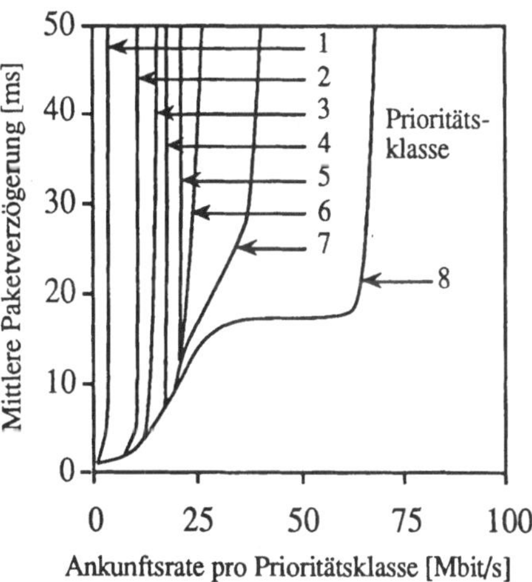

Abb. 4.6 Durchsatz und Paketverzögerung bei 8 Prioritätsklassen (nach [DYK 87])

Die Abbildung 4.6 zeigt, wie bei steigender Ankunftsrate der Durchsatz niedriger Prioritätsklassen zurückgeht, weil die Token Holding Timer die zu diesen Klassen gehörigen Schwellwerte erreichen. Bei dem hier betrachteten Beispiel haben nur die Stationen mit Priorität 7 und Priorität 8 eine Durchsatzgarantie während alle anderen Stationen schließlich nicht mehr senden dürfen.

Durchsatz und Verzögerungen werden bei FDDI u.a. durch Ankunftsraten und Zugehörigkeit zu Prioritätsklassen, Verteilung der Zwischenankunftszeiten, Anzahl der Stationen, Verteilung der Paketlänge und Wahl der Schwellwerte bestimmt. Darüber hinaus zeigen einige der in [DYK 87]

veröffentlichten Ergebnisse, daß Durchsatzreduktion und Anstieg von Verzögerungen bei Stationen niedriger Priorität stark von der Lage der betreffenden Stationen relativ zu Stationen mit höherer Priorität beeinflußt werden.

Offenbar ergibt sich eine kaum überschaubare Fülle von ggf. praxisrelevanten Szenarien, deren Untersuchung i.d.R. sehr aufwendig ist. Zwar findet das in Abb. 4.4 dargestellte Warteschlangensystem mit einem Server und zyklischer Abfertigung schon seit langer Zeit in der Forschung große Beachtung, da es nicht nur für Kommunikationssysteme sondern z.B. auch im Bereich der Betriebssysteme von großer Bedeutung ist, doch stoßen mathematische Analysen hier schnell an die Grenzen ihrer Einsetzbarkeit. Die Timer-gesteuerte Regelung des Medienzugangs in FDDI macht die analytische Behandlung exakter Modelle unmöglich und erzwingt so den Einsatz von Simulationen, welche bei HSLANs allerdings Rechenzeiten benötigen, die für Sensitivitätsanalysen kaum vertretbar sind. In vielen Fällen können aber auch wichtige Ergebnisse erzielt werden, indem die Modelle so weit vereinfacht werden, daß eine mathematische Analyse möglich wird. Dann kann auf die Vielzahl bekannter Ergebnisse zurückgegriffen werden, wie sie z.B. in [WAT 84], [TRA 86] und [TAK 86] zusammenfassend dargestellt sind.

Verzichtet man auf die Betrachtung von Prioritäten, die ja bei FDDI ohnehin nur optional vorgesehen sind, dann bleibt das Problem der Timer-gesteuerten Übertragung bei nun einer Prioritätsklasse. Der in [WEL 89a] dargestellte Vergleich von FDDI mit dem "Backbone Wideband Network (BWN)", einem im Rahmen eines ESPRIT-Projektes entwickelten Token Rings mit Einzelabfertigung, zeigt allerdings, daß bei geringer bis mittlerer Last kein nennenswerter Unterschied dieser beiden Systeme besteht. Zur näheren Untersuchung wurden umfangreiche Simulationen durchgeführt (vgl. [WEL 89b]), bei denen für unterschiedliche Werte T_OPR die mittlere Anzahl der gesendeten Pakete pro Tokenerhalt bestimmt wurde. Als Lastmodell wurde die in Kapitel 4.2. näher charakterisierte Quelle für "gemischten Verkehr" eingesetzt, die von Stationen mit jeweils 2 Mbit/s generierter Last ausgeht.

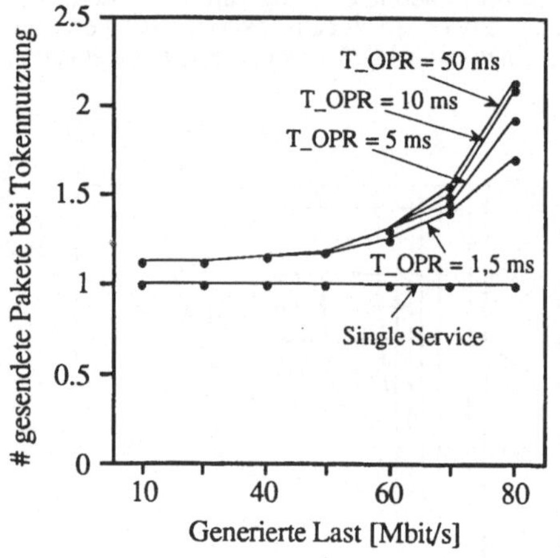

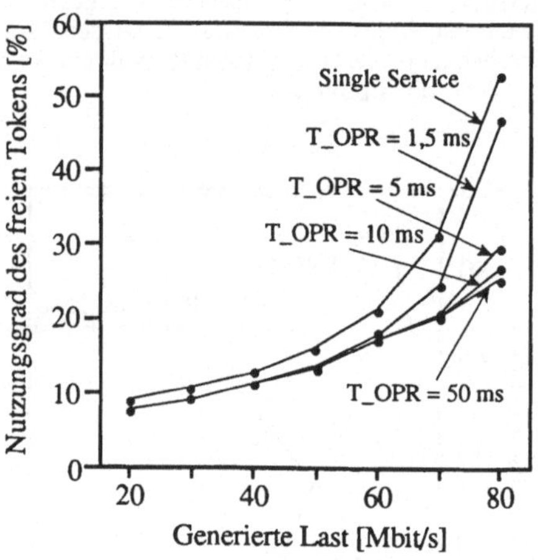

Abb. 4.7 Mittlere Anzahl gesendeter Pakete pro Tokenerhalt (nach [WEL 89b])

Die Abb. 4.7 zeigt den "Nutzungsgrad" des freien Tokens, d.h. die relative Häufigkeit der Nutzung des vorbeilaufenden freien Tokens, und die mittlere Anzahl der gesendeten Pakete bei Tokennutzung, d.h. den bedingten Erwartungswert der Anzahl gesendeter Pakete für den Fall, daß das Token für Datenübermittlung genutzt wird. Es wird deutlich, daß das Token bei geringer bis mittlerer Last nur vergleichsweise selten genutzt wird. Auch bei 50 Mbit/s Gesamtlast, d.h. 25 Stationen mit je 2 Mbit/s generierter Last, wird - im wesentlichen unabhängig von der Wahl von T_OPR - im Mittel nur etwa jedes fünfte Token "eingefangen" bzw. das Token nur an jeder fünften Station genutzt. Auch die Anzahl der gesendeten Pakete unterscheidet sich nur unwesentlich, wobei zu berücksichtigen ist, daß für die linke Graphik nur wirkliche Nutzungen des Tokens betrachtet wurden. Würde man hier die Darstellung der bedingten Erwartungswerte durch eine Darstellung der unbedingten Erwartungswerte unter Einbeziehung des Nutzungsgrades ersetzen, dann würden bei geringer bis mittlerer Last überhaupt keine Unterschiede erkennbar. Da ähnliche Ergebnisse auch für andere Lastquellen ermittelt wurden, kann man davon ausgehen, daß sich in vielen Fällen die Ergebnisse für ein System mit Single Service nur unwesentlich von den Ergebnissen für das bei FDDI eingesetzte Timer-gesteuerte Verfahren unterscheiden. Token Ringe mit Einzelabfertigung sind aber einer mathematischen Analyse zugänglich, wie sie z.B. in [MAR 88a] detailliert beschrieben und zu umfangreichen Sensitivitätsanalysen eingesetzt wurde.

Im Rahmen dieses Beitrages ist es nicht möglich, auf die Vielzahl von Veröffentlichungen zur Leistungsbewertung von FDDI detailliert einzugehen. Eine Anfrage an eine große Literatur-Datenbank ergab im Juni 1989 etwa 60 Veröffentlichungen in deutscher oder englischer Sprache, die sich mit dieser Problematik befassen. Der interessierte Leser sei auf die bekannten Fachzeitschriften und Tagungsbände verwiesen.

4.3.2. DQDB

Während der MAC-Standard von FDDI seit mehreren Jahren stabil ist, handelt es sich bei DQDB um ein neuartiges Protokoll, bei dem auch im Jahre 1989 noch Veränderungen des MAC-Protokolls vorgenommen worden sind. So ist zu erklären, daß es bis heute nur wenige Veröffentlichungen gibt, die sich mit der Leistungsfähigkeit von DQDB befassen. Leider fallen darüber hinaus noch einige der Veröffentlichungen unter die in Kapitel 4.1.3. angedeutete Klasse von Simulationen, die keine bzw. nur unzureichende Angaben zum Simulationsmodell machen. (Berechtigte?) Zweifel erscheinen hier vor allem dann zulässig, wenn - wie in [NEW 86] und [NEW 88] - die Autoren das von ihnen selbst entwickelte Protokoll mit anderen Verfahren vergleichen und zu dem Ergebnis kommen, daß das eigene Protokoll besser ist.

Die nachfolgenden Betrachtungen untersuchen Paketverzögerung und Durchsatz von DQDB, wobei der Schwerpunkt auf Fairness-Betrachtungen liegt. Auf einen detaillierten Leistungsvergleich von DQDB und FDDI soll verzichtet werden, da dieser sehr weitreichende und schwer begründbare Annahmen voraussetzt. Der in [HUB 88] enthaltene Vergleich von FDDI-II und DQDB bestätigt weitgehend die Vermutung, daß DQDB bei geringer Last den Token-Passing-Verfahren überlegen ist (Grund: DQDB ermöglicht bei geringer Last fast immer sofortigen Medienzugang; Wartezeit auf das Token entfällt). Da die Ringlatenz die Leistungsfähigkeit von FDDI wesentlich beeinflußt, ist darüber hinaus davon auszugehen, daß DQDB bei sehr großen geographischen Distanzen leistungsfähiger ist.

4.3.2.1. Paketverzögerung bei geringer Last

Da DQDB einer mathematischen Analyse nur schwer zugänglich ist, sollen in diesem Kapitel die Betrachtungen auf Simulationsergebnisse beschränkt bleiben. Dabei gilt das Interesse dem asynchronen Verkehr, da isochroner Verkehr durch das Management regelt werden soll (der entsprechende Standard liegt derzeit noch nicht vor).

Die Abb. 4.8 zeigt die mittlere Paketverzögerung (Wartezeit bis zum Medienzugang + eigentliche Sendezeit) für 25 km Medienlänge, 25 angeschlossene Stationen (äquidistant entlang des Mediums) und gleichverteilte Zieladresse. Dargestellt ist die Paketverzögerung für die Station 1 (unmittelbar am Slot Generator) und die Station 13 (in der Mitte) über der generierten Last, wobei die von jeder Station generierte Last zwischen 0,1 Mbit/s und 2 Mbit/s variiert wurde. Als Lastmodell wurde ein recht aufwendiges Modell von interaktivem Verkehr verwendet, dessen Details [DAV 89] zu entnehmen sind und das auch Paketsegmentierung berücksichtigt (die Vertrauensintervalle lagen bei ca. 2 % der Mittelwerte; sie wurden aus Gründen der Übersichtlichkeit nicht eingezeichnet). Als Datenrate des DQDB-Systems wurde 100 Mbit/s gewählt, d.h. 50 Mbit/s pro Bus.

Die Simulationsergebnisse machen deutlich, daß die mittleren Paketverzögerungen bei geringer Last von der Lage relativ zum Slot Generator abhängig sind. Dieses Ergebnis wird auch durch Simulationen bestätigt, die unter Verwendung eines völlig anderen Lastmodells in [HUB 88] für 100 km Medienlänge veröffentlicht wurden. Um den Zusammenhang zwischen Lage der Station und mittlerer Paketverzögerung deutlicher zu machen, wurde in Abb. 4.9 die mittlere Paketverzögerung für jede der angeschlossenen Stationen einzeln dargestellt, wobei die generierte Last pro Station als 0,1 Mbit/s bzw. als 1 Mbit/s gewählt wurde.

Der Grund für die unterschiedlichen Paketverzögerungen ist in der gleichmäßigen bzw. ungleichmäßigen Verteilung der Last auf die beiden unidirektionalen Busse zu sehen. Bei geringer Last ist die Wahrscheinlichkeit, daß beide Busse gleichzeitig belegt vorgefunden werden, außerst gering. Daher findet der Nachfolger eines Paketes, das auf einen freien Slot auf Bus A wartet, mit großer Wahrscheinlichkeit den Bus B frei und kann sofort gesendet werden, sofern die Zieladresse den Zugang zu Bus B vorschreibt. Stationen in der Mitte des Mediums profitieren sehr stark von diesem Effekt, da sie bei der angenommen Gleichverteilung der Zieladresse ihre Last gleichmäßig auf beide Busse (aus der Sicht der Warteschlangentheorie: schwach belastete, unabhängige Server) verteilen. Dagegen müssen die unmittelbaren Nachbarn des Slot Generators alle Pakete auf den einzigen in Frage kommenden Bus senden; für die übrigen Stationen ergeben sich Werte zwischen diesen Extremen.

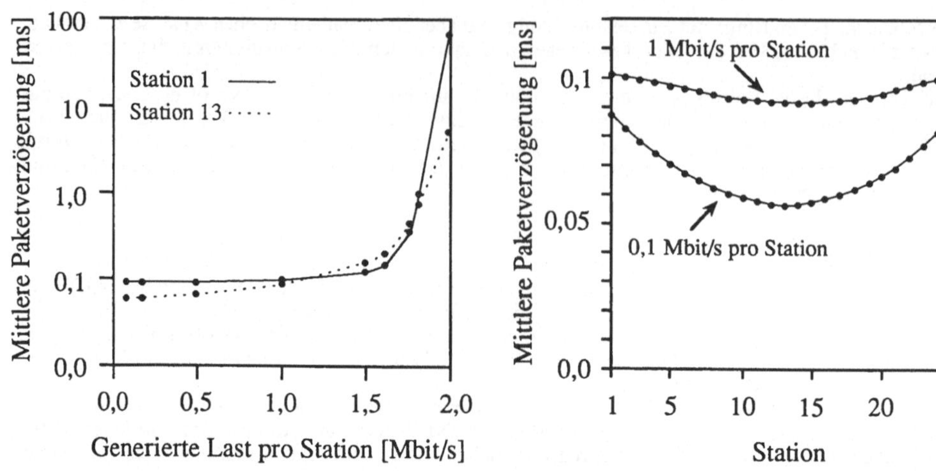

Abb. 4.8 Mittlere Paketverzögerung bei DQDB **Abb. 4.9** Stationsspezifische Paketverzögerung (nach [DAV 89])

Auch für völlig unterschiedlich gewählte Lastmodelle hat sich bei allen bisher an der RWTH Aachen durchgeführten Simulationen bei geringer Last das hier diskutierte Verhalten gezeigt. Wesentlich ist nur die Annahme der Gleichverteilung der Zieladresse. Dagegen ist der in Abb. 4.8 dargestellte Verlauf der Kurve bei höherer Last nicht als typisch zu beurteilen. Bei den hier gewählten Parametern kam es zu zwei Schnittpunkten der Kurven, welche die Werte für Station 1 bzw. für Station 13 verbinden. In dem hier betrachteten Fall war es also so, daß für die Station unmittelbar am Slot Generator größere Verzögerungen beobachtet wurden als für eine Station in der Mitte. Bei 2 Mbit/s generierter Last gingen für die Stationen unmittelbar am Slot Generator sogar ca. 10 % der Pakete verloren, obwohl jeweils 400 Pufferplätze bereitstanden und andere Stationen nicht überlastet waren. Diese starke Benachteiligung bei nur 50 Mbit/s generierter Gesamtlast (bei 100 Mbit/s nomineller Datenrate) kann nicht als typisch bezeichnet werden, da bei anderen Medienlängen und anderer Gesamtzahl der Stationen z.T. die Stationen am Slot Generator sogar bevorzugt werden. Mit der damit verbundenen Problematik befaßt sich das nachfolgende Kapitel, in dem saturierte Sender vorausgesetzt werden.

4.3.2.2. Saturierte Sender

Während bei geringer Last die Benutzer die im vorangegangenen Kapitel diskutierte Bevorzugung von Stationen in der Mitte - sie bewegt sich in Bereichen deutlich unterhalb einer Millisekunde - in der Regel nicht bemerken, zeigt DQDB bei hoher Last deutliche Schwächen, auf die nun eingegangen werden soll. Können bei geringer Last alle Sendewünsche erfüllt werden, so muß bei Hochlast die zur Verfügung stehende Bandbreite "gerecht" unter den angeschlossen Stationen aufgeteilt werden. Der DQDB Standard stellt in diesem Zusammenhang lapidar fest: 'the distributed queue is a function which allows the formation and operation of a queue of asynchronous segments which is distributed across the subnetwork, yet which gives the performance of a centralized queue. It guarantees minimum access delay and uniform sharing of capacity'. Was bedeutet in diesem Zusammenhang "uniform sharing of capacity"? In einem Netz mit n Stationen kann dieser Satz bedeuten, daß der Zugang zu jedem n-ten Slot ermöglicht wird. Da aber die Verteilung der Sendewünsche auf die beiden Busse, die ja durch die geographische Lage des Senders und des Empfängers bestimmt wird, ungleichmäßig erfolgt, folgt aus der gerechten Verteilung der Bandbreite jedes einzelnen Busses in der Regel nicht ein gerechter Zugang zum Gesamtnetz.

Die nachfolgend kurz zusammengefaßten Betrachtungen, deren Details in [MAR 89c] enthalten sind, setzen saturierte Sender voraus. Darüber hinaus generiere der Slot Generator ausschließlich Slots für asynchronen Verkehr, d.h. er generiere ausschließlich "queued-arbitrated slots". Wenn DQDB keine Reservierungsmechanismen enthalten würde, dann würden die jeweils am Busanfang gelegenen Stationen alle Slots mit eigenen Daten belegen und "stromabwärts" gelegene Stationen würden niemals sendeberechtigt. Da aber der Medienzugang durch "request counter" und "countdown counter" geregelt wird, können durch Übertragung entsprechender "Requests" die stromaufwärts gelegenen Stationen aufgefordert werden, zugunsten der stromabwärts gelegenen Stationen auf den Medienzugang zu verzichten.

Problematisch wird die Situation nun dadurch, daß für die Reservierung Signallaufzeiten zu berücksichtigen sind: Es geht Zeit verloren, bis alle Stationen über den Sendewunsch unterrichtet sind, und es geht Zeit verloren, bis ein freier Slot eine weit vom Slot Generator entfernte Station erreicht. Da die Stationen kein neues Request übertragen dürfen, bevor sie für das vorangegangene Request Zugang zu einem freien Slot erhalten haben, werden auf diese Weise Stationen benachteiligt, die weit vom Slot Generator entfernt sind.

Es ist schwer, ein Warteschlangensystem zur exakten Modellierung von DQDB zu finden, und es wäre noch schwieriger, ein derartiges Modell in angemessener Weise zu analysieren. Für Fairness-Betrachtungen genügt allerdings das relativ einfache Modell, das in Abb. 4.10 am Beispiel des Zugangs zu Bus A für insgesamt n angeschlossene Stationen dargestellt ist. Bei saturierten Sendern folgt jeder Übertragung (Slotsendezeit t_s) eine neue Reservierung, wobei die Reservierungszeit der Station i als $t_{r,i}$ bezeichnet werde, d.h. $t_{r,i}$ gibt die Zeit an, bis alle Stationen den Sendewunsch bemerkt haben.

Der Zugang zu Bus B unterscheidet sich vom Zugang zu Bus A im wesentlichen dadurch, daß Station 1 nicht auf Bus B sendet und Station n nicht auf Bus A sendet (vgl. Abb. 3.6). Daher ergibt sich ein Modell für Bus B formal durch Ersetzung von $t_{r,1}$ durch $t_{r,n}$ sowie von λ_1 durch λ_n.

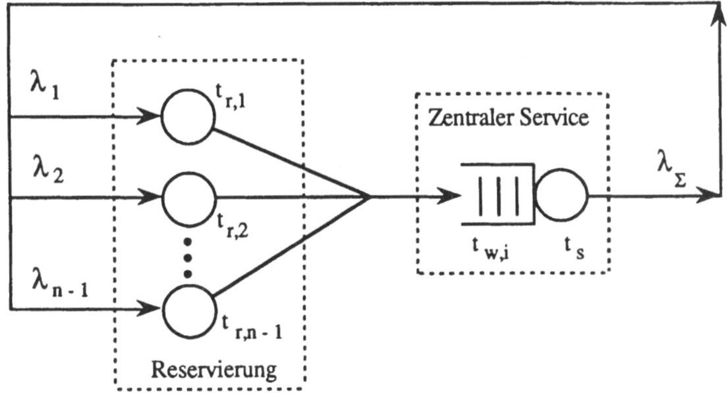

Abb. 4.10 Modellierung des Zugangs zu Bus A

Da der zentrale Service durch einen deterministischen Server erfolgt und (für saturierte Sender) die Reservierungszeiten ebenfalls deterministisch sind, muß davon ausgegangen werden, daß die Wartezeiten $t_{w,i}$ bis zum Medienzugang erheblich variieren können. Mit entsprechend gewählten Parametern und Startbedingungen können leicht Szenarios konstruiert werden, bei denen einige Stationen stets sofort bedient werden, während andere stets die maximal mögliche Zeit warten müssen. In diesem Zusammenhang ist zu beachten, daß die maximale Wartezeit durch $(n - 2) \cdot t_s$ gegeben ist, da die Abfertigung gemäß FIFO erfolgt und an jeden Bus eine Station angeschlossen ist, die niemals Daten sendet.

Für DQDB kann bei saturierten Sendern leicht gezeigt werden, daß alle Slots für Datenübertragung genutzt werden, d.h. der Durchsatz gemessen in "Slots pro Sekunde" ergibt sich als $\lambda_\Sigma = 1/t_s$. Da sich der Durchsatz aus Sicht einzelner Stationen als $\lambda_i = 1/(t_{r,i} + t_{w,i} + t_s)$ ergibt, müssen die folgenden Bedingungen erfüllt sein:

$$\frac{1}{t_s} = \sum_{i=1}^{n-1} \frac{1}{t_{r,i} + t_{w,i} + t_s} \quad \text{für Bus A} \qquad \frac{1}{t_s} = \sum_{i=2}^{n} \frac{1}{t_{r,i} + t_{w,i} + t_s} \quad \text{für Bus B}$$

Die Analyse wird nun dadurch erschwert, daß sowohl die Reservierungszeiten $t_{r,i}$ als auch die Wartezeiten $t_{w,i}$ vom Startzustand abhängen. Die sich daraus unmittelbar ergebende Folgerung, daß eine exakte Analyse für "den" stationären Zustand nicht möglich ist, wird sowohl durch Simulationen an der RWTH Aachen als auch durch die in [WON 89a] veröffentlichten Resultate bestätigt.

Im folgenden gehen wir davon aus, daß die Reservierungszeiten ausschließlich durch Signallaufzeiten vom und zum Slot Generator bestimmt sind. Da sowohl Datenübertragung als auch Reservierung auf einer Slot-Struktur basieren und die beiden Busse voneinander unabhängig sind, gehen wir ferner davon aus, daß jeweils insgesamt eine Slotzeit Verzögerung hinzukommt. Gemäß dieser Heuristik ergeben sich die Reservierungszeiten als

Bus A: $t_{r,1} = 0$ $\qquad t_{r,i} = 2 \dfrac{L_{A,i}}{V} + t_s \qquad$ für $1 < i \leq n$

Bus B: $t_{r,n} = 0;$ $\qquad t_{r,i} = 2 \dfrac{L_{B,i}}{V} + t_s \qquad$ für $1 \leq i < n$

wobei $L_{A,i}$ und $L_{B,i}$ die Entfernung zwischen Station i und dem unmittelbaren Nachbarn des Slot Generators für Bus A bzw. Bus B und V die Signalgeschwindigkeit sei.

Die nachfolgenden Betrachtungen gehen nun davon aus, daß $t_{w,1} = ... = t_{w,n} = t_w$ gilt, d.h. wir gehen von gleichen Wartezeiten auf den zentralen Server aus. Dann wird es möglich, die Wartezeit t_w (z.B. iterativ) so zu bestimmen, daß die oben angegebenen Beziehungen zwischen λ_Σ und λ_i erfüllt sind. Diese Heuristik, deren Ergebnisse mit simulativ bestimmten Werten überprüft werden müssen, läßt aufgrund der FIFO-Abfertigung im zentralen Server insbesondere dann gute Resultate erwarten, wenn die Wartezeiten im Vergleich zu den Reservierungszeiten groß sind. Dies ist dann der Fall, wenn viele Stationen an das gemeinsame Medium angeschlossen und/oder die geographischen Entfernungen klein sind. Darüber hinaus sind gute Ergebnisse aber auch dann zu erwarten, wenn die Wartezeiten im Vergleich zu t_s vernachlässigbar klein sind.

Es muß in diesem Zusammenhang klar festgestellt werden, daß es i.a. keine Garantie für die Korrektheit der hier ermittelten Lösung gibt! Diese unbefriedigende Situation ist darin begründet, daß das hier betrachtete System exakt nicht analysierbar ist und die für beliebige Parameterwahl bestimmbaren Schranken sich als unbrauchbar erweisen. Wenn auch in keinem der von uns analytisch und simulativ betrachteten Szenarien deutliche Abweichungen von Analyse und Simulation beobachtet wurden und die Resultate auch i.w. mit den simulativ bestimmten Ergebnissen in [WON 89a] übereinstimmen, so darf dies nicht als Beweis für die Allgemeingültigkeit der ermittelten Resultate mißverstanden werden. Dagegen liefert die hier diskutierte Heuristik eine gute Grundlage für Plausibilitätskontrollen von simulativ bestimmten Werten.

Da die Datenpakete gemäß der Slotstruktur segmentiert werden müssen, hängt der erzielbare Durchsatz nicht nur von der Verwendbarkeit freier Slots sondern auch von der Paketlängenverteilung ab. Im folgenden soll dieser Effekt jedoch nicht näher untersucht werden, denn wir gehen zur Vereinfachung davon aus, daß alle Slots vollständig mit Daten gefüllt sind. Dies ermöglicht es uns, den Durchsatz unmittelbar aus der Anzahl der verfügbaren Slots pro Zeiteinheit zu bestimmen.

Abbildung 4.11 zeigt den stationsspezifischen Durchsatz für 5 Stationen, wobei vorausgesetzt wurde, daß die Stationen äquidistant entlang des Mediums verteilt sind (mit Station 1 und Station 5 als Slot Generator für Bus A bzw. Bus B). Als Datenrate wurde 50 Mbit/s pro Bus gewählt.

Für 100 km Medienlänge können die Slot Generatoren die Übertragungskapazität des jeweiligen Busses fast exklusiv nutzen, ein Effekt, der bei kleinerer Medienlänge stark reduziert wird. Beachtet man, daß der Durchsatz auf Bus B ein Spiegelbild des Durchsatzes auf Bus A ist, dann ergibt sich aus dem Durchsatz auf Bus A unmittelbar der in Abb. 4.11 dargestellte Gesamtdurchsatz.

Während bei großer Medienlänge und wenigen Stationen die Slot Generatoren stark bevorzugt werden, sind diese Stationen bei geringer Medienlänge benachteiligt; im Grenzfall der Medienlänge von 0 km ist ihr Durchsatz halb so groß wie der der anderen Stationen. Diese Aufteilung der Übertragungskapazität ist fair, wenn als Topologie die in Abb. 3.6 dargestellte geschlossene Bus-Topologie verwendet wird. Sie ist dagegen unfair, wenn die offene Bus-Topologie eingesetzt wird.

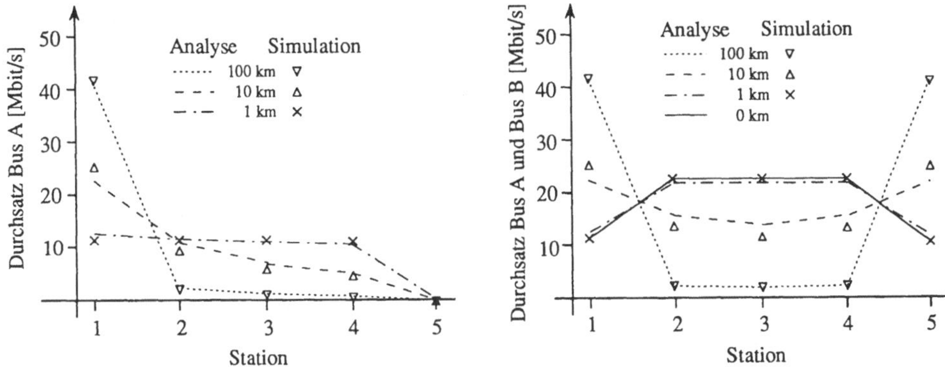

Abb. 4.11 Durchsatz pro Station für 5 angeschlossene Knoten

Bei Anschluß von nur 4 Sendern an jeden Bus wird der Durchsatz im wesentlichen durch Reservierungszeiten bestimmt, da in diesem Fall die Wartezeiten $t_{W,i}$ am zentraler Server klein sind. Weitergehende Betrachtungen in [MAR 89c] zeigen, daß selbst bei großer Medienlänge die Verteilung der Übertragungskapazität gerecht wird, wenn hinreichend viele Stationen angeschlossen sind. Je größer die Medienlänge ist, desto mehr Stationen müssen allerdings als Sender aktiv werden, damit die Bevorzugung einzelner Stationen verhindert wird; im Fall von 100 km Medienlänge bedeutet dies z.B. die Aktivität von mindestens etwa 100 Stationen.

4.4. Bedeutung höherer Protokollebenen

Die bisherigen Betrachtungen haben sich auf die Leistungsfähigkeit von Medienzugangsprotokollen beschränkt, wobei die Leistungsfähigkeit des Empfängers nicht berücksichtigt wurde. Wenn auch die Mehrzahl der Veröffentlichungen zur Leistungsbewertung von lokalen Kommunikationssystemen auf die Betrachtung des Empfängers verzichtet, so soll hier doch kurz auf diese wichtige Problematik eingegangen werden, bei der u.a. Charakteristika der Sender, die Verteilung der Zieladressen, Leistungscharakteristik des Transportkanals und Eigenschaften des Empfängers zu beachten sind. Die große Zahl relevanter Parameter und z.T. schwer begründbare aber dennoch notwendige Annahmen erschweren eine Untersuchung mittels mathematischer Analyse oder Simulation.

Bereitet schon bei konventionellen lokalen Kommunikationsystemen die Betrachtung des Empfängers außerordentlich große Probleme, so gilt dies aufgrund der höheren Datenrate erst recht für Hochgeschwindigkeitsnetze. Selbst wenn ein einzelner Sender nicht schneller Daten bereitstellt, als sie der Empfänger verarbeiten kann, so kann doch durch simultane Übertragung mehrerer Sender mit gleicher Zieladresse sehr schnell eine Situation entstehen, bei der einzelne Empfangsstationen kurzzeitig stark überlastet werden. An dieser Stelle seien die in [MAR 88b] dargestellten Simulationsergebnisse erwähnt, bei denen ein Hochgeschwindigkeits-Token-Ring für die bereits mehrfach erwähnten Quellen für gemischten Verkehr (File Transfer und interaktiver Verkehr) untersucht wurde. Obwohl bei dem untersuchten Szenario die Empfänger in 80 % der Gesamtzeit leer waren, führte simultaner File Transfer in Extremfällen dazu, daß bis zu 26 % der generierten Last am Empfänger verloren gingen, wenn - wie in der Simultation angenommen - an jedem Empfänger Pufferplätze für 100 Pakete bereitgestellt wurden. Diesem Effekt, der natürlich den Verlust von Nachrichten zur Folge hat, muß durch geeignete Maßnahmen der Fehlererkennung und der Flußkontrolle entgegengewirkt werden, die gemäß dem ISO-Referenzmodell prinzipiell auf den Ebenen 2, 3 und 4 eingesetzt werden können.

Wie Bux und Grillo in [BUX 85] zeigen, ist die Wahl adäquater Fenstergrößen keineswegs ein trivial lösbares Problem: Da die Fenstergröße eine obere Schranke für die Zahl der erforderlichen Paketwiederholungen angibt, ist i.a. davon auszugehen, daß die auf diese Weise zusätzlich generierte Last mit wachsender Fenstergröße zunimmt. Da andererseits zu kleine Fenster die tatsächlich verfügbare Datenrate durch unnötiges Warten auf Quittungen einschränken, schlagen Bux und Grillo eine dynamische Flußkontrolle vor: Bei Paketverlust soll die Fenstergröße auf 1 gesetzt und danach für jede fehlerfreie Übertragung um 1 erhöht werden, bis eine sinnvolle Maximalgröße erreicht ist. Auf diese Weise können die Sender dynamisch an die aktuelle Belastung der Empfänger angepaßt werden.

Wong und Schwartz schlagen in [WON 89b] ähnliche Verfahren vor, die speziell für den Einsatz in Hochgeschwindigkeits-Backbone-Netzen zur Kopplung lokaler Netze entwickelt wurden. Unterschiedliche Strategien wurden zwar im Hinblick auf Durchsatz und mittlere Paketverzögerung untersucht, doch basieren die in [WON 89b] veröffentlichten Ergebnisse auf weitreichenden

Annahmen, die nur aus Gründen der mathematischen Analysierbarkeit gemacht wurden. Da viele für die Praxis außerordentlich wichtige Fälle auf diese Weise nicht behandelt werden können, sehen sich auch Wong und Schwartz eher am Anfang eines langen Weges als am Ziel.

Die hier skizzierte Problematik der Flußkontrolle in Hochgeschwindigkeitsnetzen macht deutlich, daß schon unmittelbar oberhalb des Medienzugangsprotokolls viele Probleme heute nicht gelöst sind. Dies gilt auch für die darüberliegenden Protokollebenen, deren Funktionalität sich in Zukunft an die Bedürfnisse neuartiger Anwendungen anpassen muß. Während für Datenübertragung mit 100 Mbit/s die Aufgaben der Ebene 1 als prinzipiell gelöst betrachtet werden können und die Standardisierung von Protokollen zur Regelung des Medienzugangs in Hochgeschwindigkeitsnetzen weit fortgeschritten ist, gibt es nur für wenige Einsatzgebiete konkrete Vorstellungen über adäquate Protokolle der höheren Ebenen.

5. Zusammenfassung und Ausblick

In den vorangegangenen Kapiteln wurden Protokolle für Hochgeschwindigkeitsnetze vorgestellt und im Hinblick auf ihre Leistungsfähigkeit diskutiert, wobei ausschließlich Protokolle betrachtet wurden, die aus Sicht der Standardisierung derzeit von Bedeutung sind. Das wichtigste Protokoll ist heute sicher FDDI. Hier gibt es bereits erste Produkte, die dem derzeitigen Stand der Normung entsprechen, und zahlreiche Hersteller haben für die nahe Zukunft FDDI-Anschlüsse angekündigt. An vielen Orten werden derzeit insbesondere von Hochschulen und großen Industrieunternehmen FDDI-Ringe geplant, so daß FDDI das rechte Produkt zur rechten Zeit zu sein scheint. Aus Sicht der Leistungsbewertung ist FDDI als relativ unproblematisch zu betrachten, wobei allerdings bei extrem großer Medienlänge Performance-Einbußen zu beachten sind, die sich unmittelbar aus dem verwendeten expliziten Token-Passing-Prinzip ergeben. Darüber hinaus wurde gezeigt, daß bei FDDI die Wahl der angestrebten Tokenrotationszeit T_OPR gravierende Auswirkungen auf die Leistungsfähigkeit des Systems hat.

Bei DQDB handelt es sich im Gegensatz zu FDDI um ein Protokoll, das sich zwar vorwiegend an den Bedürfnissen von öffentlichen Netzbetreibern orientiert, das aber natürlich auch im privaten Bereich eingesetzt werden kann. Die Betrachtungen in Kapitel 4.3.2. haben deutlich gemacht, daß bei DQDB in gewissen Fällen Stationen aufgrund ihrer geographischen Lage benachteiligt werden. Da derartige Benachteiligungen zwar sowohl bei niedriger als auch bei hoher Last beobachtet werden können, für den Benutzer aber nur bei hoher Last erkennbar werden, bleiben zwei Lösungsansätze: Modifikation des Protokolls oder Vermeidung hoher Systemauslastung.

Die Zukunft wird zeigen, ob diejenigen Recht behalten, die FDDI als "Ethernet der Zukunft" bezeichnen und diesem Hochgeschwindigkeitsnetz ähnlich große Erfolge prophezeien wie sie CSMA/CD-basierte Produkte im Bereich der Bürokommunikation gehabt haben. Sei es nun FDDI, DQDB oder ein anderes Verfahren, Hochgeschwindigkeitsnetze können nur dann große Umsatzzahlen erreichen, wenn einerseits die entsprechenden Komponenten preislich akzeptabel angeboten werden und andererseits neue Einsatzgebiete erschlossen werden, denen herkömmliche Technik nicht gerecht wird. Derartige Hochgeschwindigkeitsnetze stellen allerdings große Anforderungen an die eingesetzten Kommunikations-Controller, so daß diese in vielen Fällen zu entscheidenden Enpässen werden können, vgl. z.B. [MAR 89a]. Aus diesem Grunde werden in zahlreichen Projekten (vgl. z.B. [JEN 88], [MAR 89d], [ZIT 89]) neue Architekturen für Hochgeschwindigkeits-Controller untersucht.

Literaturverzeichnis

[ANSI 87a] FDDI Token Ring Media Access Control, American National Standard for Information Systems, X3.139-1987

[ANSI 87b] FDDI Physical Layer Protocol (PHY), Draft Proposed American National Standard, X3T9/85-39, X3T9.5/83-15, Rev. 15, September 1987

[ANSI 88] FDDI Physical Layer Medium Dependent (PMD), Draft Proposed American National Standard, X3T9/86-71, X3T9.5/84-48, Rev. 7.3, May 1988

[BAL 86] G. Balbo et al., "Combining Queueing Networks and Generalized Stochastic Petri Net Models for the Analysis of Some Software Blocking Phenomena", *IEEE Trans. on Software Eng.*, Vol. SE-12, No. 4, April 1986, pp. 561 - 576

[BUX 85] W. Bux, D. Grillo, "Flow Control in Local-Area Networks of Interconnected Token Rings", *IEEE Trans. Commun.*, Vol. COM-33, No. 10, Oct. 1985

[DAN 88] A. Danthine, O. Spaniol, (Eds.), High Speed Local Area Networks, Proc. of the IFIP TC6/WG6.4 International *Workshop on High Speed Local Area Networks*, Participants Edition, Liège, 1988

[DAV 89] P. Davids, Th. Welzel, "Performance Analysis of DQDB Based on Simulation", Ausarbeitung eines Vortrages auf dem *Third IEEE Workshop on Metropolitan Area Networks*, San Diego, März 1989, Aachener Informatik Bericht, Nr. 89-10, ISSN 0935-3232

[DYK 87] D. Dykeman, W. Bux, "An Investigation of the FDDI Media-Access Control Protocol", *Proc. of EFOC/LAN 87*, pp. 229 - 236

[GBU 89a] P. Gburzynski, P. Rudnicki, "On Formal Modeling of Communication Channels", *Proc. of INFOCOM '89*, Ottawa (Canada) April, IEEE 1989, pp. 143 - 151

[GBU 89b] P. Gburzynski, P. Rudnicki, "A Note on the Performance of ENET II", *IEEE J. Select. Areas Commun.*, Vol. SAC-7, No. 3, April 1989, pp. 424 - 426

[GIHR 89] O. Gihr, M. Weixler, "Messung der Datenverkehrsprofile in lokalen Netzen", in *Kommunikation in verteilten Systemen*, P.J. Kühn, (Ed.), Springer 1989, pp. 861 - 877

[GRA 89] M. Graf-Siebald, R. Bordewisch, "Modellierung und Leistungsbewertung eines Local Area Networks (LAN), in *Kommunikation in verteilten Systemen*, P.J. Kühn, (Ed.), Springer 1989, pp. 539 - 553

[HAL 85] S. Halfin, "Delays in Queues, Properties and Approximations", *Proc. of ITC 11*, M. Akiyama, (Ed.), North-Holland, 1985, pp. 1.4-3-1 - 1.4-3-6

[HUB 88] M.N. Huber et al., "QPSX and FDDI-II Performance Study of High Speed LAN's", Proc. of *EFOC/LAN-88*, pp.316 - 321

[IEEE 85a] Local Area Networks: Logical Link Control, ANSI/IEEE Standard 802.2, 1985

[IEEE 85b] Local Area Networks: CSMA/CD, ANSI/IEEE Standard 802.3, 1985

[IEEE 85c] Local Area Networks: Token-Passing Bus, ANSI/IEEE Standard 802.4, 1985

[IEEE 85d] Local Area Networks: Token Ring Access, ANSI/IEEE Standard 802.5, 1985

[IEEE 89] Proposed Standard: DQDB Metropolitan Area Network, P802.6_/D7, Unapproved Draft - Published for Comment Only, IEEE, May 7, 1989

[ISO 84] Information Processing Systems - Open Systems Interconnection - Basic Reference Model, IS 7498, 1984

[JAG 82] D.L. Jagerman, "Approximate Mean Waiting Times in Transient GI/G/1 Queues", *The Bell Syst. Techn. Journal*, Vol. 61, No. 8, Oct. 1982, pp. 2003 - 2022

[JAI 86] R. Jain, S.A. Routhier, "Packet Trains - Measurements and a New Model for Computer Network Traffic", *IEEE J. Select. Areas Commun.*, Vol. SAC-4, No. 6, Sept. 1986, pp. 986 - 995

[JEN 88] M.N. Jensen et al., "VLSI-architectures Implementing Lower Layer Protocols in Very High Data Rate LAN's", Proc. of the IFIP TC6/WG6.4 International *Workshop on High Speed Local Area Networks*, Participants Edition, Liège, 1988

[KLE 76] L. Kleinrock, *"Queueing Systems, Volume II: Computer Applications"*, Wiley, 1976

[KÜHN 86] P.J. Kühn, "Modelling of New Services in Computer and Communication Networks", in *Computer Networking and Performance Evaluation*, T. Hasegawa et al., (Eds.), North-Holland, 1986, pp. 283 - 303

[MAR 88a] P. Martini, *"Leistungsbewertung von Medienzugangsprotokollen für lokale Hochgeschwindigkeitsnetze"*, Dr. Alfred Hüthig Verlag, 1988

[MAR 88b] P. Martini, O. Spaniol, Th. Welzel, 'File Transfer in High Speed Token Ring Networks: Performance Evaluation by Approximate Analysis and Simulation', *IEEE Journal on Selected Areas in Communications*, Vol. 6, No. 6, July 1988, pp. 987 - 996

[MAR 89a] P. Martini, "High Speed Bridges for High Speed Local Area Networks - Packets per Second vs. Bits per Second", *Proc. of INFOCOM '89*, Ottawa (Ontario), IEEE 1989, pp. 474 - 483

[MAR 89b] P. Martini, "High Speed Local Area Networks", *IEEE CompEuro'89 Tutorial: VLSI and Computer Peripherials*, Hamburg, IEEE 1989, pp. T4-101 - T4-125

[MAR 89c] P. Martini, "The DQDB Protocol. What about fairness?", to be published by IEEE in the *Proc. of Globecom 89*, Dallas (Texas), Nov. 1989

[MAR 89d] P. Martini, M. Rupprecht, "Designing High Speed Controllers for High Speed Local Area Networks", to be published by IEEE in the *Proc. of Globecom 89*, Dallas (Texas), Nov. 1989

[NEW 86] R.M. Newman, J.L. Hullet, "Distributed Queueing: A Fast and Efficient Packet Access Protocol for QPSX, in *New Communication Services: A Challenge to Computer Technology*, P.Kühn, (Ed.), North-Holland, 1986, pp. 294 - 299

[NEW 88] R.M. Newman et al., "The QPSX MAN", *IEEE Communications Magazine*, Vol. 26, No. 4, April 1988, pp. 20 - 28

[REI 86] M. Reiser, "Communication-System Models embedded in the OSI-Reference Model, A Survey," in *Computer Networking and Performance Evaluation*, T. Hasegawa et al., (Eds.), North-Holland, 1986, pp. 85 - 111

[SPA 87] O. Spaniol, A. Danthine, (Eds.), High Speed Local Area Networks, Proc. of the IFIP TC6/WG6.4 International *Workshop on High Speed Local Area Networks*, North-Holland, 1987

[TAK 86] H. Takagi, M. Murata, "Queueing Analysis of Scan-Type TDM and Polling Systems", in *Computer Networking and Performance Evaluation*, T. Hasegawa et al. (Eds.), North-Holland, 1986, pp. 199 - 211

[TRA 86] P. Tran-Gia, T. Raith, "Multiqueue Systems with Finite Capacity and Nonexhaustive Cyclic Service", in *Computer Networking and Performance Evaluation*, T. Hasegawa et al. (Eds.), North-Holland, 1986, pp. 213 - 225

[WAT 84] K.S. Watson, "Performance Evaluation of Cyclic Service Strategies - A Survey", in *Performance '84*, Gelenbe (Ed.), North-Holland 1984, pp. 521 - 533

[WEL 89a] Th. Welzel, "FDDI and BWN Backbone Networks: A Performance Comparison Based on Simulation", Proc. of the *Eighth Annual International Phoenix Conference on Computers and Communications*, IEEE, 1989, pp. 190 - 194

[WEL 89b] Th. Welzel, "Modellierung und Leistungsbewertung von Lokalen Hochgeschwindigkeitsnetzen", Dissertation RWTH Aachen (voraussichtliche Fertigstellung: September 1989)

[WON 89a] J.W. Wong, "Throughput of DQDB Networks under Heavy Load", *IBM Research Report*, RZ 1813 (#65107), 1989

[WON 89b] L.-N. Wong, M. Schwarz, "Flow Control in Metropolitan Area Networks", *Proc. of INFOCOM '89*, Ottawa (Ontario), IEEE 1989, pp. 826 - 833

[ZIT 89] M. Zitterbart, "A Multiprocessor Architecture for High Speed Network Interconnections", *Proc. of INFOCOM '89*, Ottawa (Ontario), IEEE 1989, pp. 212 - 218

Erfahrungen bei der Durchsatzoptimierung eines Transputer-Netzwerks für ISO-OSI-Architekturen am Beispiel der LLC-Teilschicht

R. Ulrich, R. Hinze, H. Dietsch
Informatik-Forschungsgruppe IX
(Prof. Dr.-Ing. D. Seitzer)
Universität Erlangen-Nürnberg
Martensstr. 3
D-8520 Erlangen

Zusammenfassung

Für den Bereich zwischen physischem Medium und MAC-Schnittstelle der IEEE 802.X-Normen gibt es leistungsfähige VLSI-Bausteine. Der im Vergleich dazu geringe Durchsatz an der TRANSPORT-Schnittstelle ISO-OSI-konformer Kommunikationsarchitekturen ist zum großen Teil darauf zurückzuführen, daß die Möglichkeit zum Parallelismus, die den Funktionen der (Teil-) Schichten LOGICAL LINK CONTROL (LLC), NETWORK und TRANSPORT innewohnt, in bisherigen Implementierungen nicht genutzt wird. Diesem Beitrag liegt das Konzept zugrunde, ISO-OSI-Funktionen bis hinauf zur Anwendungsschicht mittels Transputer-Netzwerken zu verwirklichen. Dargestellt und diskutiert werden Varianten einer in OCCAM formulierten Implementierung der LLC-Teilschicht. Parameter sind Typ und Kopplungsart der Transputer sowie Zahl, Kommunikationsverhalten und Zuordnung nebenläufiger Prozesse. Ein Verfahren zur Zeitmessung in Transputer-Netzwerken wird skizziert. Bei einer kombinierten Speicher/Link-Kopplung der Transputer zeigen die Ergebnisse einen um das 10-fache höheren Durchsatz als bei bisherigen Lösungen.

1. Einleitung: Stand der Technik - Eigene Ziele

Für den Bereich zwischen physischem Medium und MAC-Schnittstelle der IEEE-802.X-Normen werden leistungsfähige VLSI-Bausteine angeboten, die zu den marktgängigen Mikrorechnerfamilien kompatibel sind. Solch massiver Hardware-Unterstützung zum Trotz ergeben sich bei herkömmlichen Implementierungen unbefriedigende Durchsatzraten an der TRANSPORT-Schnittstelle (/ARONOFF87/). Einen Überblick über aktuelle Leistungsdaten gibt /STRAUSS87/: Der maximale TRANSPORT-Durchsatz von 1,8 MBit/s bei Paketgrößen von 1 KByte wird den Software-Implementierungen der ISO-OSI-(Teil)-Schichten LLC, NETWORK und TRANSPORT angelastet. Diese Ergebnisse stützen sich auf das übliche Konzept eines Standard-Mikrorechners, der mittels Speicherkopplung die TRANSPORT-

Dienste für einen "Host"-Rechner erbringt. Als größtes Durchsatzhemmnis werden Kopiervorgänge innerhalb der Schichten und beim Übergang zwischen Schichten, d.h. bei der Inanspruchnahme von Diensten, ausgemacht (/WOODSIDE86/).

Diesem Beitrag liegt die Vorstellung zugrunde, ISO-OSI-Funktionalität zwischen MAC- und Anwendungsschnittstelle auf der Basis von Transputer-Netzwerken (/INMOS87/) zu verwirklichen. Transputer-Implementierungen sind ein Kompromiß zwischen üblichen reinen Software-Lösungen und andernorts angestrebten reinen Hardware-Lösungen (/DANG88/), wobei letzteren sicherlich die Flexibilität fehlen wird, die zur Eingliederung in die Praxis "offener Systeme" immer noch vonnöten ist.

Um Leistung und Grenzwerte eines Transputer-basierten "Kommunikationswerks" abschätzen zu können, wurde die ISO/OSI-Teilschicht "Logical Link Control (LLC)" (/ISO8802-2/) in OCCAM (/INMOS88/) in einer flexiblen Entwicklungsumgebung für Transputer implementiert. Die verbindungsorientierten LLC-Dienste werden zwar in existierenden lokalen Netzen kaum verwendet, doch läßt ihre Untersuchung Schlüsse auf die Leistungsdaten einer ähnlich gestalteten TRANSPORT-Schicht zu. Die Schlußfolgerung gilt vor allem bei kleinen Datenpaketen mit ausgeschalteter Prüfsummen-Sicherung, die in typischen Anwendungen den Hauptteil des Durchsatzes ausmachen (/GIHR89/).

Für die Evaluation von Varianten der vorgestellten Lösung wurde ein Verfahren zur Zeitmessung in Transputer-Netzwerken geschaffen, das auf der OCCAM-spezifischen synchronen Interprozeßkommunikation gründet. Hauptziel für die Aktivitäten, die diesem Beitrag zugrunde liegen, ist der Einsatz Transputer-basierter "Kommunikationswerke" in MAP- und High-Speed-LAN-Umgebungen.

2. Materialien

2.1. Transputer: Eigenschaften, Verbindungsstrukturen, Programmierwerkzeuge

Transputer sind Mikrorechner, die speziell für den Aufbau von leistungsfähigen lose gekoppelten Multiprozessorsystemen geeignet sind. Transputer zeichnen sich vor allem durch folgende Eigenschaften aus:

- Befehlssatz und Hardware sind für die Ausführung von parallelen Prozessen mit zwei Prioritätsklassen optimiert: Niederpriore Prozesse werden nach dem Time-Slicing-Verfahren bearbeitet und können von hochprioren Prozessen verdrängt werden. Hochpriore Prozesse unterliegen nicht dem Zeitscheibenverfahren und rechnen so lange, bis sie sich durch eine Ein- oder Ausgabeoperation blockieren.
- Taskwechsel benötigen weniger als eine Mikrosekunde.
- Einer der beiden eingebauten Timer hat eine Auflösung von einer Mikrosekunde. Die Scheduler-Hardware ist in der Lage, Prozesse beim Erreichen eines bestimmten Zeitpunkts zu aktivieren.
- Vier schnelle serielle Kommunikationskanäle mit eigener DMA-Einheit ("Links") können parallel zur Aktivität der CPU Daten transferieren.

Mittels der Links lassen sich Transputer zu größeren, leistungsfähigeren Verarbeitungseinheiten verknüpfen. Die Verbindungsstrukturen können entweder statisch den Anforderungen angepaßt oder durch einen Kreuzschienenverteiler dynamisch geändert werden. In Fällen, bei denen einem hohen Bedarf an Rechenleistung ein relativ geringer Kommunikationsbedarf gegenüber steht, ist die Kopplung über Links sicher optimal. Bei Anwendungen, die schnellen Zugriff auf große Datenmengen erfordern, stellt die maximale Link-Übertragungsgeschwindigkeit von 1.8 MByte/s eine Beschränkung dar, die nur durch den Einsatz von Multiportspeichern umgangen werden kann.
Als Programmiersprache für Transputer-Netzwerke bietet sich OCCAM an, mit dem sich auf mehrere Transputer verteilte Prozeßsysteme übersichtlich und effizient implementieren lassen.

2.2. Zur LLC-Teilschicht im ISO-OSI-Modell

Das IEEE-Normungsprojekt 8Ø2 für LAN-Protokolle teilt die "Data Link"-Schicht des ISO-OSI-Referenzmodells auf in eine

- medienabhängige MAC- ("Medium Access Control")- Teilschicht und eine
- medienunabhängige LLC- ("Logical Link Control")- Teilschicht.

Das LLC-Dokument (/ISO88Ø2-2/) spezifiziert die Dienste an den beiden medienunabhängigen Schnittstellen, legt die Struktur der LLC-Protokolldateneinheiten fest und beschreibt das Verhalten der LLC-Funktionen mittels Automaten.
Auf der Basis zweier Dienstklassen werden zwei Betriebsarten definiert: Die Dienstklasse "Class 1" enthält ungesicherte verbindungslose Datagramm-Dienste, im folgenden als "Type 1"-Dienste bezeichnet. Die Dienstklasse "Class 2" bietet zusätzlich zu den "Type 1"-Diensten "Type 2"-Dienste an, die aus logischen Verbindungen mit Fehlererkennung (über Sequenznummern und Timer), einfachen Fehlerbehebungsmechanismen sowie Flußkontrolle bestehen. Aufgrund der niedrigen Fehlerrate bei lokalen Netzen wird dort allgemein LLC Class 1 bevorzugt, während bei öffentlichen Netzen nach CCITT-X.25 LLC Class 2 zum Einsatz kommt.
Der Verarbeitungsaufwand bei Type 1 ist sehr gering, da am Service Access Point (SAP) eintreffende Datenpakete lediglich mit einem drei Byte langen Header versehen und dann an die MAC-Teilschicht weitergereicht werden.
Die verbindungsorientierte Variante muß für jede Verbindung Empfangs- und Sendezähler führen, vier Timer überwachen und Quittungen für empfangene Datenpakete geben. Der daraus resultierende Arbeitsaufwand ist durchaus mit dem in der TRANSPORT-Schicht zu vergleichen, wenn in dieser weder Segmentierung noch Prüfsummenberechnung stattfinden (was in heute existierenden Netzen die Regel ist.). Bei einer Implementierung der ISO-OSI-Architektur auf einem als Pipeline verschaltetem Transputersystem sollte sich deshalb auch der Durchsatz an der Transportschnittstelle in ähnlichen Größenordnungen bewegen wie der von LLC Type 2.

3. Die Transputer-Implementierung der LLC-Teilschicht

Bei der Abbildung von Anwendungsaufgaben auf ein Transputer-Netzwerk ist die erzielte Leistung stark von der gewählten Verteilung der Prozesse auf die einzelnen Transputer abhängig. Bei Programmen, die nur relativ einfache Operationen an Datenströmen durchführen, kann schon bei einer kleinen Prozessoranzahl die begrenzte Link-Übertragungsgeschwindigkeit dazu führen, daß durch den Einsatz eines weiteren Transputers keine Leistungssteigerung erzielt wird. Zusätzlich zwingt z.B. die Implementierung eines als Automat beschreibbaren Prozesses auf mehr als einem Transputer zum Austausch von Zustandsnachrichten zwischen den Teil-Automaten, was das Datenaufkommen weiter vergrößert. Speziell für die LLC-Teilschicht bedeutete dies, daß ein Transputer bei der Bearbeitung von LLC-Type-2-Paketen zwar durchaus ausgelastet ist, aber der Einsatz eines weiteren bei den derzeit möglichen Link-Geschwindigkeiten keinen Vorteil bringt. In den folgenden Untersuchungen wurde deshalb versucht, die Parallelarbeit zwischen Prozessen und Link-DMA innerhalb des Bausteins zu optimieren. Als Test- und Meßanordnung wurde das in Bild 1 skizzierte 4-Transputer-System verwendet:

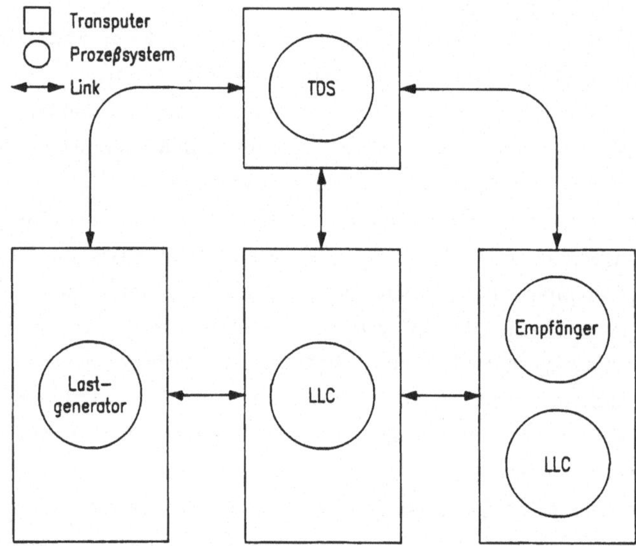

Bild 1 Struktur der LLC-Testanordnung

Die Transputer sind nur über Links miteinander verbunden. Die vom Lastgenerator (=NETWORK-Schicht) erzeugten Pakete werden über einen Link-Kanal an einen LLC-Transputer übergeben, dort bearbeitet und an einen Transputer, auf dem ebenfalls LLC und ein Empfänger implementiert sind, übergeben. Die durch den Empfänger zusätzlich erzeugte Last kann vernachlässigt werden, weil dafür das Weiterleiten der Datenpakete an die NETWORK-Schicht entfällt. Am Lastgenerator können Paketgrößen, deren Verteilung, sowie Ankunftsrate und Ankunftsratenverteilung der Pakete in weiten Grenzen eingestellt werden. Der Transputer, auf dem das Entwicklungssystem (TDS) läuft, besitzt

als einziger Zugriff zur Peripherie des PCs und tritt bei Ein- und Ausgaben der anderen Prozessoren als Vermittler auf.
Bild 2 zeigt den inneren Aufbau der LLC-Teilschicht. Die Pfeile symbolsieren OCCAM-Kanäle, über die Datenpakete laufen. Die LLC-Protokollmaschine ist als niederpriorer Prozeß realisiert. Das obere Sender/Empfängerprozeß-Paar bedient die Link-Schnittstelle zum Lastgenerator, das untere Paar die Schnittstelle zur Partnerinstanz. Die Warteschlangen vor den Senderprozessen bestehen aus einer Anzahl von Wechselpuffern und sorgen dafür, daß die Links bei unterschiedlich langen Paketen nicht leerstehen. Empfänger und LLC-Prozeß kommunizieren über einen Wechselpuffer. Free Space Manager, Sender und Empfänger sind hochpriore Prozesse, um die unter DMA-Steuerung ablaufenden Link-Übertragungen zum frühestmöglichen Zeitpunkt zu starten. Sender- und Empfängerprozesse sind nur aktiv, wenn sie die Übertragung eines Datenpaketes anstoßen. Dabei wird die niederpriore LLC-Protokollmaschine stets verdrängt. Während die Link-Hardware die Datenpakete überträgt, kann die LLC-Protokollmaschine nahezu unbeeinträchtigt weiterarbeiten.

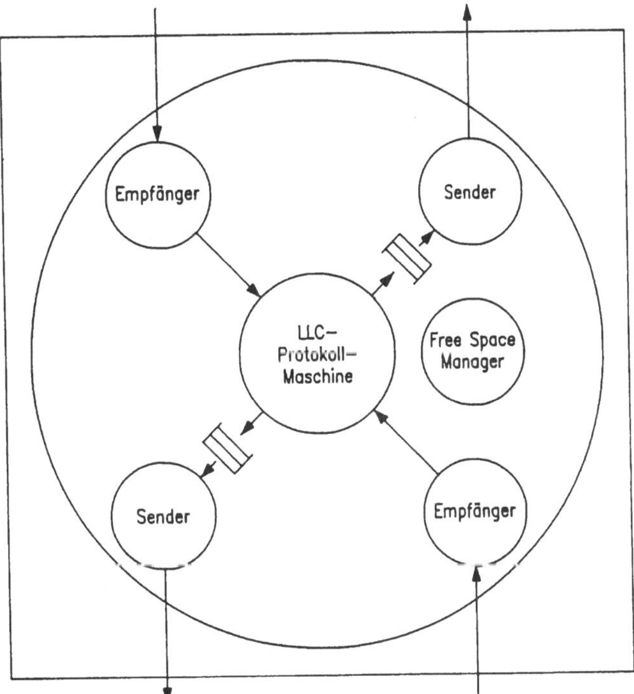

Bild 2 Innerer Aufbau der LLC-Teilschicht

Die erste Version der LLC-Instanz wurde in "reinem" OCCAM auf einem Transputer des Typs T414 implementiert, in dem Prozesse keinerlei gemeinsame Daten besitzen. Als Ursache für die geringe Leistung wurden die Kopiervorgänge zwischen den Prozessen vermutet, was durch eine Vergleichsmessung mit einem "leeren" LLC-Prozeß bestätigt wurde. Eine verbesserte T414-Version verzichtet innerhalb des Transputers auf Kopiervorgänge

und tauscht zwischen den Prozessen lediglich Zeiger aus. Damit verstößt das Programm allerdings gegen den durch OCCAM auferlegten Verzicht auf gemeinsame Daten, was die Fehlersuche erschwert und einen Korrektheitsbeweis nahezu unmöglich macht. Der um 50% höhere Durchsatz rechtfertigt jedoch diese Vorgehensweise.
Bei Paketlängen größer als 256 Bytes wird der Durchsatz von der Übertragungsgeschwindigkeit der Links des T414 begrenzt. Die Nachfolgetypen T800 und T425 erzielen infolge eines besseren Link-Protokolls trotz gleicher Brutto-Übertragungsrate annähernd die doppelte Nettodatenrate. Durch Umstellung auf die derzeit höchste Übertragungsgeschwindigkeit von 20 MBit/s (brutto) konnte bei großen Paketen insgesamt eine Vervierfachung des Durchsatzes erzielt werden.

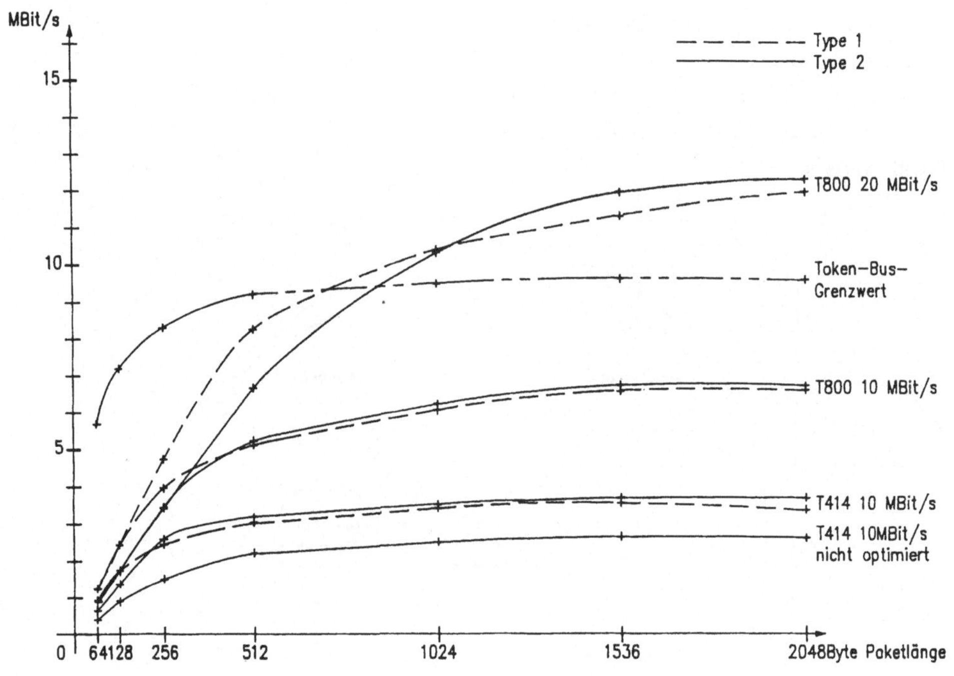

Bild 3 LLC-Durchsatz im Überblick

Bild 3 liefert eine Gesamtschau der Meßergebnisse. Zu beachten sind die Unterschiede zwischen den Betriebsarten "Type 1" und "Type 2". Der errechnete Grenzwert für den Token Bus (/ISO8802-4/) ist ebenfalls eingezeichnet. Aufgetragen sind die Ergebnisse für die nicht optimierte T414-Version, die optimierten Versionen für T414 mit 10 MBit/s, T800 mit 10 MBit/s sowie T800 mit 20 MBit/s Link-Übertragungsrate. Aus den Kurven ist deutlich der Einfluß der Paketgrößen auf den Durchsatz zu erkennen. Bei Paketlängen über 1 KByte besitzen Datagrammdienst und verbindungsorientierter Dienst nahezu den gleichen Durchsatz, weil die Übertragung eines Pakets über ein Link länger dauert als seine Bearbeitung. (Die etwas geringeren Werte für den verbindungslosen Dienst resultieren aus dem besonderen spezifischen Verhalten des Free Space Managers.) Bei geringen Datenlängen stellt die Verarbeitungsleistung des Prozessors den Engpaß dar:

Die Durchsatzwerte für Type 1 sind deutlich höher als für Type 2, während sich unterschiedliche Link-Übertragungsraten beim T800 nicht bemerkbar machen.
Das geplante Einsatzgebiet der Transputer-Implementierung des ISO/OSI-Protokollstacks liegt neben HS-LAN-Anschaltungen (FDDI) im Bereich der Fertigungsautomatisierung (MAP). Hier ist der Vergleich mit dem Token-Bus-Durchsatz (10 MBit/s BROADBAND) interessant. Die folgende Berechnung geht davon aus, daß sich eine Anzahl n_S von Stationen an einem relativ kurzen Bus (< 200m) befindet, aber nur eine Station sendewillig ist und n_d Datenbytes in n_P Frames mit minimalem Frame-Abstand sendet, bevor das Token einen kompletten Umlauf macht. Mit dem Frame-Overhead n_F und der in Bytes ausgedrückten Reaktionszeit n_R einer Station vom Erhalt des Tokens bis Tokenweitergabe bzw. Sendebeginn ergibt sich der Wirkungsgrad u als

$$u = \frac{n_d}{n_d + n_P \cdot n_F + n_S \cdot (n_F + n_R)}$$

Die während eines Token-Umlaufs übertragbare Datenmenge ist dabei von der maximal zulässigen Token Rotation Time (TRT) abhängig. Bei fünf Stationen, TRT= 10 ms, 22 Bytes Frame-Overhead und einer für heute erhältliche VLSI-Controller (/MOTOROLA87/) typischen "Reaktionszeit" von 15 Bytes ergibt sich der im Bild 3 eingezeichnete Grenzwert. Es zeigt sich, daß für die im MAP-Bereich üblichen Datagrammlängen von ca. 256 Bytes nach wie vor die Protokollsoftware den Engpaß darstellt.

4. Ansätze zur weiteren Optimierung des LLC-Durchsatzes

4.1. Ein Verfahren zur Zeitmessung in Transputersystemen

Als Ansatzpunkt für weitere Optimierungen sind genauere Untersuchungen über die zeitlichen Abläufe im Testsystem notwendig.
Da jeder Transputer zwei Timer mit unterschiedlicher Zeitauflösung besitzt (einen für hochpriore, einen für niederpriore Prozesse), wurden an allen relevanten Stellen im Code entsprechende Timer-Abfragen eingefügt. Um die Messung nicht zu verfälschen, wurden die Werte zuerst lokal gespeichert und erst nach dem Ende des Experiments an das Entwicklungssystem zur Auswertung übertragen. Probleme ergaben sich aus der Tatsache, daß hochpriore Prozesse nur den Timer mit der Auflösung 1 µs lesen können, niederpriore nur den mit der Auflösung 64 µs. Die Timer werden zwar beide vom Systemtakt des Transputers angesteuert, starten aber nicht gleichzeitig. Uhren auf anderen Transputern laufen zusätzlich mit einer im Rahmen der Quarz-Toleranzen anderen Frequenz. Es besteht nun die Aufgabe, die mit den lokalen Zeitbasen registrierten Ereignisse in einen globalen Zeitmaßstab einzuordnen.

Aufgrund der Tatsache, daß ein Kommunikationsvorgang immer eine Synchronisation nach dem Rendezvous-Verfahren bedeutet, kann ein Verfahren angegeben werden, mit dem ein Uhrenvergleich möglich ist:

1) Hochpriorer Prozeß startet Kommunikation, wird blockiert.
2) Niederpriorer Prozeß liest Timer T_L.
3) Niederpriorer Prozeß startet Kommunikation.
4) Dadurch deblockierter hochpriorer Prozeß liest Timer T_H.

Durch diese Reihenfolge wird garantiert, daß beide Prozesse fast gleichzeitig ihre zugeordneten Timer lesen. Der Zeitversatz ist die für Kommunikation und Prozeßumschaltung benötigte Zeit und kann vernachlässigt werden, weil er weniger als eine Mikrosekunde beträgt. Da diese Synchronisation nur einmal durchgeführt werden muß, kann durch zeitweilige Blockierung aller anderen Prozesse dafür gesorgt werden, daß das Verfahren genau in der gewünschten Reihenfolge abläuft. Nach dem gleichen Prinzip werden die Transputer untereinander synchronisiert. Um von den unterschiedlichen (aber als konstant angenommenen) Taktfrequenzen auf eine gemeinsame Zeitbasis umrechnen zu können, wird nach Ende des Experiments nochmals eine Synchronisation vorgenommen. Dieser einfache Synchronisations-Algorithmus setzt also nur lokale Zeitmarken, mit deren Hilfe nach Ende der Messungen auf eine gemeinsame Zeitbasis umgerechnet werden kann. Gegenüber anderen Verfahren, die zur Laufzeit eine gemeinsame Zeitbasis errechnen (z. B. /CARLINI88/), verfälscht er wegen seines geringen Aufwandes während der Messung nicht die zu messenden Durchsatzwerte.

Das Ergebnis einer solchen Messung zeigt Bild 4 am Beispiel von 128-Byte-Paketen auf einem Transputer IMS T800 mit einer Link-Geschwindigkeit von 20 Mbit/s.

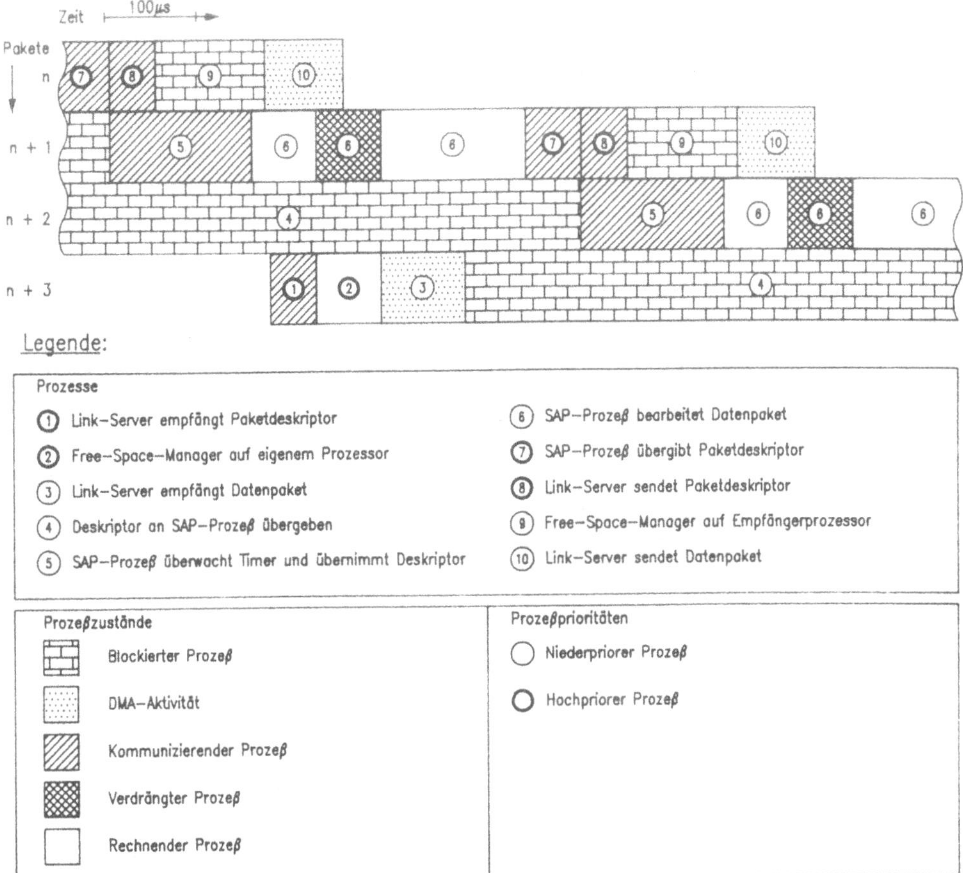

Bild 4 Verschränkung der LLC-Prozesse bei 128-Byte-Paketen (Type 2)

Ein Datenpaket durchläuft nacheinander die Bearbeitungsschritte 1 bis 10. Nach dem Empfang des Paketdeskriptors (1), der Parameter des Pakets enthält (Quell- und Zieladresse, Länge, Dienstklasse,...), kann über den Free Space Manager Puffer für die Daten angefordert werden (2). Die Daten werden per Link-DMA übertragen (3). Danach wird das Paket vom Empfänger an die LLC-Protokolleinheit übergeben (4). Sobald die LLC-Protokolleinheit bereit ist, übernimmt sie das Paket (5) und bearbeitet es (6). In (7) wird das Paket an den Sender-Prozeß übergeben, der bei leerer Warteschlange sofort den Paketdeskriptor an den Empfängerprozeß auf dem anderen Transputer verschickt (8). Nachdem dort Puffer bereitgestellt wurde (9), kann das Datenpaket übertragen werden (10).

Um die Verschränkung der Prozesse zu verdeutlichen, wurden die Zeitspuren von vier aufeinanderfolgenden Paketen in Beziehung gesetzt. Während Paket n per Link-DMA zur Partnerinstanz übertragen wird, wird Paket n+1 in der LLC-Instanz bearbeitet und Paket n+3 über Link empfangen. Paket n+2 steht zur Bearbeitung durch die LLC-Protokolleinheit bereit. Zu erkennen ist die Verdrängung der niederprioren LLC-Protokollmaschine durch den hochprioren Free Space Manager. In den Zeiten, in denen ein hochpriorer Prozeß Kommunikation über ein Link durchführt, können andere Prozesse rechnen (allerdings durch häufige Taskwechsel stark beeinträchtigt). An der Zeit, in der sich (4) im Zustand "blockiert" befindet, ist zu sehen, wie weit die Paketgröße erhöht werden kann, ohne das Verhalten des Systems wesentlich zu ändern. Da in der LLC-Teilschicht keine Segmentierung oder Prüfsummenbildung stattfindet, sind - bis auf die Link-Übertragungszeiten (3) und (10) - alle Zeiten von der Paketlänge unabhängig.

Ansatzpunkte für weitere Verbesserungen zeigten vor allem die relativ langen Kommunikationszeiten (1), (5), (7) und (8). Während dieser Zeiten finden viele Prozeßwechsel statt, weil Paketdeskriptoren nicht im Ganzen, sondern als eine Anzahl von einzelnen Werten ausgetauscht werden. Auch die Laufzeit des Free Space Managers erschien sehr lang.
Durch die Übertragung eines Paketdeskriptors als Byte-Array in einen einzigen Kommunikationsvorgang und durch Optimierungen im Free Space Manager konnte der Datendurchsatz nochmals um 25% gesteigert werden.

4.2 Emulationen von Kopplungsvarianten

Bisher wurde nur die Kopplung von Transputern über Links betrachtet (Variante A). Im folgenden sollen zwei (in dieser Untersuchung emulierte) Kopplungsmöglichkeiten dargestellt und bewertet werden.
Variante B geht davon aus, daß ein üblicher VLSI-LAN-Controller in der MAC-Schicht eingesetzt wird. Statt dem Versenden von LLC-Datenpaketen über ein Link werden die Pakete in eine verzeigerte Liste in einem Dual-Port-Speicher eingereiht.
Variante C nimmt das Vorhandensein eines Multiport-Speichers für alle Transputer des Kommunikationswerks an. Der globale Speicher reduziert in dieser Variante lediglich die Kopiervorgänge, die Interprozeßkommunikation erfolgt nach wie vor über Links. Zugriffskonflikte im Dual-Port-Speicher bei Variante B sind wegen des Zugriffsverhaltens des LAN-Controllers relativ selten und werden vernachlässigt. Bei Variante C wurden die durch den Zugriff anderer Transputer auf den Multiport-Speicher entstehenden Wartezyklen durch einen Last-Prozeß auf dem LLC-Transputer nachgebildet. Bild 5 zeigt eine Gegenüberstellung der Varianten.

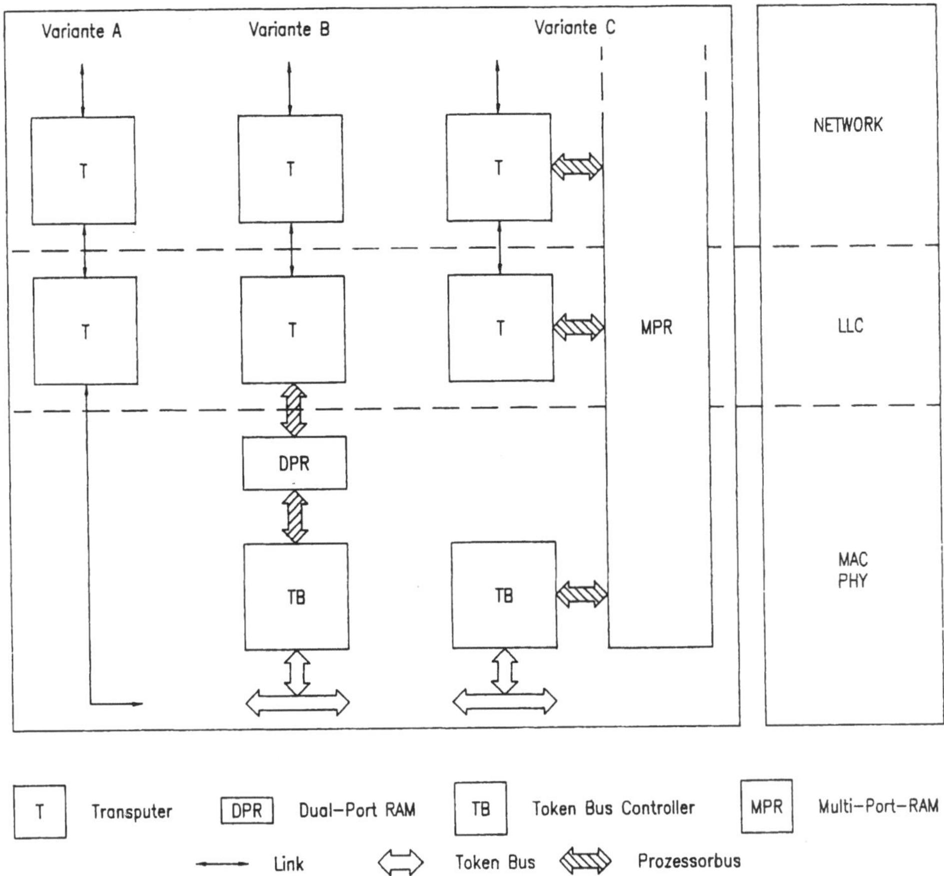

Bild 5 Kopplungsvarianten im Überblick

Die Ergebnisse dieser Emulationen zeigen die Bilder 6 und 7. Die Werte für Variante A entsprechen dabei den Meßergebnissen für den Transputer T800 mit 20 MBit/s Link-Übertragungsrate (Bild 3) nach nochmaliger Optimierung.

Die von der Hardware her erheblich aufwendigere Multiport-Variante ist bei Type-1-Paketen ab einer Datenlänge von 512 Bytes deutlich überlegen. Die Varianten A und B zeigen sich gleichwertig.

Durchsatzwerte, die auf einer Token-Bus-Anschaltung auf Transputerbasis (/SCHENK89/) erzielt wurden, sind unter Variante B2 in Bild 7 eingezeichnet. Da hier das Systemverhalten auf höchste Leistung bei kurzen Datenpaketen abgestimmt wurde, liegen die Durchsatzwerte in diesem Bereich deutlich über den durch die Emulation abgeschätzten.

Bild 8 zeigt am Beispiel für Type-2-Pakete den Zusammenhang zwischen Bearbeitungszeiten und Link-Übertragungsrate. Als Bearbeitungszeit wird hierbei die Taktzeit der Pipeline bezeichnet, und nicht die Zeit, die vom Empfang des Paketdeskriptors bis zum

Senden des letzten Datenbytes vergeht. Wie zu erwarten, steigt die Paketbearbeitungszeit bei den Varianten A und B linear mit der Link-Übertragungszeit.

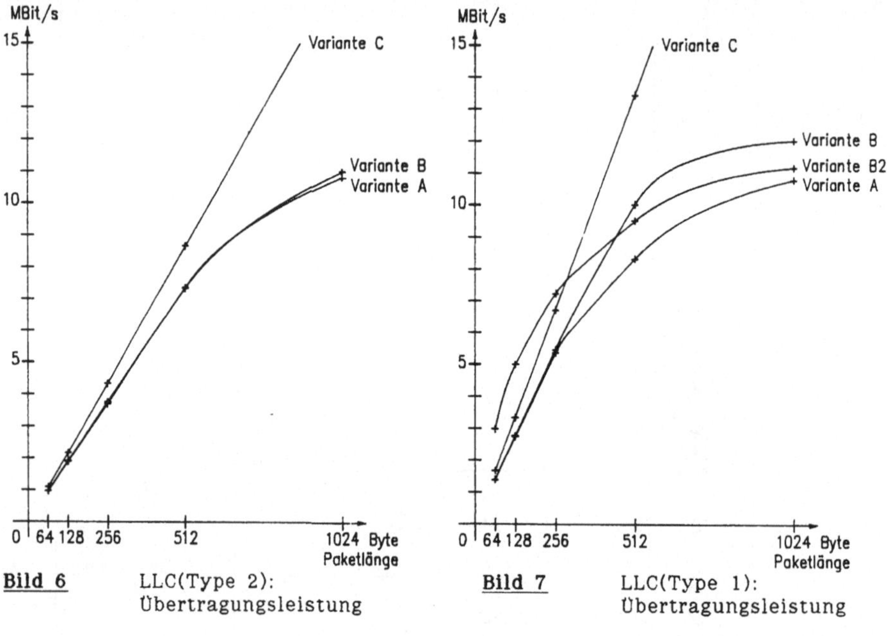

Bild 6 LLC(Type 2): Übertragungsleistung

Bild 7 LLC(Type 1): Übertragungsleistung

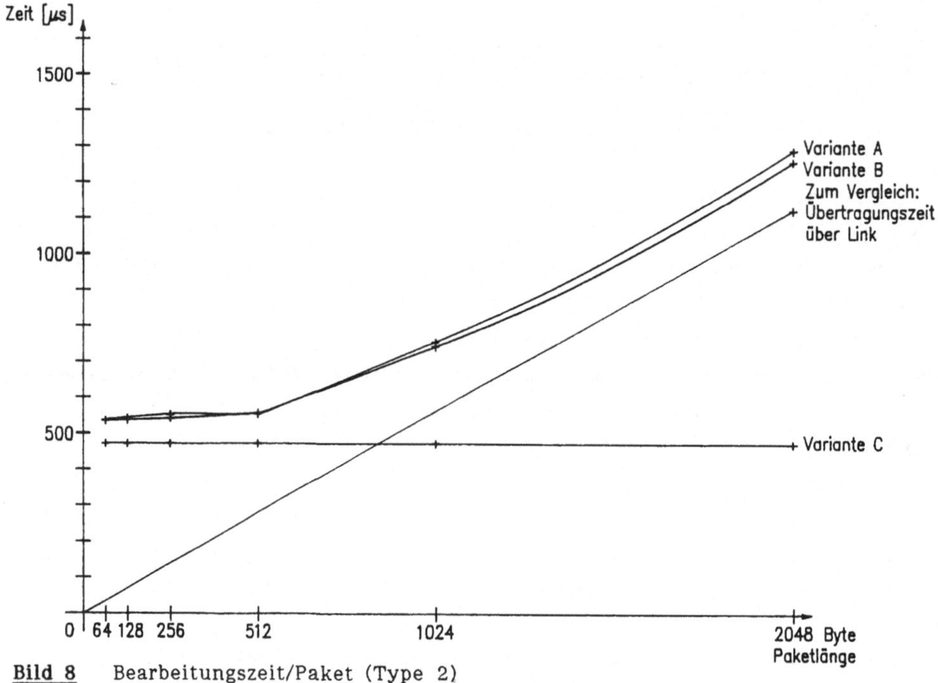

Bild 8 Bearbeitungszeit/Paket (Type 2)

In Bild 9 ist die mittlere Durchlaufzeit eines Pakets für den Hochlastfall (gefüllte Pipeline) und für den Fall geringer Last (leere Pipeline) aufgetragen. Durch Addition dieser als typisch für eine Schicht erachteten Zeiten für jede durchlaufene Schicht erhält man in etwa die an der TRANSPORT-Schnittstelle zu erwartende Durchlaufzeit für kurze Datenpakete. Verglichen mit den bei Token-Bus-Systemen auf der MAC-Ebene üblichen Latenzzeiten von ca. 10 ms sind auch die von der Variante B erzielten Zeiten durchaus tragbar.

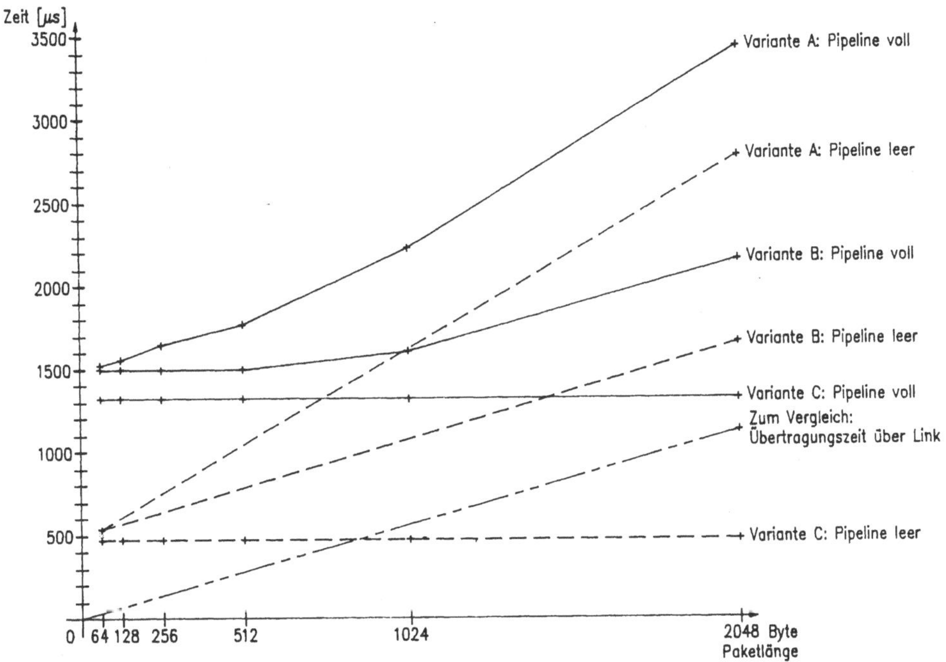

Bild 9 Durchlaufzeit/Paket (Type 2)

5. Beurteilung der Ergebnisse

Zieht man als Anwendungsumgebungen für die vorgestellten Architekturvarianten einerseits die MAP-Welt und andererseits aktuelle HS-LANs (wie z.B. FDDI) in Betracht, dann können auf Grund der Ergebnisse die folgenden Aussagen gemacht werden:

- Bei kurzen Datenpaketen, wie sie in prozeßnahen Teilbereichen von MAP-Anwendungen vorherrschen, stellt die Rechenleistung des LLC-Prozessors den Engpaß dar, so daß die im Vergleich zur Link-Kopplung wesentlich aufwendigere Speicherkopplung keine Durchsatzsteigerung bringt. Ein der Variante B entsprechendes Transputer-Kommunikationswerk weist um eine Größenordnung bessere Durchsatzwerte auf als herkömmliche LAN-Anschaltungen.

- Eine File-Server-Anschaltung an einen FDDI-Ring über ein speichergekoppeltes Transputer-Kommunikationswerk könnte deutlichen Nutzen ziehen aus den bei 4 KByte großen Paketen erreichbaren TRANSPORT-Übertragungsraten von 30-40 MBit/s.

Bei der Suche nach Lösungen für das (bisher weitgehend unbeachtet gebliebene) Problemfeld leistungsfähiger Übergänge zwischen HS-LANS bzw. zwischen HS-LANs und Standard-LANs stellt die (Speicherkopplungs-)Variante C einen erfolgversprechenden Ansatz dar.

Die bei der Implementierung der Logical Link Control-Teilschicht gewonnenen Ergebnisse sind übertragbar auf jede Schicht einer Protokollhierarchie. Aufgrund der guten Leistungsdaten eines auf Transputern basierenden Kommunikationswerks scheint es angebracht, die bisher oberhalb der Transport-Schicht befindliche Grenze zwischen Kommunikationsrechner und Host zu höheren Protokollschichten hin zu verlagern.

Literatur

/ARONOFF87/ Aronoff,R.,Mills,K.,Wheatley,M.:
Transport Layer Performance - Tools and Measurement.
IEEE Networks 1, 3 (1987), S.21-31.

/CARLINI88/ Carlini,U.,Villano,U.:
A Simple Algorithm for Clock Synchronization in Transputer Networks.
Software-Practice an Experience 18,
4 (1988), S. 321-347

/DANG88/ Dang,M.,Diot,C.,Sabouni,I.,Sponga,L.:
Specific Data Structure Intended for the Implementation of High Level ISO Standards: Associated Algorithms and Dedicated Hardware.
Proceedings "Euromicro '88", Microprocessing and Microprogramming 24, Nr. 1-5 (1988), North-Holland.

/GIHR89/ Gihr,O.,Weixler,M.:
Messung der Datenverkehrsprofile in lokalen Netzen.
Proceedings "Kommunikation in verteilten Systemen"
Informatik-Fachbericht 205, Springer, 1989.

/HINZE89/ Hinze,R.:
Implementierung der IEEE 802.2 "Logical Link Control"-Funktionen für ein Transputersystem.
Diplomarbeit am Lehrstuhl für Technische Elektronik, FAU Erlangen-Nürnberg, 1989.

/INMOS87/ INMOS Limited:
Product Information: The Transputer Family.
1987.

/INMOS88/ INMOS Limited:
OCCAM 2 Reference Manual.
Prentice Hall, 1988.

/ISO7809/ ISO IS 7809:
HDLC - Consolidation of Classes of Procedures.
1984.

/ISO8802-2/ ISO DIS 8802-2 (IEEE 802.2):
Local Area Networks, Part 2: Logical Link Control.
1985.

/ISO8802-4/ ISO DIS 8802-4 (IEEE 802.4):
Local Area Networks, Part 4: Token-Passing Bus Access Method and Physical Layer Specifications.
1985.

/MOTOROLA87/ Motorola:
MC68824 Token Bus Controller Data Sheet, 1987.

/SCHENK89/ Schenk,A.:
Entwurf,Aufbau und Inbetriebnahme einer IEEE 802.4 Token-Bus-Anschaltung auf Transputerbasis.
Diplomarbeit am Lehrstuhl für Technische Elektronik, FAU Erlangen-Nürnberg, 1989.

/STRAUSS87/ Strauss,P.:
OSI throughput performance: Breakthrough or bottleneck ?
Data Communications 16, 5(1987), S.53-56.

/WOODSIDE86/ Woodside,C.M.,Montealegre,J.R.:
On Packet Buffering and Protocol Performance.
In: Diaz, M. (Hrsg.): Protocol Specification,
Testing and Verification, V.
North-Holland, 1986.

Applying the SRPT Principle to Improve the Performance of the CSMA/CD Media Access Control Protocol

Leonhard Schmickler Klaus Scholten*

Lehrstuhl Datenfernverarbeitung
RWTH Aachen
Kopernikusstr. 16
D-5100 Aachen

Abstract

The optimal preemptive strategy SRPT[1] reduces the mean delay time by preferring short jobs or messages especially in systems with a high coefficient of variation of the message length. In packet switching networks the service time, i.e. the transmission time, is known in advance, so that this strategy can be applied. The CSMA/CD[2] media access control (MAC) protocol of the widely used Ethernet LAN's with their heterogeneous mixed traffic load is a promising object for a first application of SRPT. In this paper the optimized CSMA/CD/SRPT protocol based on this strategy is introduced. Access to the channel is allowed only to messages which are shorter than the rest of the message currently transmitted. Short messages can interrupt the transmission of a longer one by provoking a collision. After such a collision stations start their transmission according to the message length in reserved slots. This is similar to the hyperchannel priorities for station numbers. Simulation results for a realistic network topology and user scenario show that the mean transfer time of interactive terminal traffic and paging traffic as well as that of the total traffic can be substantially reduced.

1 Introduction to the SRPT/RR strategy

The SRPT[1] strategy was introduced by SCHRAGE and MILLER in 1966 [18]. In an M/G/1 system it offers the smallest mean number of jobs and the lowest mean waiting and delay time among all conceivable strategies by preferring the jobs with the shortest remaining processing time. Therefore it is called the *optimal* strategy [17]. If all messages are of the same length (coefficient of variation $c = 0$) SRPT and FIFO show the same behavior. But when the difference between short and long messages increases, SRPT leads to a significant performance gain.

Further investigations have been made by GÖRG [6], [7] concerning the overhead caused by preemptions and on a combination of SRPT with the Round Robin (RR) time slice strategy. Even with an overhead of $10 - 20$ % of the mean service time the combined SRPT/RR strategy keeps the mean waiting time remarkably shorter than standard strategies, such as FIFO and PS (Processor Sharing). In [15] it is shown that the advantage of SRPT can also be used in priority systems.

*Since November 1, 1988, Mr. Scholten is with IBM Germany in Boeblingen

[1]SRPT: Shortest Remaining Processing Time first

[2]CSMA/CD: Carrier Sense Multiple Access with Collision Detection

A profitable field of application for the SRPT/RR combined strategy are packet switching networks where the service time of any message is known by its length. The time to transmit a single maximum length packet of a message represents the maximum time slice. Interrupts of transmissions should only take place after a successfully transmitted packet. GÖRG proposed in [7] a *reversed numbering* of the packets of a message to indicate the remaining number of packets, i.e. the remaining processing time.

In order to apply the SRPT/RR strategy in a station transmission queue, the arriving messages have to be placed in the queue according to their length. The actual length, i.e. the remaining processing time of partially transmitted messages is approximated by the rest packet number RPN. The packet with the smallest RPN in the queue is transmitted next[3] (packet $C2$ in Fig. 1).

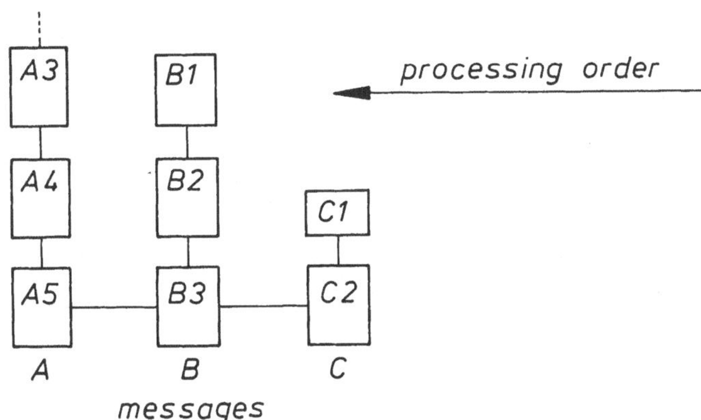

Figure 1 Example of an SRPT organized station queue

BUX considered in [3] local area networks based on bus systems with ordered access. One strategy investigated was with a "shortest-packet-first" discipline where the packets of each frame were scheduled according to their length. But the achieved performance gain was relatively small, because the discipline could not be performed among all waiting packets.

In this paper a modification of the CSMA/CD MAC protocol is introduced as an application of SRPT/RR. The main idea is to favor short messages according to their RPN in the station queues and in the access protocol. This will reduce the mean transmission time significantly for different traffic classes.

First we will give a brief review of traffic measurements in Ethernets. Then the new protocol will be introduced as CSMA/CD/SRPT and will be investigated by means of simulation of an Ethernet with a typical topology and user scenario.

2 Workload characteristics of CSMA/CD Ethernet LAN's

Among local communication systems developed in the last decade, the most widely used is the Ethernet LAN [12], with the standardized CSMA/CD protocol [1]. A relatively high data rate of 10 *Mbits/s* seems to satisfy communication requirements of many attached users. Office automation and partly manufacturing automation are the typical fields of application.

First measurement results were published by SHOCH and HUPP [20]. At that time the bit rate was 2.49 *Mbits/s* and the maximum packet length was 554 *bytes*. The network was only lightly

[3] If the original order of the messages generated by a process must be guaranteed, SRPT output queue sorting is only applied to the first message of the different processes.

loaded (average utilization 2.7 % with a maximum of 32.4 %). Foreseeing dramatically increasing loads, they investigated the maximum possible throughput for their low speed Ethernet. In an experiment they could reach a network utilization of 98 % and therefore concluded that complex optimizations were not needed. SHOCH and HUPP considered only the networks point of view, disregarding the strongly increasing response times for high load conditions.

The more recent paper of GUSELLA [9] describes a completely changed workload situation. The standardized 10-Mbits/s-Ethernet provides a maximum packet length of about 1500 *bytes*. The network analyzed contains file- and printer-servers as well as terminal servers. Network wide file systems (e.g. Network File System NFS, and Network Disk Protocol ND) are installed, where diskless workstations produce a high rate of paging traffic by virtual file access. Faster network interfaces make file transfers more bursty with an interarrival time close to the minimum data link layer interframe spacing of 9.6 μs. This scenario leads to a very unbalanced load among the different stations and a high coefficient of variation of the message length transmitted on the channel. The regular communication between a single client and a file server can substantially load an Ethernet of the current generation. GUSELLA observed a mean network utilization of 6.5 % with short-term peaks above 30 %.

It is known that the performance of the standardized Ethernet with 10 $Mbits/s$ decreases remarkably if the traffic load is raised above about 50 % [4]. The average network utilization in most installations is less than 10 %, but there are short-term peaks in the range of 30 – 40 %. Many of the rapidly growing Ethernet installations are operating near the point of temporary overload situations.

Therefore it is useful to develop an optimization of Ethernet's MAC protocol to reduce the response times with relatively small changes (such as a new controller chip) at the already installed systems. Due to the high coefficient of variation of the message length in these systems, they could be a promising application for the SRPT principle.

3 The new protocol CSMA/CD/SRPT

3.1 General requirements

In this section we introduce the new protocol as a modification and optimization of the standardized CSMA/CD [1]. The system remains unchanged with respect to the distributed buffering capabilities and does not require control from a central monitor station. Within the new protocol most of the elements of the standardized CSMA/CD remain unchanged, for example: jam signal, binary exponential backoff (BEB), and the slot time of 51.2 μs.

Using the broadcast nature of an Ethernet, stations are able to completely monitor the activities on the channel. But in present implementations they ignore all received packets except those containing their own address. For the new protocol described below distributed monitoring of all packet headers is essential.

For the use of the SRPT/RR strategy a mechanism is needed which allows to find the shortest among all messages in the local network waiting for transmission. Therefore we introduce an additional field in the packet headers for the rest packet number of a message (RPN). It indicates the number of packets (*time slices*) needed to complete the transmission of a message. The size of this field is fixed according to the XNS-protocol [2].
The combined SRPT/RR strategy is applied at two levels:

- In the *station output queue* by sorting the messages in the queue according to the number of packets needed to complete their transmission indicated by the RPN of its first packet;
- In the *network access mechanism* by implementing a controlled access mechanism.

3.1.1 SRPT in the station output queue

Every message arriving in a station is sorted into the station's output queue according to the RPN of its first packet (this is the one with the largest number). If a new message is shorter than the waiting ones it is placed ahead of the queue.

The additional time necessary for the organization of the station's output queue according to the RPN's has not been taken into account. Two packets are assumed to wait in the data encoder of the controller. The first of these packets is never replaced until it is transmitted. The second packet in the encoder can be replaced by the first packet of an arriving message with a lower RPN, if the first packet in the data encoder is not presently in service. Thus conflicts are avoided if the station gets channel access during the reorganization of the output queue. The transmission time of one packet can be taken as an approximate lower bound for the time available to reorganize the station queue.

3.1.2 The controlled access mechanism

In standard CSMA/CD any station is allowed to start transmission after the interpacket gap of 9.6 μs following the end of carrier of a transmission. Under higher network load many collisions occur significantly reducing system performance.

Sensing the introduced $RPN = R_l$ of the last or currently transmitted packet a station can decide whether to provoke a collision or not. If its own RPN is greater than or equal to the packet expected next with $RPN = R_l - 1$ of the other station, it does not access the channel. The other station can continue with its next packet. Collisions are therefore avoided. Only if the channel is idle for more than a slot time due to an error or overload situation in the other station the value of R_l can be ignored. If the stations RPN is smaller than $R_l - 1$, it starts its transmission after the interpacket gap and a collision occurs. During the contention period following this provoked collision time slots for the start of transmission of packets with an RPN in a certain range are reserved. This is similar to the hyperchannel [21] priorities with slots reserved for a certain station number. If more than one station starts a transmission in one of these slots, a backoff algorithm can be used to solve this conflict. To reduce the number of collisions a message having sent its packet with $RPN = 3$ is not interrupted. If the last packet on the channel with $RPN = 1$ completed a message and no other message is known from an observed provoked collision any station can access the channel.

In our model we investigated a mechanism with $k = 4$ time slots and the following RPN intervals:

$\leq I(1) = 1$	$\leq I(2) = 3$	$\leq I(3) = 7$	$\leq I(4) = R_l - 1$

The small number of four time slots limits the additional overhead (cf. Fig. 11). The upper bounds of the intervals can be changed depending on a presently observed load on the network. For the simulations we used the standardized binary exponential backoff (BEB) to solve a collision occurring during such a time slot.

3.2 Protocol requirements

The new media access control protocol requires four additional registers for every station interface, a slot counter, and a slot timer for the contention periods.

- R_o *Output-RPN-register*; stores the RPN of the packet at the top of the station's output queue in the data encoder ($R_o \geq 0$; initial[4] value $R_o = 0$);
- R_l *last packet register*; it stores the RPN of the last successfully transmitted packet in the system ($R_l \geq 1$; initial value $R_l = 1$);

[4]The register holds its initial value after each setup of the station and when the whole network is "empty".

R_c *collision register*; stores $R_l - 1$ of the expected packet if a provoked collision occurred, resp. during the contention period the upper bound R_u of the slot interval in which the collision occurred ($R_c \geq 0$; initial value $R_c = 0$);

R_u *upper bound register*; stores the upper bound $I(i),\ i = 1, ..., k$ of the k intervals during the contention period ($R_u \geq 1$; initial value $R_u = 1$). For the investigated model we chose $k = 4$ with $I(1) = 1$, $I(2) = 3$, $I(3) = 7$, and $I(4) = R_l - 1$.

It is important to note, that the values in the registers, except register R_o, are equal in every station due to the distributed monitoring of the activities on the channel.
Every station, idle or ready for transmission, is always sensing the channel including

- frame detection to store the *RPN* of the currently transmitted packet in register R_l,
- controlled access mechanism (see 3.1.2 and 3.3: P3),
- and collision detection.

The station's slot timer is started after each successful transmission, after the jam signal and during the contention period.

3.3 Protocol Description

The following formalized description of the new CSMA/CD/SRPT protocol will later be explained by an example (cf. section 3.4).

P1 `IF sensing a successful transmission of a packet with number` RPN `THEN`

P1.1 `IF` $R_l = 1$ `THEN` $R_c := 0$;
/* packet remembered from a collision is transmitted */

P1.2 `store` $R_l := RPN$;
/* store the *RPN* of the header field of this packet */

P2 `IF ready for transmission THEN`

P2.1 `IF channel busy THEN wait UNTIL end of carrier;`

P2.2 `IF channel idle THEN`

P2.2.1 `IF (last successfully transmitted packet was from this station)`

P2.2.2 `OR` $((R_o < R_l - 1)$ `AND` $(R_l > 3))$
/* output *RPN* is smaller than next expected *RPN*: provoke a collision */

P2.2.3 `OR` $(R_l = 1$ `AND` $R_c = 0)$
/* a transmission is completed and no other message is known from a collision */

P2.2.4 `OR` $(R_l = 1$ `AND` $R_o \leq R_c)$
/* a transmission is completed and output *RPN* is less than or equal to that of a message known from a collision: provoke a collision */

P2.2.5 `OR` $(R_o \leq R_u)$
/* own *RPN* lies in the current interval of the controlled access mechanism */

P2.2.6 `OR (channel still idle after one slot)`
`THEN start transmission`
`ELSE GOTO P2;`

P3 `IF transmission started AND collision detected THEN`

P3.1 `terminate transmission and send jam signal;`

P3.2 `IF` $R_l > 1$ `THEN` $R_c := R_l - 1$;
/* if last transmitted packet didn't complete a transmission then store the expected *RPN* of its next packet */

P3.3 `set slot counter` $i := 1$; /* first time slot of contention period */

```
P3.4 reset collision counters;
P3.5 set R_u := I(i);
P3.6 reset and start slot timer;
P3.7 IF R_o ≤ R_u THEN
       • start transmission in this slot;
       • IF collision detected THEN
         store R_c := R_u; use BEB;
P3.8 IF channel idle during this time slot THEN
  P3.8.1 set slot counter i := i + 1;
  P3.8.2 continue with P3.5;
```

3.4 Example

We illustrate this protocol in Fig. 2 and 3 by displaying the activities on the channel. In these figures the downward arrow represents an arrival of a new message in the system and the double downward arrow indicates a transmission attempt of more than one station in the same time slot.

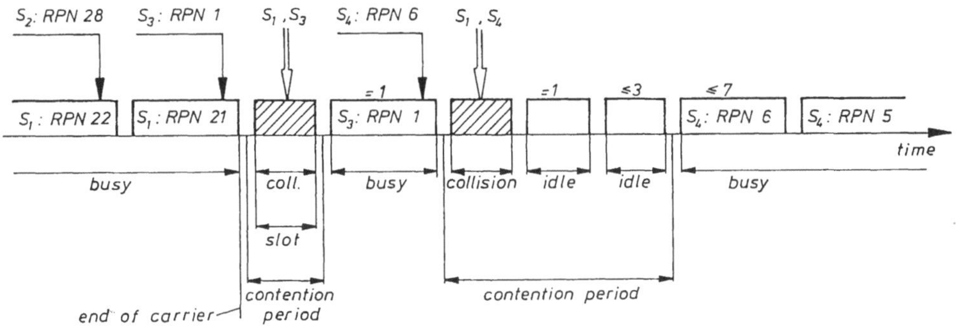

Figure 2 Explanation of the CSMA/CD/SRPT protocol (part I)

It is assumed that during the transmission of a packet with $RPN = 22$ of station S_1 a message with $RPN = 28$ arrived in the empty queue of the station S_2. Station S_2 gets no channel access after the interpacket gap because its RPN is greater than $R_l - 1$ (cf. P2.2.2). A collision is avoided. While station S_1 continues with the packet $RPN = 21$ a one-packet-message arrives in station S_3. Therefore station S_3 provokes a collision after the completion of the current transmission (cf. P2.2 and P2.2.2).

After detecting the collision all stations in the system store the RPN of the interrupted message by setting $R_c = R_l - 1 = 20$ (cf. P3.2), whereas the registers R_l still hold the value 21 of the last successfully transmitted packet. The collision initiates the contention period beginning with a slot time reserved for packets with $RPN = I(1) = 1$. In this example station S_3 starts the transmission of its one-packet-message (cf. P3.7).

During the transmission of station S_3 a message with $RPN = 6$ arrives in station S_4. After the detection of end of carrier, only stations having a message with $RPN \leq R_c = 20$ in their queue are allowed to access the channel (cf. P2.2.4). This condition is matched by the RPN's of the stations S_1 and S_4 and again a collision takes place in the time slot after the end of carrier.

Now the first two slots of the contention period are idle, because no message with an $RPN \leq 3$ arrived. In the third time slot after the collision station S_4 successfully starts its transmission of packet $RPN = 6$.

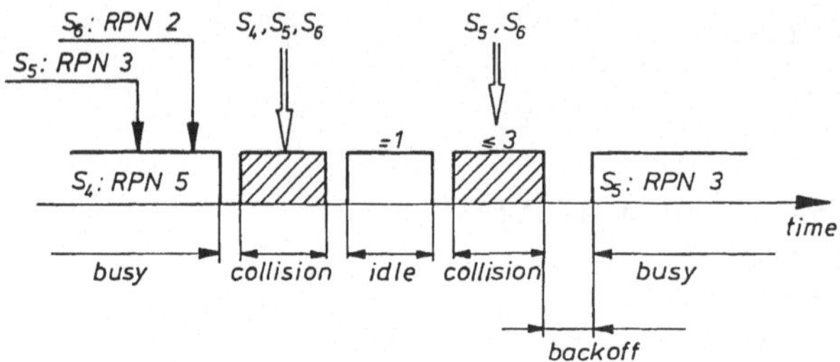

Figure 3 Explanation of the CSMA/CD/SRPT protocol (part II)

The example is continued in Fig. 3 where two short messages arrive in stations S_5 and S_6. After end of carrier a collision takes place between station S_4, S_5 and S_6. After the first idle slot time S_5 and S_6 collide again. The stations use the backoff algorithm and S_5 sends the packet $RPN = 3$.

All stations in the network have stored the $R_c = 3$ of the last collision during the contention period (cf. P3.7). Therefore the other stations wait, observing the activities on the channel, and comparing the two registers R_l and R_c with the RPN of the messages arriving in their own queue (cf. P2.).

4 Traffic Assumptions

For the investigation of the new MAC protocol by means of simulation, a detailed LAN topology as shown in Fig. 4 with traffic assumptions based on recent measurement results has been designed.

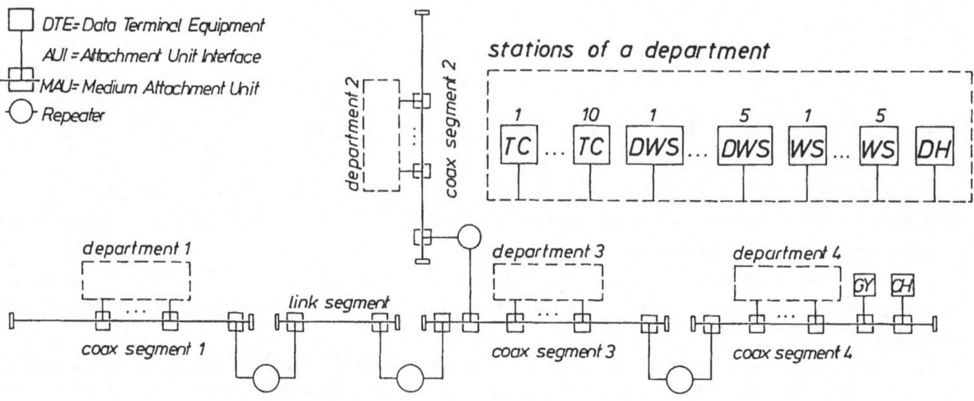

Figure 4 Configuration of the simulated Local Area Network; segment length: 500 m; distance between two stations in one segment: 23.3 m; data rate: 10 $Mbits/s$

The simulated network consists of 4 coax segments and one link segment which are connected by repeater sets. In each coax segment, representing e.g. a department of an administration or a research institution, 21 stations are attached to the cable: 10 terminal concentrators (TC), 5 workstations (WS), 5 diskless workstations (DWS) and 1 department host computer (DH). In segment 4 a central host (CH) and a gateway (GY) to other local or wide area networks are installed in addition to the 21 stations of department 4. The traffic between stations in different segments and to the central host and the gateway can be so intense that routers instead of repeaters do not solve the problem. The maximum delay in the stations and in the media access units as defined in the standard [1] have been taken into account in the simulation; see section 5. The binary backoff algorithm and the jam time have not been changed. The mean propagation delay for each transmission event has been assumed as half of the networks end to end transmission delay, i.e. 9.59 μs.

All traffic sources in the simulation model are independent Poisson processes. For the message generation rates of the stations we choose a basic unit $\lambda_0 = 1\ h^{-1}$. The message lengths are hyperexponentially distributed, see Table 1. The distribution parameters have been chosen on the basis of recent measurements: Marshall/Morgan [11], Welzel [22] and Pawlita [13] for interactive terminal traffic, and Schmickler [16] for file traffic. The constant page size was fixed to the value of a SUN 3 station. The protocol overhead according to the XNS-protocol, supporting numbered packets [2], is 72 *bytes* and the maximum information field length in layer 4 is 1454 *bytes*. If a messages requires more than one packet, the interpacket gap was set to 9.6 μs.

Table 1 shows the traffic generation rates for the different types of stations and the message length distributions.

Table 1 Stations and their input rates; basic rate $\lambda_0 = 1h^{-1}$
packet length: 72 *bytes* = 576 *bits* overhead, 1454 *bytes* information field length

Message length Mean value with Overhead Coeff. of variation #Stations	$\overline{x} = 100\ bits$ $\overline{Q}_{TI} = 676\ bits$ $c = 1.3$ Terminal Input	$\overline{x} = 1500\ bits$ $\overline{Q}_{TR} = 2076\ bits$ $c = 6.1$ Terminal Response	$\overline{x} = 40000\ bytes$ $\overline{Q}_F = 336128\ bits$ $c = 6.0$ File	$\overline{x} = 8192\ bytes$ $\overline{Q}_P = 68992\ bits$ $c = 0$ Paging
40 TC	4750 λ_0	-	-	-
20 WS	-	2500 λ_0	60 λ_0	-
20 DWS	750 λ_0	-	-	360 λ_0
4 DH	-	31500 λ_0	296 λ_0	1800 λ_0
1 GY	-	21250 λ_0	128 λ_0	-
1 CH	-	55000 λ_0	832 λ_0	-
86 *Sum*	λ_{TI}=205000λ_0	λ_{TR}=252250λ_0	λ_F=3344λ_0	λ_P=14400λ_0

The assumptions for the generation rates and the message lengths yield a ratio of 1-packet-messages (terminal input and response) to multi-packet-messages (paging and file traffic) of 96 % to 4 %. But only 16 % of the bytes are carried in the 1-packet-messages.

The network utilization ρ (throughput rate/transmission rate) depends on the traffic generation rates, the mean message lengths (including overhead; Table 1), and the transmission rate v = 10 *Mbits/s*.

$$\rho = \frac{\lambda_{TI}\,\overline{Q}_{TI} + \lambda_{TR}\,\overline{Q}_{TR} + \lambda_P\,\overline{Q}_P + \lambda_F\,\overline{Q}_F}{v} \tag{1}$$

With the basic input rate of $\lambda_0 = 1\ h^{-1}$ we obtain a utilization $\rho = 7.7$ %. During the simulation runs the load was raised by increasing the basic input rate from $\lambda_0 = 1\ h^{-1}$ up to $\lambda_0 = 9\ h^{-1}$, keeping the ratio of terminal to paging resp. file traffic constant. For the curves depicted in section 5 the exact value of ρ was determined from the simulation.

5 Simulation results

Since analytical solutions of a model of such a complexity are not tractable, the model in section 4 with the new access control protocol CSMA/CD/SRPT and SRPT in the station queues was investigated by means of simulation. The simulation program has been written in SIMPAS (**SIM**ulation system in **PAS**cal) [10] running on a computer SIEMENS 7.561. The CSMA/CD-scheme was modelled according to [5]. As shown in Fig. 5 the messages generated in the stations (generation time by GEN and message length by MSG-LEN) are written into their waiting queues WQ. For the simulation of IEEE 802.3 the queues are sorted according to the arrival times (FIFO) and for CSMA/CD/SRPT according to the packet numbers (SRPT/RR). The first packet of every WQ is forwarded into the carrier sense waiting queue CSWQ, which is always organized as a FIFO queue. The collision window of the first packet is computed. All packets with a current carrier sense time[5] falling into this intervall are transmitted into the collision detection queue CDQ. In case of CSMA/CD/SRPT the packets with $RPN \geq R_l$ are retransmitted into the CSWQ. The other packets are further analyzed. Depending on the time when the other stations receive the signal of the first packet in the CDQ (with the lowest current carrier sense time) they sense the channel busy or a collision occurs. In case of a collision the current carrier sense times of the concerned packets are changed according to the BEB. Otherwise the first packet of the CDQ is transmitted and leaves the system. Afterwards all CDQ-packets are sent back and sorted into the CSWQ, according to their new current carrier sense times.

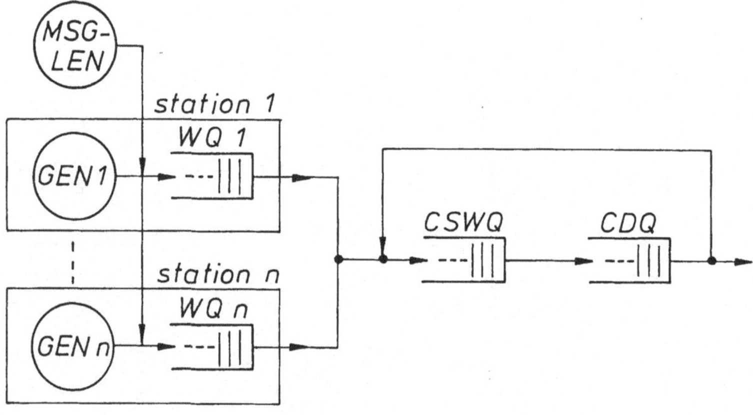

Figure 5 CSMA/CD model

The random number generation (MSG-LEN and GEN in Fig. 5) is based on the quasi-ideal uniform table generator PURAN [8] which was extracted from a physical source.

The transfer time τ_T includes waiting time, propagation and repeater delay and transmission time τ_B of the packetized message. In the following figures the mean transfer times $\overline{\tau_T}$ of the four traffic types are depicted with their symmetric absolute errors $\sigma_{\mu+} = \sigma_{\mu-} = \sigma_\mu/\sqrt{2}$ [19], representing the average deviation below and above the value of $\overline{\tau_T}$. To obtain sufficient statistical significance for the simulated mean value of the traffic with the lowest generation rate (file transfer), the relative error of its mean value was limited to $\sigma_\mu/\overline{\tau_T} \leq 0.2$ (for exact values cf. Fig. 6-11). The standardized CSMA/CD and the new CSMA/CD/SRPT protocol applying SRPT in the station queues and on the network as a whole are compared.

Fig. 6 shows the mean transfer time for the short terminal input messages generated in terminal concentrators (TC) and diskless workstations (DWS). A significant performance gain of more than a factor of 10 could be obtained, with absolute values in the range of $2 - 10\ ms$. The 1-packet-messages immediately provoke a collision and get channel access in the first time slot of the contention period.

The situation is quite similar for the terminal response messages generated in the other stations (WS, DH, CH, GY), see Fig. 7. But there is an important difference to the standard: SRPT in the station queues immediately puts any arriving 1-packet-message at the top of the queue. Furthermore

[5]This is the time when the controller starts to sense the carrier in order to transmit this packet.

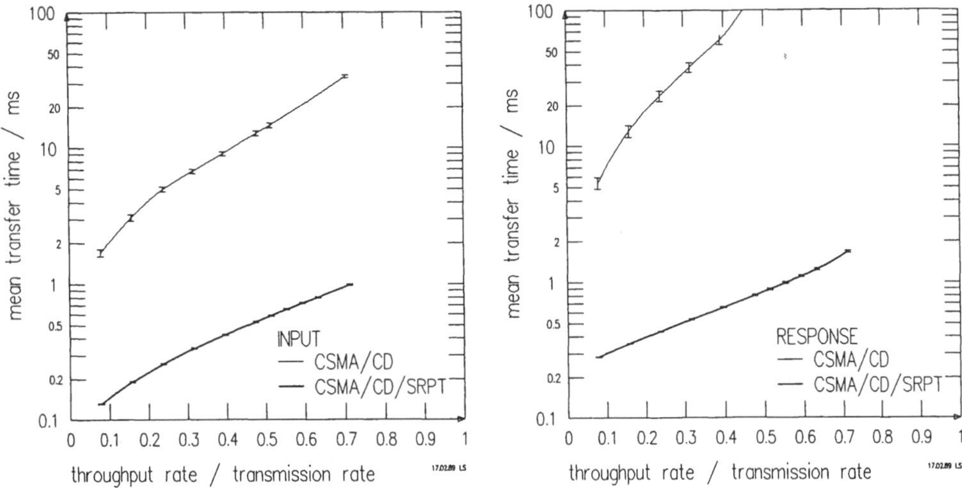

Figure 6 Mean transfer time of *terminal input* messages (TC, DWS).

Figure 7 Mean transfer time of *terminal response* messages (WS, DH, GY, CH).

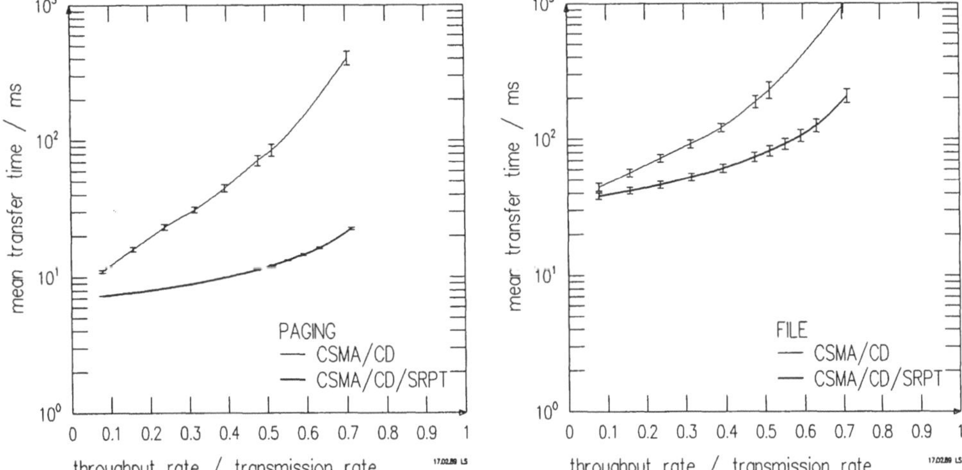

Figure 8 Mean transfer time of *paging* messages (DWS, DH).

Figure 9 Mean transfer time of *file* messages (WS, DH, GY, CH).

terminal responses do not have to wait until the completion of a file transfer. The improvement of their mean transfer time reaches $10 - 50\ ms$. Terminal input and response messages representing a common traffic class are nearly independent of the file and paging traffic. Their mean transfer time is reduced nearly down to their pure transmission time.

The constant paging messages are more frequent but much shorter than files. They too take advantage of the sorted output queue, especially in the department hosts, s. Fig. 8. Because of their constant length they often collide with each other in the same slot during the contention period and the conflict must be solved by the binary exponential backoff. Therefore their relative performance gain is smaller than for the response traffic, but nevertheless about $20 - 40\ ms$.

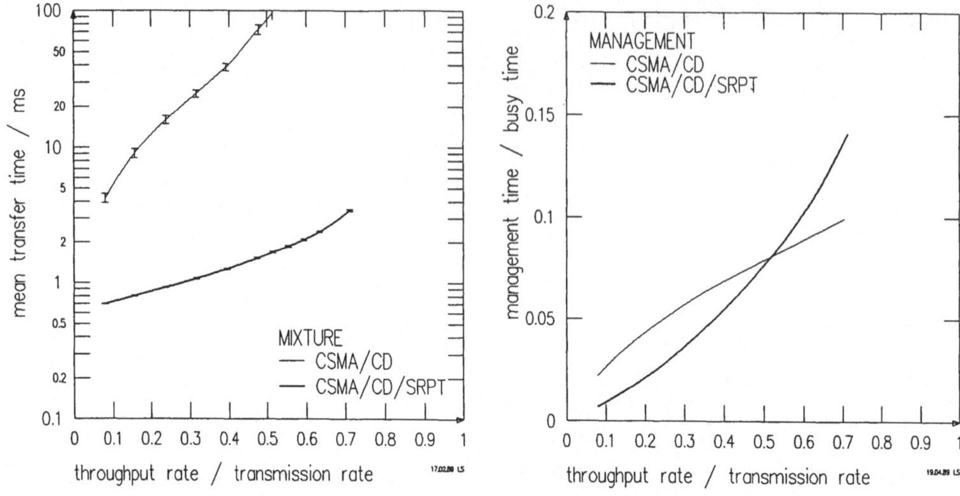

Figure 10 Mean transfer time of all messages.

Figure 11 Management time (contention time) normalized by the busy time (total transmission time).

Without knowing the behavior of SRPT one would say that the long files must be the victims of preferring the shorter messages. But looking at Fig. 9 we see that even the mean transfer time for files could be improved. This is due to the fact that the files are also queued in the station and transmitted over the network according to the lowest rest packet number. Therefore on an average the SRPT principle is of advantage for files as well.

The overall mean transfer time, Fig. 10, shows the great improvement for the whole system obtained by the CSMA/CD/SRPT protocol. Under light traffic the channel access nearly behaves like standard CSMA/CD, while the SRPT organized station queues are responsible for the improvement. If the network load increases, it is the controlled access mechanism which leads to a significant performance gain. The mean response time as the main measure for the performance seen by the users can be kept relatively low, even under higher load conditions.

The prejudice that SRPT causes a large value of the standard deviation of the transfer time could be disproven. The simulation results show that the standard deviation of the transfer time normalized by the mean transmission time is remarkably smaller for the CSMA/CD/SRPT protocol than for the standard CSMA/CD. This result is in agreement with new analytical investigations of PERERA [14] for the M/G/1 system.

Fig. 11 shows, as expected, that the overhead time, i.e. the duration of the contention periods and backoff times, decreases under light load for the new protocol, because collisions are avoided. The main part of the overhead is due to empty slot times (mean value $\approx$ 90 μs i.e. 1.8 slots). Above 50 % network load, the number of interrupts (provoked collisions) increases and the overhead becomes greater than for standard CSMA/CD. But nevertheless the performance improvement is still significant because of the strong reduction of the waiting times. This verifies the analytical investigations of GÖRG [6] concerning the influence of overhead time in the system M/G/1/SRPT.

6 Conclusion

In this paper the first practical application of the SRPT principle to an existing packet switching network, namely the Ethernet LAN with the CSMA/CD protocol, has been investigated. Under realistic conditions with a high coefficient of variation of the message lengths it is shown that the new CSMA/CD/SRPT protocol with the SRPT principle applied to the output queue and the network as a whole considerably reduces the mean transfer time especially under a high traffic load. Using the rest packet number collisions are avoided. Short messages, e.g. terminal traffic or protocol packets, are favored similar to the way the hyperchannel protocol gives priority to certain station numbers. Their performance is not affected by longer file or paging transmissions. This means that the total delay including higher protocol layers is also reduced. Implementing the new protocol, the network provider can allow a remarkably higher network load without increasing the mean transfer times for the users. The additional functions of the new protocol as well as the output queue management can be realized by a minor extension to the Ethernet controller hardware and firmware.

The SRPT principle is applicable to other network types, for example by using already installed priority mechanisms. Further investigations concerning Token Bus, Token Ring, and High Speed LAN's have been started.

Acknowledgement

We would like to thank Prof. Dr.-Ing. F. Schreiber for his valuable support and for carefully reading this paper, Dipl.-Ing. N. Niebert for many helpful discussions and Dr. rer. nat. C. Görg for reviewing this paper.

References

[1] *IEEE 802.3: Carrier Sense Multiple Access with Collision Detection (CSMA/CD); Access Method and Physical Layer Specifications.* IEEE, New York, 1985.

[2] *Xerox System Integration Standard: Internet Transport Protocols.* Xerox, December 1981.

[3] W. Bux. *Analysis of a Local-Area Bus System with Controlled Access.* In *Performance of Data Communication Systems and their Application*, pp. 11–22, North-Holland, 1981.

[4] W. Bux. *Local-Area Subnetworks: A Performance Comparison.* IEEE Trans. Commun., Vol. COM-29, No. 10, pp. 1465–1473, 1981.

[5] N. Dahmen, U. Killat, and R. Stecher. *Performance Analysis of Token Bus and CSMA/CD Protocols Derived from FORCASD Simulation Runs.* In *Performance of Computer Communication Systems*, Elsevier North-Holland, Amsterdam, 1984.

[6] C. Görg. *Evaluation of the Optimal SRPT Strategy with Overhead.* IEEE Trans. Commun., Vol. COM-34, No. 4, pp. 338–344, 1986.

[7] C. Görg. *Further Results on a New Combined Strategy Based on the SRPT-Principle.* (to appear in IEEE Trans. Commun.).

[8] M. Gude. *Concept for a High Performance Random Number Generator Based on Physical Random Phenomena.* Frequenz, Vol. 39, No. 7/8, pp. 187–190, 1985.

[9] R. Gusella. *The Analysis of Diskless Workstation Traffic on an Ethernet.* Technical Report UCB/CSD 87/379, Computer Science Dev., University of California, Berkeley, California, 1987.

[10] M. Kochan, C. Görg, H. Tuchel, and N. Niebert. *SIMPASII. Ereignisorientiertes Simulationssystem in der Programmiersprache PASCAL.* Technical Report, Lehrstuhl DFV, RWTH Aachen, 1984.

[11] W.T. Marshall and S.P. Morgan. *Statistics of Mixed Data Traffic on a Local Area Network.* Computer Networks and ISDN Systems, No. 10, 1985.

[12] R.M. Metcalfe and D.R. Boggs. *Ethernet: Distributed Packet Switching for Local Computer Networks.* Communications of the ACM, Vol. 19, No. 7, pp. 395–404, 1976.

[13] P. Pawlita. *Traffic Measurements in Data Networks, Recent Measurement Results, and Some Implications.* IEEE Trans. Commun., Vol. COM-29, No. 4, pp. 525–535, 1981.

[14] R. Perera. *Contributions to the Theory of Queueing Systems with the Optimal Strategies SPT and SRPT.* Ph. D. Thesis, RWTH Aachen, 1989. (in German).

[15] X.H. Pham and F. Schreiber. *The Optimal Strategy SRPT in Priority Systems of Digital Communication Networks.* AEÜ, Vol. 43, No. 3, pp. 129–134, 1989.

[16] L. Schmickler. *Representation of Data Network Measurement Results by the Mixed Erlang Approximation Method MEDA.* In *Proc. 12th Int. Teletraffic Congress (ITC)*, Turin, 1988.

[17] L.E. Schrage. *A Proof of the Optimality of the Shortest Remaining Processing Time Discipline.* Operations Research, Vol. 16, pp. 678–690, 1968.

[18] L.E. Schrage and L.W. Miller. *The Queue M/G/1 with the Shortest Remaining Processing Time Discipline.* Operations Research, Vol. 14, pp. 670–684, 1966.

[19] F. Schreiber. *Improved Simulation by Application of the Objective Bayes-Statistics.* AEÜ, Vol. 34, pp. 243–249, 1980.

[20] J.R. Shoch and J.A. Hupp. *Measured Performance of an Ethernet Local Network.* Communications of the ACM, Vol. 23, No. 12, pp. 711–721, 1980.

[21] O. Spaniol. *Analysis and Performance Evaluation of HYPERchannel Access Protocols.* Performance Evaluation, Vol. 1, pp. 170–179, 1981.

[22] T. Welzel. *Analyse und Messung der Workload einer technisch-wissenschaftlichen Programmentwicklungsumgebung als Grundlage der Leistungsbewertung Lokaler Netze.* In *Informatik Fachberichte: Kommunikation in verteilten Systemen, GI/NTG Fachtagung Erlangen*, Springer-Verlag, Berlin, 1987.

S-ALOHA-MULTI-HOP-NETZE MIT ADAPTIVER SENDEREICHWEITE BEI OMNIDIREKTIONALEM UND DIREKTIONALEM SENDEN

Claus Gotthardt, Hans-Jürgen Perz
Fachbereich Elektrotechnik - Datenverarbeitungstechnik
FernUniversität-GH Hagen, D-5860 Iserlohn

Kurzfassung

Es wird ein Multi-hop Paket-Funk-Netz mit dem zufallsorientierten Kanalzugriffsprotokoll S-ALOHA untersucht, bei dem alle Stationen die Entfernungen zu ihren Nachbarstationen kennen und in der Lage sind, ihre Sendereichweite zwischen 0 und der maximalen Sendereichweite R zu variieren. Die Sendeleistung und damit die Sendereichweite wird derart eingestellt, daß die jeweilige Nachbarstation gerade noch erreicht werden kann, wobei es keine Rolle spielt, ob sie Ziel- oder Relaisstation des Datagramms ist. Durch diese Sendereichweiten-Anpassung wird die Anzahl durch eine Übertragung gestörter Stationen deutlich reduziert. Es wird gezeigt, daß dadurch eine deutliche Verbesserung des Durchsatzes und des mittleren One-hop Fortschrittes in Richtung Ziel gegenüber einer festen Sendereichweite erzielt werden kann.

Die Durchsatzgleichung sowie die Fortschrittgleichung wird für ein Netz mit zufällig verteilten Stationen hergeleitet, numerisch ausgewertet und mit Ergebnissen des S-ALOHA Protokolls bei fester Sendereichweite verglichen.

Zur Validierung der Analyseergebnisse wurde ein Simulationsprogramm erstellt, welches den Modellannahmen entspricht. Im Simulationsprogramm sind die Stationen zufällig auf einer Kugeloberfläche angeordnet um die Annahme eines unendlich ausgedehnten Netzes, wie es im analytischen Modell unterstellt wird, nachzubilden. Die Anordnung der Stationen auf der Kugeloberfläche hat gegenüber anderen möglichen Anordnungen, z.B. in einer Kreisfläche, einige Vorteile, welche im Kapitel 4 beschrieben werden.

1. EINLEITUNG

Zur Kommunikation lokal verteilter evtl. mobiler Teilnehmer kann man paketvermittelte Datenkanäle verwenden. Bei geringer Datenmenge pro Übertragung und stochastischem Verkehrsaufkommen sind zufallsorientierte Vielfachzugriffsprotokolle vom ALOHA- oder CSMA-Typ besonders geeignet. Für das Slotted-ALOHA (S-ALOHA) Zugriffsverfahren wird der Kanal durch einen Takt in äquidistante Zeitschlitze (Slots) unterteilt. Die Slotlänge entspricht dabei genau einer Paketlänge. Auf einem derartigen Kanal kann eine Station übertragen und alle anderen Stationen im Sendebereich dieser Station können empfangen. Eine sendewillige Station überträgt ein Paket in einem Slot ohne Rücksicht auf andere Kanalbenutzer. Übertragen bei einem sog. vollvermaschten System zwei oder mehr Stationen gleichzeitig, dann liegt eine Kollision vor. Keine der beiden Stationen ist erfolgreich und beide Pakete müssen erneut übertragen werden. Damit nicht beide Stationen ihr Paket in gleichen, zeitlich später liegenden Slots wiederholen (2.Versuch) und wieder kollidieren, muß jede Station eine Pausendauer, d.h. eine bestimmte Anzahl Slots, die sie ausläßt bevor sie wiederholt, zufällig auswählen.

Die heute bekannten S-ALOHA-Systeme sind vollvermascht, d.h. die Sendereichweite der Stationen ist netzweit, jede Station des Netzes kann jede andere direkt erreichen. Hier soll aber ein System betrachtet werden, bei dem die Sendereichweite, bedingt durch übertragungstechnische Gegebenheiten, wie z.B. die Anwendung im Gigahertz-Bereich, stark begrenzt ist. Es entsteht ein teilvermaschtes S-ALOHA-System, ein sog. Multi-hop System, bei dem die Verhältnisse erheblich komplizierter sind. Die reduzierte Sendeleistung hat zur Folge, daß jede Station nur noch einen Teil der Stationen des Netzes direkt empfangen kann. Datenpakete müssen u.U. über Zwischenstationen übertragen werden, um zu ihrem Ziel zu gelangen, d.h. die Pakete müssen für eine Quelle-Ziel-Übertragung mehrmals übertragen werden.

Andererseits erreicht man aber auch eine räumliche Trennung der einzelnen Stationen, denn Stationen die sich nicht hören können, können sich auch nicht stören. Ein übertragenes Paket wird nur von einem Teil der Stationen des Netzes empfangen. Es besteht die Möglichkeit, daß eine Station in einem anderen Teil des Netzes zur gleichen Zeit überträgt und trotzdem erfolgreich ist. Diese Tatsache wird als räumliche Wiederbenutzung des ALOHA-Kanals bezeichnet.

Eine weitere Möglichkeit zur Verringerung der Interferenzen ist die Einführung einer adaptiven Sendereichweite. Nach dem Auswählen der jeweiligen Empfangsstation wird die Sendeleistung derart reduziert, daß die ausgewählte Station gerade noch erreichbar ist. Dadurch werden weiter entfernte Stationen, an die das Paket nicht adressiert ist auch nicht unnötigerweise gestört. Für dieses System sollen im folgenden einige Leistungsmerkmale untersucht werden.

Für Multi-hop Systeme sind bereits Ergebnisse für den Durchsatz bekannt. Es wurde für S-ALOHA /SILV.78/ und CSMA /GOTTH.89/ gezeigt, daß dieser sich gegenüber einem vollvermaschten System nicht verringert, sondern sogar deutlich größer sein kann als 1/e bei vollvermaschten S-ALOHA-Systemen bzw. 0.85 beim vollvermaschten CSMA-System.

In einem teilvermaschten Netz müssen Pakete mehrmals erfolgreich übertragen werden, bevor sie ihr Ziel erreichen. Demzufolge ist der One-hop Durchsatz kein ausreichendes Leistungsmaß. Ein geeigneteres Leistungsmaß ist der Quelle-Ziel-Durchsatz, der One-hop Durchsatz dividiert durch die mittlere Anzahl benötigter Hops. Die mittlere Anzahl benötigter Hops ist aber nicht bekannt. Als Leistungsmaß wird daher der mittlere One-hop Fortschritt benutzt. In /TAKA.84/ wird gezeigt, daß der mittlere One-hop Fortschritt pro Übertragung maximal ist, wenn der S/E-Radius so eingestellt ist, daß sich im Mittel W=8 Nachbarstationen im Übertragungsbereich einer Station befinden.

Eine weitere interessierende verkehrstheoretische Größe ist die Paketverzögerung. Über Paketverweilzeiten in teilvermaschten S-ALOHA-Systemen sind bisher kaum Aussagen gemacht worden.

Wie zu vermuten wird sich zeigen, daß großer Durchsatz und kleine Verzögerungsdauer sich gegenseitig ausschließen. Ein Maß für die Paketverzögerung ist die sog. Backlogzeit pro erfolgreichem Hop. Im vorliegenden Aufsatz soll versucht werden diese mittlere Backlogzeit eines Paketes zu berechnen. Die Backlogzeit eines Paketes ist definiert als die Zeit vom 1. Übertragungsversuch bis zur erfolgreichen Übertragung.

2. BESCHREIBUNG DES MODELLS

In diesem Abschnitt sollen die hier benutzten Modelle mit den adaptiven Sendereichweiten für den S-ALOHA-Kanal beschrieben werden.

Sende-/Empfangsradius R

Alle Stationen haben die gleiche Sendereichweite R. Stationen, die sich in einem Kreis mit Radius R um eine Station herum befinden, können diese Station hören. Stationen außerhalb dieses Kreises können das nicht, sie haben keine direkte Verbindung. Empfängt eine Station in einem Slot zwei oder mehr Pakete gleichzeitig, d.h. sie hat in diesem Slot mehr als eine sendende Station in ihrem Empfangsbereich, dann entsteht eine Kollision. Eine Kollision entsteht auch wenn der Empfänger sendet.

Die Besonderheit in diesem Modell ist die Reduktion der Interferenzen durch die Adaption der Sendereichweite. Die Stationen sind in der Lage, nach der Auswahl der jeweiligen Empfangsstation den Sendebereich zwischen 0 und R so zu variieren, daß die ausgewählte Station gerade noch erreichbar ist. Befindet sich die ausgewählte Empfangsstation Q im Abstand r von der Sendestation P, so wird der Senderadius auf r eingestellt. Dies vermindert die Anzahl der Stationen, die dieses Paket auch empfangen und dadurch gestört werden können.

MODELL 1:

rundumstrahlende Antenne bei adaptivem Sende-Radius. Die rundumstrahlende Antenne hat ein kreisförmiges Abstrahlfeld mit dem Radius r zur Folge.

MODELL 2:

richtungsselektierende Antenne. Die richtungsselektierende Antenne besitzt ein keulenförmiges Abstrahlfeld. Im Modell wird die Funktion der Lemniskate als Abstrahlfeld verwendet. Die Empfangsstation liegt bei $c*R$ ($c \geq 1$). Für eine begrenzte Anzahl von Antennen muß c größer 1 sein. Empfangen wird rundum.

a) Lemniskate: $r_1^3 = R_1 (x^2 - y^2)$ b) Lemniskate: $r_1^4 = R_1^2 (x^2 - y^2)$

Räumliche Verteilung der Stationen

Die Stationen sind zufällig in der Ebene angeordnet, wobei die möglicherweise mobilen Stationen als quasi stationär angesehen werden. Die Mobilität hat keinen Einfluß auf den Durchsatz. Der Einfluß auf die Backlogzeit ist ausführlich in /BRAGO.87/ diskutiert. Die Stationen sind homogen in einem zweidimensionalen Raum mit der Dichte τ verteilt. Die Anordnung repräsentiert eine Momentaufnahme eines Mobilfunknetzes. Die Anzahl Stationen in einer

Kreisfläche mit Radius R genügt somit einer Poisson-Verteilung mit Mittelwert $W = \tau\pi R^2$. W ist auch ein Maß für die Netzkonnektivität.

Die charakteristischen Systemgrößen Durchsatz, Fortschritt und Backlogzeit sollen für ein System mit einer großen Anzahl Stationen M (M>>W) in einem flächenmäßig homogenen Bereich bestimmt werden. Randeffekte bleiben für die Analyse unberücksichtigt.

Kanalzugriffsprotokoll S-ALOHA

Das Kanalzugriffsprotokoll ist S-ALOHA, wobei die Slotdauer einer Paketlänge plus der maximalen Ausbreitungsdauer entspricht.

Verkehrsaufkommen

Jede Station besitzt zu jedem Zeitpunkt ein Paket. Gesendet wird gemäß einer Poisson-Verteilung mit einem Mittelwert von p Paketen/Slot, solange die Station nicht blockiert ist, d.h. solange sie kein früher kollidiertes Paket mehr zu übertragen hat. Zu wiederholende Pakete sind in p nicht enthalten.
Ein kollidiertes Paket wird mit der Wahrscheinlichkeit α in jedem der auf die Kollision folgenden Slots wiederholt übertragen. Es existieren somit zwei voneinander unabhängige Ankunftsprozesse, die beide abhängig sind von der Anzahl zurückgelegter, früher kollidierter Pakete. Die Anzahl Slots, die ein Paket bis zur erneuten Übertragung ausläßt oder überspringt (Wartezeit) ist geometrisch verteilt. Die mittlere Wartezeit ist:

$$\sum_{k=1}^{\infty} \alpha k(1-\alpha)^{k-1} = 1/\alpha \qquad (1)$$

Quittungsverkehr wird nicht betrachtet. Es wird angenommen, daß der erfolgreiche Empfang eines Paketes auf einem eigens dafür vorgesehenen Kanal sofort bestätigt wird.

Verteilung der Ziele eines Paketes

Die Verkehrsbeziehungen untereinander sind gleichmäßig verteilt. Jede Station sendet zu allen anderen M-1 Stationen mit gleicher Wahrscheinlichkeit.

Routing-Verfahren MFR

Pakete, die für ein vorgegebenes Ziel F im Netz bestimmt sind, werden weiter geroutet, indem das Paket an den Nachbarn übertragen wird, der am weitesten in Richtung Ziel liegt. Befindet sich keine Station in Vorwärtsrichtung (Richtung Ziel), so muß das Paket zurück übertragen werden. Befindet sich keine weitere Station im Kreis mit Radius R um P, dann wird in diesem Slot nicht übertragen.

In dem beschriebenen Modell sind folgende Leistungsgrößen von Bedeutung:

1) $S(p,\alpha,N)$ = One-hop Durchsatz

Der One-hop Durchsatz ist definiert als die mittlere Anzahl erfolgreicher One-hop Übertragungen einer Station/Slot. Da für eine vollständige Übertragung zwischen Quelle und Ziel im allgemeinen mehrere One-hop Übertragungen nötig sind, erhält man den wirklichen Netzdurchsatz, indem man den One-hop Durchsatz durch die mittlere Anzahl benötigter Hops dividiert. Die mittlere Anzahl benötigter Hops pro Quelle-Ziel Übertragung ergibt sich aus dem mittleren Abstand zwischen einer beliebigen Quelle und einem zufälligen Ziel. Diese beiden Größen sind Zufallsvariable, die aber hier nicht betrachtet werden. Anstatt dessen wird als Maß für den wirklichen Netzdurchsatz der mittlere Fortschritt eines Paketes/Slot definiert.

2) $Z(p,\alpha,N)$ = mittlerer Fortschritt eines Paketes in Richtung Ziel/Slot.

Der mittlere One-hop Fortschritt ist der physikalische Weg der pro Hop in Richtung Ziel zurückgelegt wird. Um jedoch eine einheitenfreie Größe zu erhalten wird üblicherweise mit der Wurzel aus der Dichte multipliziert. Die Größe Z soll im wesentlichen nur als Vergleichsmaß dienen um unterschiedliche Modelle zu vergleichen.

3) $D(p,\alpha,N)$ = One-hop Backlog

Der One-hop Backlog ist definiert als die Backlogzeit eines Paketes zuzüglich dessen Übertragungszeit (1 Slot). Die Paketverzögerung entsteht aufgrund von Kollisionen. Beträgt die Wartezeit eines kollidierten Paketes bis zur wiederholten Übertragung im Mittel $1/\alpha$ und das Paket kollidiert im Mittel k-mal, so entsteht eine mittlere Verzögerung von k/α. Die zufällige Verzögerungszeit eines Paketes nach einer Kollision wird hier als geometrisch verteilt angenommen. In /LAM.74/ wurde gezeigt, daß Durchsatz und Backlogzeit bei S-ALOHA-System im wesentlichen nur vom Mittelwert dieser Verteilung abhängen, so daß das hier verwendete Modell näherungsweise doch sehr allgemeingültig ist.

3. ANALYSE

Im folgenden gilt $\alpha=p$, d.h. neue und kollidierte Pakete werden mit gleicher Wahrscheinlichkeit in einem Slot übertragen. p ist somit die Wahrscheinlichkeit dafür, daß eine Station unabhängig von vorhergehenden Kollisionen ein Paket im nächsten Slot überträgt.

3.1 Modell 1: Übertragung mit einer rundumstrahlenden Antenne bei adaptivem Sende-Radius

Die am einfachsten bestimmbare Größe ist der One-hop Durchsatz, die mittlere Anzahl erfolgreicher Übertragungen pro Slot. Da für alle Stationen ein identisches Verkehrsverhalten angenommen wird, kann man eine beliebige Station auswählen, und deren mittlere Anzahl erfolgreich empfangener Pakete/Slot berechnen. Der Gesamtdurchsatz ergibt sich aus der Summation über alle Stationen.

Der Durchsatz einer beliebig ausgewählten Station Q, die außer dem jeweiligen Sender P noch i andere Stationen empfängt, ist

$$S_1=p(1-p)(1-p)^i \qquad (2)$$

Dieser Durchsatz gilt unter der Bedingung, daß sich ein Sender P und i andere Stationen im Sende-/Empfangsbereich des Empfängers Q befinden. Da nun der Sendebereich in diesem Modell so eingestellt wird, daß der ausgewählte Empfänger Q im Abstand r von P gerade erreicht wird, müssen sich diese i weiteren Stationen in einer Kreisfläche mit Radius r um Q befinden. Für die Wahrscheinlichkeit, daß sich i weitere Stationen im Abstand r um Q befinden gilt wegen der Markoff-Eigenschaft der Poisson-Verteilung

$$P\{x=i\,|\,\tilde{r}=r\} = \frac{(\tau\pi r^2)^i}{i!}\, e^{-\tau\pi r^2} \tag{3}$$

Mit der folgenden Beziehung aus der Wahrscheinlichkeitsrechnung

$$E[S_1\,|\,\tilde{r}=r] = \sum_{i=0}^{\infty} E[S_1\,|\,i,\tilde{r}=r]\; P\{i\,|\,\tilde{r}=r\} \tag{4}$$

folgt somit für den One-hop Durchsatz einer Station unter der Bedingung, daß der Abstand zwischen Sender und Empfänger r ist

$$E[S_1\,|\,\tilde{r}=r] = \sum_{i=0}^{\infty} p(1-p)(1-p)^i\, \frac{(\tau\pi r^2)^i}{i!}\, e^{-\tau\pi r^2} \tag{5}$$

Um die Bedingung aufzulösen wird nun die Verteilung der Positionen des Empfängers Q in Bezug auf den Sender P benötigt. Der Abstand r zwischen Sender und Empfänger bzw. der Winkel δ ergeben sich aus dem Auswahlverfahren des nächsten Empfängers, d.h. aus dem gewählten Routingverfahren. Zur Modellierung wird hier ein sog. MFR-Routing angenommen. Dabei wird immer derjenige Empfänger ausgewählt der sich am weitesten in Richtung Ziel befindet (Bild 1). Praktisch wird dies jedoch anders gelöst, es wird die Anzahl Hops minimiert. Dieses ist jedoch kaum zu modellieren. Simulationsergebnisse haben jedoch gezeigt, daß das Modell zu dem in der Praxis verwendeten Verfahren fast identische Aussagen liefert. Es gilt nach /GOTTH.89/:

BILD 1

MFR-Routing

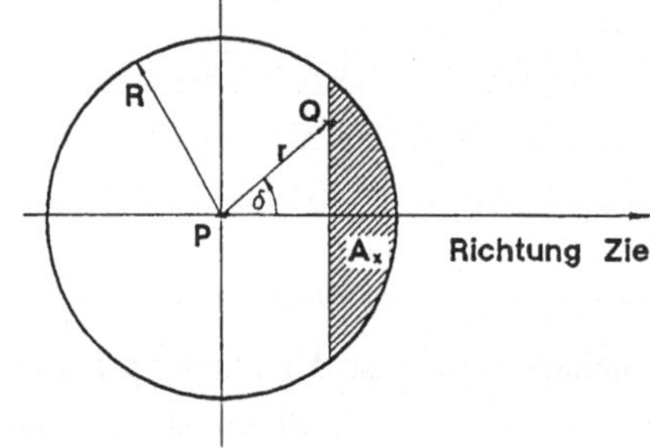

$$\begin{aligned} P\{r<\tilde{r}\le r+dr,\ \delta<\tilde{\delta}\le\delta+d\delta\} &= P\{\text{DER EMPFÄNGER BEFINDET SICH IM PUNKT } (r,\delta)\}\\ &= P\{\text{KEINE STATION IN } A_x\}\\ &\quad *P\{\text{MINDESTENS EINE STATION BEI } (r,\delta)\}\\ &= e^{-\tau A_x}\ \delta\ r\ dr\ d\delta \end{aligned} \tag{6}$$

$$A_x = R^2[\arccos(r/R\,\cos\delta) - r/R\,\cos\delta\,(1-(r/R\,\cos\delta)^2)^{1/2}] \tag{7}$$

Da es sich hier bei r und δ um kontinuierliche Zufallsvariable handelt, kann man die Bedingung auflösen, indem man über die betrachtete Kreisfläche integriert. Es gilt somit für den One-hop Durchsatz $E[S_1]$

$$E[S_1] = \int_0^R \int_0^{2\pi} E[S_1 | \tilde{r}=r] \; P\{r<\tilde{r}<r+dr, \delta<\tilde{\delta}<\delta+d\delta\}$$

$$E[S_1] = p(1-p) \int_0^R \int_0^{2\pi} e^{-p\tau\pi r^2} \; e^{-\tau Ax} \; 2 \; \delta \; r \; d\delta \; dr$$

$$E[S_1] = p(1-p) \; 2W/\pi \int_0^1 \int_0^{\pi} e^{-pWt^2} \; e^{-a(t)W/\pi} \; t \; d\delta \; dt \qquad (8)$$

mit $a(t) = \arccos(t \cos\delta) - t \cos\delta \; (1-(t \cos\delta)^2)^{1/2}$

$W = \tau\pi R^2$

Mit $x = r*\cos\delta$ gilt für den mittleren Fortschritt eines Paketes pro Slot in Richtung Ziel:

$$E[Z_1](\tau)^{1/2} = \frac{2W^{3/2} p(1-p)}{\pi^{3/2}} \int_0^1 \int_0^{\pi} e^{-pWt^2} \; e^{-a(t)W/\pi} \; t^2 \cos\delta \; d\delta \; dt \qquad (9)$$

Eine weitere interessante Leistungsgröße ist die Backlogzeit. Bei der Untersuchung von Paketverzögerungen hat die Beweglichkeit der Stationen einen Einfluß. Der Grund dafür ist, daß man zur Bestimmung von Backlogzeiten den Kanal zu mehreren Zeitpunkten, zu denen ein Paket wiederholt wird, betrachten muß, denn in der Zwischenzeit kann sich wegen der Mobilität die Anzahl Nachbarn einer Station verändert haben. Da die Beweglichkeit der Stationen im Verhältnis zur zufälligen Wartezeit zwischen zwei Übertragungsversuchen sehr gering ist, kann das System als quasi stationär betrachtet werden /GOTTH.89/. Es wird angenommen, daß die Anzahl Nachbarstationen sich bis zur erfolgreichen Übertragung eines Paketes nicht ändert.

Die Backlogzeit eines Paketes erhält man, indem zunächst die Backlogzeit des vollvermaschten Systems mit i Stationen berechnet wird. Um die Backlogzeit des teilvermaschten oder Multi-hop Systems unter der Annahme der Quasi-Stationarität zu erhalten, muß die Backlogzeit des vollvermaschten Systems mit i Stationen mit den entsprechenden Auftrittswahrscheinlichkeiten gewichtet und von 0 bis aufsummiert werden /BRAGO.87/.

$$E[D | i, r=r] = \frac{1-(1-p)^{i-1}}{p(1-p)^{i-1}}$$

$$E[D | \tilde{r}=r] = \sum_{i=0}^{\infty} E[D | i+2, \tilde{r}=r] \; P\{i | \tilde{r}=r\}$$

$$E[D | \tilde{r}=r] = 1/p \; \{1/(1-p) \; e^{p\tau\pi r^2/(1-p)} - 1\} \qquad (11)$$

Nach der Integration erhält man somit für die Backlogzeit in einem Multi-hop System mit adaptivem Sende-Radius:

$$E[D] = \int_0^R \int_0^{2\pi} E[D|\tilde{r}=r]\ P\{r<\tilde{r}<r+dr, \delta<\tilde{\delta}<\delta+d\delta\}$$

$$E[D] = \frac{2W}{\pi p} \int_0^1 \int_0^{\pi} \left(\frac{e^{p\tau t^2/(1-p)}}{1-p} 1\right) e^{-a(t)W/\pi}\ t\ d\delta\ dt \qquad (12)$$

3.2 Modell 2: Übertragung mit richtungsselektierender Antenne bei adaptivem Sende-Radius

Im folgenden wird ein Multi-hop System betrachtet, bei dem die Stationen in der Lage sind, eine richtungsselektierende Antenne auf den jeweiligen Empfänger auszurichten. Im Modell wird angenommen, daß das Abstrahlfeld der Antenne einer Lemniskate entspricht.

Modell 2a - Gleichung der Lemniskate a) $r_1^4 = R_1^2\ (x^2-y^2)$

$$= R_1^2\ r_1^2 \cos 2\delta$$

Zunächst ist die Fläche der Lemniskate zu bestimmen, sie wird der Einfachheit halber in Polarkoordinaten berechnet.

$$A_a(r) = \iint_A dy\ dx = \iint_A r\ dr\ d\delta$$

Für die Grenzen gilt:

$$-\pi/4 < \delta \leq \pi/4 \qquad 0 < r_1 \leq R_1(\cos 2\delta)^{1/2}$$

Nach dem Lösen der Gleichung ergibt sich die Fläche zu:

$$A_a(r) = \frac{\pi R_1^2}{2\pi}$$

R ist der Abstand zwischen Sender P und Empfänger Q . Da sich der Empfänger nicht in der Spitze der Lemniskate befinden soll, es gilt $R_1 = cR$ ($c \geq 1$).

$$A_a(r) = \frac{\pi r^2}{\pi}$$

Modell 2b - Gleichung der Lemniskate b) $r_1^3 = R_1\ (x^2-y^2)$

$$= R_1\ r_1^2 \cos 2\delta$$

analog ergibt sich die Sendefläche zu:

$$A_b(r) = \frac{\pi r^2}{4}$$

Für den Durchsatz unter der Bedingung des Abstrahlfeldes in der Form einer Lemniskate gilt:

$$E[S_1] = p(1-p) \int_0^R \int_0^{2\pi} e^{-p\tau A(r)} e^{-\tau Ax} \; 2r\delta \; d\delta \; dr \tag{13}$$

$$E[S_1] = p(1-p)2W/\pi \int_0^1 \int_0^{\pi} e^{-p\tau A(t)} e^{-\tau a(t)} \; 2r\delta \; d\delta \; dr$$

$$A_a(t) = t^2/\pi \qquad \text{(Modell 2a)}$$

$$A_b(t) = t^2/4 \qquad \text{(Modell 2b)}$$

Analog zu Gleichung (9) ergibt sich der mittlere Fortschritt eines Paketes pro Slot in Richtung Ziel:

$$E[Z1](\tau)^{1/2} = \frac{2W^{3/2}p(1-p)}{\pi^{3/2}} \int_0^1 \int_0^{\pi} e^{-pWA(t)} e^{-a(t)W/\pi} \; t^2 \cos\delta \; d\delta \; dt \tag{14}$$

Für die Berechnung der Backlogzeit kann die Gleichung (11) und die dafür getroffenen Annahmen auch für das gerichtete Senden herangezogen werden. Durch das Integrieren über die Kreisfläche erhält man somit für die Backlogzeit in einem Multi-hop System mit adaptivem und lemniskatenförmigem Sende-Feld:

$$E[D] = \int_0^R \int_0^{2\pi} E[D|\tilde{r}=r] \; P\{r<\tilde{r}<r+dr, \delta<\tilde{\delta}<\delta+d\delta\}$$

$$E[D] = \frac{2W}{\pi p} \int_0^1 \int_0^{\pi} \left(\frac{e^{p\tau A(t)/(1-p)}}{1-p} - 1\right) e^{-a(t)W/\pi} \; t \; d\delta \; dt \tag{15}$$

4. DAS SIMULATIONSMODELL

Zunächst soll das Simulationsmodell kurz beschrieben werden. Auf der Oberfläche einer Kugel mit dem Radius $r_k = 2$ km befinden sich 30 Stationen, deren Positionen über einen Zufallszahlengenerator bestimmt werden. Der Vorteil der Kugel liegt in der Tatsache begründet, daß Randeffekte wie bei einem kreisförmigen Modell nicht auftreten, d.h. die Stationen auf der Kugel verhalten sich, bzgl. der Größen Durchsatz und Fortschritt, wie Stationen in einem unendlich ausgedehnten Netz. Die Routingmatrix ist derart aufgebaut, daß der jeweils kürzeste Weg in Hops zum Empfänger ausgewählt wird. Die maximale Senderreichweite beträgt im Modell 1 1,9 km, im Modell 2a 1.8 km und im Modell 2b 2,1 km.

Es wird vorrausgesetzt, daß jede Station die Entfernung zu ihren Nachbarn sowie den Routingpartner für Pakete zu anderen Stationen kennt. Für das Modell 2 muß jede Station zusätzlich die Richtung zu ihren Nachbarn kennen, damit die Antenne ausgerichtet werden kann.

Jede Station besitzt zu jedem Zeitpunkt ein Paket. In jedem Slot wird mit der Wahrscheinlichkeit p gesendet. Es gilt $\alpha = p$, somit spielt es keine Rolle, ob ein Paket zuvor kollidiert ist.

5. NUMERISCHE ERGEBNISSE

Für das Kanalzugriffsprotokoll S-ALOHA wird für teilvermaschte Netze und festen Senderadius in /TAKL.84/ gezeigt, daß das Optimum erreicht wird, wenn der Senderadius so eingestellt wird, daß jede Station im Mittel W = 8 Nachbarstationen besitzt. Der maximale Fortschritt bei angepaßter Sendereichweite wird für eine mittlere Anzahl von W = 7 Nachbarstationen erreicht.

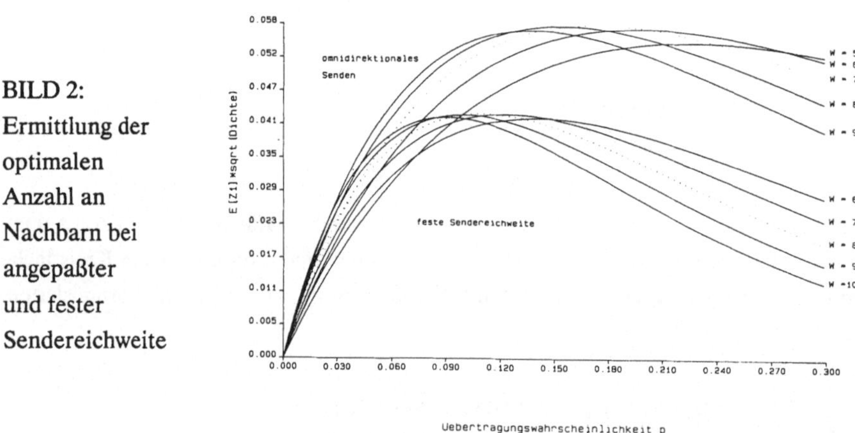

BILD 2: Ermittlung der optimalen Anzahl an Nachbarn bei angepaßter und fester Sendereichweite

Der maximale Fortschritt bei angepaßter Sendereichweite für das direktionale Senden wird für eine mittlere Anzahl von W = 13 (Modell 2a) und W = 14 (Modell 2b) Nachbarstationen erreicht. Die optimale mittlere Anzahl Nachbarn unterscheidet sich also wesentlich von der eines System mit konstantem Senderadius R. Die erhöhte optimale Anzahl resultiert aus der geringeren Störung der Nachbarstationen.

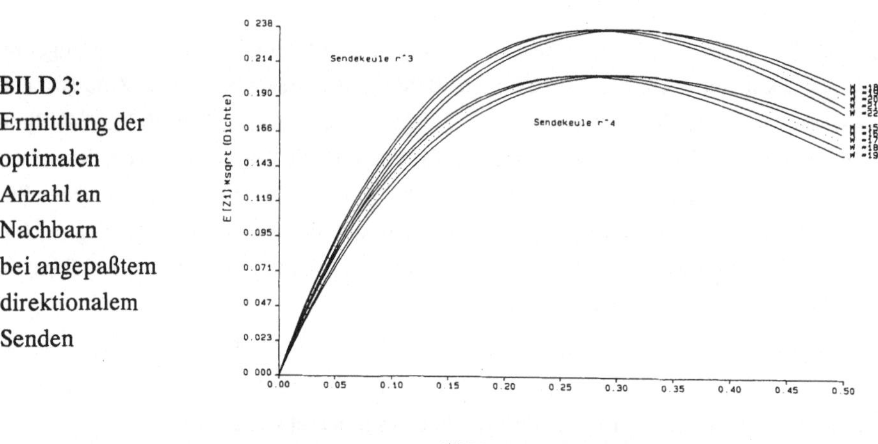

BILD 3: Ermittlung der optimalen Anzahl an Nachbarn bei angepaßtem direktionalem Senden

Bild 4 zeigt einen Durchsatzvergleich der Modelle mit festem und angepaßtem Senderadius

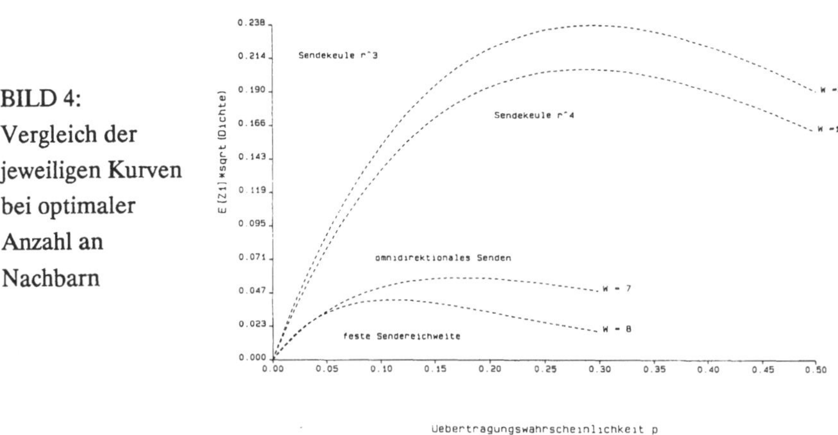

BILD 4: Vergleich der jeweiligen Kurven bei optimaler Anzahl an Nachbarn

Im Bild 5 sind Durchsatzkurven gegenübergestellt. Die erste Kurve entspricht der Analyse, die zweite Kurve zeigt Simulationsergebnisse mit dem 95% Konfidenzintervall. Die Kurven zeigen eine gute Übereinstimmung.

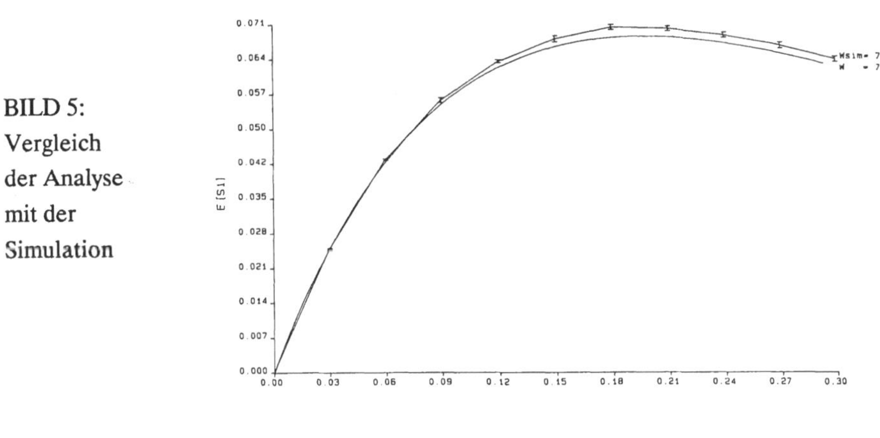

BILD 5: Vergleich der Analyse mit der Simulation

In Bild 6 sind die Ergebnisse für die Backlogzeit dargestellt. Aus den Kurven ist leicht ersichtlich, daß eine zu große Übertragungswahrscheinlichkeit zu sehr langen Paketlaufzeiten führt.

Die Kurve zeigt im Bereich zwischen der Übertragungswahrscheinlichkeiten p = 0.01 bis p = 0.18 keine nennenswerte Änderung der Backlogzeit. Das System ist stabil und relativ unempfindlich gegenüber zufälligen Verkehrsschwankungen. Für eine größere Übertragungswahrscheinlichkeit steigt die Backlogzeit deutlich an. Der Grund dieses Anstiegs liegt in der Zunahme der Kollisionen und der daraus resultierenden Instabilität.

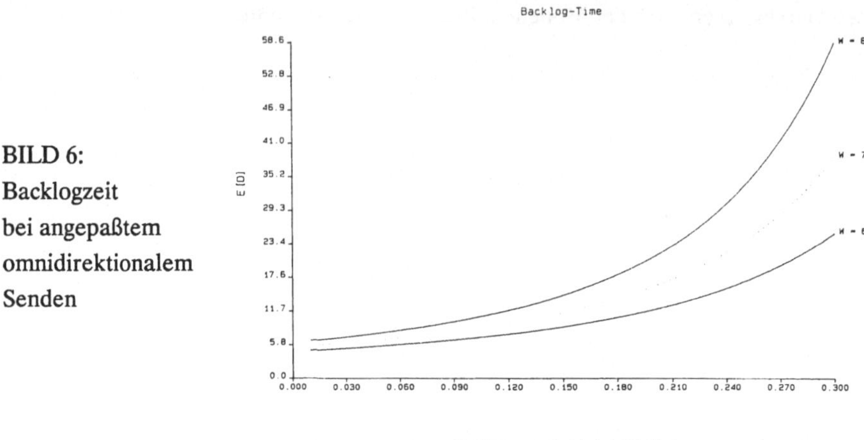

BILD 6:
Backlogzeit
bei angepaßtem
omnidirektionalem
Senden

6. ZUSAMMENFASSUNG

Es wurde die Analyse eines S-ALOHA Multi-hop Systems mit adaptiver Sendereichweite bei omnidirektionalem und direktionalem Senden untersucht. Als wichtigstes Leistungskriterium wurde der Durchsatz betrachtet.

Die Ergebnisse der Analyse und der Simulation haben gezeigt, daß sich mit Stationen, die in der Lage sind die Sendereichweite zu adaptieren, bzw. richtungsselektierend zu senden, ein erheblich größerer Netzdurchsatz erzielen läßt als bei dem Modell mit festem Senderadius R.

7. LITERATUR

/BRAGO.87/ V. Brass, C. Gotthardt:
On Throughput and Delay in S-ALOHA Multihop Networks
Informatik-Fachberichte 154
Springer Verlag Berlin/Heidelberg, S. 236-249, Sep. 1987

/GOTTH.89/ C. Gotthardt:
Durchsatz in CSMA-Multi-Hop-Netzen
Informatik-Fachberichte 205
Springer Verlag Berlin/Heidelberg, S. 309-323, Feb. 1989

/HOULI.86/ T.-C. Hou, V.O.K. Li:
Transmission Range Control in Multihop Packet Radio Networks.
IEEE Trans. Comm., Vol.COM-34, No. 1, January 1986, pp. 38-44

/LAM.74/ S. Lam, L. Kleinrock:
On stability of packet switching in a random multi-access broadcast channel
7th Hawaii Int.Conf. on Systems Science, Honolulu, January 1974
Proc. Special Subconf. on Computer Nets 1974

/SIKL.80/ J.A. Silvester, L. Kleinrock:
On the Capacity of One-Hop S-ALOHA Packet Radio Networks
jwith Adjustable Transmission Power.
Pacific Telecomm. Conf. 1980, pp. 1A.25-1A.32

/TAKL.84/ H. Takagi, L. Kleinrock:
Optimal Transmission Ranges for Randomly Distributed
Packet Radio Terminals.
IEEE Trans. Comm., Vol.COM-32, No. 3, March 1984, pp. 246-257

Untersuchung expansiver Techniken des Verkehrsmanagements für leitungsvermittelte Netze

Wolfgang Koops
Forschungsinstitut der Deutschen Bundespost
Postfach 5000, D-6100 Darmstadt, F.R.G.

Kurzfassung

In dem vorliegenden Beitrag werden zwei Verfahren zur Durchsatzerhöhung in leitungsvermittelten Netzen untersucht. Es sind dies das Verfahren 'Trunk Reservation', das dem Schutz einer Verkehrs- oder Dienstart dient, sowie das Verkehr auf verschiedene Wege aufteilende 'Proportional Bidding'.

Es wird eine Modellbildung zur Berechnung von lastabhängigen Blockierungswahrscheinlichkeiten und Belastungen in kleinen Netzen beschrieben. Die Verfahren werden in symmetrischen und unsymmetrischen Netzen anhand von Beispielen mit unterschiedlichen Lastsituationen untersucht und bewertet.

Abstract

Two methods for advanced throughput in circuit switched networks are investigated in the present paper. These methods are 'Trunk Reservation', which protects a traffic or service, and 'Proportional Bidding', that shares out traffic on different paths.

The modeling for computing load-depending blocking probabilities and carried load in small network will be described. The methods will be examined and evaluated for symmetrical and unsymmetrical circuit switched networks using different offered loads.

1 Einleitung

Datennetze werden neben anderen Grundsätzen auch nach dem Gesichtspunkt entstehender und verursachender Kosten geplant, aufgebaut und weiter entwickelt. Ein Maximalausbau ist unrealistisch, so daß die weitgehend zufällig auftretenden Verbindungswünsche unter ungünstigen Bedingungen zu Engpaßsituationen führen, die auf Teilnehmer, Netzbetreiber oder Netzelemente nachteilig wirken.

Ziel der Überlastabwehrstrategie (Congestion Control, Traffic Management) ist es, ungewünschte Situationen nicht entstehen zu lassen oder, wenn dies nicht möglich war, die Schadenshöhe und -dauer zu minimieren. Zwei wesentliche Kategorien sind hierbei zu unterscheiden; Verkehrsflußkontrolle

(Flow Control), die die Menge von Forderungen reguliert und begrenzt, und Verkehrslenkung (Routing), die den Weg der einzelnen Nachricht im Netz aufgrund freier Kapazitäten zwischen Quelle und Senke spezifiziert.

Die Verkehrsflußkontrolle soll die Belastung des Netzes herabsetzen, indem sowohl Forderungswünsche, die nicht an ihr Ziel geleitet werden können, sofort abgewiesen werden als auch anderen Verkehr beeinträchtigende Forderungen nicht bedient werden. Das Routing zielt auf eine Verbesserung der Netzverfügbarkeit unter verschiedensten Lastsituationen ebenso wie bei Störungen von Übertragungsstrecken und Vermittlungsstellen (VSt).

Der vorliegende Beitrag beschäftigt sich mit der Leistungsfähigkeit zweier dem Routing zuzurechnenden Verfahren zur Erhöhung des Verkehrsdurchsatzes in leitungsmittelten Netzen. Die Verfahren 'Trunk Reservation' und 'Proportional Bidding' werden in kleinen Modellnetzen hinsichtlich resultierender Blockierungswahrscheinlichkeiten sowie der Belastungen auf Verbindungsabschnitten untersucht.

Nach einer kurzen Erläuterung der zwei untersuchten Verfahren in Kapitel 2, beschreibt Kapitel 3 die Modellbildung. Anschließend werden in Kapitel 4 eine Netzanalyse und Verfahrensbewertung vorgenommen.

2 Verfahren zur Erhöhung des Verkehrsdurchsatzes

Die Verfahren 'Trunk Reservation' (TR - Leitungszahlreservierung) und 'Proportional Bidding' (PB - Verkehrsflußteilung) sollen in ihren Auswirkungen auf Verkehrsflüsse untersucht werden.

2.1 Trunk Reservation

TR für einen bestimmten Verkehr (oder Dienst) schützt diesen, da es von der Gesamtheit freier Leitungen immer eine vorgegebene Anzahl an Leitungen für den privilegierten Verkehr freihält.

Aus dem Modell eines Geburts-Sterbe-Prozesses erhält man für zwei Verkehrstypen und für Poissonankünfte an ein Bündel mit n Leitungen die u. a. in /4/ berechneten Wahrscheinlichkeiten P_i, dafür daß i Leitungen belegt sind. Ab der m-ten Leitung, $m \leq n$, sei nur noch privilegierter Verkehr zugelassen, d.h. m-1 Leitungen befördern beide Verkehrsarten und n-m nur eine. Die P_i berechnen sich zu:

$$P_i = \frac{A^i}{i!} \cdot P_0, \qquad \text{für} \quad i \leq m-1$$

$$P_i = \frac{A^{m-1} A_D^{i-(m-1)}}{i!} \cdot P_0, \qquad \text{für} \quad i > m-1$$

$$mit \quad P_0 = \left(\sum_{i=0}^{m-1} \frac{A^i}{i!} + \sum_{i=m}^{n} \frac{A^{m-1} A_D^{i-(m-1)}}{i!} \right)^{-1} .$$

In Kenntnis der Angebote A_D (privilegierter Verkehr) und A (Summe der Verkehre) lassen sich die Blockierungswahrscheinlichkeiten für den privilegierten und nicht privilegierten Verkehr berechnen. Für den Fall, daß keine Leitungszahlreservierung stattfindet, d. h. m=n, ergibt sich folgerichtig die Erlangsche Verlustformel für $P_n = E(n, A)$.

Die Priorisierung eines Verkehrs führt zur Benachteiligung anderer. Diese Benachteiligung ist z.B. bei Überlast für Überlaufverkehr, der zuerst anderen Wegen angeboten wurde, erwünscht. Ein größerer Anteil zurückgewiesenen Überlaufverkehrs erhöht für den den kürzesten Weg nehmenden Direktverkehr die Wahrscheinlichkeit einer Bedienung, woraus unmittelbar geringere Überlaufverkehrsanteile für andere Abschnitte resultieren.

Der Frage nach der Anzahl zu reservierender Leitungen kommt eine besondere Bedeutung zu, weil bei hochausgelasteten Leitungen die jeweils wenigen nicht belegten nur dem bevorzugten Verkehr verfügbar sind. In Abhängigkeit von der Last und der Anzahl reservierter Leitungen kann ein Verdrängungsprozeß durch diesen entstehen, der den nicht bevorzugten Verkehr in einem ungewollten Ausmaß trifft.

2.2 Proportional Bidding

Bei dem Verfahren der Verkehrsflußteilung wird der Verkehr im Ursprungsknoten auf verschiedene Wege als Direktverkehr aufgeteilt, ohne vorher jeweils anderen Wegen angeboten worden zu sein. In einem bemessenen Netz verspricht dieses Prinzip dann Vorteile, wenn tatsächliche Angebote Leitungen unterlasten, aber auch überlasten. Bild 2.1 zeigt das Prinzip des PB: Das Direktangebot A_{12} wird in zwei Anteile $A_{12}s_{12}$ und $A_{12}(1 - s_{12})$ aufgeteilt, die als Direktangebot der Kante 1 bzw. 2 angeboten werden.

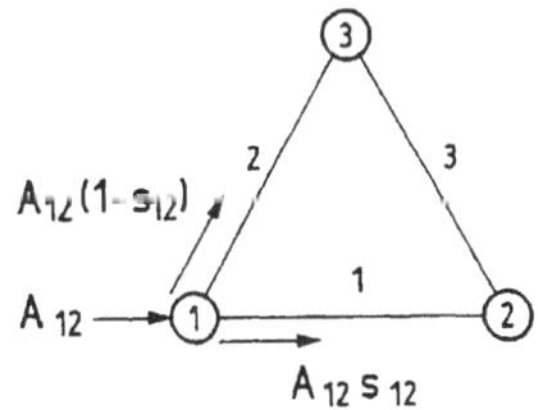

Bild 2.1: Darstellung der Verkehrsflußteilung

Eine Verkehrsflußteilung zur Auslastung wenig belasteter und zur Entlastung stark belasteter Wege kann zur besseren Gesamtauslastung des Netzes beitragen. Die richtige Wahl der Aufteilungsfaktoren s , $0 \leq s \leq 1$, für den Erstweg bzw. (1-s) für den Zweitweg ist besonders wichtig, da bei falscher Einstellung der Schaden den Nutzen übersteigt. PB muß deshalb als dynamisches Verfahren eingerichtet werden.

3 Modellbildung

Die in dieser Arbeit betrachteten Netze bilden Teile eines realen Netzes in einer oder in zwei Hierarchieebenen nach. Ein Modell davon besteht aus drei Knoten (Vermittlungsstellen) und drei Kanten (Leitungsbündeln, Verbindungsabschnitten) zwischen ihnen. Die Anzahl der Leitungen in einem Leitungsbündel sind beliebig. An jedem Knoten kann Angebot für jeden anderen Knoten vorhanden sein.

Die Ankunftsprozesse werden als Poissonströme und unabhängig angenommen. Die Belegungsdauern seien unabhängig vom Systemzustand negativ-exponential verteilt mit gleichem Mittelwert $1/\mu = 1$. Überlaufverkehre werden ebenfalls als Poissonströme aufgefaßt, entsprechend der in /2/ angegebenen Näherungsmethode des 'Ersatz-Poisson-Verkehrs'. Hierbei werden die statistischen Eigenschaften des Überlaufverkehrs durch eine geeignete Erhöhung seines Angebotswerts reflektiert. Dieser Zuschlagfaktor wurde in den hier gezeigten Beispielen einheitlich auf 1.25 gesetzt.

Zunächst werden gemäß einer Angebotsmatrix unter Berücksichtigung der Verkehrslenkungsmöglichkeiten und einzuhaltender Blockierungswahrscheinlichkeiten die Leitungszahlen berechnet.

Das so berechnete Netz wird in einer weiteren Bearbeitungsstufe mit neuen, anderen Angebotsmatrizen konfrontiert als deren Folge andere Blockierungswahrscheinlichkeiten und Ende-zu-Ende-Blockierungen (EEB) auftreten. In Abschnitt 2 beschriebene Schutzmechanismen für Verkehrsarten oder Dienste wurden in dem Programmpaket zur Netzberechnung realisiert, so daß Auswirkungen unterschiedlicher Maßnahmen studiert werden können.

Die Bemessung des Netzes muß ebenso wie die spätere Berechnung von Leistungsgrößen iterativ bis zum Erreichen einer vorzugebenden Genauigkeitsgrenze erfolgen, da das System wegen Alternativ- bzw. Überlaufverkehre rückkopplungsbehaftet ist. Die Berechnung der Leitungszahlen zwischen zwei Knoten wird für beide Richtungen gemeinsam auf wechselseitigen Leitungen vorgenommen. Es besteht Wahlmöglichkeit zwischen Leitungseinsparung bei einzuhaltender Dienstgüte (Grade Of Service - GOS) und Erhöhung des GOS für die je Richtung berechneten aber wechselseitig betriebenen Leitungen.

Bild 3.1 zeigt die Verkehrsbelastung eines Abschnitts in einem 3-Knoten-Netz. Je Richtung treten Direktverkehre A_D und Überlaufverkehre $A_{\ddot{U}}$ auf. Das Gesamtangebot A_i je Kante i ist die Summe der Einzelangebote, so daß mithilfe der Erlangschen Verlustformel die Blockierungswahrscheinlichkeit berechnet werden kann.

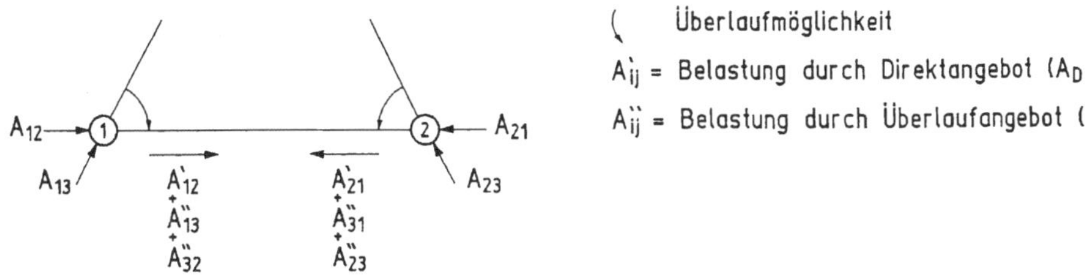

Bild 3.1: Verkehre eines 3-Knoten-Netzes auf einem Verbindungsabschnitt

Die Betrachtung der Blockierungswahrscheinlichkeiten ist von elementarer Bedeutung. Neben den wichtigen EEB für beliebige Verbindungen sind die Einzelblockierungswahrscheinlichkeiten x_{ij} der Verkehre je Kante für spätere Auswertungen zu analysieren.

Um die EEB einer beliebigen Verbindung zu berechnen, sind aus der Gesamtblockierung die individuellen Blockierungswahrscheinlichkeiten $x_{ij}, i = 1,2,3, j = A, D$, die gemäß Gleichung 3.1 voneinander abhängig sind, zu bestimmen.

$$E(n_i, A_i) \cdot A_i = E(n_i, A_i) \cdot (A_{iD} + A_{i\ddot{U}}) = x_{iD} \cdot A_{iD} + x_{i\ddot{U}} \cdot A_{i\ddot{U}} \tag{3.1}$$

Vergleichende Berechnungen mit den in /2/ vorgestellten Berechnungsvarianten wiesen nachstehende einfache Näherungen als hinreichend genau aus.

$$x_{iD} = E(n_i, A_i) \cdot (A_{iD} + A_{i\ddot{U}}) \cdot (A_{iD} + 1.8 A_{i\ddot{U}})^{-1},$$

$$x_{i\ddot{U}} = 1.8 \cdot x_{iD} \qquad \mathit{f\ddot{u}r} \quad A_{i\ddot{U}} > 0.$$

Die EEB werden nach /1/ ermittelt:

$$EEB_{kl} = 1 - \sum_i \text{ Wahrscheinlichkeit (Angebot an Weg i und Weg i frei)}$$

k, l Ursprungs- bzw. Zielknoten

Für das einfache Modellnetz (Bild 4.1) ergibt sich die EEB für die Verkehrsbeziehung (1,2) oder (2,1)

$$EEB_{12} = x_1 x_2 + x_1 x_3 - x_1 x_2 x_3$$

und unter Berücksichtigung von (3.1)

$$EEB = x_{1D} x_{2\ddot{U}} + x_{1D} x_{3\ddot{U}} - x_{1D} x_{2\ddot{U}} x_{3\ddot{U}}.$$

Das vorgenannte Verfahren zur Bestimmung von Leistungsgrößen kann wegen der vereinfachenden Näherungen ebenso wie andere bekannte Verfahren keine exakten Ergebnisse liefern. Im Hinblick auf eine phänomenologische Betrachtung, die in späteren Arbeiten auf größere Modelle schwenkt, und wegen der oben angedeuteten Vergleichsrechnungen, scheint das Vorgehen gerechtfertigt zu sein.

4 Netzanalyse

Das 3-Knoten Netz wird hinsichtlich der Verkehrslenkung mit verschiedenen Randbedingungen untersucht. Die Lastsituationen

- generelle Erhöhung
- lokale Erhöhung
- konstante aber schiefe Last

werden als eingehende, steuernde Systemparameter verwendet. Die Untersuchungen beziehen sich zunächst auf die Angebote

$$A_I(Erl) = \begin{pmatrix} -- & 20 & 15 \\ 20 & -- & 10 \\ 15 & 10 & -- \end{pmatrix} \; und \;\; A_{II}(Erl) = \begin{pmatrix} -- & 200 & 150 \\ 200 & -- & 100 \\ 150 & 100 & -- \end{pmatrix}$$

unter Nennlast (= 100% Angebot) bei einzuhaltenden Blockierungswahrscheinlichkeiten B = 0.2 bzw. B = 0.01 auf den Kanten für alle Möglichkeiten des Überlaufs. Tabelle 4.1 zeigt die berechneten Leitungszahlen für die Kanten 1 bis 3 mit $N = (n_1, n_2, n_3)$. Nachfolgende prozentuale Angaben beziehen sich immer auf Nennlast.

A	A_I		A_{II}	
B	0.2	0.01	0.2	0.01
N	42,35,29	53,42,31	389,325,260	431,331,230

Tabelle 4.1: Leitungszahlen in Abhängigkeit von A und B

4.1 Allgemeine Überlastproblematik

Eine erste Leistungsbewertung wird anhand des in Bild 4.1 dargestellten Modells vorgenommen. Die für die Leitungsbemessung maßgebende Blockierungswahrscheinlichkeit sei B = 0.2 bei wechselseitig betriebenen Leitungen mit Leitungseinsparung, die Angebotsmatrix A_I gelte, Routing sei in jedem Knoten erlaubt.

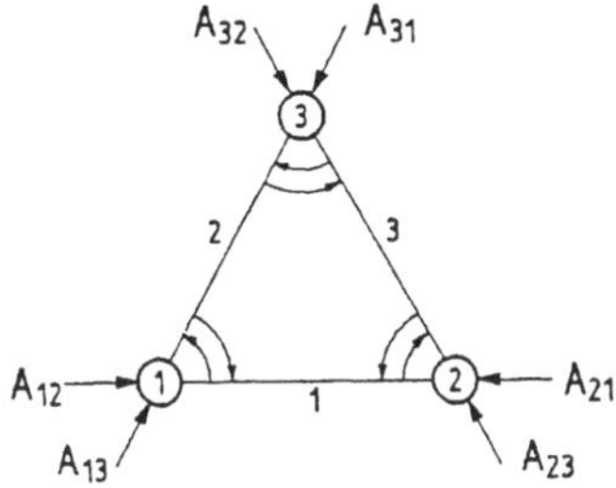

Bild 4.1: Modellnetz mit 3 Knoten und alternativer Verkehrslenkung

Bild 4.2 zeigt die Abhängigkeit maßgebender Größen vom Angebot auf der Kante 1. Dabei bedeuten: Y_{ij} - Belastung auf Kante i infolge Direktangebots $[j = D]$ bzw. infolge Summenangebots $[j = S]$. Das Angebot $A_1.$ an die Kante 1 setzt sich aus zwei Direktangeboten und vier Angeboten von Überlaufverkehren zusammen. Die Belastung $Y_1 = A_1(1 - B_1)$ der Kante 1 geht bei dem dem Netz dargebrachten Gesamtangebot $A_I > 90\%$ zurück und erreicht seinen Wert erst wieder ab $A_I > 110\%$. Die Blockierungswahrscheinlichkeiten an den Kanten wächst für $A_I > 90\%$ in eine Größenordnung, die einen verstärkten Anstieg des Überlaufverkehrs verursacht und so eine weitere netzinterne Angebotsausbreitung nach sich zieht. Bei $A_I = 95\%$ beträgt der Anteil des Überlaufverkehrs bereits mehr als 15 %.

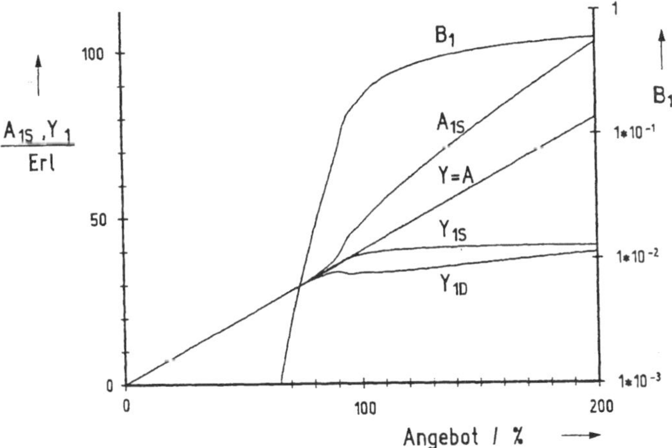

Bild 4.2: Angebote und Belastungen sowie Blockierungswahrscheinlichkeit eines Verbindungsabschnitts

Mit Leitungszahlen $N = (53, 42, 31)$, die unter Nennlast Blockierungswahrscheinlichkeiten der Kanten von $B = 0.01$ einzuhalten erlauben, ergibt sich eine annähernd gleiche Qualitätsaussage. Der Belastungseinbruch ist wiederum bei $B \approx 0.2$ bemerkbar, d. h. $110\% < A_{II} < 140\%$.

4.2 Charakterisierung des Trunk Reservation

TR für einen Verkehr mit optimaler Anzahl freizuhaltender Leitungen $n_{Opt}, 0 \leq n_{Opt} \leq n$ je Kante, ist in jedem Fall besser im Sinne höherer möglicher Belastungen als starres alternatives Routing. Es bedeutet aber eine diffizile lastabhängige Einstellung der n_{Opt} im gesamten Netz unter Echtzeitbedingungen. Mit zunehmender Last steigt die Anzahl von Abweisungen alternativ gelenkter Verkehre und der damit korrespondierende Zeitaufwand, der für die Informationsgewinnung durch Belegungsprüfung notwendig ist, zumal Änderungen dann dynamisch erfolgen. Ständig wiederholte Messungen müssen netzweit ausgewertet werden. Für Verkehrsmatrizen $A > 130\%$ besteht zudem keine praxisrelevante Abweichung zwischen TR mit optimaler Leitungszahl und starrer Verkehrslenkung für einen Weg.

4.2.1 TR im symmetrischen Netz

- Hohe Blockierungswahrscheinlichkeit

Für beide Lastsituationen zeigt sich, daß mit zunehmender Last ab etwa $A > 95\%$ TR geboten ist. Tabelle 4.2 weist die Belastung ohne TR sowie N_{Opt} aller Kanten für höchstmögliche Belastungssituationen Y_{TR} aus. N_{Opt} ist sowohl von der Leitungszahl als auch von dem Angebot abhängig.

Im Nennlastbereich und bis zu 30 % darüber verschafft TR eine höhere Belastung des Netzes. Sehr viel höhere Angebote erfordern eine Leitungszahlreservierung für den Direktverkehr, die einer absoluten Blockierung für Überlaufverkehr entspricht. TR mit hohem N und feste Verkehrslenkung für einen Weg sind identisch in ihren Auswirkungen, wobei deutliche Unterschiede an Verkehrsmanagementtätigkeiten festzustellen sind.

A / %	95	100	110	130	200
A_I					
Y / %	92.7	93.4	85.7	73.1	48.6
Y_{TR} / %	97.6	95.4	90.6	81.2	56.7
N_{Opt}	1,1,1	3,2,2	5,4,4	5,5,5	8,8,8
A_{II}					
Y / %	100.0	92.9	82.8	69.5	45.5
Y_{TR} / %	100.0	98.5	92.7	81.4	53.9
N_{Opt}	0,0,0	5,5,6	10,10,10	11,10,10	20,20,20

Tabelle 4.2: Auswirkungen des TR (B = 0.2)

- Geringe Blockierungswahrscheinlichkeit

Tabelle 4.3 zeigt die Auswirkungen des TR im Modellnetz bei kleiner Blockierungswahrscheinlichkeit

(B = 0.01). Der Bedarf an TR setzt erst bei höheren Lastsituationen ein. In beiden Modellen ist nur ein schmaler Lastbereich vorhanden innerhalb dessen TR sinnvoll Anwendung finden kann. Das Modell II (hohe Last) erfordert einen früheren Einsatz des TR, da die Erlangsche Verlustformel bei höheren Angeboten relativ weniger Leitungen liefert, diese aber höhere Belastungen erfahren.

A / %	100	105	110	130	200
A_{III}					
Y / %	100.0	99.9	99.5	87.5	57.9
Y_{TR} / %	100.0	99.9	99.5	92.5	66.9
N_{Opt}	0,0,0	0,0,0	0,0,0	6,5,5	10,10,10
A_{IV}					
Y / %	100.0	90.3	85.5	71.6	46.8
Y_{TR} / %	100.0	98.4	95.8	83.5	54.9
N_{Opt}	0,0,0	12,12,12	21,21,21	21,21,21	21,21,21

Tabelle 4.3: Auswirkungen des TR (B = 0.01)

4.2.2 TR im unsymmetrischen Netz

Bild 4.3 zeigt einen Ausschnitt eines möglichen leitungsvermittelten Fernnetzes. Darin sind die Angebots- (A_{III}) und Verkehrsmatrizen durch Pfeile und Werte dargestellt; $A_{IV} = 10A_{III}$. Die Knoten 1, 2 und 3 können in dieser Reihenfolge beispielsweise die zentralen Netzknoten der Städte Braunschweig, Hannover und Frankfurt bedeuten. Der Verkehr von und nach Knoten 3 (Fernverkehr) habe immer Erst- und Zweitwegemöglichkeit, dagegen habe Verkehr von 1 nach 2 und umgekehrt (Nahverkehr) nur eine Wegemöglichkeit. Es soll TR für Direktverkehr im Fernbereich angewendet werden.

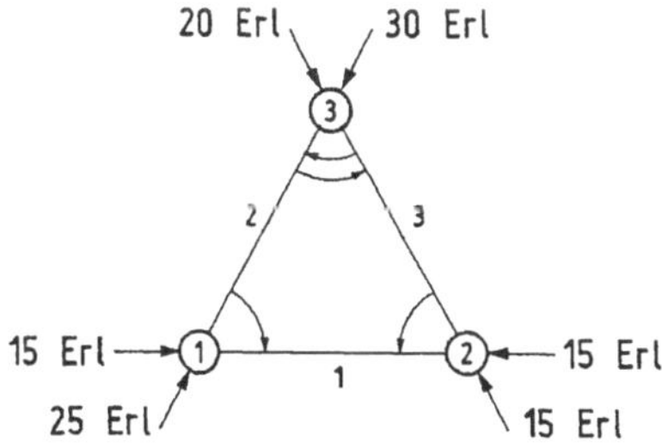

Bild 4.3: Modellnetz mit Angebotsaufkommen und alternativer Verkehrslenkung

Die durch TR geschützten Verkehre auf den Kanten 2 und 3 ermöglichen nur wenig mehr (< 2%) Belastung. Indem der jeweils andere Forderungsstrom einer höheren Blockierung unterworfen ist, sinkt die Blockierungswahrscheinlichkeit des Erstwegs bzw. steigt die des Zweitwegs. Die individuelle

EEB des Fernverkehrs nimmt deshalb nur geringfügig ab. Die die Nahverbindung repräsentierende Kante 1 erhält weniger Überlaufverkehr, wodurch deren Auslastung für Direktverkehr steigt.

TR für Direktverkehr auf Kante 1 (Nahverkehr) leistet nicht dasselbe, da die Direktangebote an die Kanten 2 und 3 nicht mehr bevorzugt werden.

Eine bedeutende Steigerung des Durchsatzes an Fernverkehr ließe sich dann erzielen, wenn der überlaufende Fernverkehr keine erhöhte Blockierung infolge Schutzmaßnahmen erleiden würde.

Die Ergebnisse werden gleichermaßen für hohe wie für niedrige Blockierungswahrscheinlichkeiten und Angebote erzielt.

Im folgenden sollen die Auswirkungen des TR für Überlaufverkehre, die die Kante 1 nutzen, untersucht werden. D. h. der Fernverkehr auf dieser Kante darf, weil für ihn Leitungen freigehalten werden, mehr von diesen benutzen als der nur den Direktweg nehmende Nahverkehr. Die Bilder 4.4 und 4.5 geben die Verkehrsbelastungen für Angebote A_{III} und A_{IV} bei $B = 0.2$ wider.

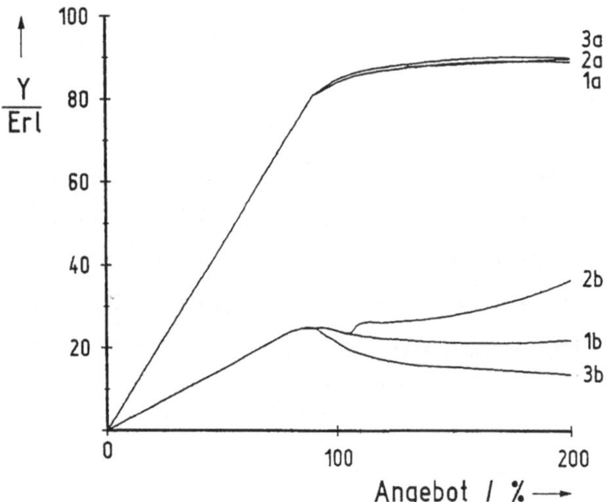

Bild 4.4: Belastungen im Nah- und Fernbereich
(B=0.2) - geringe Angebote, kleine Leitungszahlen -

Die Belastungskurven gelten für starres alternatives Routing, starres alternatives Routing mit optimaler TR sowie für TR für Verkehr von und nach Knoten 3 auf der Kante 1, welches einem Schutz des Überlaufverkehrs im Nahbereich des beschriebenen Modellnetzes entspricht. Die Fernverkehrswerte können nicht bedeutend erhöht werden. Auf Kante 1 können zwar für den Fernverkehr alle Leitungen reserviert werden, aber es ist für diesen nur einer von zwei seriell geführten Verbindungsabschnitten der Alternativroute, so daß die positiven Auswirkungen begrenzt bleiben. Die Sensibilität für den Nahbereichsverkehr ist anhand der Kurven 3b, die die Belastungen für Direktverkehr zwischen Knoten 1 und 2 kennzeichnen, ausgedrückt. Jeweils nur eine für Fernverkehr reservierte Leitung läßt die

Belastung schon drastisch fallen. Blockierungswahrscheinlichkeiten von B = 0.01 im Netz ziehen die gleichen qualitativen Aussagen nach sich.

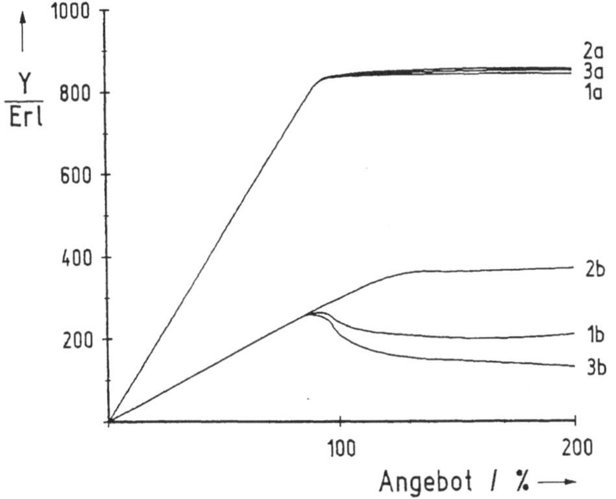

Bild 4.5: Belastungen im Nah- und Fernbereich
(B=0.2) - hohe Angebote, große Leitungszahlen -

4.3 Charakterisierung des Proportional Bidding

In einem stimmigen Netz, in dem Prognose, technischer Ausbau und Verkehrsangebote harmonieren, wird PB keinen Vorteil bringen, da durch eine Verkehrsflußaufteilung Unstimmigkeiten zwischen Angeboten und Leitungsanzahlen entstehen. Dagegen kann u. U. mithilfe des PB eine verbindungsspezifische Angebotsveränderung und Belastungserhöhung erreicht werden, wenn die aktuelle Angebotsmatrix nicht mehr mit der für den Netzausbau maßgeblichen Angebotsmatrix übereinstimmt. Notwendige Voraussetzung einer erfolgversprechenden Anwendung des PB ist die Kenntnis der Belastungen im Nenn- und Istzustand auf den Abschnitten, die den Auswirkungen von PB-Maßnahmen unmittelbar unterliegen. Sämtliche seriellen Abschnitte des Wegs, der Verkehrsanteile zur Entlastung eines anderen Wegs aufnehmen soll, müssen geringere als für die Bemessung vorgegebene Blockierungswahrscheinlichkeiten aufweisen. Der Verkehrsaufteilungsfaktor s_j. in einem Knoten j mit beliebigem Ziel wird nach Gleichung 4.1 bestimmt, wobei A_{kDNenn} dem Nennangebot an dem eigentlichen Direktweg entspricht:

$$s_i^* = \begin{cases} 1 & falls\ A_{iDNenn} \leq A_{iDIst} \\ 1 - \frac{A_{iDNenn} - A_{iDIst}}{A_{KDNenn}} & sonst \end{cases} \tag{4.1a}$$

für alle i Kanten der Alternativverbindung und

$$s_j = \max_i s_i^*. \tag{4.1b}$$

Der zunächst ohne Verkehrslenkung dem Alternativweg angebotene Verkehrsanteil orientiert sich an der Aufnahmefähigkeit der Kante mit der geringsten Bewertung und belastet alle Kanten der Alternativverbindung.

4.3.1 PB im symmetrischen Netz

Für das in Bild 4.1 dargestellte und für A_I bzw. A_{II} bemessene Netz gelten die Angebotsmatrizen A_V bzw. A_{VI} unter Nennlast.

$$A_V(Erl) = \begin{pmatrix} -- & 17 & 20 \\ 20 & -- & 8 \\ 15 & 10 & -- \end{pmatrix} \qquad und \quad A_{VI}(Erl) = 10A_V$$

Die Angebote seien hier bei konstanter Netzlast derart verändert, daß die Kanten 1 und 3 weniger, Kante 2 mehr Belastung erfahren. PB soll das Angebot A an die Kanten 1 und 3 zur Erzielung einer gleichmäßigeren Kantenbelastung und annähernd gleich großer Blockierungswahrscheinlichkeiten umlegen.

Die Netzauslastung sinkt unter den genannten Annahmen immer, wenn das Verfahren PB eingesetzt wird. Dabei ist keine Abhängigkeit zu großen oder kleinen Lastsituationen und hohem oder geringen Blockierungswahrscheinlichkeiten zu erkennen. Der Gewinn einer geringeren Blockierungswahrscheinlichkeit auf dem Direktweg wird durch eine höhere auf den Abschnitten des Alternativwegs erkauft. Da die letztgenannten zugleich Direkt- und Alternativwege für andere Angebote darstellen, wird auch diese Belastung geringer. Die individuelle EEB kann nicht gesenkt werden.

4.3.2 PB in unsymmetrischem Netz

Die nachfolgende Untersuchung bezieht sich wieder auf Bild 4.3 mit den veränderten Angebotsmatrizen

$$A_{VII}(Erl) = \begin{pmatrix} -- & 12 & 31 \\ 15 & -- & 12 \\ 30 & 20 & -- \end{pmatrix} \qquad bzw. \quad A_{VIII}(Erl) = 10A_{VII}$$

bei Blockierungswahrscheinlichkeiten von B = 0.2 bzw. B = 0.01. In Tabelle 4.4 sind die Ergebnisse zusammengestellt.

Es wurde der Frage nachgegangen, ob eine Belastungssteigerung in dem Bereich möglich ist, in dem nur alternativ zu lenkender Verkehr existiert (Fernverkehr). Der Einsatz des PB bietet sich hiernach in Netzen mit allgemein hoher Blockierungswahrscheinlichkeit der einzelnen Verbindungsabschnitte an, und dies in Netzen, die für hohe Angebotswerte ausgelegt sind, umso mehr.

Netze, die für geringe Blockierungswahrscheinlichkeiten ausgelegt sind, erzielen per se unter Nennlast eine hohe Belastung, so daß bei ihnen weniger Spielraum zu noch höheren Belastungen als bei Netzen mit höherer Blockierungswahrscheinlichkeit vorhanden ist. Die Berechnung der

$$EEB_{13} = EEB_{31} = x_2 \cdot (x_1 + x_3 - x_1 x_3)$$

für Verkehr von Knoten 1 nach 3 bzw. umgekehrt zeigt, daß für höhere Blockierungswahrscheinlichkeiten der Kanten 1 und 3 der Einfluß des Subtrahends wächst und zu einer relativen Verringerung der EEB beiträgt.

Last	A_{VII}		A_{VIII}		A_{VII}		A_{VIII}	
Blws	0.2		0.2		0.01		0.01	
A / %	dY	s	dY	s	dY	s	dY	s
90	j	0.90	j	0.78	n	1.00	n	1.00
95	j	0.90	j	0.82	n	1.00	n	1.00
100	j	0.93	j	0.87	n	1.00	j	0.93
105	j	0.97	j	0.92	n	1.00	n	1.00
110	n	1.00	j	0.98	n	1.00	n	1.00

Blws - Blockierungswahrscheinlichkeit

dY - Belastungssteigerung in dem Bereich möglich, in dem nur alternativ zu lenkender Verkehr existiert (Ferntrasse) ?

j, n - ja, nein

Tabelle 4.4: Einfluß des PB im unsymmetrischen Netz

Von Netzen mit großen Leitungszahlen (hohe Belastungen) wird gegenüber solchen mit geringen Leitungszahlen grundsätzlich ein höherer prozentualer Anteil an Verkehrsaufkommen verkraftet werden können, weil der sog. Bündelungsgewinn je Abschnitt zum Tragen kommt. Eine abschnittsweise Überlastung kann deshalb in größerem Maße verteilt werden, wodurch sich die kleineren erreichbaren Werte von s erklären. Die Belastungserhöhungen, sofern sie erzielbar sind, bewegen sich im Bereich $< 1\%$.

5 Zusammenfassung

Leitungsvermittelte Netze werden verstärkt mit digitaler Technik ausgebaut und erneuert, sie werden u. a. flexibler hinsichtlich geänderter Nutzungsvarianten und neu einzuführender Dienste. Gleichzeitig können Überlastsituationen entstehen, die Teile des Netzes oder auch das ganze Netz in ihrer Funktionsweise bis hin zur Funktionsunfähigkeit beeinträchtigen.

Der Beitrag stellt nach einer kurzen Beschreibung der zwei untersuchten Verfahren 'Trunk Reservation' und 'Proportional Bidding' die Modellbildung und Leistungsbewertung dieser zur Durchsatzerhöhung in leitungsvermittelten Netzen dar.

Es zeigten sich von den Lastsituationen abhängige unterschiedliche Anwendungsgegebenheiten, die qualitativ und quantitativ beschrieben wurden.

6 Literaturverzeichnis

/1/ Chan, Wan S.:
Recursive Algorithms for Computing End-to-End Blocking in a Network with Arbitrary Routing Plan
IEEE Transactions in Communications, Vol. COM-28, S. 153-164, No. 2, 1980

/2/ Kleinewillinghöfer-Kopp, R.:
Ein Modell für die kostengünstige Bündelbemessung in Fernsprechnetzen.
Zeitschrift für Operations Research Vol. 29 , Nr. 4, S. B73- B100, 1985.

/3/ Schwartz, M.:
Telecommunication Networks.
Addison-Wesley Publishing Company, November 1987.

/4/ Yum, T.-K.G.; Schwartz, M.:
Comparison of Routing Procedures for Circuit-Switched Traffic in Nonhierarchical Networks.
IEEE Transactions on Communications, Vol. COM-35, S. 535-544, No. 5, Mai 1987.

Band 169: G. Heyer, J. Krems, G. Görz (Hrsg.), Wissensarten und ihre Darstellung. VIII, 292 Seiten. 1988.

Band 170: A. Jaeschke, B. Page (Hrsg.), Informatikanwendungen im Umweltbereich. 2. Symposium, Karlsruhe, 1987. Proceedings. X, 201 Seiten. 1988.

Band 171: H. Lutterbach (Hrsg.), Non-Standard Datenbanken für Anwendungen der Graphischen Datenverarbeitung. GI-Fachgespräch, Dortmund, März 1988, Proceedings. VII, 183 Seiten. 1988.

Band 172: G. Rahmstorf (Hrsg.), Wissensrepräsentation in Expertensystemen. Workshop, Herrenberg, März 1987. Proceedings. VII, 189 Seiten. 1988.

Band 173: M. H. Schulz, Testmustergenerierung und Fehlersimulation in digitalen Schaltungen mit hoher Komplexität. IX, 165 Seiten. 1988.

Band 174: A. Endrös, Rechtsprechung und Computer in den neunziger Jahren. XIX, 129 Seiten. 1988.

Band 175: J. Hülsemann, Funktioneller Test der Auflösung von Zugriffskonflikten in Mehrrechnersystemen. X, 179 Seiten. 1988.

Band 176: H. Trost (Hrsg.), 4. Österreichische Artificial-Intelligence-Tagung. Wien, August 1988. Proceedings. VIII, 207 Seiten. 1988.

Band 177: L. Voelkel, J. Pliquett, Signaturanalyse. 223 Seiten. 1989.

Band 178: H. Göttler, Graphgrammatiken in der Softwaretechnik. VIII, 244 Seiten. 1988.

Band 179: W. Ameling (Hrsg.), Simulationstechnik. 5. Symposium. Aachen, September 1988. Proceedings. XIV, 538 Seiten. 1988.

Band 180: H. Bunke, O. Kübler, P. Stucki (Hrsg.), Mustererkennung 1988. 10. DAGM-Symposium, Zürich, September 1988. Proceedings. XV, 361 Seiten. 1988.

Band 181: W. Hoeppner (Hrsg.), Künstliche Intelligenz. GWAI-88, 12. Jahrestagung. Eringerfeld, September 1988. Proceedings. XII, 333 Seiten. 1988.

Band 182: W. Barth (Hrsg.), Visualisierungstechniken und Algorithmen. Fachgespräch, Wien, September 1988. Proceedings. VIII, 247 Seiten. 1988.

Band 183: A. Clauer, W. Purgathofer (Hrsg.), AUSTROGRAPHICS '88. Fachtagung, Wien, September 1988. Proceedings. VIII, 267 Seiten. 1988.

Band 184: B. Gollan, W. Paul, A. Schmitt (Hrsg.), Innovative Informations-Infrastrukturen. I. I. I. – Forum, Saarbrücken, Oktober 1988. Proceedings. VIII, 291 Seiten. 1988.

Band 185: B. Mitschang, Ein Molekül-Atom-Datenmodell für Non-Standard-Anwendungen. XI, 230 Seiten. 1988.

Band 186: E. Rahm, Synchronisation in Mehrrechner-Datenbanksystemen. IX, 272 Seiten. 1988.

Band 187: R. Valk (Hrsg.), GI – 18. Jahrestagung I. Vernetzte und komplexe Informatik-Systeme. Hamburg, Oktober 1988. Proceedings. XVI, 776 Seiten.

Band 188: R. Valk (Hrsg.), GI – 18. Jahrestagung II. Vernetzte und komplexe Informatik-Systeme. Hamburg, Oktober 1988. Proceedings. XVI, 704 Seiten.

Band 189: B. Wolfinger (Hrsg.), Vernetzte und komplexe Informatik-Systeme. Industrieprogramm zur 18. Jahrestagung der GI, Hamburg, Oktober 1988. Proceedings. X, 229 Seiten. 1988.

Band 190: D. Maurer, Relevanzanalyse. VIII, 239 Seiten. 1988.

Band 191: P. Levi, Planen für autonome Montageroboter. XIII, 259 Seiten. 1988.

Band 192: K. Kansy, P. Wißkirchen (Hrsg.), Graphik im Bürobereich. Proceedings, 1988. VIII, 187 Seiten. 1988.

Band 193: W. Gotthard, Datenbanksysteme für Software-Produktionsumgebungen. X, 193 Seiten. 1988.

Band 194: C. Lewerentz, Interaktives Entwerfen großer Programmsysteme. VII, 179 Seiten. 1988.

Band 195: I. S. Bátori, U. Hahn, M. Pinkal, W. Wahlster (Hrsg.), Computerlinguistik und ihre theoretischen Grundlagen. Proceedings. IX, 218 Seiten. 1988.

Band 197: M. Leszak, H. Eggert, Petri-Netz-Methoden und -Werkzeuge. XII, 254 Seiten. 1989.

Band 198: U. Reimer, FRM: Ein Frame-Repräsentationsmodell und seine formale Semantik. VIII, 161 Seiten. 1988.

Band 199: C. Beckstein, Zur Logik der Logik-Programmierung. IX, 246 Seiten. 1988.

Band 200: A. Reinefeld, Spielbaum-Suchverfahren. IX, 191 Seiten. 1989.

Band 201: A. M. Kotz, Triggermechanismen in Datenbanksystemen. VIII, 187 Seiten. 1989.

Band 202: Th. Christaller (Hrsg.), Künstliche Intelligenz. 5. Frühjahrsschule, KIFS-87, Günne, März/April 1987. Proceedings. VII, 403 Seiten. 1989.

Band 203: K. v. Luck (Hrsg.), Künstliche Intelligenz. 7. Frühjahrsschule, KIFS-89, Günne, März 1989. Proceedings. VII, 302 Seiten. 1989.

Band 204: T. Härder (Hrsg.), Datenbanksysteme in Büro, Technik und Wissenschaft. GI/SI-Fachtagung, Zürich, März 1989. Proceedings. XII, 427 Seiten. 1989.

Band 205: P. J. Kühn (Hrsg.), Kommunikation in verteilten Systemen. ITG/GI-Fachtagung, Stuttgart, Februar 1989. Proceedings. XII, 907 Seiten. 1989.

Band 206: P. Horster, H. Isselhorst, Approximative Public-Key-Kryptosysteme. VII, 174 Seiten. 1989.

Band 207: J. Knop (Hrsg.), Organisation der Datenverarbeitung an der Schwelle der 90er Jahre. 8. GI-Fachgespräch, Düsseldorf, März 1989. Proceedings. IX, 276 Seiten. 1989.

Band 208: J. Retti, K. Leidlmair (Hrsg.), 5. Österreichische Artificial-Intelligence-Tagung, Igls/Tirol, März 1989. Proceedings. XI, 452 Seiten. 1989.

Band 209: U. W. Lipeck, Dynamische Integrität von Datenbanken. VIII, 140 Seiten. 1989.

Band 210: K. Drosten, Termersetzungssysteme. IX, 152 Seiten. 1989.

Band 211: H. W. Meuer (Hrsg.), SUPERCOMPUTER '89. Proceedings, 1989. VIII, 171 Seiten. 1989.

Band 212: W.-M. Lippe (Hrsg.), Software-Entwicklung. Fachtagung, Marburg, Juni 1989. Proceedings. IX, 290 Seiten. 1989.

Band 213: I. Walter, Datenbankgestützte Repräsentation und Extraktion von Episodenbeschreibungen aus Bildfolgen. VIII, 243 Seiten. 1989.

Band 214: W. Görke, H. Sörensen (Hrsg.), Fehlertolerierende Rechensysteme / Fault-Tolerant Computing Systems. 4. Internationale GI/ITG/GMA-Fachtagung, Baden-Baden, September 1989. Proceedings. XI, 390 Seiten. 1989.

Band 215: M. Bidjan-Irani, Qualität und Testbarkeit hochintegrierter Schaltungen. IX, 169 Seiten. 1989.

Band 216: D. Metzing (Hrsg.), GWAI-89. 13th German Workshop on Artificial Intelligence. Eringerfeld, September 1989. Proceedings. XII, 485 Seiten. 1989.

Band 218: G. Stiege, J. S. Lie (Hrsg.), Messung, Modellierung und Bewertung von Rechensystemen und Netzen. 5. GI/ITG-Fachtagung, Braunschweig, September 1989. Proceedings. IX, 342 Seiten. 1989.